AF352866

Advancing Sustainability in Geotechnical Engineering

Advancing Sustainability in Geotechnical Engineering

Editors

Shuren Wang
Chen Cao
Hong-Wei Yang

Basel • Beijing • Wuhan • Barcelona • Belgrade • Novi Sad • Cluj • Manchester

Editors

Shuren Wang	Chen Cao	Hong-Wei Yang
School of Civil Engineering	School of Mining	School of Civil Engineering
Henan Polytechnic University	Liaoning Technical University	Sun Yat-sen University
Jiaozuo	Fuxin	Zhuhai
China	China	China

Editorial Office
MDPI AG
Grosspeteranlage 5
4052 Basel, Switzerland

This is a reprint of articles from the Special Issue published online in the open access journal *Sustainability* (ISSN 2071-1050) (available at: www.mdpi.com/journal/sustainability/special_issues/ WD20A552B4).

For citation purposes, cite each article independently as indicated on the article page online and as indicated below:

Lastname, A.A.; Lastname, B.B. Article Title. *Journal Name* **Year**, *Volume Number*, Page Range.

ISBN 978-3-7258-1636-1 (Hbk)
ISBN 978-3-7258-1635-4 (PDF)
doi.org/10.3390/books978-3-7258-1635-4

Contents

Preface . . . . **vii**

Shuren Wang, Chen Cao and Hongwei Yang
Advancing Sustainability in Geotechnical Engineering
Reprinted from: *Sustainability* 2024, *16*, 4757, doi:10.3390/su16114757 . . . . **1**

Xiaowei Lu, Jingyu Jiang, Wen Wang, Xuewen Cao and Lei Hong
Laboratory Experimental Study on the Pressure Relief Effect of Boreholes in Sandstone under
High-Stress Conditions
Reprinted from: *Sustainability* 2023, *15*, 15557, doi:10.3390/su152115557 . . . . **5**

Minhe Luo, Ding Wang, Xuchun Wang and Zelin Lu
Analysis of Surface Settlement Induced by Shield Tunnelling: Grey Relational Analysis and
Numerical Simulation Study on Critical Construction Parameters
Reprinted from: *Sustainability* 2023, *15*, 14315, doi:10.3390/su151914315 . . . . **19**

Yuxi Hao, Mingliang Li, Wen Wang, Zhizeng Zhang and Zhun Li
Study on the Stress Distribution and Stability Control of Surrounding Rock of Reserved
Roadway with Hard Roof
Reprinted from: *Sustainability* 2023, *15*, 14111, doi:10.3390/su151914111 . . . . **40**

Wen Wang, Lei Hong, Xuewen Cao, Xiaowei Lu, Fan Wang, Tong Zhang and Weibing Zhu
Experimental Study of the Mechanical and Acoustic Emission Characteristics of Sandstone by
Using High-Temperature Water-Cooling Cycles
Reprinted from: *Sustainability* 2023, *15*, 13358, doi:10.3390/su151813358 . . . . **61**

Shuren Wang, Zhixiang Wang, Jian Gong and Qianqian Liu
Damage Mode and Energy Consumption Characteristics of Paper-Sludge-Doped Magnesium
Chloride Cement Composites
Reprinted from: *Sustainability* 2023, *15*, 13051, doi:10.3390/su151713051 . . . . **78**

**Lin Hu, Zhijun Zhang, Lingling Wu, Qing Yu, Huaimiao Zheng, Yakun Tian and Guicheng
He**
Experimental Study on the Solidification of Uranium Tailings and Uranium Removal Based on
MICP
Reprinted from: *Sustainability* 2023, *15*, 12387, doi:10.3390/su151612387 . . . . **94**

Samad Narimani, Seyed Morteza Davarpanah, Neil Bar, Ákos Török and Balázs Vásárhelyi
Geological Strength Index Relationships with the Q-System and Q-Slope
Reprinted from: *Sustainability* 2023, *15*, 11233, doi:10.3390/su151411233 . . . . **109**

Nan Zhou and Jianhui Yang
Spatial Effect Analysis of a Long Strip Pit Partition Wall and Its Influence on Adjacent Pile
Foundations
Reprinted from: *Sustainability* 2023, *15*, 10409, doi:10.3390/su151310409 . . . . **125**

Ying Chen, Zhiwen Wang, Qianjia Hui, Zhaoju Zhang, Zikai Zhang, Bingjie Huo, et al.
Influence of Gas Pressure on the Failure Mechanism of Coal-like Burst-Prone Briquette and the
Subsequent Geological Dynamic Disasters
Reprinted from: *Sustainability* 2023, *15*, 7856, doi:10.3390/su15107856 . . . . **142**

Rashad Alsirawan, Edina Koch and Ammar Alnmr
Proposed Method for the Design of Geosynthetic-Reinforced Pile-Supported (GRPS) Embankments
Reprinted from: *Sustainability* **2023**, *15*, 6196, doi:10.3390/su15076196 . . . . **155**

Kuidong Gao, Jihai Liu, Hong Chen, Xu Li and Shuan Huang
Dynamic Characteristics of Rock Holes with Gravel Sediment Drilled by Bit Anchor Cable Drilling
Reprinted from: *Sustainability* **2023**, *15*, 5956, doi:10.3390/su15075956 . . . . **175**

Jinxiang Deng, Mengjie Li, Yakun Tian, Zhijun Zhang, Lingling Wu and Lin Hu
Using Electric Field to Improve the Effect of Microbial-Induced Carbonate Precipitation
Reprinted from: *Sustainability* **2023**, *15*, 5901, doi:10.3390/su15075901 . . . . **192**

Mingtao Ji, Xuchun Wang, Minhe Luo, Ding Wang, Hongwei Teng and Mingqing Du
Stability Analysis of Tunnel Surrounding Rock When TBM Passes through Fracture Zones with Different Deterioration Levels and Dip Angles
Reprinted from: *Sustainability* **2023**, *15*, 5243, doi:10.3390/su15065243 . . . . **214**

Jincheng Hua, Xinwu Wang, Huanhuan Liu and Haisu Sun
Nonlinear Finite Element Analysis of a Composite Joint with a Blind Bolt and T-stub
Reprinted from: *Sustainability* **2023**, *15*, 4790, doi:10.3390/su15064790 . . . . **229**

Ammar Alnmr, Richard Paul Ray and Rashad Alsirawan
A State-of-the-Art Review and Numerical Study of Reinforced Expansive Soil with Granular Anchor Piles and Helical Piles
Reprinted from: *Sustainability* **2023**, *15*, 2802, doi:10.3390/su15032802 . . . . **247**

Grzegorz Kacprzak, Artur Zbiciak, Kazimierz Józefiak, Paweł Nowak and Mateusz Frydrych
One-Dimensional Computational Model of Gyttja Clay for Settlement Prediction
Reprinted from: *Sustainability* **2023**, *15*, 1759, doi:10.3390/su15031759 . . . . **283**

Xin Huang, Tong Wang, Yanbin Luo and Jiaqi Guo
Study on the Influence of Water Content on Mechanical Properties and Acoustic Emission Characteristics of Sandstone: Case Study from China Based on a Sandstone from the Nanyang Area
Reprinted from: *Sustainability* **2023**, *15*, 552, doi:10.3390/su15010552 . . . . **297**

Preface

Geotechnical engineering links engineering construction with the earth, including foundation engineering, slope engineering, tunnel engineering, mining engineering, etc. To promote the sustainable development of geotechnical engineering, it is necessary to make full use of the bearing capacity of rock and soil mass; reduce the amount of building materials with high carbon emissions, such as concrete and reinforcement; and reduce the quantity of work. To achieve engineering construction objectives, it has become a hot topic of global concern to transform rock and soil mass; improve the anti-floating, anti-flood, and anti-seismic capabilities of geotechnical engineering; and avoid polluting, damaging, and otherwise influencing the ecological environment around geotechnical engineering sites.

The aim of 'Advancing Sustainability in Geotechnical Engineering' was to require geotechnical engineers to actively respond to various natural disasters, implement energy conservation and emission reduction, and achieve friendly, coordinated and sustainable development of geotechnical engineering and the ecological environment during the investigation, design, and construction stage. In addition, in this Special Issue, we hope to showcase original and innovative papers highlighting the most challenging new methods, materials, equipment, and techniques relevant to promoting sustainable development in geotechnical engineering.

Shuren Wang, Chen Cao, and Hong-Wei Yang
Editors

Editorial

Advancing Sustainability in Geotechnical Engineering

Shuren Wang [1,*], **Chen Cao** [2] **and Hongwei Yang** [3]

1 School of Civil Engineering, Henan Polytechnic University, Jiaozuo 454003, China
2 School of Mining, Liaoning Technical University, Fuxin 123000, China; caochen@lntu.edu.cn
3 School of Civil Engineering, Sun Yat-sen University, Zhuhai 519000, China; yanghw9@mail.sysu.edu.cn
* Correspondence: w_sr88@163.com

1. Introduction

Geotechnical engineering is a key element for all engineering construction that establishes contact with the earth, including foundation engineering, slope engineering, tunnel engineering, mining engineering, etc. To promote the sustainable development of geotechnical engineering, it is necessary to fully utilize the bearing capacity of rock and soil masses, reduce the amount of building materials with high carbon emissions, such as concrete and reinforcement, and minimize the workload [1,2]. To achieve engineering construction objectives, it has become a hot topic of global concern to modify the rock and soil masses, improve the anti-floating, anti-flood, and anti-seismic capabilities of geotechnical engineering, and avoid the pollution of, damage to, and influence on the ecological environment around the geotechnical engineering site [3,4].

The aim of this Special Issue is to enable geotechnical engineers to actively respond to various natural disasters, implement energy conservation and emission reduction, and achieve the friendly, coordinated, and sustainable development of geotechnical engineering and the ecological environment during the investigation, design, and construction processes. In addition, in this Special Issue, we hope to showcase original and innovative research papers highlighting the most important new methods, materials, equipment, and techniques relevant to promoting sustainable development in geotechnical engineering.

2. Progress in the Research Topic

We provided a forum for professionals and academics to communicate their impactful research, resulting in the publication of 17 papers in this Special Issue, mainly covering the following five aspects: the physical and mechanical properties of sandstone samples under different conditions, the prediction and control methods of surface subsidence, the solidification of wastes and their safe utilization, safety analysis and its control for engineering structures, and the application of innovative technology and methods in geotechnical engineering. Please visit the following website for more information: https://www.mdpi.com/journal/sustainability/special_issues/WD20A552B4 (accessed on 15 August 2023).

The physical and mechanical properties of sandstone samples under different conditions: To study the effects of deep-rock drilling and pressure relief under high stress conditions in enhanced geothermal systems, Lu et al. conducted two types of sandstone drilling–pressure relief experiments: the staged drilling of pressure relief holes before peak stress and one-time drilling. To reveal the physical and mechanical properties of sandstone under high-temperature water-cooling cycling conditions, Wang et al. conducted uniaxial compression acoustic emission tests on sandstone after high-temperature water-cooling cycles using an RMT-150B electrohydraulic servo rock testing system and a DS-5 acoustic emission detection system. They analyzed the deformation, strength, and acoustic emission characteristics of sandstone under different temperatures and cycle times. To investigate the influence of the water content on the mechanical properties of sandstone and the evolution

Citation: Wang, S.; Cao, C.; Yang, H. Advancing Sustainability in Geotechnical Engineering. *Sustainability* **2024**, *16*, 4757. https://doi.org/10.3390/su16114757

Received: 22 May 2024
Accepted: 26 May 2024
Published: 3 June 2024

of crack propagation, Huang et al. carried out laboratory compression tests and discrete element method numerical simulations of sandstone under different conditions.

The prediction and control methods of surface subsidence: Excessive surface settlement poses significant challenges to shield tunneling construction, resulting in damage to adjacent buildings, infrastructure, and underground pipelines. Luo et al. investigated the surface settlement induced by shield tunneling during the construction of the Qingdao Metro Line 6 between Haigang Road Station and Chaoyang Road Station, China. In fracture zones, tunneling with a double-shield Tunnel Boring Machine (TBM) presents significant challenges. To assure the tunnel construction's safety and efficiency, it was necessary to conduct a stability analysis of the tunnel's surrounding rock when a TBM passed through the fracture zones. Ji et al. analyzed the stability of the tunnel's surrounding rock when the TBM passed through fracture zones in the Qingdao Metro Line 8 (China) with different deterioration levels and dip angles. Working out appropriate load–settlement relationships is considered to be a very difficult geomechanical task. Kacprzak et al. proposed a 1-D rheological model, which accounted for the complex behavior of soil related to the settlement process.

The solidification of wastes and their safe utilization: To reduce the pollution caused by paper sludge in the environment and overcome issues of poor water resistance and brittleness in Magnesium Oxychloride Cement (MOC), Wang et al. modified the MOC by adding different dosages of paper sludge and studied the mechanical properties and damage modes of composite MOC materials containing paper sludge through uniaxial compression tests. They verified the effect of paper sludge on the mechanical properties of the MOC materials and the change in damage modes from the perspective of energy dissipation. The governance of uranium tailings aims to improve their stability and reduce radionuclide uranium release. In order to achieve this goal, Hu et al. successively carried out a uranium removal solution test and a uranium tailings grouting test using microbially induced calcium carbonate precipitation (MICP) technology. They discussed the effect of the MICP on the reinforcement of the uranium tailings and the synchronous control of radionuclide uranium in the tailings. This study provides valuable insights and references for safely disposing of uranium tailings.

Safety analysis and its control for engineering structure: Laboratory testing of raw coal samples is the dominant research method for disaster prediction. Chen et al. studied the failure characteristics and dynamic disaster propensities of coal-like burst-prone briquettes under different gas pressures. They developed a self-made multi-function rock–gas coupling experimental device and synthesized burst-prone briquettes. Hao et al. revealed the distribution law of the surrounding rock stress and the displacement of the surrounding rock after roof cutting and pressure relief of a reserved roadway with a hard roof using numerical simulation. Zhou and Yang studied the influence of the distance from the pile foundation to the partition wall as well as the stiffness of the retaining wall on the displacement and bending moment of the pile foundation during excavation. Soft soils with unfavorable properties can be improved using various ground-improvement methods. To overcome the limitations of current design methods that do not fully account for all interactions, Alsirawan et al. developed a simplified design method for geosynthetic-reinforced pile-supported embankments.

The application of innovative technology and methods in geotechnical engineering: Gravel and sediment frequently build up in holes during the anchor cable installation process, which makes it harder to install the anchor cable and causes the reinforcement to fail. Gao et al. proposed a bit–anchor setup approach to set up a drill bit at the front of the anchor cable to aid the anchor cable drilling. The feasibility of this approach was established with the DEM-MBD joint simulation and proved the correctness of the model. The precipitation of calcium carbonate induced by Bacillus subtilis has garnered considerable attention as a novel rock and soil reinforcement technique. The content and structure of calcium carbonate produced through this reaction play a crucial role in determining the rocks' and soil's reinforcement effects. Deng et al. investigated the effects of electrification on the physical and chemical characteristics, such as the growth and reproduction of Pasteurii. Using datasets obtained from igneous, sedimentary, and metamorphic rock slopes from various regions worldwide, Narimani et al. investigated different relationships between the geological strength index and the Q-system and Q-slope.

We hope that these articles provide readers with valuable information on recent developments in science, technology, and related research for achieving the goals of this Special Issue.

Author Contributions: The idea for and concept of this research topic came from discussion among the guest editors. S.W. organized and wrote the manuscript. C.C. and H.Y. provided critical feedback. All authors have read and agreed to the published version of the manuscript.

Funding: This work was financially supported by Key Project of Natural Science Foundation of Henan Province (232300421134), First-Class Discipline Implementation of Safety Science and Engineering (AQ20230103), Zhongyuan Science and Technology Innovation Leading Talent Program (244200510005), China.

Acknowledgments: We thank the various authors for submitting their work to this Special Issue and the reviewers who agreed to review the individual contributions.

Conflicts of Interest: The authors declare that the research was conducted in the absence of any commercial or financial relationships that could be construed as a potential conflict of interest.

List of Contributions

1. Huang, X.; Wang, T.; Luo, Y.; Guo, J. Study on the Influence of Water Content on Mechanical Properties and Acoustic Emission Characteristics of Sandstone: Case Study from China Based on a Sandstone from the Nanyang Area. *Sustainability* **2022**, *15*, 552. https://doi.org/10.3390/su15010552.
2. Kacprzak, G.; Zbiciak, A.; Józefiak, K.; Nowak, P.; Frydrych, M. One-Dimensional Computational Model of Gyttja Clay for Settlement Prediction. *Sustainability* **2023**, *15*, 1759. https://doi.org/10.3390/su15031759.
3. Alnmr, A.; Ray, R.P.; Alsirawan, R. A State-of-the-Art Review and Numerical Study of Reinforced Expansive Soil with Granular Anchor Piles and Helical Piles. *Sustainability* **2023**, *15*, 2802. https://doi.org/10.3390/su15032802.
4. Hua, J.; Wang, X.; Liu, H.; Sun, H. Nonlinear Finite Element Analysis of a Composite Joint with a Blind Bolt and T-stub. *Sustainability* **2023**, *15*, 4790. https://doi.org/10.3390/su15064790.
5. Ji, M.; Wang, X.; Luo, M.; Wang, D.; Teng, H.; Du, M. Stability Analysis of Tunnel Surrounding Rock When TBM Passes through Fracture Zones with Different Deterioration Levels and Dip Angles. *Sustainability* **2023**, *15*, 5243. https://doi.org/10.3390/su15065243.
6. Deng, J.; Li, M.; Tian, Y.; Zhang, Z.; Wu, L.; Hu, L. Using Electric Field to Improve the Effect of Microbial-Induced Carbonate Precipitation. *Sustainability* **2023**, *15*, 5901. https://doi.org/10.3390/su15075901.
7. Gao, K.; Liu, J.; Chen, H.; Li, X.; Huang, S. Dynamic Characteristics of Rock Holes with Gravel Sediment Drilled by Bit Anchor Cable Drilling. *Sustainability* **2023**, *15*, 5956. https://doi.org/10.3390/su15075956.
8. Alsirawan, R.; Koch, E.; Alnmr, A. Proposed Method for the Design of Geosynthetic-Reinforced Pile-Supported (GRPS) Embankments. *Sustainability* **2023**, *15*, 6196. https://doi.org/10.3390/su15076196.
9. Chen, Y.; Wang, Z.; Hui, Q.; Zhang, Z.; Zhang, Z.; Huo, B.; Chen, Y.; Liu, J. Influence of Gas Pressure on the Failure Mechanism of Coal-like Burst-Prone Briquette and the Subsequent Geological Dynamic Disasters. *Sustainability* **2023**, *15*, 7856. https://doi.org/10.3390/su15107856.
10. Zhou, N.; Yang, J. Spatial Effect Analysis of a Long Strip Pit Partition Wall and Its Influence on Adjacent Pile Foundations. *Sustainability* **2023**, *15*, 10409. https://doi.org/10.3390/su151310409.
11. Narimani, S.; Davarpanah, S.M.; Bar, N.; Török, Á.; Vásárhelyi, B. Geological Strength Index Relationships with the Q-System and Q-Slope. *Sustainability* **2023**, *15*, 11233. https://doi.org/10.3390/su151411233.
12. Hu, L.; Zhang, Z.; Wu, L.; Yu, Q.; Zheng, H.; Tian, Y.; He, G. Experimental Study on the Solidification of Uranium Tailings and Uranium Removal Based on MICP. *Sustainability* **2023**, *15*, 12387. https://doi.org/10.3390/su151612387.

13. Wang, S.; Wang, Z.; Gong, J.; Liu, Q. Damage Mode and Energy Consumption Characteristics of Paper-Sludge-Doped Magnesium Chloride Cement Composites. *Sustainability* **2023**, *15*, 13051. https://doi.org/10.3390/su151713051.
14. Wang, W.; Hong, L.; Cao, X.; Lu, X.; Wang, F.; Zhang, T.; Zhu, W. Experimental Study of the Mechanical and Acoustic Emission Characteristics of Sandstone by Using High-Temperature Water-Cooling Cycles. *Sustainability* **2023**, *15*, 13358. https://doi.org/10.3390/su151813358.
15. Hao, Y.; Li, M.; Wang, W.; Zhang, Z.; Li, Z. Study on the Stress Distribution and Stability Control of Surrounding Rock of Reserved Roadway with Hard Roof. *Sustainability* **2023**, *15*, 14111. https://doi.org/10.3390/su151914111.
16. Luo, M.; Wang, D.; Wang, X.; Lu, Z. Analysis of Surface Settlement Induced by Shield Tunnelling: Grey Relational Analysis and Numerical Simulation Study on Critical Construction Parameters. *Sustainability* **2023**, *15*, 14315. https://doi.org/10.3390/su151914315.
17. Lu, X.; Jiang, J.; Wang, W.; Cao, X.; Hong, L. Laboratory Experimental Study on the Pressure Relief Effect of Boreholes in Sandstone under High-Stress Conditions. *Sustainability* **2023**, *15*, 15557. https://doi.org/10.3390/su152115557.

References

1. von der Tann, L.; Basu, D.; Ritter, S.; Capellen, P.S.; Stordal, I.F. Sustainability in geotechnical engineering: What does it mean and why does that matter? *Proc. Inst. Civ. Eng.-Eng. Sustain.* **2023**, *176*, 247–259. [CrossRef]
2. Pathmanandavel, S.; MacRobert, C.J. Digitisation, sustainability, and disruption-promoting a more balanced debate on risk in the geotechnical community. *Georisk Assess. Manag. Risk Eng. Syst. Geohazards* **2020**, *14*, 246–259. [CrossRef]
3. Jorge-Arturo, P.J.; César-Augusto, G.U.; Rodrigo-Elías, E.R. Assessment of geotechnical hazard due to deep excavations in Bogota clays: A contribution for sustainability in urban environments. *Rev. Fac. Ing.* **2020**, *29*, e11373. [CrossRef]
4. Wang, Y.; Tang, Y.F.; Zhang, F.; Guo, J.L. Laboratory model tests of seismic strain response of anti-seismic anchor cables. *Front. Earth Sci.* **2022**, *10*, 863511. [CrossRef]

sustainability

MDPI

Article

Laboratory Experimental Study on the Pressure Relief Effect of Boreholes in Sandstone under High-Stress Conditions

Xiaowei Lu [1,2], Jingyu Jiang [1,*], Wen Wang [2], Xuewen Cao [3] and Lei Hong [2]

[1] School of Safety Engineering, China University of Mining and Technology, Xuzhou 221116, China; luxiaowei@hpu.edu.cn
[2] School of Energy Scinence and Engineering, Henan Polytechnic University, Jiaozuo 454003, China; wangwen2006@hpu.edu.cn (W.W.); 15606486753@163.com (L.H.)
[3] School of Resources and Civil Engineering, Northeastern University, Shenyang 110819, China; xuewenxmail@163.com
* Correspondence: jiangjingyu@cumt.edu.cn

Abstract: To study the effects of deep rock drilling pressure relief under high stress conditions in enhanced geothermal systems, two kinds of drilling pressure relief experiments were conducted on sandstone—the staged drilling of pressure relief holes before peak stress and one-time drilling. Pressure relief experiments were carried out on sandstone with two borehole methods of the stage-by-stage drilling and one-time drilling of pressure relief boreholes ahead of the experiments. FLAC3D was used to analyze the plastic zone evolution during drilling and the relationship between stress and plastic zone volume. The results reveal the pre-peak stress change characteristics and pressure relief features of non-prefabricated boreholes under high stress. The experiments show that the staged drilling of pressurized samples involves stages of rapid and gradual decreases in stress, with total relief amplitudes increasing but single-borehole relief decreasing with more holes. Under the same conditions, staged drilling has better relief effects and results in greater energy dissipation, indicating that incremental pre-peak pressure relief is beneficial for reducing the surrounding rock's impact tendency and improving stability. The research results can provide good guidance and reference for the long-term stability analysis of borehole-containing rock and rock burst hazard control.

Keywords: rock mechanics; drilling under pressure; uniaxial compression; numerical simulation; energy dissipation

Citation: Lu, X.; Jiang, J.; Wang, W.; Cao, X.; Hong, L. Laboratory Experimental Study on the Pressure Relief Effect of Boreholes in Sandstone under High-Stress Conditions. *Sustainability* **2023**, *15*, 15557. https://doi.org/10.3390/su152115557

Academic Editors: Syed Minhaj Saleem Kazmi and Kaihui Li

Received: 31 July 2023
Revised: 5 October 2023
Accepted: 27 October 2023
Published: 2 November 2023

1. Introduction

Due to its cleanliness and widespread spatial distribution, geothermal resources have become a key clean energy source prioritized by countries around the world for research and development [1–4]. China has abundant geothermal resources widely distributed across the country. Sedimentary basin-type hot dry rock resources are an important type of geothermal energy, mainly distributed in northwestern, northern, and northeastern China. Hot dry rocks refer to deep high-temperature, low-permeability rock formations without naturally existing water bodies inside. Artificial geothermal reservoirs need to be formed through measures such as artificial water injection and hydraulic fracturing to achieve geothermal energy extraction and utilization. Enhanced geothermal systems are engineered geothermal systems that economically extract deep geothermal energy from low-permeability rocks by artificially creating geothermal reservoirs, which means developing geothermal energy from hot dry rocks [2–5].

Drilling pressure relief is commonly used for surrounding rock and rock burst hazards and is one of the important measures for preventing rock bursts. It has the advantages of simple operation, safe construction and a good relief effect [6–9]. Meanwhile, drilling pressure relief can also be used to develop the heat storage layer for enhanced geothermal systems, but there are also some technical difficulties and risks involved. Drilling pressure relief can generate or activate large-scale fracture networks in hot dry rocks, which can be accessed through seismic monitoring and directional drilling. This can improve the permeability and hydraulic connectivity of the heat storage layer, enabling fluid circulation and heat exchange in the fracture networks. However, drilling pressure relief may also lead to induced seismicity that poses threats to the environment and humans. Therefore, accurate parameter design and stringent safety control are needed for drilling pressure relief to achieve optimal relief effects and minimize seismic impacts.

Scholars have carried out many experiments, theoretical derivations and numerical simulations on borehole pressure relief for rock burst hazards. W. D. Ortlepp and T. R. Stacy [10] pointed out that stress relief during excavation under high-stress conditions may cause strain rock bursts. P K KAISER [11] pointed out that the pressure relief effect and free surface generated by roadway excavation change the triaxial compression equilibrium state of deep rock and provide the conditions necessary for the occurrence of rock bursts. Liu Honggang et al. [12] uncovered the mechanism of borehole pressure relief by simulating engineering examples using the RFPA and ANSYS software. Zhang Ying et al. [13] established an elastoplastic model of the coal body around boreholes using COMSOL software and found that coal strength, burial depth and borehole diameter are the main factors controlling the pressure relief range of boreholes. Zheng He et al. [14] carried out a numerical simulation of deep roadways and found that borehole pressure relief technology can effectively release the deformation and destruction energy of surrounding rock and improve the stress environment. Zhu Staou et al. [15] proposed the concept of the energy dissipation index through rock mechanics experiments and deduced the relationship between the energy dissipation index and the impact energy index to determine the anti-impact drill hole parameters. Wang Shuwen et al. [16] proposed the concept of borehole energy dissipation rate from the perspective of energy by analyzing the failure mechanism of borehole-surrounding rock. The pressure relief effect is exponentially related to borehole diameter and negatively correlated with coal strength. Wang Meng et al. [17] inverted the mechanical parameters of rock mass using FLAC3D and proposed a method for deter-mining the key parameters affecting the pressure relief effect. Jia Chuanyang et al. [18] found that borehole parameters have a great influence on the pressure relief effect based on indoor tests and PFC software simulations. The larger the borehole diameter and depth and the smaller the spacing, the better the pressure relief effect. The fundamental reason for borehole pressure relief is the stress release caused by fracture penetration. Qin Zihan et al. [19] detected the pressure relief effect zone using electromagnetic wave CT detection technology and found that coal strength, borehole diameter and spacing are the main factors affecting the pressure relief effect. Ma Binwen et al. [20] deduced a boundary equation for the borehole pressure relief zone based on the theory of impact ground pressure stress control. Zhao Zhenhua et al. [21] carried out a stress relaxation test using amphibolite samples containing pressure relief holes and determined the mechanism underlying pressure relief holes during the stress relaxation process. Lin, P. et al. [22] explored the crack evolution mechanism of granite samples with prefabricated boreholes of different sizes, spacing and layouts using a uniaxial compression test. The test results show that, under the interaction of stresses, three typical cracks are generated around the borehole: shear cracks, tensile cracks and mixed (shear–tensile) cracks. Zhao, T.B. et al. [23] further explored the failure characteristics and mechanisms of samples with different borehole arrangements via PFC and found that shear accompanied by splitting failure is the main failure mode for samples with different borehole diameters, while, for samples with different numbers of boreholes in a row, splitting accompanied by shear failure is the main failure mode. However, splitting or shear failure modes may occur for samples containing a different number of borehole rows. Huang, B. et al.'s [24] study on the acoustic emission characteristics of samples with different borehole diameters also confirmed that, when the borehole diameter is larger, the AE activity before the peak strength point is stronger, and the bearing capacity of the sample is significantly weaker.

In summary, most of the current research is limited to theoretical analysis and numerical simulation, with less systematic experimental verification. The understanding of borehole pressure relief effects and laws under high-stress conditions is still insufficient, and the study of the evolution law of the stress field during dynamic drilling is insufficient. Therefore, it is of great significance for research on borehole pressure relief in rock burst hazards to analyze the peak stress change characteristics by drilling multiple pressure relief boreholes in stressed sandstone samples.

2. Specimen Test Methods

2.1. Specimen Preparation

Sandstone was selected as the research object for the experiments. The samples were taken from the same sandstone block and processed by cutting and grinding them into cubic specimens with length 80 mm, width 80 mm and height 80 mm. The flatness of the end faces was less than 0.02 mm, meeting the standards for rock mechanics testing. The geometrical parameters are shown in Table 1.

Table 1. Specimen parameters.

Group	No.	Unloader Hole Quantity	Length /mm	Width /mm	Height /mm	Notes
I	A11	/	80.05	79.96	80.02	Full specimens
	A12	/	80.89	80.73	79.80	
II	B11	1	80.05	79.83	80.02	Staged drilling of pressure relief holes
	B12	1	80.28	80.14	80.32	
	B21	2	80.22	79.77	80.19	
	B22	2	79.92	81.13	80.21	
	B31	3	80.13	80.21	80.17	
	B32	3	80.08	79.87	79.79	
III	C11	1	80.81	80.83	79.87	Pressure relief holes
	C12	1	80.12	80.07	79.92	
	C21	2	79.65	79.76	80.23	
	C22	2	80.21	80.17	79.89	One drilling event
	C31	3	80.07	80.26	80.18	
	C32	3	79.93	80.12	80.08	

2.2. Test Methods

The RMT-150-B electrohydraulic servo test system was used for the uniaxial compression tests. The axial load was measured using a 1000 kN force sensor with a precision of 1.0×10^{-3} kN. The axial compressive deformation was measured using a 5.0 mm displacement sensor, and the circumferential deformation was measured using two 2.5 mm displacement sensors. The deformation precision was 1.0×10^{-3} mm. Displacement loading was applied throughout the test at a loading rate of 0.002 mm/s, as shown in Figure 1.

Preload stress: Uniaxial compression tests on intact samples yielded an average peak stress of 67.73 MPa. A preload stress that is too low cannot simulate the stress environment of deep rock, while excessive stress may cause sample failure during borehole drilling under pressure, making it difficult to conduct pressurized borehole tests. The preload stress was determined to be 38.50 MPa.

Pressure relief borehole parameters: Loading was stopped upon reaching the preload stress, with the hydraulic cylinder displacement maintained. Pressure relief boreholes were then drilled into the samples. The boreholes were drilled vertically into the end faces of the samples using a fixed electric drill. Reference [17] indicates that the relief effect is most significant when the borehole spacing is 2–6 times the borehole diameter. Thus, 8 mm diameter through-holes with 80 mm depth were adopted, with 40 mm spacing between the borehole centers in the multi-borehole samples, as shown in Figure 2.

Figure 1. RMT-150-B electrohydraulic servo rock test system.

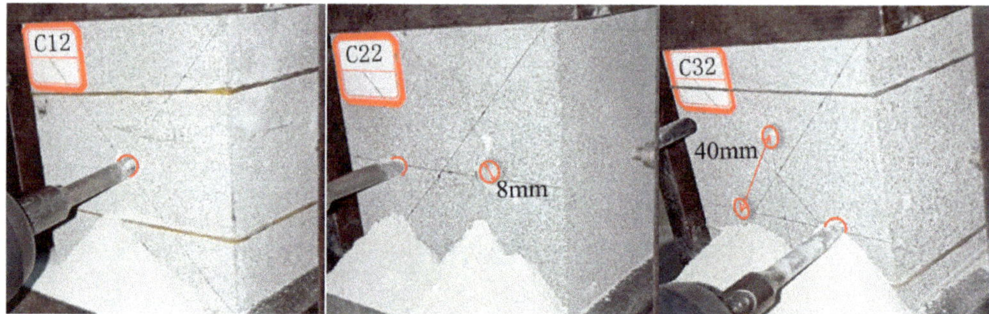

Figure 2. Sample loading diagram.

Two borehole drilling modes were implemented: staged drilling and one-off drilling. Staged drilling refers to drilling the next borehole in multi-borehole samples after the stress of the previous borehole has stabilized. One-off drilling means immediately drilling subsequent boreholes upon completing the first without waiting for stress stabilization (Figure 3).

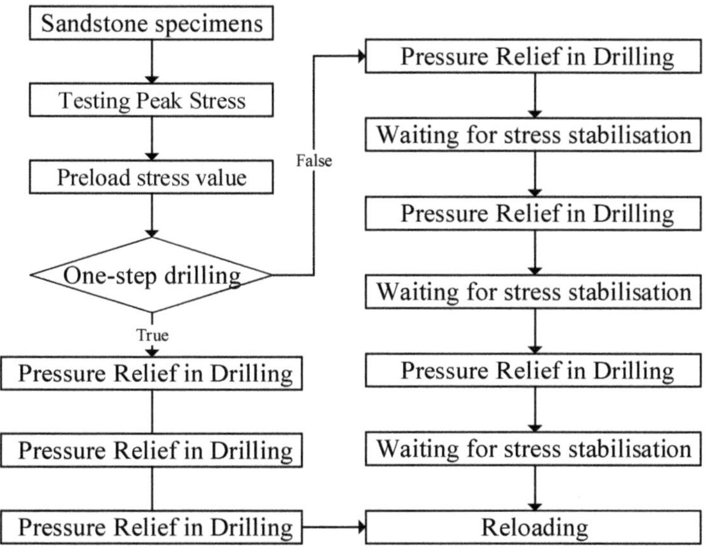

Figure 3. Flowchart of the experiment.

3. Test Results and Analysis

3.1. Staged Drilling Stress Evolution Characteristics for Pressure Relief Holes

Figure 4 shows the stress–time curves for staged pressure relief borehole drilling in the samples. It took 461 s, 554 s and 581 s for the stresses in B11, B21 and B32 to reach the preload stress, respectively. Figure 4 indicates that B11 has a shorter compaction stage and faster rate of increase in stress, meaning fewer pores in the sample. In the first stage of pressure relief borehole drilling, the sample stresses dropped sharply under the action of pressure relief holes: the stresses decreased to 30.75 MPa, 30.92 MPa and 29.94 MPa in B11, B21 and B32, with reductions of 20.1%, 19.7% and 23.4%, respectively. In the second drilling stage, the stress reduction was slower than that in the first stage, and the relief amplitude decreased: the B21 stress decreased to 27.68 MPa with a total reduction of 28.1%, 10.5% less than that in the first stage; the B32 stress decreased to 26.70 MPa with a total reduction of 30.6%, 10.8% less than that in the first stage. In the third drilling stage of B32, the stress decreased to 25.15 MPa with a total reduction of 34.6%, 5.8% less than that in the second stage. The samples were reloaded after completing borehole drilling, with lower peak stresses compared to intact samples. B11 and B32 showed brittle failure, while B21 exhibited ductile failure.

Taking B32 as an example, we characterize the stress evolution of the staged unpressurized drilling. From 0 to 581 s, the slope and increase rate of the stress–time curve rose under uniaxial compression, indicating pore compaction in the sample. The stress increased to the preload stress at 581 s, and then borehole drilling started. At 642 s after drilling was completed, the stress dropped sharply to 29.49 MPa due to mechanical disturbance and stress concentration, decreasing by 23.4%. From 642 s to 1210 s, the stress remained unchanged, suggesting that the pressure relief of the first borehole had reached its limit. The second borehole was drilled at 1211 s and completed at 1282 s. The stress decreased to 26.7 MPa with a reduction of 30.6%. Unlike the cliff-like drop in the first stage, the second stage showed a more gradual stress reduction, indicating microcrack propagation in the sample. The stress stabilized until 1817 s. The third borehole was drilled at 1818 s and completed at 1921 s, reducing the stress to 25.15 MPa with an amplitude of 34.6%. The stress remained stable from the completion of drilling to 2411 s, marking the end of the third stage of pressure relief. Reloading at 2412 s increased the stress to 49.38 MPa at 2596 s, at which point brittle failure occurred with stress dropping to 0 MPa. This suggests that the post-relief peak stress decreased but the sample retained some load-bearing capacity before brittle failure.

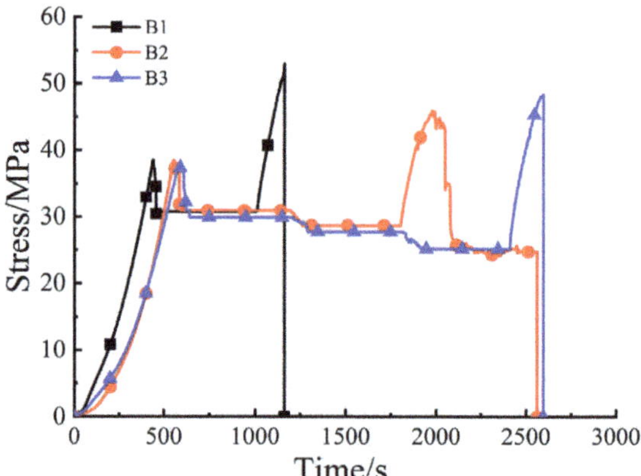

Figure 4. The stress-time curve for the staged drilling of the pressure relief hole of the specimen.

After pressure relief in each stage, the stresses of the samples showed cliff-like drops, but the stress drop amplitudes in subsequent stages gradually decreased. It is analyzed that with the increasing number of pressure relief holes, the effective load-bearing area inside the samples gradually decreased, leading to a saturation of the pressure relief effect. After completing all stages of hole drilling, the three samples were reloaded to failure, and their failure loads were lower than those of the intact samples, indicating that the drilling of pressure relief holes did reduce the load-bearing capacity of the samples. Analysis of sample B32: during the compaction stage, the stress increase rate grew larger, suggesting a compaction of voids. After pressure relief in each stage, stress decrease and stabilization occurred, and the pressure relief amplitudes in the second and third stages were smaller than the first stage, consistent with the gradually saturated relief effect. Upon reloading, brittle failure occurred, demonstrating that the drilling of pressure relief holes decreased the load-bearing capacity of the sample.

3.2. Evolution Characteristics of One-Off Pressure Relief Borehole Drilling

Figure 5 presents the stress–time curves of one-off pressure relief drilling. The stress gradually increased, with C32 having a shorter compaction stage and faster increase rate, indicating fewer internal pores. Upon reaching the preload stress, borehole drilling caused varying degrees of stress reduction. The stresses in C12, C22 and C32 decreased to 30.61 MPa, 29.85 MPa and 27.37 MPa, with reductions of 20.5%, 22.4% and 28.9%, respectively.

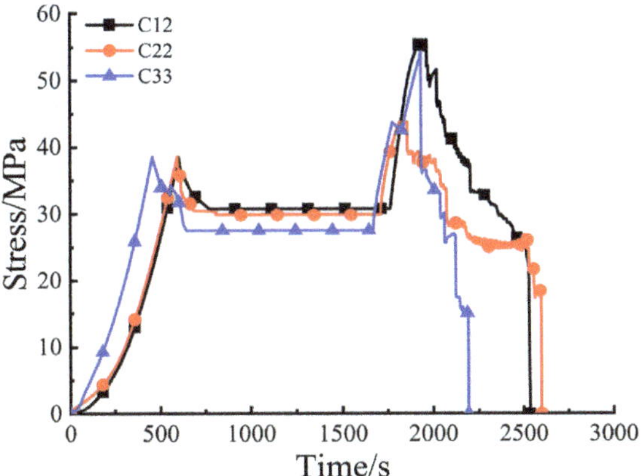

Figure 5. The stress-time curve for one drilling of the specimen pressure relief hole.

The samples failed upon reloading, exhibiting fluctuating and decreasing stress with visible crack expansion, compaction and re-expansion, unlike the brittle failure of the staged drilling samples. For single and double-borehole samples, axial pressure induced fractures at borehole edges propagating through the sample, causing failure. In the triple-borehole sample, the top borehole connected with the other two via penetrating cracks. Further propagation to the edges led to sample failure and the loss of load bearing capacity, as shown in Figure 6.

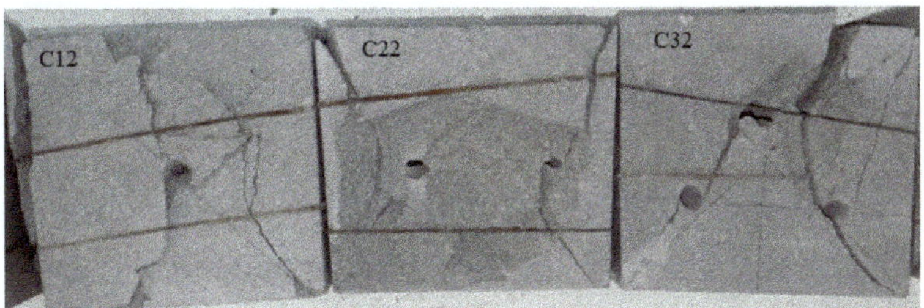

Figure 6. Specimen loading diagram.

Let us take C32 as an example to analyze the stress evolution in one-off drilling. Initially, the stress gradually increased under axial compression. At 460 s, the stress reached the preload value, and then three sequential boreholes were drilled without interruption while maintaining the loading displacement. From 460 s to 644 s, the stress decreased by 28.9% to 27.37 MPa. It remained relatively stable from 645 s to 1667 s, suggesting the pressure relief limit had been reached. Reloading at 1667 s led to a sudden drop in stress, indicating violent crack expansion. The subsequent compaction increased the stress until failure. In the post-peak region, the sample showed features of ductile failure. The time from peak stress to final failure was 259 s, with large drops and small rises in between.

After the one-time drilling of pressure relief holes, the stresses of the samples showed different degrees of cliff-like drops, but the drop amplitudes were lower than those of staged drilling. This is believed to be because the one-time drilling led to concentrated stress and excessively rapid release. Upon reloading to failure, the samples exhibited fluctuating and decreasing stress ductile failure modes rather than brittle failure, which was due to the large number of cracks generated inside the samples during the one-time pressure relief. Analysis of sample C32: after pressure relief, the stress decreased by 28.9%, then stabilized. Upon reloading, large fluctuations of stress occurred, indicating the process of crack propagation, compaction, and re-propagation. Failure occurred after 259 s. For single-hole and double-hole samples, failure occurred at the edges of pressure relief holes, while for the triple-hole sample, failure occurred between the holes. This shows that the number and layout of pressure relief holes affect the failure modes of the samples. Overall, one-time pressure relief led to a large amount of crack generation and stress redistribution in the samples, making the failure more ductile but also decreasing the load bearing capacity, so caution should be exercised during the operation.

3.3. Test Result Comparison and Analysis

Figure 7 presents the stress evolution of staged pressure relief drilling under high stress. With increasing borehole quantity, the stress reductions were as follows: B11 decreased to 30.75 MPa, with a reduction of 20.1%; B21 decreased to 27.68 MPa, with a reduction of 28.1%; and B32 decreased to 25.15 MPa, with a reduction of 34.6%.

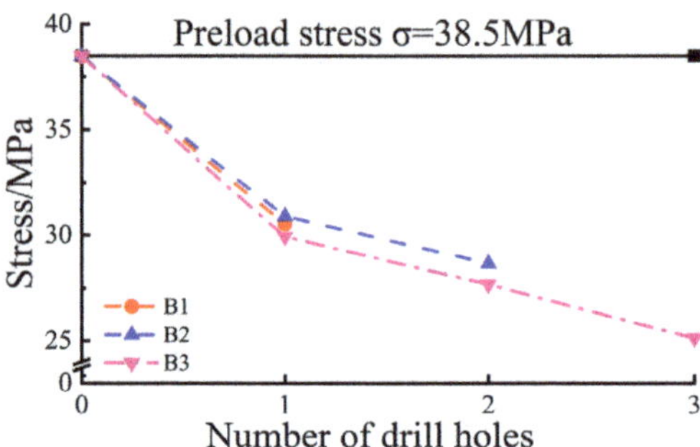

Figure 7. The variation in stress in the specimen pressure relief hole drilled in stages.

Figure 8 shows the stress changes in one-off drilling: the stress of C12 decreased from 38.50 MPa to 30.61 Mpa, with a reduction of 20.5%; the stress of C22 decreased from 38.50 Mpa to 29.85 Mpa, with a reduction of 22.4%; and the stress of C32 decreased from 38.50 MPa to 27.37 MPa, with a reduction of 28.9%.

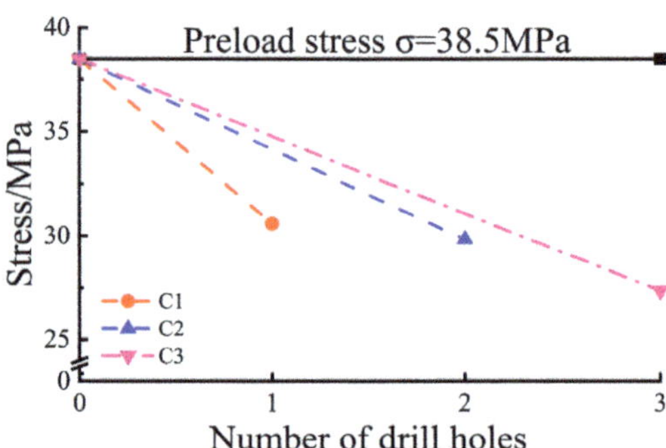

Figure 8. The variation of stress in one drilling of the specimen pressure relief hole.

Comparing the test results reveals that the pressure relief amplitude of B32 was 6.5% higher than that of B21, and that of B21 was 8.0% higher than that of B11; the pressure relief amplitude of C32 was 6.5% higher than that of C22, and that of C22 was 1.9% higher than that of C12. The stress–time curves indicate that the samples retained considerable strength after pressure relief and could still provide good load-bearing capacity. This suggests that, with a spacing of 40 mm and up to three boreholes, more boreholes lead to better pressure relief ahead of peak stress.

Figures 9 and 10 present the stress changes for samples with the same number of boreholes but different drilling modes under uniaxial compression. For the two-borehole samples, the stress decreased to 27.68 MPa for B21 and 29.85 MPa for C22, with the pressure relief amplitude of staged drilling (B21) being 5.7% higher than that of one-off drilling (C22). As shown in Figure 9, the stresses were reduced to 25.15 MPa for B32 and 27.37 MPa for C32, giving a 5.7% higher relief amplitude for staged drilling (B32) over one-off drilling (C32). The results demonstrate that staged drilling is more effective than one-off drilling for the same number of pressure relief boreholes in sandstone samples under high stress, especially with more boreholes: the increased relief of three-hole samples was 1.9 times that of two-hole sample.

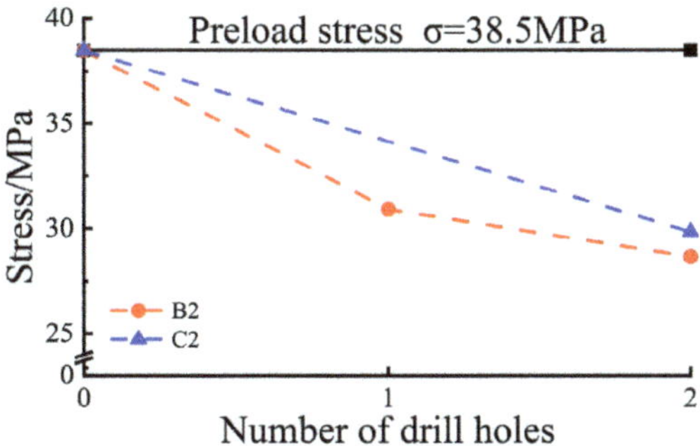

Figure 9. Characteristics of variation in pressure relief stress for a sample with two pressure relief holes.

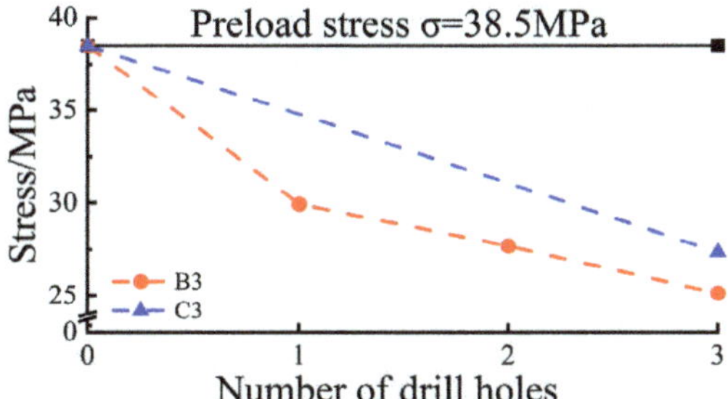

Figure 10. Characteristics of variation in pressure relief stress for a sample with three pressure relief holes.

When the number of pressure relief holes increases, both staged drilling and one-time drilling lead to varying degrees of improvement in the stress drop amplitudes of the samples. This indicates that increasing the number of pressure relief holes is advantageous for enhancing the pressure relief effect. Under the same number of pressure relief holes, the stress drop amplitude of staged drilling is significantly higher than that of one-time drilling. For instance, the three-hole sample B32 has a drop amplitude of 34.6%, while C32 has 28.9%. This demonstrates that staged drilling provides better pressure relief than one-time drilling. It is analyzed that staged drilling can gradually release the internal energy stored in the rock, which facilitates the control of the pressure relief process. In contrast, one-time drilling tends to cause stress concentration and abrupt release. Under high stress conditions, the more pressure relief holes drilled sequentially, the more pronounced the improvement in pressure relief. The pressure relief amplitude increase of the three-hole sample is 1.9 times that of the two-hole sample. After reloading, the sample still possesses considerable load bearing capability, indicating the staged drilling of pressure relief holes can effectively reduce the stress level of the rock mass, given proper control over the number and layout of the holes. In summary, for rock masses under high stress, adopting the staged drilling of pressure relief holes and gradually increasing the hole quantity can achieve superior pressure relief effects, benefiting the stable control of the rock mass.

4. Numerical Simulation

4.1. Building of the Model

A FLAC3D model was built based on the sample dimensions, using the Mohr–Coulomb constitutive model. The model parameters are as follows: density $2700/(\text{kg·m}^{-3})$, bulk modulus $19.79/\text{GPa}$, shear modulus $19.86/\text{GPa}$, internal friction angle $34°$, cohesion $4.8/\text{MPa}$ and tensile strength $6.4/\text{MPa}$. The bottom boundary displacement was constrained, and 38.50 MPa vertical stress was applied in the Z direction. Circular boreholes with an 8 mm diameter and 80 mm depth were excavated in the model. The calculation convergence criteria required the unbalanced force ratio to reach 1×10^{-5}.

4.2. The Analysis of Numerical Simulation Results Compared with Experimental Results

The borehole pressure relief effect in sandstone samples under high stress was simulated, obtaining the plastic zone evolution as shown in Figure 11. Taking B32 as an example, the plastic zone volume was extracted using the Fish language and compared with the stress changes.

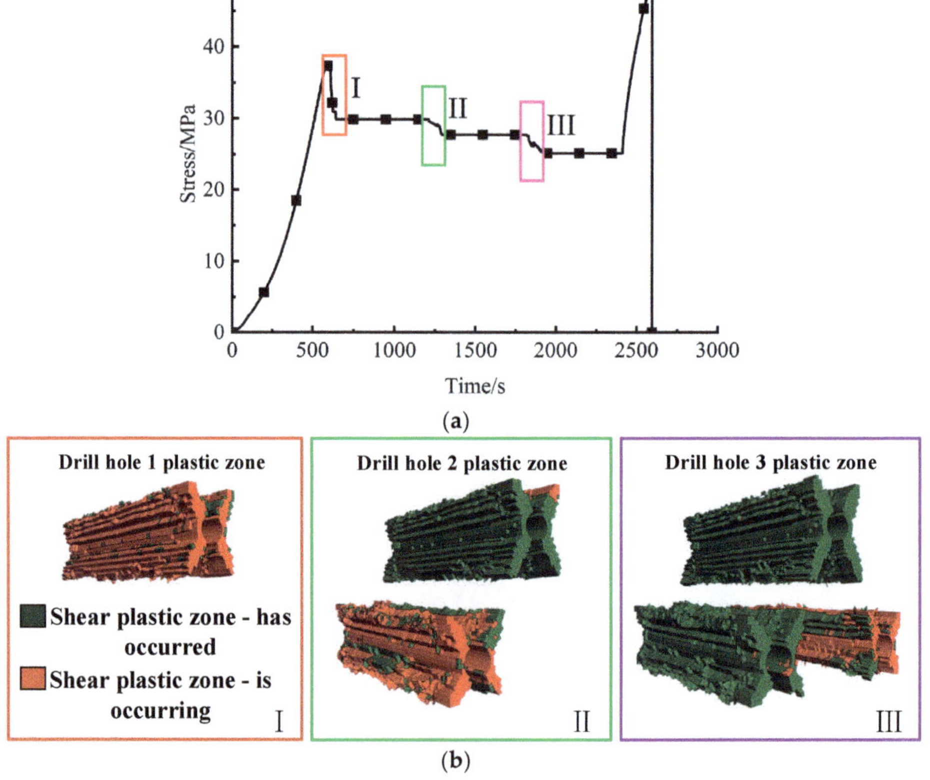

Figure 11. *Cont.*

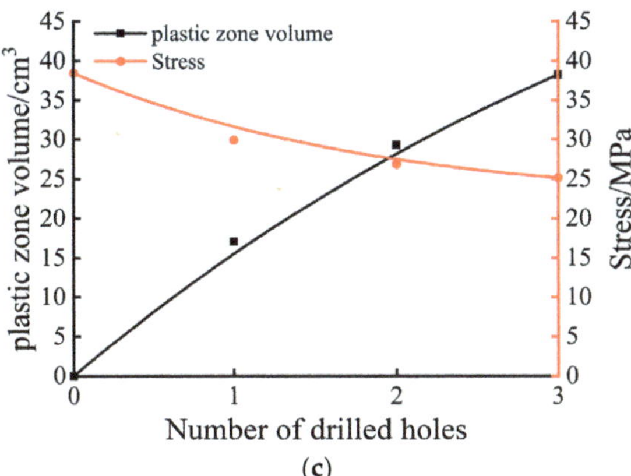

(c)

Figure 11. Characteristic diagram of the plastic zone of the sample. (**a**) B32 stress–time curve. (**b**) Cloud view of the plastic zone of the specimen. (**c**) B32 plastic zone versus stress relationship.

As shown in Figure 11b, after calculating to equilibrium, two types of plastic zones were observed—occurred and occurring shear plastic zones. The minimum distance between the zones was 15 mm. Comparing the plastic zone clouds at each stage shows that the occurring shear plasticity was concentrated around the new borehole during drilling, while the small volume increased its concentration at the edges of previous boreholes. This indicates that the influence of new boreholes on previous plastic zones is minor at 40 mm spacing, with stable borehole-surrounding rock. It implies that, with increasing stress, cracks would propagate from the plastic zone edges to the sample boundaries, penetrating the boreholes and causing failure, consistent with experimental observations. Figure 11c shows a close relationship between pressure relief amplitude and plastic zone volume. For the first borehole, the volume was 17.06 cm^3 with a stress of 29.94 MPa. For the second borehole, the volume increased to 29.29 cm^3 with a stress of 26.70 MPa. For the third borehole, the volume reached 38.27 cm^3 with a stress of 25.15 MPa. This demonstrates that more boreholes lead to larger plastic zone volumes and greater stress reductions under high stress, indicating better pressure relief. The occurring shear plastic zone volumes for B32 were 16.28 cm^3, 11.55 cm^3 and 8.98 cm^3 for each stage, with corresponding stress reductions of 22.2%, 6.6% and 5.8%. This suggests that the single-borehole plastic zone volume is negatively correlated with borehole quantity, and the relief effect diminishes gradually.

5. Discussion

The energy transformation of rock samples under uniaxial compression can be divided into four stages—compaction energy input, elastic energy accumulation, crack expansion energy dissipation and post-peak failure energy release [25]—as shown in Figure 12.

This study focuses on pre-peak borehole pressure relief in sandstone. Since the preload stress is much lower than the failure stress of intact samples, there is no yielding before drilling relief boreholes. Thus, the energy process involves only compaction input and elastic accumulation. The total energy absorption increases continuously under loading, with most of the energy stored as elastic strain energy. At this point, the sample elastic strain energy is the peak elastic strain energy without the limit being reached, so macroscopic failure does not occur [26].

Drilling the first relief borehole causes mechanical disturbances that induce microcracks in the surrounding rock. The released elastic strain energy promotes microcrack growth. Crack propagation expands the plastic zone, decreasing the stress [11]. After the first borehole, loading stops; therefore, the lost elastic strain energy cannot be replenished without external work. The remaining elastic energy is insufficient to extend the cracks further, and the sample stabilizes. Now, the retained elastic strain energy is less than that of intact samples. The second borehole provides less driving energy for microcrack growth, with a smaller plastic zone increase and stress reduction compared to the first stage. The third stage works on the same principle.

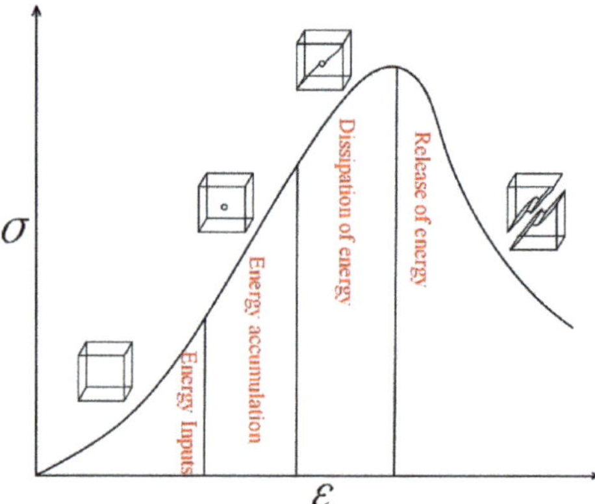

Figure 12. Schematic diagram of energy evolution during rock deformation and failure.

Xie Heping et al. [27] stated that energy dissipation is directly related to strength loss and that the original strength decay reflects the dissipated energy amount. The test results show that the pressure relief amplitude in the first stage is larger than that in the second stage, which is larger than that in the third stage under the same energy input. This demonstrates decreasing energy dissipation over the stages. The efficiency of single-borehole pressure relief ahead of peak stress decreases with more drilling events, indicating limited relief and close correlation with retained elastic energy. Comparing staged and one-off drilling reveals that more boreholes lead to greater total unloading amplitudes, suggesting increased total energy dissipation. Staged drilling has slower dissipation over longer cycles but larger relief amplitudes, reducing impact tendencies. One-off drilling dissipates energy faster over shorter cycles and is therefore suitable for rock burst hazards.

Due to experimental constraints and the specificity of pressurized borehole tests, this study did not accurately calculate the dissipated energy values and input energy differences between drilling events. Further research is needed to obtain accurate energy quantifications.

6. Conclusions

The following conclusions were obtained from the borehole pressure relief experiments on sandstone samples under high stress:

(1) For non-prefabricated boreholes in sandstone samples under high stress, staged drilling resulted in pressure relief amplitudes of 20.1%, 28.1% and 34.6%, while one-off drilling led to relief amplitudes of 20.5%, 22.4% and 28.9%. The pressure relief effect is positively correlated with the number of boreholes. Staged drilling is more effective than one-off drilling.

(2) The numerical simulation results show that the surrounding rock forms plastic zones under stress. In staged drilling tests with 40 mm spacing, the influence of new boreholes on existing plastic zones is minor, with each stage dominated by the individual borehole plasticity.

(3) Combining the results of the lab tests and simulations shows that the volume of staged drilling plastic zones is an important factor influencing pressure relief. The volumes are 16.2808 cm^3, 11.5506 cm^3 and 8.9823 cm^3 for each stage, with corresponding stress reductions of 22.2%, 6.6% and 5.8%. This indicates that the single-borehole relief amplitude is negatively correlated with the number of existing boreholes.

Author Contributions: Conceptualization, J.J. and W.W.; methodology, X.L.; software, X.C.; validation, W.W., X.C. and L.H.; formal analysis, X.L.; investigation, X.C.; resources, J.J.; data curation, X.L.; writing—original draft preparation, X.L.; writing—review and editing, J.J.; visualization, X.C.; supervision, W.W.; project administration, W.W.; funding acquisition, W.W. All authors have read and agreed to the published version of the manuscript.

Funding: This study was supported by the Program for National Major Achievement Cultivation, Theory and Key Technology of Natural Gas and Coal Resources Cooperative Mining, NSFRF230202; Theory and Technology of Natural Gas-Coal-Uranium Mining Synergy; Research on the Theory and Key Technology of Coordinated Natural Gas and Coal Mining, 23HASTIT011; Research on the Mechanism of Coordinated Coal and Natural Gas Exploitation and Disaster Warning in Ordos Basin, T2022-2; National Natural Science Foundation of China funded project (52174109).

Data Availability Statement: Due to the nature of this research, participants of this study did not agree for their data to be shared publicly, so supporting data is not available.

Conflicts of Interest: The authors declare no conflict of interest.

References

1. Wang, J.Y.; Hu, S.B.; Pang, Z.H.; He, L. Estimate of geothermal resources potential for hot dry rock in the continental area of China. *Sci. Technol. Rev.* **2012**, *30*, 25–31.
2. Xu, T.F.; Zhang, Y.J.; Zeng, Z.F.; Bao, X. Technology progress of enhanced geothermal system (hot dry rock). *Sci. Technol. Rev.* **2012**, *30*, 42–45.
3. Brown, D. The US hot dry rock program-20 years of experience in reservoir testing. In Proceedings of the World Geothermal Congress, Florence, Italy, 18–31 May 1995; pp. 2607–2611.
4. Lu, C.; Wang, G.L. Current status and prospect of hot dry rock research. *Sci. Technol. Rev.* **2012**, *33*, 13–21.
5. Zhai, H.Z.; Su, Z.; Wu, N.Y. Development experiences of the Soultz enhanced geothermal systems and inspirations for geothermal development in China. *Adv. New Renew. Energy* **2014**, *2*, 286–294.
6. Lu, Y.; Zou, X.; Liu, C.; Xing, S.J. Technology of Digging Stress-Relax Entry by the Roadside and Its Application. *J. Min. Saf. Eng.* **2006**, *23*, 329–332.
7. Guo, B.; Lu, T. Analysis of Floor Heave Mechanism and Cutting Control Technique in Deep Mines. *J. Min. Saf. Eng.* **2008**, *25*, 91–94.
8. Wang, L.; Jiang, F.; Yu, Z. Similar Material Simulation Experiment on Destressing Effects of the Deep Thick Coal Seam with High Burst Liability after Mining Upper and Lower Protective Seams. *Chin. J. Geotech. Eng.* **2009**, *31*, 442–446.
9. Song, X.; Zuo, Y.; Zhu, W. Effect of Lateral Pressure Coefficients on Pressure-Released Hole Combined Support with Rockbolt under Dynamic Disturbance. *Chin. J. Undergr. Space Eng.* **2013**, *9*, 1076–1081.
10. Ortlepp, P.W.D.; Stacey, T.R. Rockburst Mechanisms in Tunnels and Shafts. *Tunn. Undergr. Space Technol.* **1994**, *9*, 59–65. [CrossRef]
11. Kaiser, P.; Yazici, S.; Maloney, S. Mining-Induced Stress Change and Consequences of Stress Path on Excavation Stability—A Case Study. *Int. J. Rock Mech. Min. Sci.* **2001**, *38*, 167–180. [CrossRef]
12. Liu, H.; He, Y.; Xu, J.; Han, L.-J. Numerical Simulation and Industrial Test of Boreholes Distressing Technology in Deep Coal Tunnel. *J. China Coal Soc.* **2007**, *32*, 33–37.
13. Zhang, Y.; Hao, F.; Liu, C.; Liu, W.J. Borehole Relief Range Considering Plastic Softening and Dilatancy of Coal. *J. Liaoning Tech. Univ.* **2013**, 1599–1604.
14. Zheng, H.; Wang, M.; Xu, S. Research on Surrounding Rock Pressure Relief in Deep Roadway by Borehole and Its Control Technology. *Min. Saf. Environ. Prot.* **2014**, 51–55.
15. Zhu, S.; Jiang, F.; Shi, X.; Sun, G.; Zhang, Z.; Cheng, X.; Zhang, H. Energy Dissipation Index Method for Determining Rockburst Prevention Drilling Parameters. *Rock Soil Mech.* **2015**, *36*, 2270–2276.
16. Wang, S.; Pan, J.; Liu, S.; Xia, Y.X.; Gao, X.J. Evaluation Method for Rockburst-Preventing Effects by Drilling Based on Energy-Dissipating Rate. *J. China Coal Soc.* **2016**, *41*, 297–304.
17. Wang, M.; Wang, X.; Xiao, T. Borehole Destressing Mechanism and Determination Method of Its Key Parameters in Deep Roadway. *J. China Coal Soc.* **2017**, *42*, 1138–1145.
18. Jia, C.; Jiang, Y.; Zhang, X.; Wang, D.; Luan, H.; Wang, C. Laboratory and Numerical Experiments on Pressure Relief Mechanism of Large-Diameter Boreholes. *Rock Soil Mech.* **2017**, *39*, 1115–1122.
19. Qin, Z. Study of Pressure Relief with Large Diameter Drilling Hole and Results Verified. *Coal Min. Technol.* **2018**, *23*, 82–85.
20. Ma, B.W.; Deng, Z.G.; Zhao, S.K.; Li, S.G. Analysis on Mechanism and Influencing Factors of Drilling Pressure Relief to Prevent Rock Burst. *Coal Sci. Technol.* **2020**, *48*, 35–40.
21. Zhao, Z.; Zhang, X.; Li, X. Experimental Study of Stress Relaxation Characteristics of Hard Rocks with Pressure Relief Hole. *Rock Soil Mech.* **2019**, *40*, 170–177.
22. Lin, P.; Wong, R.H.; Tang, C. Experimental Study of Coalescence Mechanisms and Failure under Uniaxial Compression of Granite Containing Multiple Holes. *Int. J. Rock Mech. Min. Sci.* **2015**, *77*, 313–327. [CrossRef]

23. Zhao, T.-B.; Guo, W.-Y.; Yu, F.-H.; Tan, Y.-L.; Huang, B.; Hu, S.-C. Numerical Investigation of Influences of Drilling Arrangements on the Mechanical Behavior and Energy Evolution of Coal Models. *Adv. Civ. Eng.* **2018**, *2018*, 1–12. [CrossRef]

24. Huang, B.; Guo, W.-Y.; Fu, Z.-Y.; Zhao, T.-B.; Zhang, L.-S. Experimental Investigation of the Influence of Drilling Arrangements on the Mechanical Behavior of Rock Models. *Geotech. Geol. Eng.* **2018**, *36*, 2425–2436. [CrossRef]

25. Zhang, L.; Gao, S.; Wang, Z. Rock Elastic Strain Energy and Dissipation Strain Energy Evolution Characteristics under Conventional Triaxial Compression. *J. China Coal Soc.* **2014**, *39*, 1238–1242.

26. Zhao, Z.; Lu, R.; Zhang, G. Analysis on Energy Transformation for Rock in the Whole Process of Deformation and Fracture. *Min. Res. Dev.* **2006**, *26*, 8.

27. Xie, H.; Ju, Y.; Li, L. Criteria for Strength and Structural Failure of Rocks Based on Energy Dissipation and Energy Release Principles. *Chin. J. Rock Mech. Eng.* **2005**, *24*, 3003–3010.

 *sustainability*

Article

Analysis of Surface Settlement Induced by Shield Tunnelling: Grey Relational Analysis and Numerical Simulation Study on Critical Construction Parameters

Minhe Luo [1], Ding Wang [2,3,*], Xuchun Wang [1,*] and Zelin Lu [1]

[1]   School of Civil Engineering, Qingdao University of Technology, Qingdao 266520, China; sbermin@163.com (M.L.); luzelin1025@163.com (Z.L.)
[2]   School of Mechanical and Electrical Engineering, China University of Mining and Technology, Beijing 100083, China
[3]   State Key Laboratory for Geomechanics and Deep Underground Engineering, China University of Mining and Technology, Beijing 100083, China
*   Correspondence: wangdingneil@outlook.com (D.W.); xcwang2008@126.com (X.W.); Tel.: +86-0532-8507-1310 (X.W.)

Abstract: Excessive surface settlement poses significant challenges to shield tunnelling construction, resulting in damage to adjacent buildings, infrastructure, and underground pipelines. This study focused on investigating the surface settlement induced by shield tunnelling during the construction of Qingdao Metro Line 6 between Haigang Road Station and Chaoyang Road Station. Firstly, the settlement data from the left line of the shield tunnel were evaluated by grey relational analysis. The relational coefficients were calculated to assess the correlation degrees of each influential parameter. Subsequently, the four critical influential parameters with the highest relational degrees were chosen to investigate their effects on surface settlement through numerical simulations under different scenarios. The results show that the four parameters with the highest relational degrees were thrust, grouting pressure, earth pressure, and strata elastic modulus. It should be noted that the strata elastic modulus significantly affects surface settlement, while the grouting pressure influences the settlement trough width in weak strata. Moreover, improper thrust magnitude can lead to an increase in surface settlement. Based on these findings, recommendations are proposed for the right-line tunnel construction and practical countermeasures for surface settlement during shield tunnelling construction are provided.

Keywords: shield tunnelling; surface settlement; grey relational analysis; numerical simulation; influential parameters

Citation: Luo, M.; Wang, D.; Wang, X.; Lu, Z. Analysis of Surface Settlement Induced by Shield Tunnelling: Grey Relational Analysis and Numerical Simulation Study on Critical Construction Parameters. *Sustainability* **2023**, *15*, 14315. https://doi.org/10.3390/su151914315

Academic Editors: Hong-Wei Yang, Shuren Wang and Chen Cao

Received: 28 August 2023
Revised: 25 September 2023
Accepted: 26 September 2023
Published: 28 September 2023

1. Introduction

Rapidly escalating urbanization calls for efficient transport systems and has made metro construction crucial to meet growing urban transportation demand. The shield tunnelling method has become the primary construction method for underground metro projects, owing to the rapid construction speed, automated operation, and minimal impact on surface traffic. However, shield tunnelling inevitably disturbs the surrounding rocks, leading to stress redistribution within the soil [1] and inducing surface settlement [2,3]. The excessive surface settlement can damage adjacent buildings and infrastructure [4,5], jeopardizing the safety of pedestrians, vehicles, and roads [6]. This in turn can lead to work stoppages and result in additional costs [7]. Moreover, it may trigger soil liquefaction and landslides, leading to severe construction accidents. Therefore, it is essential to understand and predict the behavior of surface settlement induced by shield tunnelling to ensure the safety and stability of urban areas.

An approach using Gaussian normal distribution to represent the lateral surface settlement induced by shield tunnelling was initially proposed by Peck [8]. The prediction formulas to describe surface settlement and its influential parameters were introduced and improved by subsequent researchers [9–14]. Notably, these formulas typically rely on empirical parameters and require local measurements for theoretical adjustments. In addition, some researchers have also studied surface settlement through model tests [15–17], but these tests have limitations in accurately imitating field conditions and actual construction scenarios. For greater authenticity and credibility, researchers utilized the measured data to analyze the correlation between the surface settlement and the influential construction parameters. For instance, Kannangara et al. [18] investigated the impact of shield tunnelling parameters on surface settlement. They identified the correlation between earth pressure and grouting pressure on the peak of surface settlement. Based on the construction data from EPB-TBM in the Moscow metro, Ter-Martirosyan et al. [19] studied the effect of metro tunnel construction parameters on surface settlement. The results indicated a close mutual relationship between earth pressure and surface settlement. Dongku Kim et al. [20] analyzed data from the construction section of the Hong Kong Metro and identified parameters such as boring speed, thrust, and cutterhead torque presented a modest correlation with the surface settlement. However, it cannot offer specific functional relationships or in-depth analysis of the complex interactions between the construction parameters and the surface settlement.

In recent years, numerical simulation methods have been widely employed to solve the interdisciplinary problems encountered in underground engineering, involving the surrounding rock deformation affected by novel metallic rock support materials [21,22] and the surface settlement issues induced by shield tunnelling [23–30]. Wang et al. [31] employed particle flow to simulate surface settlement in sandy soils, revealing its dynamic nature involving soil arch formation and destruction. Alzabeebee et al. [32] studied traffic load, microtunnel diameter, and backfill height using a finite element model. They found traffic loads significantly impact microtunnel-induced pavement settlement. Through finite element analysis, Yuan et al. [33] investigated the effect of parameters like deformation modulus, burial depth, and tunnel spacing on surface settlement during shield construction in weathered mudstone areas. Numerical simulations can theoretically model the entire shield construction process. However, existing simulations frequently suffer from inaccurate selection and validation of essential simulation parameters due to the multitude of construction variables. In addition, these simulations typically oversimplify the surface settlement process by focusing on a single and homogeneous stratum, leaving room for improvement to consider complex geological conditions and various influential parameters.

Existing data analysis methods have limitations in quantifying settlement impact patterns, and numerical simulations encounter challenges in selecting critical construction parameters. Hence, this study proposes a novel hybrid numerical approach combined with grey relational analysis to analyze the critical construction parameters influencing surface settlement comprehensively. The grey relational analysis serves to quantify the correlation degree of various parameters with surface settlement by calculating grey relational coefficients [34]. When coupled with numerical simulations, it facilitates optimized parameter selection and enhances the assessment of specific parameters' impact on surface settlement. The engineering background was based on the shield construction section of Qingdao Metro Line 6 between Haigang Road Station and Chaoyang Road Station. This tunnel section presents complex geological conditions with diverse soil properties and various stratigraphic characteristics, where severe settlements occurred during the construction of the left line. In this study, the settlement monitoring data from the left line were evaluated by grey relational analysis to assess the relational degrees of each influential parameter. Subsequently, numerical simulations were conducted to investigate the effect of critical construction parameters on surface settlement, and recommendations are proposed for the right-line tunnelling. Finally, practical countermeasures for surface settlement during shield tunnelling construction are provided based on the research findings to ensure the safety and stability of future urban subway projects.

2. Project Overview

The shield construction section of Qingdao Metro Line 6, between Haigang Road Station and Chaoyang Road Station, comprises two single-line single-tunnel sections, starting and ending at Y(Z)DK 31 + 248.393~32 + 039.183. The right-line section is 790.790 m in length, while the left is 791.732 m. The depth of vault ranges from 12.9 m to 25.3 m. CR Equip. Machine 179, produced by China Railway Engineering Equipment Group Co., Ltd., headquartered in Zhengzhou, China, was employed for this tunnel section. The shield machine excavated and constructed the tunnel from Chaoyang Road Station and advanced toward Haigang Road Station, featuring an excavation diameter of ϕ = 6303 mm. The cutterhead comprised six spoke-type cutters and six face plates, totaling 34 face cutters, 12 gauge cutters, and six center cutters. The main parameters of the shield machines are detailed in Table 1.

Table 1. Main parameters of the shield machine.

Parameter	CR Equip. Machine 179
Tunnel boring type	Earth pressure balance
Excavation diameter (mm)	6303
Shield diameter (mm)	6250
Cutterhead rotation speed (rpm)	0~3.7
Maximum thrust (kN)	39,000
Rated torque (kN/m)	6000
Unblocking torque (kN/m)	7200
Horizontal turning capability (m)	250
Maximum climbing capability (‰)	±50
Cutterhead opening ratio (%)	34

The topography of the construction section slopes from northwest to southeast. The YDK31 +248.393~+339.682 section exhibits an erosional slope landform, while the YDK31 +339.682~32 + 039.183 section displays an erosional accumulation and gently sloping landform. Within the section, there are six standard layers and 19 sublayers. The left line of the construction section traverses several main strata, including powdery clay, clayey gravelly sand, strongly weathered granite porphyry, moderately weathered granite porphyry, slightly weathered granite porphyry, blocky granite (fractured), strongly weathered tuff, moderately weathered tuff, and sandy tuff (fractured). The groundwater in the section mainly consists of quaternary pore water and bedrock fissure water. Additionally, the construction section is influenced by the Lingshanwei fracture and its secondary fracture. The fault's exposure along the Lingshanwei area shows a northeast strike of 40° to 55°, predominantly northwest tendency, and a dip of 70° to 88°. This section reveals three fracture zones and one severely affected zone of the Lingshanwei fault. Table 2 provides more detailed stratigraphic physical and mechanical parameters, sourced from the "Geotechnical Investigation Report of Qingdao Metro Line 6 Phase I Project, Haigang Road Station—Chaoyang Road Station," supplied by Qingdao Municipal Research Institute of Surveying and Mapping. The geological profile of the left-line construction section is illustrated in Figure 1. Figures 2 and 3 depict the layout of monitoring points and an aerial view in the vicinity of Haigang Road Station, respectively.

Table 2. Stratigraphic physical and mechanical parameters of the construction section between Haigang Road Station and Chaoyang Road Station.

Stratum	Density (kg/m³)	Cohesion (kPa)	Internal Friction (°)	Elastic Modulus [1] (MPa)	Poisson's Ratio
Plain fill soil	1650	8	14	8	0.3
Powdery clay	1980	18	16	16	0.4
Clayey gravelly sand	2050	14	28	12	0.3

Table 2. *Cont.*

Stratum	Density (kg/m³)	Cohesion (kPa)	Internal Friction (°)	Elastic Modulus [1] (MPa)	Poisson's Ratio
Sandy tuff (fractured)	2450	25	33	40	0.3
Blocky granite (fractured)	2500	48	35	100	0.3
Slightly weathered tuff	2610	710	45	1250	0.2
Moderately weathered tuff	2600	400	40	550	0.3
Strongly weathered tuff	2300	22	30	20	0.3
Slightly weathered granite porphyry	2680	530	45	1080	0.2
Moderately weathered granite porphyry	2600	300	40	400	0.3
Strongly weathered granite porphyry	2300	15	30	18	0.3

[1] The elastic modulus in the table is the tangential modulus.

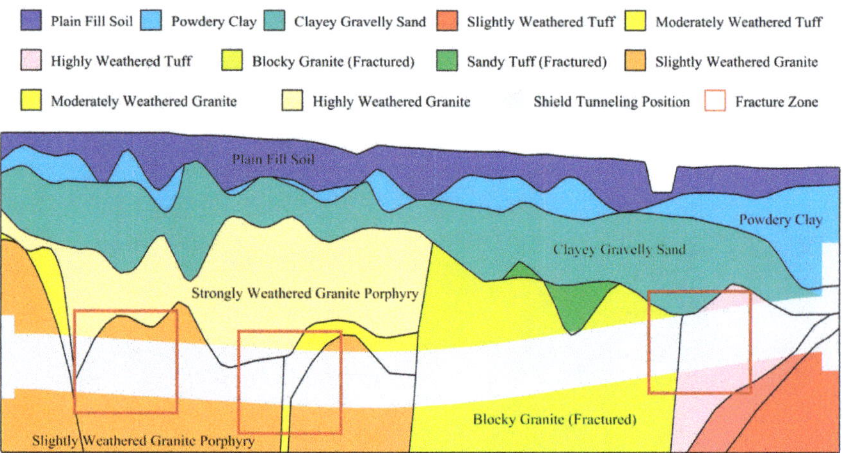

Figure 1. Geological profile of the left-line construction section between Haigang Road Station and Chaoyang Road Station.

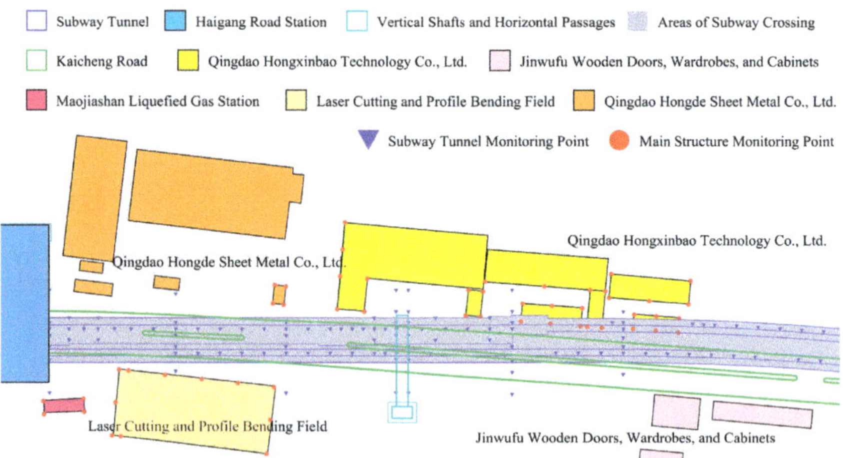

Figure 2. Layout plan of monitoring sites at Haigang Road Station.

Figure 3. Aerial photo of the proposed tunnel construction section between Haigang Road Station and Chaoyang Road Station.

3. Grey Relational Analysis

Grey relational analysis is a quantitative method used to assess the correlation of trends among different parameters in a complex system. It is particularly suitable for systems with multiple independent variables and unclear interconnections. The method involves the following key steps.

1. Determine the evaluation index sequence and the influential parameter sequence.
2. Transform the obtained sequences into dimensionless form.
3. Calculate the difference sequences and identify the maximum and minimum differences.
4. Compute the relational coefficients and determine the grey relational degree.

The above four steps enable researchers to evaluate the influence of different factors on the evaluation index and facilitate the subsequent optimization analysis.

3.1. Model Index Determination

When calculating the relational degree in the grey relational model, it is necessary to determine the evaluation index sequence (dependent variables) and the influential parameter sequence (independent variables) that characterize the system behavior. The selection of the evaluation index sequence directly affects the rationality of the relational degree calculation results:

$$X_0 = \{x_0(1), x_0(2), \cdots, x_0(n)\} \tag{1}$$

$$X_i = \{x_i(1), x_i(2), \cdots, x_i(n)\} \tag{2}$$

where X_0 is the evaluation index sequence and X_i is the influential parameter sequence.

Previous studies [35–39] have identified several influential factors for the soil movement and surface settlement induced by shield tunnelling:

1. Changes in the stress state of the excavated soil.
2. Squeezing and forward movement of the soil in front of the excavation face and the surrounding soil in contact with the shield.
3. Over-excavation during the tunnelling process.
4. Formation of gaps around the shield due to the shield shell radius being smaller than the cutterhead radius, causing the surrounding soil to move towards the void and result in surface settlement.

5. Improper grouting at the shield tail leads to soil compression and settlement.
6. Interaction between the soil and the lining, as well as between the lining and the segment.
7. Creep or secondary consolidation of the soft clay soil following construction disturbance around the shield tunnel, leading to continuous secondary consolidation settlement.

Among the above influential factors, factors 1, 2, and 3 affect the stability of the excavation face and can be controlled by adjusting tunnelling parameters during construction. Factors 4, 5, and 6 are related to grouting quality and can be controlled by adjusting grouting parameters. Factor 7 can be considered stratum parameters.

Based on the above correlations between the influential factors and the surface settlement, this study selected the surface settlement as the evaluation index sequence in the relational degree calculation. To ensure the measuring data's coherence, surface settlement data were obtained from the monitoring points above the centerline of the left-line tunnel section. The influential parameter sequences selected were earth pressure, tunnelling speed, and thrust as tunnelling parameters; grouting pressure and grouting volume as grouting parameters; and the weighted average of cohesion, internal friction angle, elastic modulus, and Poisson's ratio as stratum parameters. The impact on surface settlement was evaluated by analyzing the relational degrees of these parameters. Table 3 summarizes the influential construction parameters and corresponding surface settlement data collected and collated for each ring during the left-line tunnelling.

Table 3. Influential construction parameters and corresponding surface settlement.

Data ID	Settlement (cm)	Tunnelling Parameters			Grouting Parameters		Stratum Parameters			
		Thrust (kN)	Earth Pressure (kPa)	Tunnelling Speed (mm/min)	Grouting Pressure (kPa)	Grout Volume (m^3)	Cohesion (kPa)	Internal Friction (°)	Poisson's Ratio	Elastic Modulus (MPa)
0	−3.10	10,424	83	15	272	3.4	237	39.7	0.23	471
1	−2.34	9817	94	33	244	3.1	180	37.6	0.22	477
2	−4.29	7656	69	26	207	3.1	237	38.9	0.21	518
3	−1.54	11,460	121	13	298	5.0	223	39.9	0.22	542
4	−0.61	10,138	113	15	244	4.8	164	38.2	0.22	489
5	−6.27	10,438	18	12	260	3.8	213	37.6	0.21	448
6	−2.44	9855	90	16	223	3.2	230	42.9	0.20	533
7	−3.12	9440	81	12	300	3.8	237	42.9	0.20	585
8	−5.70	6886	17	22	194	5.3	172	40.9	0.21	391
9	−7.09	7441	37	33	185	5.7	145	38.7	0.21	323
10	−9.28	7791	21	20	203	5.6	160	38.4	0.20	238
11	−9.90	7419	21	13	203	5.3	191	35.4	0.20	158
12	−8.79	7965	69	37	214	5.9	161	39.5	0.21	150
13	−8.71	9489	82	15	220	5.6	175	40.1	0.21	207
14	−7.53	7045	73	28	188	5.9	77	35.9	0.26	220
15	−7.56	7645	82	29	270	5.8	137	37.9	0.24	298
16	−4.50	8921	76	27	334	5.3	142	40.1	0.23	383
17	−3.92	7809	62	27	223	5.0	225	42.9	0.21	499
18	−3.36	9379	54	39	190	5.2	198	40.6	0.22	475
19	−1.61	9723	108	14	295	6.1	199	40.4	0.22	516
20	−0.92	10,344	118	32	277	6.1	237	42.9	0.22	524
21	−1.61	10,074	66	24	232	5.4	177	40.4	0.23	435
22	−1.34	10,217	112	27	232	4.4	155	39.3	0.24	408
23	0.03	10,678	126	26	292	4.8	147	40.1	0.24	429
24	−0.21	10,735	125	28	283	5.6	107	35.7	0.28	586
25	−2.09	9961	64	44	264	6.2	159	39.6	0.27	447
26	−1.68	9572	104	25	274	4.1	40	32.6	0.28	495
27	−2.42	9374	125	34	274	6.2	40	33.3	0.28	461
28	−1.13	11,109	105	26	247	6.2	40	31.0	0.28	520
29	−1.25	9547	109	34	237	4.4	40	29.9	0.28	510
30	−2.30	8677	142	36	233	3.6	40	27.0	0.28	454
31	−1.09	10,429	145	30	268	3.6	40	29.5	0.28	515
32	−1.11	10,260	126	41	243	3.2	40	31.2	0.28	518
33	−3.11	9504	126	38	194	3.5	40	33.5	0.27	427
34	−6.23	10,703	74	34	221	3.1	40	25.9	0.28	160
35	−7.14	7041	64	33	200	3.2	40	25.7	0.28	147

Table 3. *Cont.*

Data ID	Settlement (cm)	Tunnelling Parameters			Grouting Parameters		Stratum Parameters			
		Thrust (kN)	Earth Pressure (kPa)	Tunnelling Speed (mm/min)	Grouting Pressure (kPa)	Grout Volume (m³)	Cohesion (kPa)	Internal Friction (°)	Poisson's Ratio	Elastic Modulus (MPa)
36	−7.76	7574	55	37	181	3.1	40	23.5	0.26	240
37	−10.39	7563	26	29	135	4.1	102	23.3	0.23	91
38	−10.37	7170	25	28	150	3.3	146	28.7	0.24	168
39	−3.08	9289	107	27	228	3.9	163	28.1	0.26	470
40	−2.82	9150	101	12	237	3.7	237	25.4	0.28	654

3.2. Grey Relational Degree Calculation

The grey relational degree calculation quantifies the relative influences and trends of selected influential parameter sequences compared to the evaluation index sequence. To account for the varying dimensions and magnitudes of the system's influential parameters and minimize errors in the calculation, mean normalization is performed on the original data columns for dimensionless standardization:

$$X_0^*(k) = X_0(k)/\overline{X}_0 (k = 1, 2, \cdots, n) \tag{3}$$

$$X_i^*(k) = X_i(k)/\overline{X}_i (k = 1, 2, \cdots, n) \tag{4}$$

where $\overline{X}_0$ and $\overline{X}_i$ represent the average values of the evaluation index sequence and influential parameter sequences, respectively.

The absolute differences between all index sequences and the influential parameter sequence are then calculated using the following formula:

$$\Delta_{0i}(k) = |X_0^*(k) - X_i^*(k)| (i = 1, 2, \cdots, n) \tag{5}$$

Subsequently, the maximum and minimum absolute differences are obtained with the following formulas:

$$\Delta_{max} = Max_i Max_k \Delta_{0i}(k) \tag{6}$$

$$\Delta_{min} = Min_i Min_k \Delta_{0i}(k) \tag{7}$$

where $\Delta_{0i}(k)$ is the absolute difference between the ith influential parameter sequence and the evaluation index sequence at the kth data point, and Δ_{max} and Δ_{min} represent the maximum and minimum values in the set, respectively.

Next, the relational coefficient and grey relational degree are calculated using the following expressions:

$$l_{0i} = \frac{\Delta_{min} + \rho\Delta_{max}}{\Delta_{0i}(k) + \rho\Delta_{max}} \tag{8}$$

$$r_{0i} = \frac{1}{n}\sum\nolimits_{k=1}^{n} l_{0i} \tag{9}$$

where l_{0i} is the relational coefficient between the evaluation index sequence and the ith influential parameter sequence at the kth data point. The resolution coefficient ρ is introduced, with smaller ρ values indicating greater resolution ($\rho \in (0,1)$, and in this paper, $\rho = 0.5$). r_{0i} stands for the relational degree between the evaluation index sequence and the kth influential parameter sequence, with a value domain of (0,1).

3.3. Results and Analysis

The calculation results of relational coefficients for each data point are shown in Figure 4 and Table 4. The overall relational degrees of influential construction parameters are shown in Table 5.

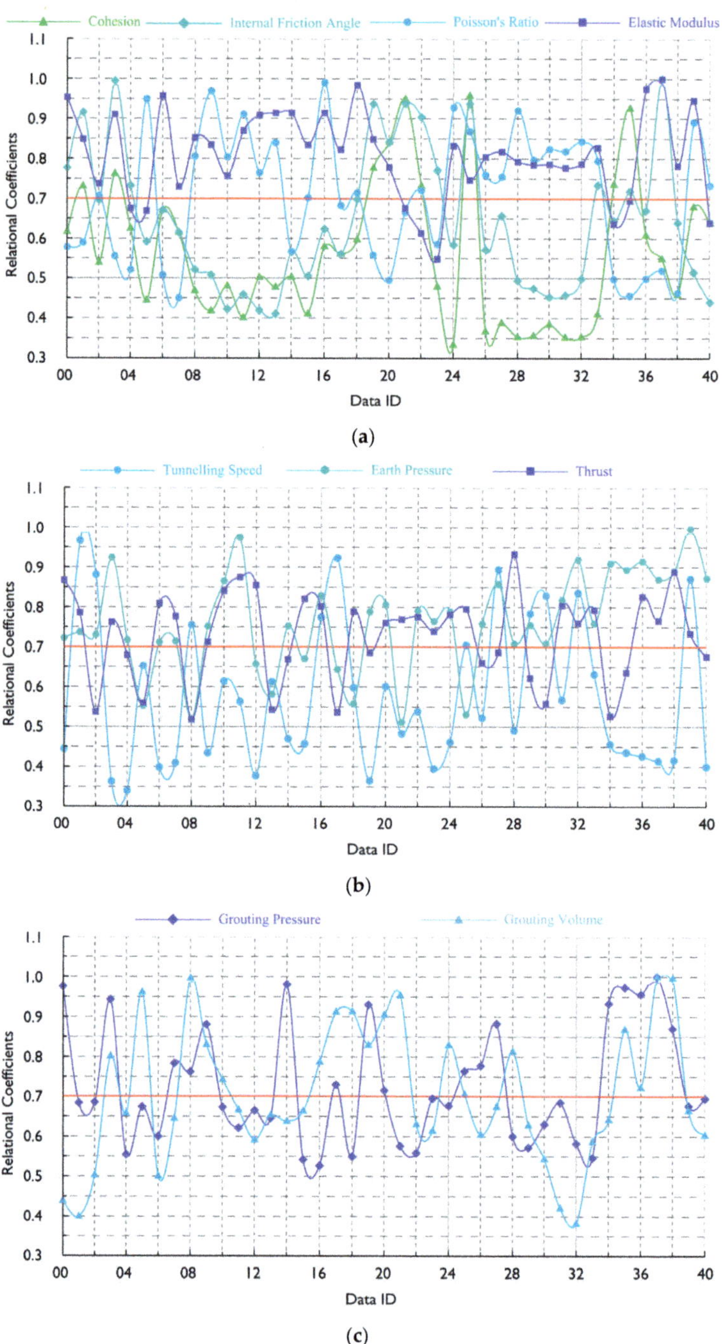

Figure 4. Calculation results of grey relational coefficents. A relational coefficient of 0.7 is placed in each subfigure to represent the correlation threshold between the influential parameters and the surface settlement: (**a**) relational coefficients of stratum parameters; (**b**) relational coefficients of tunnelling parameters; (**c**) relational coefficients of grouting parameters.

Table 4. Calculation results of grey relational coefficients.

Data ID	Relational Coefficients of Tunnelling Parameters			Relational Coefficients of Grouting Parameters		Relational Coefficients of Stratum Parameters			
	Thrust	Earth Pressure	Tunnelling Speed	Grouting Pressure	Grouting Volume	Cohesion	Internal Friction	Poisson's Ratio	Elastic Modulus
0	0.87	0.72	0.44	0.98	0.44	0.62	0.78	0.58	0.95
1	0.79	0.74	0.97	0.68	0.40	0.73	0.92	0.59	0.85
2	0.54	0.73	0.88	0.69	0.50	0.54	0.70	0.71	0.74
3	0.76	0.92	0.36	0.94	0.80	0.76	0.99	0.56	0.91
4	0.68	0.72	0.34	0.56	0.66	0.63	0.73	0.52	0.68
5	0.56	0.55	0.65	0.67	0.96	0.44	0.59	0.95	0.67
6	0.81	0.71	0.40	0.60	0.50	0.67	0.67	0.51	0.96
7	0.78	0.71	0.41	0.78	0.64	0.62	0.62	0.45	0.73
8	0.52	0.52	0.76	0.76	1.00	0.47	0.52	0.81	0.85
9	0.71	0.75	0.43	0.88	0.83	0.42	0.51	0.97	0.83
10	0.84	0.87	0.61	0.67	0.74	0.48	0.42	0.80	0.76
11	0.87	0.97	0.56	0.62	0.67	0.40	0.46	0.91	0.87
12	0.86	0.66	0.38	0.67	0.59	0.50	0.42	0.76	0.91
13	0.54	0.58	0.61	0.65	0.65	0.48	0.41	0.84	0.91
14	0.67	0.75	0.47	0.98	0.64	0.50	0.57	0.57	0.91
15	0.82	0.67	0.46	0.54	0.66	0.41	0.51	0.70	0.83
16	0.80	0.83	0.78	0.53	0.79	0.58	0.62	0.99	0.91
17	0.54	0.64	0.92	0.73	0.91	0.56	0.56	0.68	0.82
18	0.79	0.56	0.60	0.55	0.91	0.60	0.70	0.72	0.98
19	0.69	0.79	0.37	0.93	0.83	0.78	0.94	0.56	0.85
20	0.76	0.81	0.60	0.71	0.90	0.84	0.84	0.50	0.78
21	0.77	0.51	0.48	0.58	0.95	0.95	0.94	0.67	0.68
22	0.78	0.79	0.54	0.56	0.63	0.73	0.90	0.72	0.61
23	0.74	0.76	0.40	0.69	0.61	0.48	0.77	0.59	0.55
24	0.78	0.79	0.46	0.68	0.83	0.33	0.58	0.93	0.83
25	0.80	0.53	0.70	0.76	0.70	0.96	0.94	0.87	0.75
26	0.66	0.76	0.52	0.78	0.60	0.37	0.57	0.76	0.80
27	0.69	0.86	0.89	0.88	0.67	0.39	0.66	0.75	0.82
28	0.93	0.71	0.49	0.60	0.81	0.35	0.49	0.92	0.79
29	0.62	0.76	0.78	0.57	0.63	0.36	0.47	0.80	0.78
30	0.56	0.71	0.83	0.63	0.54	0.38	0.45	0.82	0.79
31	0.80	0.82	0.57	0.68	0.42	0.35	0.46	0.82	0.78
32	0.76	0.92	0.84	0.58	0.38	0.35	0.50	0.84	0.79
33	0.79	0.76	0.63	0.55	0.59	0.41	0.73	0.79	0.83
34	0.53	0.91	0.46	0.93	0.64	0.74	0.64	0.50	0.64
35	0.64	0.89	0.44	0.97	0.87	0.93	0.72	0.46	0.70
36	0.83	0.91	0.43	0.96	0.72	0.61	0.67	0.50	0.98
37	0.77	0.87	0.41	1.00	1.00	0.55	1.00	0.52	1.00
38	0.89	0.89	0.42	0.87	1.00	0.46	0.64	0.46	0.78
39	0.73	1.00	0.87	0.68	0.66	0.68	0.52	0.89	0.95
40	0.68	0.87	0.40	0.69	0.60	0.64	0.44	0.73	0.64

Table 5. Grey relational degrees of influential construction parameters.

	Parameter	Relational Degree	Ranking
Tunnelling Parameters	Earth Pressure (kPa)	0.762	2
	Tunnelling Speed (mm/min)	0.575	8
	Thrust (kN)	0.730	3
Grouting Parameters	Grouting Pressure (kPa)	0.726	4
	Grout Volume (m^3)	0.705	6
Stratum Parameters	Cohesion (kPa)	0.562	9
	Internal Friction (°)	0.648	7
	Elastic Modulus (MPa)	0.809	1
	Poisson's Ratio	0.708	5

Analysis of Figure 4, Tables 4 and 5 yield the following observations:

1. Grey relational coefficients above 0.7 indicate significant correlations between certain parameters and surface settlement. Figure 4a shows that the relation of cohesion

and internal friction angle with the surface settlement is relatively low within the Data ID range of 9–15 (corresponding to tunnelling rings 280–247). Similarly, in the interval of Data ID 34–38 (corresponding to tunnelling rings 65–41), the relation between Poisson's ratio and surface settlement is also notably low, which corresponds to fractured zone areas. This indicates a weak correlation between stratum parameters (excluding elastic modulus) and surface settlement within the fractured zone region.

2. Earth pressure exhibits the highest grey relational degree among the construction parameters. Combined with Tables 4 and 5, in Data ID 38–34, which exhibits high relational coefficients, surface settlement decreases as earth pressure gradually increases, consistent with previous studies [18]. Notably, slightly different from other findings [20], the grey correlation degree of tunnelling speed does not reach a significant level. Tunnelling speed briefly demonstrates a strong relational coefficient only when maintained near its maximum value (Data ID 25–30, corresponding to tunnelling rings 183–116). Therefore, flexibly adjusting the tunnelling speed based on the coefficients during right-line tunnelling helped mitigate surface settlement.

3. While the grouting parameters did not exhibit the highest relational degrees, they still held prominent positions [40]. Notably, grouting pressure contributes slightly more to surface settlement than grouting volume. In addition, as shown in Figure 4c, the relational coefficients between the two are higher in the fractured zone than in other intervals. Therefore, ensuring the stability of grouting pressure and controlling the grouting volume are both critical for achieving the optimal grouting effect.

4. Numerical Simulation

Grey relational analysis reveals the relational degrees of different parameters with the surface settlement. However, it cannot offer specific functional relationships or in-depth analysis of the complex interactions between the construction parameters and the surface settlement. To explore the variation patterns governing the surface settlement influenced by the critical construction parameters, numerical simulations were conducted to analyze the effect of the four parameters (thrust, grouting pressure, earth pressure, and strata elastic modulus) with the highest relational degrees. In this study, we utilized Flac3D as a numerical simulation tool, which is based on the finite difference method and designed for addressing underground geotechnical problems. It employs an implicit solution algorithm, seamlessly combining the finite difference method with the analysis of two-body dynamics to accurately simulate complex underground phenomenon.

4.1. Simulation Setup

The following simplifying assumptions were made to simulate the deformation behaviors of rock and soil zones in the model.

1. The tunnel structure and surrounding rock were assumed uniform and isotropic, exhibiting typical elastoplastic behavior.
2. The initial stress field was induced only by the self-weight of the surrounding rock, regardless of the potential influence of groundwater on the stress distribution.
3. The deformation of the rock and soil zones follows the Mohr–Coulomb strength criterion. Throughout the simulation, no progressive damage, nonlinear bending behavior, or loading–unloading cyclic loading occurred. Therefore, the lining and grouting zones were assumed as elastic material, and the shield shell zone was treated as a rigid material.

Besides the above simplifying assumptions, the following adjustments were also made to enhance the relevance to the actual conditions.

1. Normal constraints were set on the surroundings and bottom of the model to simulate actual constraints on the surrounding rocks. No constraints were applied on the top.

2. A fixed time-step value was implemented in the simulation, ensuring that the parameters of the grouting layer within the simulation varied in sync with the time step (time).

3. The simulation considered the standard gravitational acceleration (9.8 m/s^2), and the convergence threshold of the computational iterations was set at 10^{-5} to maintain numerical simulation accuracy.

4.2. Simulation Modelling

Based on the geological profile of Haigang Road Station to Chaoyang Road Station, the tunnelling rings from 140 to 180 (corresponding to Data ID 27–23) were selected as the representative interval to construct the numerical model. The stratum of the selected interval consists of powdery clay, clayey gravelly sand, strongly weathered tuff, moderately weathered tuff, slightly weathered tuff, and blocky granite (fractured). This interval lies within the transition zone between granite and weathered tuff, adjacent to fractured zones, encompassing both weak and stable strata. Such transitional zones frequently exhibit nonuniform physical properties like particle size distribution, porosity, and crack distribution, presenting intricate rock traits and mechanical behavior. Therefore, selecting this interval for simulation can provide comprehensive influence patterns of parameter variations under different conditions.

According to Saint-Venant's principle, the width of the model along the X direction was set to 3–5 times the tunnel diameter to avoid boundary effects, resulting in a total width of 30 m. Similarly, the height of the model along the Z direction was 30 m, and the length of the model along the Y direction was 60 m. Additionally, to make the simulation more representative of the actual tunnelling process, grouting, lining, and segment zones were introduced to consider the influence of support processes and grouting parameters on the strata. The thickness of the shield shell zones was set to 60 mm, and the thickness of the lining zones was set to 300 mm.

To determine the optimal meshing scheme, we conducted simulations with different grid densities. The scheme that showed no significant result changes with further refinement was chosen. The final mesh scheme divided the model into grids with a 0.375 m interval, with additional refinement at the geological transitions and tunnel excavation sections. The three-dimensional numerical model is shown in Figure 5 and consists of 373,722 zones, encompassing 274,797 grid points.

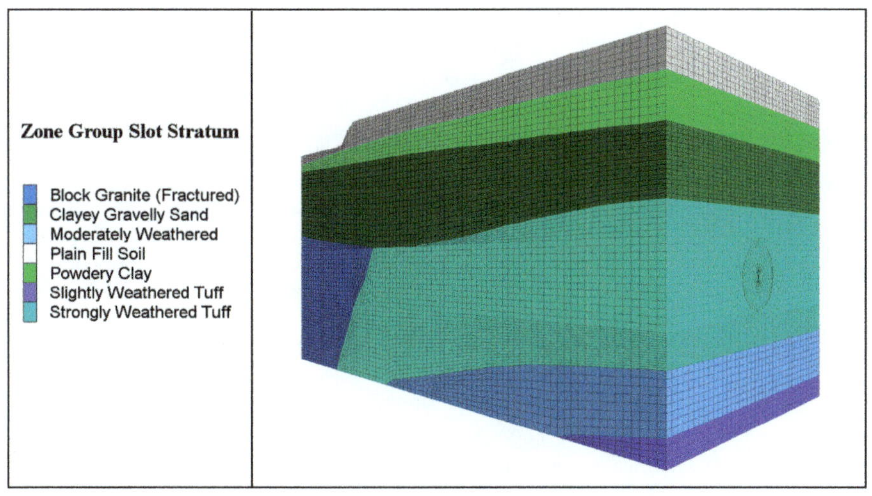

Figure 5. Three-dimensional numerical model.

4.3. Simulation Scheme

Based on the maximum and minimum values of actual construction parameters during shield tunnelling, the range for the four critical parameters were set as follows.

1. The earth pressure ranges in stress from 20 to 138 kPa.
2. The grouting pressure ranges in stress from 134 to 334 kPa.
3. The thrust ranges in force from 6886 to 11,460 kN.
4. The elastic modulus changes from 50% to 150% (compared with the exploration data).

Each parameter was set at five levels, resulting in 17 construction simulation schemes, which shown in Table 6. Each one of critical parameters varied within the specified range, while the remaining parameters were kept at the mean values. In each simulation scheme, the entire numerical simulation process was conducted using a ring (1.5 m) as the smallest circulation unit. The procedure involved the following steps.

Table 6. Simulation scheme sets for different levels of construction parameters.

Schemes	Thrust (kN)	Earth Pressure (kPa)	Elastic Modulus (%)	Grouting Pressure (kPa)
1	6886	79	100	234
2	8030	79	100	234
3	9173	79	100	234
4	10,317	79	100	234
5	11,460	79	100	234
6	9173	20	100	234
7	9173	50	100	234
8	9173	109	100	234
9	9173	138	100	234
10	9173	79	50	234
11	9173	79	75	234
12	9173	79	125	234
13	9173	79	150	234
14	9173	79	100	134
15	9173	79	100	184
16	9173	79	100	284
17	9173	79	100	334

1. Initially, the shield shell zones of the shield machine were advanced by 1.5 m to simulate the initiation of a new shield tunnelling loop.
2. Subsequently, the tunnel zones, segment zones, and grouting zones were removed to mimic the excavation process of shield construction.
3. After the release of surrounding rock stresses, the earth pressure, the thrust, and the grouting pressure were applied to replicate the grouting process of shield construction.
4. Following the grouting process, the grouting pressure was canceled, and material parameters were assigned to grouting and the segment zones to simulate the assembly process of shield construction.
5. The parameters of the grouting zones were defined as a function of time, gradually transitioning from pre-solidification parameters to post-solidification parameters with each calculation step.
6. The calculation was then allowed to converge, completing one cycle of excavation.
7. Subsequently, the shield shell zones of the shield machine were advanced by another 1.5 m, initiating the next tunnelling cycle.

This cyclic tunnelling process continued until the model was fully constructed. The geotechnical zone's parameters in the model were adopted from the geological survey report (refer to Table 2), while the material parameters for the tunnel and shield shell are presented in Table 7.

Table 7. Material parameters of the tunnel and shield zones.

Zone Name	Density (N/m³)	Elastic Modulus (MPa)	Poisson's Ratio
Segment rings	3250	3000	0.2
Shield shell	9500	200,000	0.3
Grouting layer (pre-solidification)	2100	240	0.3
Grouting layer (post-solidification)	2100	720	0.3

4.4. Model Validation and Analysis

Simulation scheme 3, where all critical construction parameters were given the mean values, was as per the initial choice numerical simulation. Subsequently, the simulation results were compared with the actual monitoring data. The regression coefficient (R^2) and root mean square error ($RMSE$) were used as evaluation metrics to assess the accuracy of the simulation, which is calculated as follows:

$$R^2 = 1 - \frac{\sum_{i=1}^{N}\left(Y_i - Y_i'\right)^2}{\sum_{i=1}^{N}\left(Y_i - \bar{Y}\right)^2} \tag{10}$$

$$RMSE = \sqrt{\frac{1}{N}\sum_{i=1}^{N}\left(Y_i - Y_i'\right)^2} \tag{11}$$

where N denotes the number of data samples, Y_i represents the observed surface settlement values, Y_i' represents the predicted surface settlement values, and $\bar{Y}$ is the mean of the observed surface settlement values.

Figure 6 displays the contour map of surface settlement, while Table 8 presents the assessment results of the simulation compared to the actual. In Table 8, the R^2 between the simulation data and the actual monitoring data is 0.911, and the RMSE is 0.235. This high consistency between the simulation results and the monitoring settlement data not only validates the reliability of numerical simulation but also confirms the effectiveness of grey relational analysis in capturing the critical construction parameters.

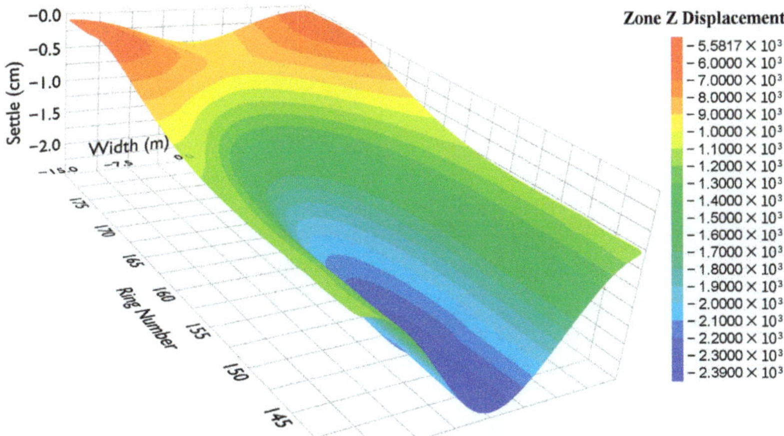

Figure 6. Contour map of surface settlement.

Further analysis of Figure 6 yields:

1. Lateral settlement varies with strata zones, reaching up to 24.2 mm with a broad trough (approximately 20 m wide) in weak strata (140–170 rings). In more stable strata (170–180 rings), maximum surface settlement decreases to 2.1 mm with a narrower trough (about 12 m).

2. Longitudinal settlement peaks at 24 mm in the 140–160 rings, with local minima at 140–145 and 152–155 rings. In stable strata, maximum surface settlement rapidly decreases from 24.2 mm to 10.3 mm.

Table 8. Validation results of the simulation compared to the actual.

Data Name	Actual Data	Simulation Data	R^2	RMSE
ZDBC65 (Data ID 27)	−2.42	−2.37		
ZDBC64 (Data ID 26)	−1.68	−2.04		
ZDBC63 (Data ID 25)	−2.09	−2.23		
ZDBC62 (Data ID 24)	−0.21	−0.56		
ZDBC61 (Data ID 23)	0.03	−0.36	0.911	0.235
ZDBC60 (Data ID 22)	−1.34	−1.09		
DBC10-01	−0.79	−0.92		
DBC10-01	−1.86	−2.01		
DBC10-01	−1.95	−2.02		
DBC10-04	−0.88	−1.01		

4.5. Simulation of Influence Patterns

To further investigate the influence patterns of critical construction parameters on surface settlement, systematic numerical simulations were conducted following the simulation scheme described in Section 4.3. Figures 7 and 8 present the results of maximum lateral and longitudinal settlements under different schemes of varying parameters.

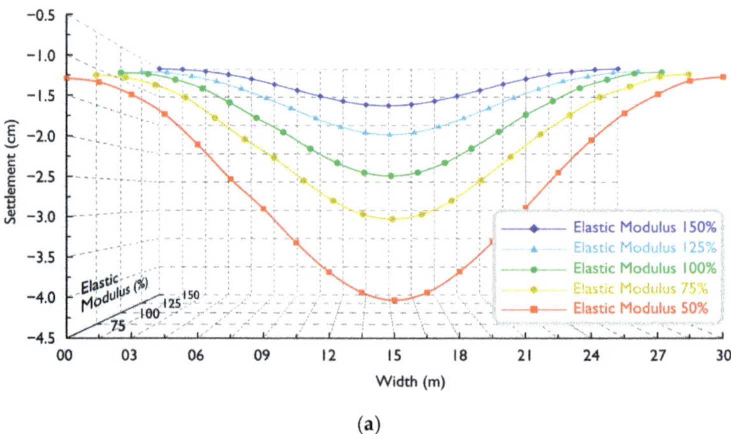

(a)

Figure 7. *Cont.*

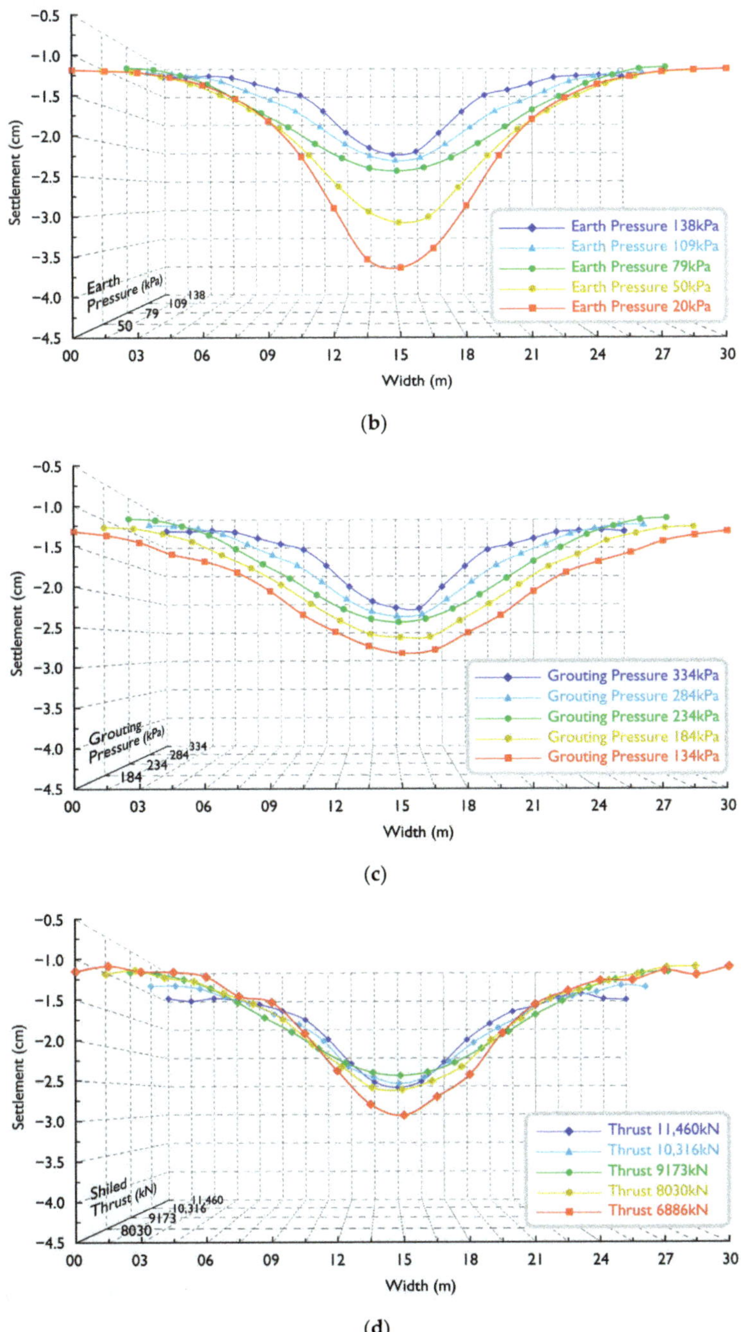

(**b**)

(**c**)

(**d**)

Figure 7. Maximum lateral surface settlement under different variations of critical construction parameters: (**a**) maximum lateral surface settlement under different elastic moduli; (**b**) maximum lateral surface settlement under different earth pressure; (**c**) maximum lateral surface settlement under different grouting pressure; (**d**) maximum lateral surface settlement under different thrust.

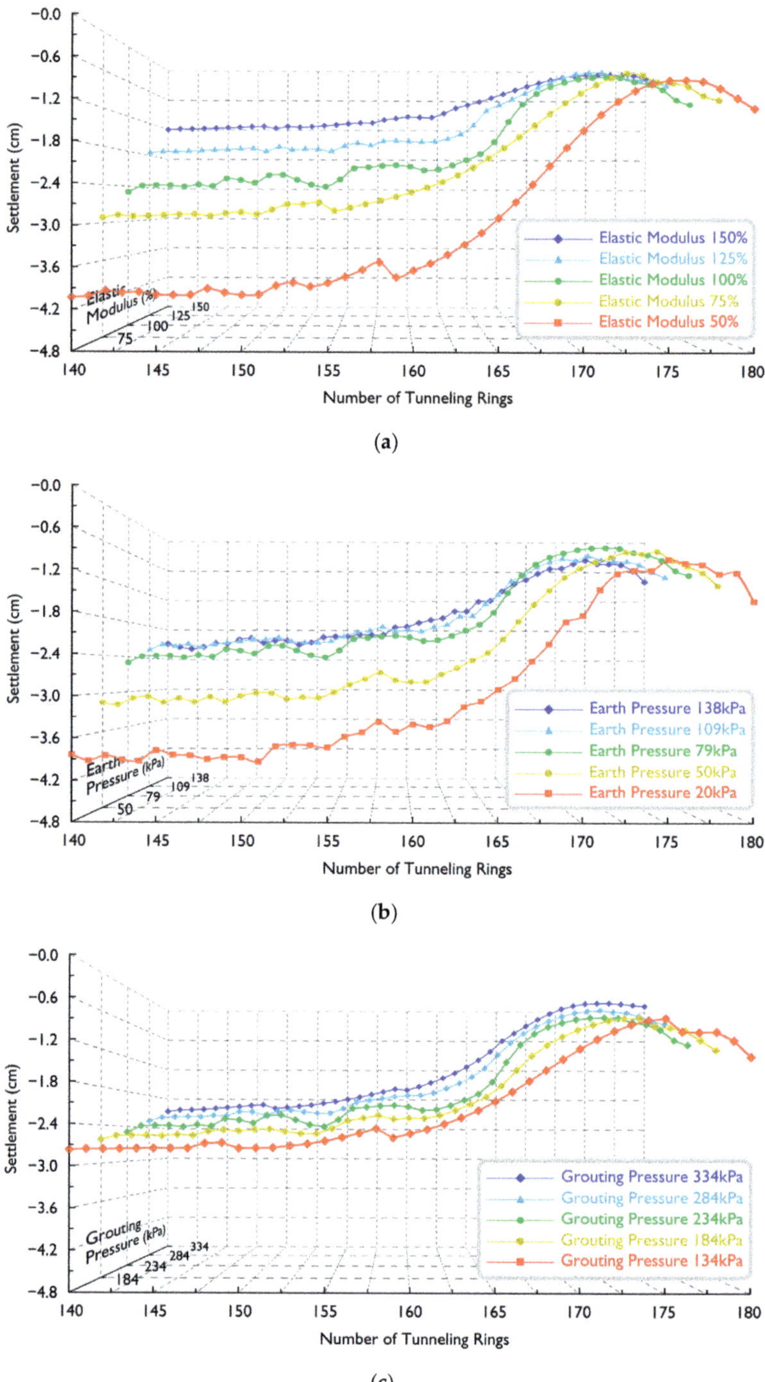

(**a**)

(**b**)

(**c**)

Figure 8. *Cont.*

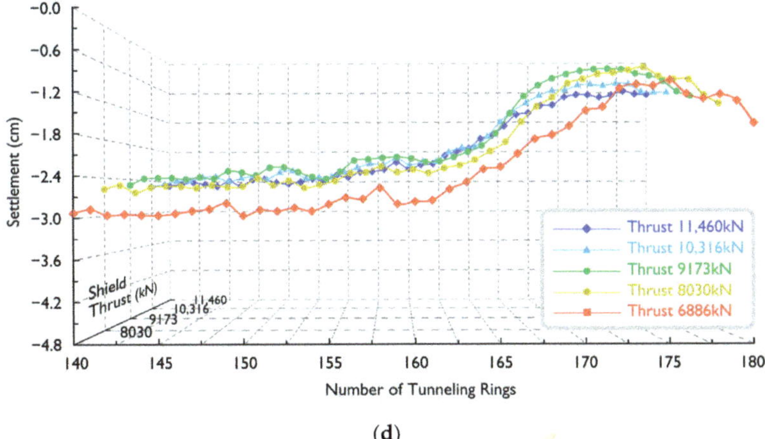

(d)

Figure 8. Maximum longitudinal surface settlement under different variations of critical construction parameters: (**a**) maximum longitudinal surface settlement under different elastic modulis; (**b**) maximum longitudinal surface settlement under different earth pressure; (**c**) maximum longitudinal surface settlement under different grouting pressure; (**d**) maximum longitudinal surface settlement under different thrust.

From Figures 7 and 8, it can be seen that:

1. Elastic modulus significantly impacts surface settlement, showing a nonlinear relationship (Figure 7a). Strata with smaller elastic modulus demonstrate increased sensitivity to shield tunnelling, resulting in larger settlement and wider troughs. In Figure 8a, reducing the elastic modulus from 150% to 50% leads to a 248% increase in maximum settlement (from 6.2 mm to 21.6 mm) in intervals with weak strata. Meanwhile, the impact on stable strata intervals is less than 50%. The actual settlement data across fractured zones (Table 3, Data ID 73–78) also coincide with this observation.

2. Earth pressure's influence on surface settlement parallels elastic modulus trends (Figures 7b and 8b). In stable strata, it has a moderate impact, but in poorer geological conditions, reducing earth pressure from 138 kPa to 20 kPa increases maximum settlement from 8.6 mm to 18.3 mm, consistent with previous findings [41,42].

3. As indicated in previous studies [18,36], the influence of grouting pressure on surface settlement is evident. Moreover, it greatly influences the settlement trough width in weak strata (Figure 7c). Increasing grouting pressure from 134 kPa to 334 kPa decreases the settlement trough from 10 m to 6 m. Somewhat differently from a previous study [40], no surface uplift occurs, likely due to reasonable grouting pressure within the soil's bearing capacity.

4. Thrust magnitude directly affects the final surface settlement [33] (Figure 8d). Excessive thrust that is beyond the strata's resistance during tunnelling leads to a significant surface settlement with surface uplift and subsequent sinking after shield tunnelling. Similarly, insufficient thrust results in incomplete compensation of strata displacement, leading to an increase in surface settlement.

5. Discussion

This study investigated the surface settlement in the construction section of Qingdao Metro Line 6 between Haigang Road Station and Chaoyang Road Station through a hybrid numerical approach comprised of grey relational analysis and numerical simulation. The research findings provide scientific evidence and engineering guidance for mitigating surface settlement during shield tunnel construction. Based on the results of grey relational analysis and numerical simulations, the following recommendations are proposed for the right-line tunnelling and future urban subway project.

As illustrated in Figures 7 and 8, it is important to note that the elastic modulus of the stratum has a significant influence on the surface settlement. Given its inherent property as a critical stratum parameter, precise characterization of the elastic modulus can provide a direct link to the potential surface settlement and offer a scientific basis for developing effective mitigation strategies. Geotechnical investigations and predictive modelling before shield tunnelling are required. In addition, it is advisable to conduct geological predictions in advance and promptly adjust construction parameters based on real-time stratum conditions during tunnel evacuation. In regions with small elastic moduli, suitable reinforcement measures like pre-excavation grouting or enhanced ground stabilization are also recommended.

Similarly, the influence of earth pressure on the surface settlement should also be considered to mitigate the settlement fluctuations during shield tunnelling. Careful adjustments in earth pressure in sections with weak geological conditions are vital to prevent excessive surface settlement, and utilizing air pressure can enhance tunnelling efficiency in strata with stable geological conditions. In light of the notable influence of grouting pressure and thrust magnitude on surface settlement, it becomes crucial to determine suitable magnitude levels of these parameters through numerical modelling and field validation. According to the simulation results, maintaining the grouting pressure between 250 and 300 kPa and keeping the thrust force matched with the initial earth pressure can effectively mitigate surface settlement. Findings (Table 5) highlight that maintaining stable and reasonable tunnelling speed throughout the process can mitigate its impact on surface settlement. For the right-line tunnelling, maintaining a tunnelling speed similar to the left line can help control the surface settlement.

The research findings offer scientific insights and practical guidance in mitigating the surface settlement during shield tunnel construction. However, there are also some limitations and constraints in this study. One constraint lies in the assumption of the initial stress field, which considers only the self-weight of the surrounding rock and overlooked parameters like rock displacement and external influences such as tectonic shifts, geological history, and seismic activity. In addition, there are also limitations when employing the Mohr–Coulomb strength criterion for the deformation of rock and soil. It cannot fully capture the complexities of various geological conditions and mechanical phenomena, especially the deformation and fracture behavior of soft rocks. Furthermore, due to limitations in computational resources, simulation time, and data availability, consideration of influential parameters like the groundwater level shifts and soil permeability's impact on surface settlement was omitted in this study. There are also certain insufficiencies in model dimensions created through numerical simulation.

To address these limitations, future study will concentrate on three main areas. Firstly, more accurate data of material strength will be acquired through model testing and field monitoring. Consequently, a more realistic initial in situ stress field model will be constructed based on these measuring data, thus enhancing simulation accuracy. Secondly, advanced mechanical models will be developed to comprehensively represent the variations in material strength, especially in describing the deformation and fracture behavior of soft rocks. Finally, efforts will be made to extend the numerical model size to better fit the engineering context. Additional factors will be taken into account, including groundwater levels, infiltration coefficients, fracture development, and the stratigraphic properties of other intervals. By incorporating diverse engineering scenarios and geological contexts, more precise engineering recommendations and effective control strategies for surface settlement will be provided.

6. Conclusions

This study investigated surface settlement in the urban subway construction section of Qingdao Metro Line 6 between Haigang Road Station and Chaoyang Road Station. By utilizing a hybrid numerical approach, the main conclusions are as follows.

1. Earth pressure emerges as the most influential tunnelling parameter, while tunnelling speed has minimal impact. Grouting pressure has a greater effect on surface settlement than grouting volume, and strata elastic modulus significantly influences settlement.
2. Numerical simulations reveal that strata with lower elastic moduli are more sensitive to shield tunnelling, resulting in larger settlement. Earth pressure and grouting pressure show similar trends, with pronounced effects in weak strata.
3. For stable strata, it is feasible to consider a judicious increment in tunnelling speed and employ compressed air pressure for excavation. Keeping grouting pressure between 250 and 300 kPa and adjusting thrust with initial earth pressure can mitigate settlement.

Author Contributions: Conceptualization, M.L. and X.W.; methodology, M.L. and D.W.; field testing, Z.L.; data curation, M.L. and Z.L.; data analysis, M.L.; writing—original draft preparation, M.L.; writing—review and editing, D.W. and X.W.; supervision, X.W. and D.W.; project administration, X.W. and D.W.; funding acquisition, X.W. and D.W. All authors have read and agreed to the published version of the manuscript.

Funding: This research was financially supported by the National Natural Science Foundation of China (grant 52204115), the Foundation of Research Institute for Deep Underground Science and Engineering (grant XD2021022), and the Shield/TBM Construction Risk Consultation Project in Qingdao Metro Line 6.

Institutional Review Board Statement: Not applicable.

Informed Consent Statement: Not applicable.

Data Availability Statement: All data included in this study are available upon request by contact with the corresponding author.

Acknowledgments: Thanks are due to Guangzhao Zhang (Qingdao University of Technology, Qingdao) for assistance with the numerical simulation, and Zelin Lu (Qingdao University of Technology, Qingdao) for valuable advice.

Conflicts of Interest: The authors declare no conflict of interest.

References

1. Ji, M.; Wang, X.; Luo, M.; Wang, D.; Teng, H.; Du, M. Stability Analysis of Tunnel Surrounding Rock When TBM Passes through Fracture Zones with Different Deterioration Levels and Dip Angles. *Sustainability* **2023**, *15*, 5243. [CrossRef]
2. Tian, Y.; Motalleb Qaytmas, A.; Lu, D.; Du, X. Stress Path of the Surrounding Soil during Tunnel Excavation: An Experimental Study. *Transp. Geotech.* **2023**, *38*, 100917. [CrossRef]
3. Liu, C.; Yang, S.; Liu, W.; Wang, Z.; Jiang, Y.; Yang, Z.; Jiang, H. Three-Dimensional Numerical Simulation of Soil Deformation during Shield Tunnel Construction. *Math. Probl. Eng.* **2022**, *2022*, 5029165. [CrossRef]
4. Peng, S.; Huang, W.; Luo, G.; Cao, H.; Pan, H.; Mo, N. Failure Mechanisms of Ground Collapse Caused by Shield Tunnelling in Water-Rich Composite Sandy Stratum: A Case Study. *Eng. Fail. Anal.* **2023**, *146*, 107100. [CrossRef]
5. Seol, H.; Won, D.; Jang, J.; Kim, K.Y.; Yun, T.S. Ground Collapse in EPB Shield TBM Site: A Case Study of Railway Tunnels in the Deltaic Region near Nak-Dong River in Korea. *Tunn. Undergr. Space Technol.* **2022**, *120*, 104274. [CrossRef]
6. Yao, Q.; Di, H.; Ji, C.; Zhou, S. Ground Collapse Caused by Shield Tunneling in Sandy Cobble Stratum and Its Control Measures. *Bull. Eng. Geol. Environ.* **2020**, *79*, 5599–5614. [CrossRef]
7. Bayati, M.; Hamidi, J.K. A Case Study on TBM Tunnelling in Fault Zones and Lessons Learned from Ground Improvement. *Tunn. Undergr. Space Technol.* **2017**, *63*, 162–170. [CrossRef]
8. Peck, B.B. Deep excavation and tunnelling in soft ground, State of the art volume. In Proceedings of the 7th ICSMFE, Mexico City, Mexico, 25–29 August 1969; pp. 225–290. Available online: https://www.issmge.org/publications/publication/deep-excavations-and-tunneling-in-soft-ground (accessed on 27 August 2023).
9. Ren, Y.; Zhang, C.; Zhu, M.; Chen, R.; Wang, J. Significance and Formulation of Ground Loss in Tunneling-Induced Settlement Prediction: A Data-Driven Study. *Acta Geotech.* **2023**, *18*, 4941–4956. [CrossRef]
10. Jin, H.; Yuan, D.; Jin, D.; Wu, J.; Wang, X.; Han, B.; Mao, J. Shield Kinematics and Its Influence on Ground Settlement in Ultra-Soft Soil: A Case Study in Suzhou. *Can. Geotech. J.* **2022**, *59*, 1887–1900. [CrossRef]

11. Lou, P.; Li, Y.; Tang, X.; Lu, S.; Xiao, H.; Zhang, Z. Influence of Double-Line Large-Slope Shield Tunneling on Settlement of Ground Surface and Mechanical Properties of Surrounding Rock and Segment. *Alex. Eng. J.* **2023**, *63*, 645–659. [CrossRef]

12. Sun, F.; Jin, Z.; Wang, C.; Gou, C.; Li, X.; Liu, C.; Yu, Z. Case Study on Tunnel Settlement Calculations during Construction Considering Shield Disturbance. *KSCE J. Civ. Eng.* **2023**, *27*, 2202–2216. [CrossRef]

13. Ahn, C.-Y.; Park, D.; Moon, S.-W. Analysis of Surface Settlement Troughs Induced by Twin Shield Tunnels in Soil: A Case Study. *Geomech. Eng.* **2022**, *30*, 325–336. [CrossRef]

14. Wang, F.; Du, X.; Li, P. Predictions of Ground Surface Settlement for Shield Tunnels in Sandy Cobble Stratum Based on Stochastic Medium Theory and Empirical Formulas. *Undergr. Space* **2023**, *11*, 189–203. [CrossRef]

15. Sohaei, H.; Hajihassani, M.; Namazi, E.; Marto, A. Experimental Study of Surface Failure Induced by Tunnel Construction in Sand. *Eng. Fail. Anal.* **2020**, *118*, 104897. [CrossRef]

16. Bel, J.; Branque, D.; Camus, T. Physical Modelling of EPB TBM in Dry Sand and Greenfield Conditions. *Eur. J. Environ. Civ. Eng.* **2023**, *27*, 3236–3259. [CrossRef]

17. Berthoz, N.; Branque, D.; Wong, H.; Subrin, D. TBM Soft Ground Interaction: Experimental Study on a 1 g Reduced-Scale EPBS Model. *Tunn. Undergr. Space Technol.* **2018**, *72*, 189–209. [CrossRef]

18. Kannangara, K.K.P.M.; Ding, Z.; Zhou, W.-H. Surface Settlements Induced by Twin Tunneling in Silty Sand. *Undergr. Space* **2022**, *7*, 58–75. [CrossRef]

19. Ter-Martirosyan, A.Z.; Cherkesov, R.H.; Isaev, I.O.; Shishkina, V.V. Surface Settlement during Tunneling: Field Observation Analysis. *Appl. Sci.* **2022**, *12*, 9963. [CrossRef]

20. Kim, D.; Pham, K.; Park, S.; Oh, J.Y.; Choi, H. Determination of Effective Parameters on Surface Settlement during Shield TBM. *Geomech. Eng.* **2020**, *21*, 153–164. [CrossRef]

21. Wang, D.; He, M.; Tao, Z.; Guo, A.; Wang, X. Deformation-Softening Behaviors of High-Strength and High-Toughness Steels Used for Rock Bolts. *J. Rock Mech. Geotech. Eng.* **2022**, *14*, 1872–1884. [CrossRef]

22. Wang, D.; He, M.; Jia, L.; Sun, X.; Xia, M.; Wang, X. Energy Absorption Characteristics of Novel High-Strength and High-Toughness Steels Used for Rock Support. *J. Rock Mech. Geotech. Eng.* **2023**, *15*, 1441–1456. [CrossRef]

23. Chen, R.; Meng, F.; Ye, Y.; Liu, Y. Numerical Simulation of the Uplift Behavior of Shield Tunnel during Construction Stage. *Soils Found.* **2018**, *58*, 370–381. [CrossRef]

24. Jia, J.; Gao, R.; Wang, D.; Li, J.; Song, Z.; Tan, J. Settlement Behaviours and Control Measures of Twin-Tube Curved Buildings-Crossing Shield Tunnel. *Struct. Eng. Mech.* **2022**, *84*, 699–706. [CrossRef]

25. Jallow, A.; Ou, C.-Y.; Lim, A. Three-Dimensional Numerical Study of Long-Term Settlement Induced in Shield Tunneling. *Tunn. Undergr. Space Technol.* **2019**, *88*, 221–236. [CrossRef]

26. Eskandari, F.; Goharrizi, K.G.; Hooti, A. The Impact of EPB Pressure on Surface Settlement and Face Displacement in Intersection of Triple Tunnels at Mashhad Metro. *Geomech. Eng.* **2018**, *15*, 769–774. [CrossRef]

27. Miliziano, S.; De Lillis, A. Predicted and Observed Settlements Induced by the Mechanized Tunnel Excavation of Metro Line C near S. Giovanni Station in Rome. *Tunn. Undergr. Space Technol.* **2019**, *86*, 236–246. [CrossRef]

28. Shahin, H.M.; Nakai, T.; Okuno, T. Numerical Study on 3D Effect and Practical Design in Shield Tunneling. *Undergr. Space* **2019**, *4*, 201–209. [CrossRef]

29. Bahri, M.; Mascort-Albea, E.J.; Romero-Hernández, R.; Koopialipoor, M.; Soriano-Cuesta, C.; Jaramillo-Morilla, A. Numerical Model Validation for Detection of Surface Displacements over Twin Tunnels from Metro Line 1 in the Historical Area of Seville (Spain). *Symmetry* **2022**, *14*, 1263. [CrossRef]

30. Anato, N.J.; Chen, J.; Tang, A.; Assogba, O.C. Numerical Investigation of Ground Settlements Induced by the Construction of Nanjing WeiSanLu Tunnel and Parametric Analysis. *Arab. J. Sci. Eng.* **2021**, *46*, 11223–11239. [CrossRef]

31. Wang, Z.; Li, G.; Wang, A.; Pan, K. Numerical Simulation Study of Stratum Subsidence Induced by Sand Leakage in Tunnel Lining Based on Particle Flow Software. *Geotech. Geol. Eng.* **2020**, *38*, 3955–3965. [CrossRef]

32. Alzabeebee, S.; Keawsawasvong, S. Numerical Assessment of Microtunnelling Induced Pavement Settlement. *Geotech. Geol. Eng.* **2023**, *41*, 2173–2184. [CrossRef]

33. Yuan, C.; Zhang, M.; Ji, S.; Li, J.; Jin, L. Analysis of Factors Influencing Surface Settlement during Shield Construction of a Double-Line Tunnel in a Mudstone Area. *Sci. Rep.* **2022**, *12*, 22606. [CrossRef] [PubMed]

34. Wang, Y.; Jiang, W.; Wang, M.; Li, Y. Risk Assessment and Implementation of Deformation Disaster for Operation Tunnel Based on Entropy Weight-Grey Relational Analysis. *Geomat. Nat. Hazards Risk* **2022**, *13*, 1831–1848. [CrossRef]

35. Ma, J.; He, S.; Cui, G.; He, J.; Liu, X. Construction Stability and Reinforcement Technology for the Super-Large Rectangular Pipe-Jacking Tunnel Passing beneath the Operational High-Speed Railway in Composite Stratum. *Geomat. Nat. Hazards Risk* **2023**, *14*, 2208720. [CrossRef]

36. Broere, W.; Festa, D. Correlation between the Kinematics of a Tunnel Boring Machine and the Observed Soil Displacements. *Tunn. Undergr. Space Technol.* **2017**, *70*, 125–147. [CrossRef]

37. Goh, A.T.C.; Zhang, W.; Zhang, Y.; Xiao, Y.; Xiang, Y. Determination of Earth Pressure Balance Tunnel-Related Maximum Surface Settlement: A Multivariate Adaptive Regression Splines Approach. *Bull. Eng. Geol. Env.* **2018**, *77*, 489–500. [CrossRef]

38. Shi, Y.-J.; Xiao, X.; Li, M.-G. Long-Term Longitudinal Deformation Characteristics of Metro Lines in Soft Soil Area. *J. Aerosp. Eng.* **2018**, *31*, 04018080. [CrossRef]

39. Pan, Y.; Ou, S.; Zhang, L.; Zhang, W.; Wu, X.; Li, H. Modeling Risks in Dependent Systems: A Copula-Bayesian Approach. *Reliab. Eng. Syst. Saf.* **2019**, *188*, 416–431. [CrossRef]
40. Moghtader, T.; Sharafati, A.; Naderpour, H.; Gharouni Nik, M. Estimating Maximum Surface Settlement Caused by EPB Shield Tunneling Utilizing an Intelligent Approach. *Buildings* **2023**, *13*, 1051. [CrossRef]
41. Vinoth, M.; Aswathy, M.S. 3D Evolution of Soil Arching during Shield Tunnelling in Silty and Sandy Soils: A Comparative Study. *Int. J. Numer. Anal. Methods Geomech.* **2023**, *47*, 299–322. [CrossRef]
42. Wang, J.; Feng, K.; Wang, Y.; Lin, G.; He, C. Soil Disturbance Induced by EPB Shield Tunnelling in Multilayered Ground with Soft Sand Lying on Hard Rock: A Model Test and DEM Study. *Tunn. Undergr. Space Technol.* **2022**, *130*, 104738. [CrossRef]

Article

Study on the Stress Distribution and Stability Control of Surrounding Rock of Reserved Roadway with Hard Roof

Yuxi Hao [1], Mingliang Li [1], Wen Wang [2,*], Zhizeng Zhang [1] and Zhun Li [1]

[1] School of Architectural Engineering, Zhongyuan University of Technology, Zhengzhou 450007, China; 6548@zut.edu.cn (Y.H.); 2021109363@zut.edu.cn (M.L.); 6175@zut.edu.cn (Z.Z.); 2022109405@zut.edu.cn (Z.L.)

[2] College of Energy Science and Engineering, Henan Polytechnic University, Jiaozuo 454000, China

* Correspondence: wangwen2006@hpu.edu.cn; Tel.: +86-0391-3987973

Abstract: According to field observation and theoretical analysis, the failure of the 1523103 reserved roadway is mainly affected by the lateral support pressure, rock mass strength, and support mode. With the mining of the 152309 working face, the lateral pressure of coal pillars on both sides of the reserved roadway increases, and since the lithology of the two sides and the floor of the roadway is weak, the reserved roadway experiences spalling and floor heave. Through numerical simulation, the distribution law of surrounding rock stress and the displacement of surrounding rock are obtained after the roof cutting and pressure relief of the reserved roadway with hard roof. According to the cause of surrounding rock failure of a reserved roadway, the combined control technology of roof cutting and pressure relief, grouting anchor cable support, and bolt support is put forward. After cutting the roof and releasing the pressure on the working face, the lateral support pressure of the two sides of the roadway is significantly reduced, the deformation of the two sides of the roadway is small, the maximum shrinkage rate of the section is reduced from 70% to 11%, and the deformation of the surrounding rock of the 1523103 reserved roadway is effectively controlled. The successful control of the surrounding rock in the 1523103 tunnel reduces the number of coal pillars to be installed, improves the coal extraction rate, and is conducive to the sustainable utilization of limited natural resources and the sustainable development of the coal industry.

Keywords: reserved roadway; roof cutting and pressure releasing; surrounding rock damage; hard roof; grouting anchor

Citation: Hao, Y.; Li, M.; Wang, W.; Zhang, Z.; Li, Z. Study on the Stress Distribution and Stability Control of Surrounding Rock of Reserved Roadway with Hard Roof. *Sustainability* **2023**, *15*, 14111. https://doi.org/10.3390/su151914111

Academic Editor: Jianjun Ma

Received: 10 August 2023
Revised: 20 September 2023
Accepted: 21 September 2023
Published: 23 September 2023

1. Introduction

In the field of coal mine safety prevention and control, the control process of the surrounding rock of a reserved roadway is a complex process that runs through each deformation stage of the surrounding rock [1], which has an important impact on the sustainable development of the coal mining industry. The sustainable development of the coal mining industry needs to realize the effective utilization of coal resources, the effective protection of coal miners, and the effective protection of the environment through scientific technology and reasonable management methods. This requires us to analyze the main factors causing roadway deformation according to the basic principle of roadway surrounding rock deformation control and put forward a reasonable surrounding rock deformation control scheme under the premise of sustainable development of the environment and industry. In response to such problems, many scholars have studied this phenomenon in an attempt to uncover solutions.

In recent years, coal mine resources have been depleted and the coal mining rate is low, which is similar to the problems faced by the construction industry as analyzed in the article by Zhimin Wang [2] and others. These phenomena prompt us to transform traditional technology into intelligent technology to extend the service life of mines. The traditional roadway surrounding rock control process is complicated, and the three-dimensional platform construction of BIM [3] technology can achieve three-dimensional visualization of mine construction through digital geological models, realize roadway safety monitoring, and improve the efficiency and effect of construction management. Feng Wan [4] used roof cutting and pressure relief combined with concrete walls to control the deformation of surrounding rock in the tunnel, and Jianjun Ma [5] used a dynamic load concrete thermo-elastoplastic damage coupling model to study the dynamic load effect of concrete. In terms of the numerical simulation of tunnel surrounding rock, Jianjun Ma [6] used a fast particle accumulation generation algorithm with discontinuous deformation and controllable gradient to simulate the mechanical properties of the rock. Regarding the impact of advanced blasting on surrounding rock, Jianjun Ma [7] applied the M-JHB4DLSM model to study the mechanical behavior of tunnels under close-range blasting. In recent years, Academician Hongpu Kang, in view of the supporting problems such as the large deformation of surrounding rock of kilometer-deep roadways, and on the basis of studying the deformation and failure mechanism of the surrounding rock of roadways, put forward the cooperative control theory of roadway surrounding rock support + modification + pressure relief "trinity". A complete set of technology systems for the control of surrounding rock of deep roadways was developed, which integrates the active support of high prestressed anchor cable, active modification of high-pressure fracturing grouting, and active pressure relief via hydraulic fracturing. The new material of NPR anchor steel, developed by Academician Manchao He, has the characteristics of being non-magnetic, with high strength, high toughness, and high uniform extension, which solves the local deformation and fracture of ordinary anchors and is the first kind of material with these characteristics at home and abroad. Up to now, in the field of tunnel surrounding rock control, there are mainly five surrounding rock control solutions: surface support technology, anchoring technology, improving the self-supporting capacity of surrounding rock, pressure relief technology, and joint control technology. In order to adapt to various construction environments, joint control technology is widely used. Joint control technology can give full play to the advantages of various support methods and achieve complementary advantages. In 2008, He [8–12] proposed the theory of "Roof cutting short-wall beam" via field investigation and indoor experimental research and formed the "longwall mining 110 method" on this basis. Then, by comparing the traditional gob-side entry retaining technology, a new technology involving roof cutting and pressure relief gob-side entry retaining was proposed. Roof cutting and pressure relief technology can not only reduce the stress of surrounding rock, ensure safety in the coal mining process, reduce the risk of accidents, and reduce environmental pollution, but also improve coal recovery and reduce waste so as to prolong the service life of coal resources. All of these advantages help to protect coal resources, provide more resources for the future, realize the effective use of resources, and promote the sustainable development of geotechnical engineering. He believes that the gangue that is produced after roof pre-splitting blasting can play a supporting role in the overlying strata due to its own fragmentation and expansion, so that the roof forms a short-arm beam structure in the lateral direction. And the filling mining method of disposing coal gangue in the goaf of the working face can not only reduce the discharge of solid waste in a coal mine, but also reduce the disaster of mining subsidence and improve the recovery rate of mine resources. It is one of the key technical ways to realize green mining in coal mines.

The fracture characteristics and activity laws of the overlying rock layer along the empty alley are closely related to the fault structure characteristics and activity laws of the overlying rock layer during the working face of the upper section and the working face of this section, but they also have their own characteristics and rules. Zhu et al. [13] analyzed the variation law of the horizontal stress distribution of hard and thick roof plates using a mechanical formula combined with a numerical simulation research method and obtained the basic top first pressure step formula, which achieved a better early warning effect and provided protection for the safety of miners as much as possible. Chen et al. [14] studied the parameters of the deep hole pre-cracked cutting roof of the hard roof and narrow coal column protection alley with respect to the excessive deformation of the surrounding rock of the hard roof roadway. Reducing the loss of coal pillar while ensuring safety, high yield, and high efficiency is an important part of coal mining. The application of the technology of retaining a narrow coal pillar in pre-splitting blasting can effectively reduce the roof compression strength, reduce the gas concentration in the goaf, reduce the coal spontaneous combustion rate, effectively improve the recovery rate of coal resources, and promote the sustainable development of the coal industry. Wang et al. [15] studied the key parameters of the coal column combined with the existing theory and numerical simulation software and obtained the breaking position of the roof plate of the goaf area, and they believed that there was a trapezoidal internal stability zone around the coal column.

Due to the variability of surrounding rock properties, quarry stress, and roadway section size, the mechanism of roadway deformation and failure is diverse and complex, resulting in different methods and countermeasures for roadway support [16]. Different roadway support methods also have different effects on the sustainable development of coal mines. The traditional roadway support mainly uses a steel frame support and wood support, which have many problems, such as the fact that a steel frame support will produce a lot of scrap steel and a wood support will waste a lot of wood resources. At the same time, these support methods also have safety risks. Modern roadway support methods pay more attention to sustainable development, mainly using new materials and technologies such as bolt support, anchor wire mesh support, shotcrete support, and so on. These support methods can not only improve the stability and safety of the roadway, but also reduce the impact on the environment, save resources, and reduce costs, which is conducive to the sustainable development of coal mines. Qin et al. [17] analyzed the distribution characteristics of the surrounding rock-bearing structure of the roadway using the existing theory, used the FLAC3D 6.0 finite element software to perform a numerical simulation, and proposed a reinforcement scheme according to the analysis results, and the experimental results proved that the bearing capacity of the shallow surrounding rock of the roadway was significantly improved after the application of the new support scheme. Taking a coal mine as the research background, Yang [18] and others used the research method of field practice + theoretical analysis + numerical simulation to analyze the pressure relief effect of severe roadway deformation on surrounding rock, and verify the feasibility of using the pressure relief effect to excavate a new roadway near the damaged roadway, which reduces the waste of coal resources and improves the efficiency of coal mining. Wang et al. [19] took a coal mine in the western part of the Qingyang mining area as the engineering background, used the method of field practice + numerical calculation to reveal the distribution law of the stress field and the displacement field of the surrounding rock of each section, and optimized the roadway support parameters. Xie [20] proposed "external anchor-internal unloading" collaborative control technology according to the deformation of the surrounding rock of a certain mine, and the new technology achieved a good surrounding rock control effect in a field application. A steel arch + bolt support system is widely used in roadway support [21], but different types of bolts have different ways of connecting steel arches, and their supporting effects are also different [22]. Strengthening the retractable steel support with the steel strand bolt passing through the steel beam can maintain the stability of the roadway even under mining stress. However, this method requires the use of a large amount of steel, which will inevitably be wasted in the support process. According to the geological conditions of a steeply inclined coal seam in a coal mine, Tu [23] developed a similar simulation system for a steeply inclined coal seam. The simulation results prove the effectiveness of roof caving and gangue filling characteristics in the process of steeply inclined coal seam mining. Filling goaf with gangue can not only reduce the discharge of solid waste from a coal mine, but also reduce the disaster of mining subsidence and improve the recovery rate of mine resources. Taking the deformation problem of a deep-buried soft rock roadway as the background, Guo et al. [24] studied the causes of roadway deformation and the failure of the original support, and put forward a new full-section combined support system involving an anchor cable, mesh shotcreting, and U-shaped steel support. Shavarskyi et al. [25] studied the influence of longwall face advance on the stress–strain state of a rock mass using GeoDenamics, determined the physical and geometric parameters of the reference pressure and the parameters of a lower sandstone filling body during the subsidence of an old roof, and analyzed the influence of additional pressure on a direct roof. In view of the surrounding rock deformation of a reserved roadway under the condition of a hard roof, most scholars study how to strengthen the surrounding rock, but there is little research on the pressure relief of the surrounding rock and the sustainable development of the environment and economy.

In this paper, by cutting the roof of the working face and blasting along the strike, the influence of the goaf on the roadway roof is reduced, the stress caused by the goaf roof to the roadway is cut off, and pressure relief is achieved. According to the failure reasons of the reserved roadway with a hard roof, the combined control technology of roof cutting and pressure relief, grouting anchor cable, and bolt support is put forward, which reduces the number of protected coal pillars and increases the coal mining rate, and can prolong the service life of the coal mine. It is beneficial to the sustainable utilization of non-renewable resources and the sustainable development of the coal industry, and can provide a reference for similar surrounding rock control work.

2. Project Overview

2.1. Geological Overview

The Gushuyuan Mine is located in Jincheng Mining Area, Shanxi, China, 1 km from Jincheng District. The No. 15 coal seam is the main coal mining seam, located in the lower part of the Taiyuan Formation. The coal seam burial depth is about 420 m; the mining width is 181 m; the mining length is 1588 m; the coal thickness is 0.81~3.56 m with an average of 2 m; and the coal seam inclination angle is 1~9° with an average of 5°. The average compressive strength of the coal seam was measured to be 22.68 MPa, which was relatively low. The direct roof of the No. 15 coal seam is made of K2 limestone, with a thickness of about 9.00 m and an average compressive strength of 92.41 MPa, which is relatively high. The floor is mainly composed of black-gray mudstone and bauxite mudstone. The average compressive strength is 46.88 MPa, and the strength is low. The maximum horizontal principal stress of the No. 15 coal seam is 9.45 MPa, the maximum–minimum horizontal principal stress is 7.11 MPa, and the lateral pressure coefficient is greater than 1, which is mainly horizontal stress and belongs to the type of tectonic stress field.

2.2. Overview of the Roadway

The No. 15 coal seam of the Gushuyuan Mine is the main coal mining seam, and the working face adopts the layout form of "two into one round", with the 1523092 alley and 1523103 alley as the inlet alleys, and the 1523091 alley as the return alley, as shown in Figure 1. The 1523091 roadway and the 1523092 roadway collapsed after mining. The 1523103 reserved roadway was retained during the mining process to serve the 152310 working face. However, due to the influence of the mining process, the surrounding rock of the 1523103 reserved roadway was serious. The failure depth is 4.54 m~5.96 m, and the maximum floor heave is 1.5 m. The roof of the roadway has almost no deformation but shows an overall sinking trend. The maximum section shrinkage rate is more than 70%, and the deformation is shown in Figure 2.

The section of the 1523103 roadway is designed as a rectangular roadway with dimensions of 5.6 m × 2.6 m. The original support method is a combined support of a bolt and anchor cable. One anchor cable with a length of 5400 mm and with row spacing of 800 mm is installed on the roof of the roadway, and two high-strength screw steel bolts without longitudinal reinforcement with a diameter of 18 mm and a length of 2000 mm are installed. The row spacing is 3600 mm × 800 mm. Three high-strength screw steel bolts without longitudinal reinforcement with a diameter of 18 mm and a length of 2000 mm are installed on each side, and the row spacing is 800 mm × 800 mm, as shown in Figure 3.

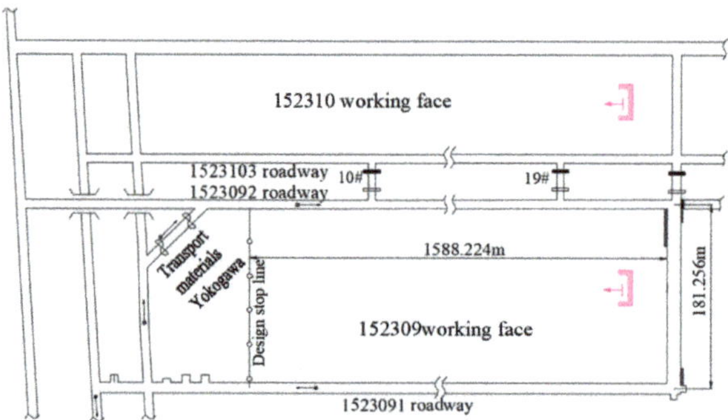

Figure 1. The layout diagram of reserved roadway and working face.

(a)   (b)

Figure 2. Real pictures of surrounding rock deformation of 1523103 roadway. (**a**) Roof subsidence; (**b**) bottom drum situation.

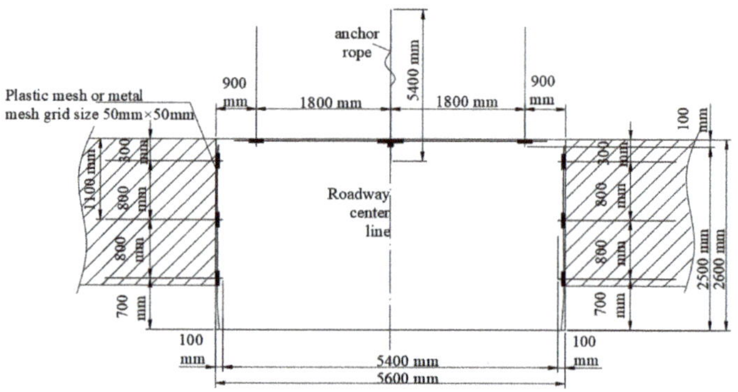

Figure 3. The original support method of 1523103 roadway.

3. Cause Analysis of Surrounding Rock Failure of Reserved Roadway

3.1. Justification of the Geomechanical Model

According to the actual production conditions of the 152309 working face, a 280 m × 280 m × 70 m model was established via FLAC3D. According to the actual situation, the dip angle of the roof and floor strata was set to 0°; the thickness of the coal seam mined by the model was 2 m; the length of the working face was 170 m; the size of the reserved roadway was 5 m × 2.5 m, measured from the bottom of the fixed model, the front and back, and the left and right; the advancing direction of the working face was measured from bottom to top; the reserved roadway was located at 25 m on the right side of the working face; and the grid near the roadway was encrypted. An analysis of the statistical data of the measured vertical stress from all over the world shows that the vertical stress increases linearly in the range of depth of 25–2700 m, which is roughly equivalent to the gravity calculated according to the average bulk density. According to the data from the geological exploration report, the average bulk density of rock mass within 320 m below the surface is 2.4×10^4 N/m^3. The upper boundary of the model was 320 m from the surface, the vertical ground stress was 7.9 MPa, and the lateral pressure coefficient was 1.2. The model is shown in Figure 4.

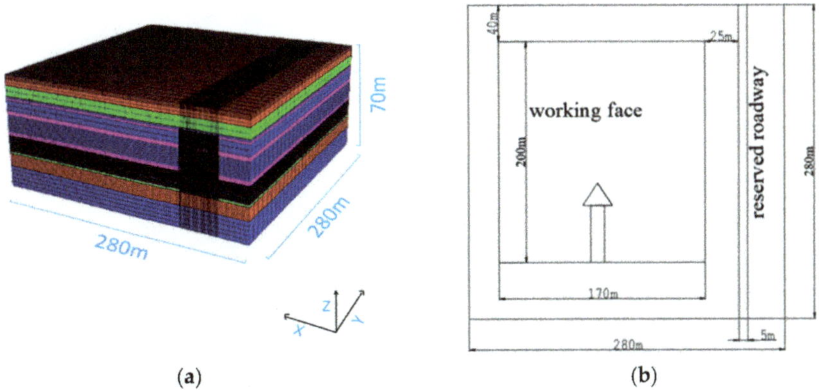

(a)

(b)

Figure 4. Numerical model and working face and reserved roadway layout diagram. (**a**) Numerical model. (**b**) Working face and reserved roadway layout diagram.

The physical and mechanical properties of a coal–rock mass are affected by joints, cracks, and other structural planes, and sometimes, the influence of structure on the properties of a coal–rock mass is much greater than that of the material itself, so, in general, the properties of a coal–rock mass and a rock block are very different. In a numerical simulation, the mechanical parameters such as the elastic modulus, cohesion, and tensile strength of a rock mass are generally 1/5~1/3 of the corresponding mechanical parameters of a rock block, and sometimes, the difference may be even greater, with the ratio reaching 1/20~1/10. The Poisson's ratio of a rock mass is generally 1.2~1.4 times greater than that of a rock block, and the joint stiffness is 0.1~0.9 times greater than that of a normal rock block. The rock mechanics parameters used in the numerical simulation were obtained via reduction based on the geological report of the Gushuyuan Mine and the experimental results of rock mechanics related to the #15 coal seam combined with the rock scale effect [26–28]. The specific parameters are shown in Table 1 below.

Table 1. Mechanical parameters of rock mass after reduction.

Name of Rock Mass (GPa)	Bulk Modulus (Gpa)	Shear Modulus (MPa)	Tensile Strength (MPa)	Cohesion (MPa)	Angle of Internal Friction (°)	Density (Kg/m^3)
limestone	24.32	12.50	1.82	4.23	38	2760
15# coal seam	4.41	2.23	0.56	0.82	23	1540
fine sandstone	13.16	4.92	1.61	4.12	36	2628
grit stone	10.17	3.26	1.60	4.05	36	2672
mudstone	5.52	2.54	1.05	2.03	26	2570
sandy mudstone	6.87	2.93	3.35	1.18	28	2707

In order to study the influence of lateral abutment pressure on the stability of roadway surrounding rock, the influence of the working face advancing process on the surrounding rock stress distribution is simulated using FLAC3D 6.0 software. The Mohr–Coulomb constitutive model is used in the numerical simulation. The normal displacement of the surface around the model is zero, and the normal and horizontal displacements at the bottom of the model are zero. The cloud map of the support pressure distribution of the working face advancing 200 m is obtained, as shown in Figure 5.

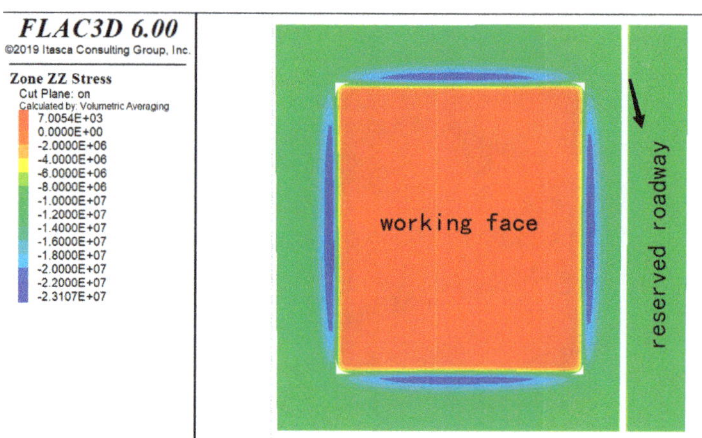

Figure 5. Working face advancing 200 m support pressure distribution cloud map.

In Figure 5, it can be seen that after the working face advances to 200 m, the areas of increased stress in the left and right coal bodies are large, the peak value of the abutment pressure reaches 20.5 MPa, the peak position is about 13 m from the coal wall of the goaf, the pressure relief area in the coal pillar is about 2 m, the area of increase stress is about 44 m, and the depth of the coal wall is 44 m. Outside of the original rock stress state, the coal pillar between the reserved roadway and the goaf in the #15 coal seam is only 25 m, and the excavation position is within the area of increased stress, as shown in Figure 6. The roadway is greatly affected by the abutment pressure of the working face during the mining period of the working face. After excavating the working face, the roof loses support in the vertical direction, resulting in the transfer of the roof load to the coal pillars on both sides, thus causing the deformation and failure of the 1523103 reserved roadway.

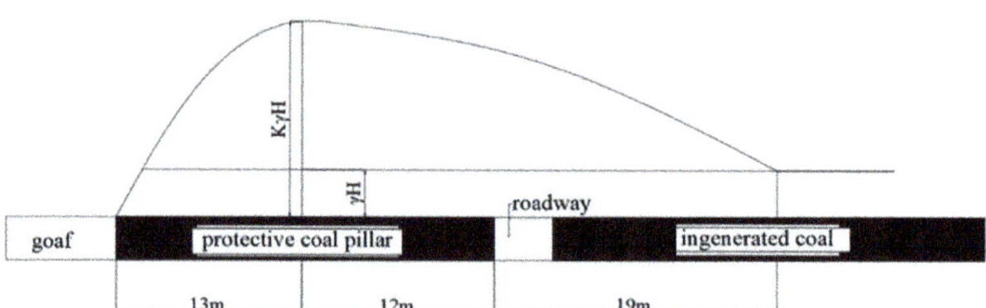

Figure 6. Relationship diagram of reserved roadway position and abutment pressure zone.

3.2. The Influence of Lateral Support Pressure on Working Face

After excavating the working face, the distribution of stress and the displacement of the roadway surrounding rock at different positions differ. Based on the upper section model, the FLAC3D 6.0 software was used to simulate the stress and displacement distribution of the surrounding rock at 25 m from the working face (stress increase area). The simulation results are shown in Figure 7.

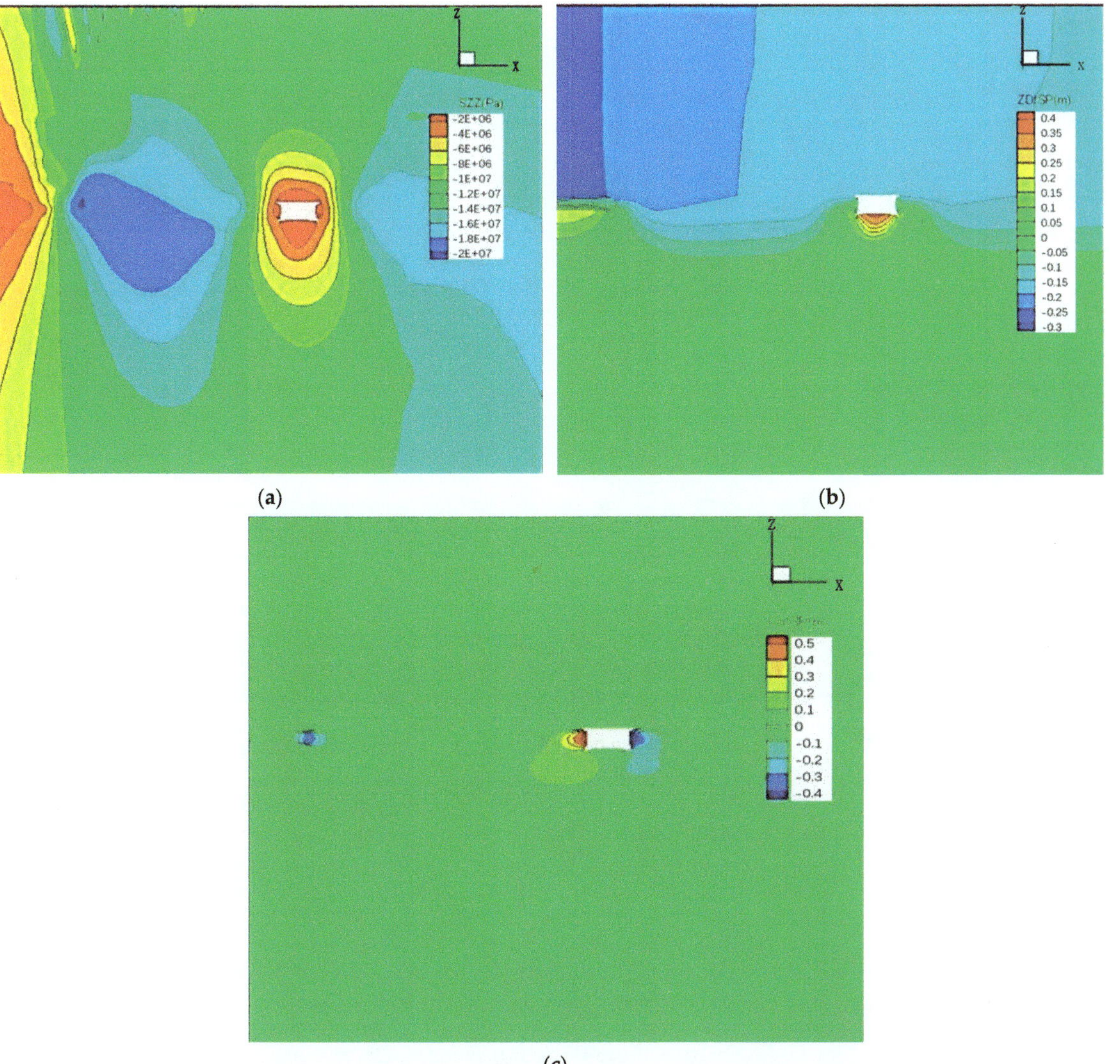

Figure 7. Cloud map of stress and displacement distribution in surrounding rock at a distance of 25 m from the working face (stress increasing area) in the roadway. (**a**) The vertical stress distribution cloud map of surrounding rock at 25 m (stress increase area) from the working face of roadway. (**b**) The vertical displacement cloud map of the roadway 25 m (stress increase area) from the working face. (**c**) The horizontal displacement cloud diagram of the roadway 25 m (where the stress increases) away from the working face.

It can be seen in Figure 7a that there is a large amount of vertical stress on the two sides of the roadway and the on roof and floor in the stress-increasing area, and the maximum vertical stress is about 2 MPa. The vertical stress of the two sides of the roadway changes violently from the surface to the depth, but the stress change in the roof and floor of the roadway from the surface to the depth is relatively gentle, which is due to the loss of vertical support in the roof after the excavation of the working face, resulting in the transfer of roof load to the coal pillars on both sides. The stress near the protective coal pillar is larger, and the maximum stress can reach 20 MPa. It can be seen in Figure 7b that when the roadway is 25 m away from the working face (stress increase area), the roof subsidence is about 0.17 m, and the bottom displacement is about 0.30 m. The vertical displacement of the roadway roof is less than the vertical displacement of the floor. This is because the difference in the lithologic strength of the roof and floor of the roadway leads to inconsistency in the deformation of the roof and floor, and the roof of the roadway is thick and hard, which causes it to sink as a whole, and the subsidence is small. As can be seen from Figure 7c, the displacement of the protective coal pillar side is about 0.51 m, and the displacement of the solid coal side is about 0.48 m. The horizontal displacement of the roadway coal pillar side is larger than that of the solid coal side. This is because the abutment pressure in the protective coal pillar is greater than that in solid coal. The internal stress of a protective coal pillar is formed by the superposition of the working face lateral abutment pressure, and the stress concentration in the coal pillar is caused by roadway excavation. To sum up, the shrinkage rate of the roadway section is about 35.3%.

3.3. Influence of Rock Mass Strength

According to the experimental results of the physical and mechanical parameters of rock, the compressive strength of roof limestone is 92.41 MPa, the natural density is 2618 kg/m^3, the elastic modulus is 25.64 GPa, and the firmness coefficient is 9.2, which belongs to hard rock. It is difficult to completely collapse the hard roof during the mining process of the working face, but it is inclined to the goaf at a certain angle on the whole end face. The collapsed gangue has a certain supporting effect on the roof, and the direct roof forms a cantilever beam on the side of the protective coal pillar. The cantilever beam takes the coal wall as the fulcrum and transfers the pressure to the protective coal pillar while carrying the pressure of itself and the overlying strata in the goaf. As the pressure value continues to rise, when it is greater than the supporting strength of the protective coal pillar, the coal wall near the goaf begins to be destroyed, resulting in the transfer of stress to the inside of the coal pillar, which, in turn, increases the internal stress of the surrounding rock of the reserved roadway, resulting in roadway damage.

According to the field conditions, the calculation formula of the caving zone under hard conditions is selected [29–31].

$$Hm = \frac{100\sum M}{2.1\sum M + 16} \pm 2.5 \tag{1}$$

$\sum M$—cumulative mining thickness. The single-layer mining thickness is 1~3 m, and the cumulative mining thickness is not more than 15 m.

The thickness of limestone strata in the roof of the #15 coal seam is 10.13 m, and the mining height is 2 m; that is, the thickness of the roof strata is greater than three times the mining height. It can be seen that the height of the caving zone is 5.7 ~7.6 m, and the coefficient of bulking is generally 1.33~1.5. Combined with Formula (1), it is finally determined that the height of the roof caving zone of the #15 coal seam is 7 m~12 m.

The compressive strength of floor mudstone is 26.88 MPa, with a firmness coefficient of 4.7, and the compressive strength of coal seam is 22.68 MPa, with a firmness coefficient of 2.3. Due to the existence of clay minerals, the state and physical and chemical properties of mudstone are easily affected by water, pressure, temperature, and other environmental factors. It is easy to expand it in water, shrink it after water loss, and disintegrate it [32,33]. The two sides of the reserved roadway at 1523103 are mostly fractured bodies or even broken bodies. The mine pressure characteristics of the surrounding rock of the reserved roadway mainly depend on the occurrence and development of the deformation and failure of the two sides of the coal body. If the control is not good, it can easily cause the following sequence of events to occur: the two sides are squeezed and broken → the two sides experience spalling and collapse → the roof support of the two sides is weakened → floor heave is caused by the extrusion of the floor → the vicious cycle of the intensification of the damage of the two sides begins.

3.4. The Influence of Supporting Method

After the excavation of the roadway, the stress of the surrounding rock of the roadway will be redistributed; the stress state of the surrounding rock will change from three-dimensional to approximately two-dimensional; and the elastic zone, plastic zone, and broken zone will be generated in the surrounding rock of the roadway over time. The change in the "three zones" will gradually increase over time and eventually stabilize. According to the field geological parameters of the Gushuyuan Mine, the radius of the plastic softening zone, $R_p = 3.4$ m, and the radius of fracture zone, $R_t = 2.74$ m, can be calculated. Given the serious deformation and failure of the loose coal on the two sides of the roadway, some theoretical formulas from at home and abroad [34–36], such as the A. H. Wilson formula, large plate fracture theory formula, and limit equilibrium theory formula, can be used to calculate the failure depth of the two sides of the reserved roadway in combination with the parameters of the on-site geological conditions. It is known that the failure depth of the two sides is 4.54 m~5.96 m.

Roadway 1523103 only has three high-strength rebar bolts without longitudinal ribs with a diameter of 18 mm and a length of 2000 mm on each side of the reserved roadway, and the supporting strength can only control the deformation of the surrounding rock during the excavation of the reserved roadway, but it is far from meeting the deformation of the surrounding rock caused by the mining of the working face. During the mining process of the working face, the damage range of the surrounding rock of the roadway is further expanded, the length and strength of the original bolt cannot meet the requirements, and its supporting effect is lost.

4. Control Countermeasures of Roadway Stability

Based on the above analysis, the lateral support pressure produced by mining in the working face has an adverse impact on the roadway, and it is not easy to collapse the hard roof, which aggravates the mine pressure of the rock mass on both sides of the roadway and leads to the destruction of the rock mass and weak floor on both sides of the roadway, resulting in the failure of the original support. In order to effectively control the stability of the surrounding rock, the mine pressure of the roadway should be reduced, the stress condition of the two sides and floor should be improved, and the support of the rock mass and floor on both sides of the roadway should be strengthened. We decided to adopt the cooperative control technology of pressure relief, improving the mechanical properties of the surrounding rock and high-strength support to control the large deformation of the surrounding rock of the reserved roadway with a hard roof in the mine.

The blasting method was used to cut off the connection between the roadway roof and the goaf so as to reduce the transmission of mine pressure in the goaf and reduce the mine pressure of the roadway. The method of grouting was used to improve the mechanical properties of the rock mass on both sides of the roadway, improve its strength and bearing capacity, and enhance the anchorage performance of the surrounding rock. An anchor cable was used to strengthen the support of the roof and both sides of the roadway. The anchor cable can increase the support strength, expand the support range, and make the rock mass on both sides of the roadway form a larger range of a bearing carrier. The cooperative support technology can reduce the roof pressure of the roadway, improve the stress state of the surrounding rock, increase the bearing capacity of the surrounding rock, expand the bearing range of the rock mass on both sides of the roadway so as to reduce the destruction of the rock mass and floor on both sides of the roadway, and control the large deformation of the roadway.

5. Design of Surrounding Rock Control Scheme of Reserved Roadway 1523103

The main reasons for the deformation of the 1523103 reserved roadway include a large amount of lateral support pressure, the weak strength of the original support, the small strength of the rock mass on both sides of the roadway and floor rock mass, and the small strength of the supporting bolt. In view of the main causes of deformation, the surrounding rock control scheme of roof cutting, the pre-splitting blasting of the roadway roof, and the grouting anchor cable providing high-strength support for the roadway roof and rock mass on both sides of the roadway are adopted to reduce the stress concentration in the coal pillars on both sides of the goaf and improve the bearing capacity of the rock mass and floor on both sides of the roadway so as to realize the long-term stability of the roadway surrounding rock.

5.1. 152309 Face-Cutting Top Pressure Relief Technology

5.1.1. Cutting Top Pressure Relief Principle

With regard to the #15 coal seam hard roof, due to its large thickness, high strength, good integrity, and being directly endowed in the upper part of the #15 coal seam, in the working face mining process, the hard roof will experience plastic bending with the advancement in the working face, showing the characteristic of slow sinking at the end of the working face to form a long cantilever beam and showing mine pressure through the cantilever beam to the coal column, thereby causing the roadway protection coal column to experience a high stress concentration and serious deformation damage. In this regard, the roof can be pre-cracked and blasted on the side of the coal column to protect the working face, the roof structure can be cut off between the 1523092 and 1523103 alleys, the length of the basic roof cantilever beam and the hanging area can be reduced, the roof energy storage conditions can be destroyed, the continuous force generated by the direct top rotation sinking inside the surrounding rock of the reserved roadway can be alleviated, the roof pressure along the empty side of the reserved roadway can be released, and the stress concentration degree of the lateral pressure in the surrounding rock of the roadway can be reduced. The stress distribution is shown in Figure 8.

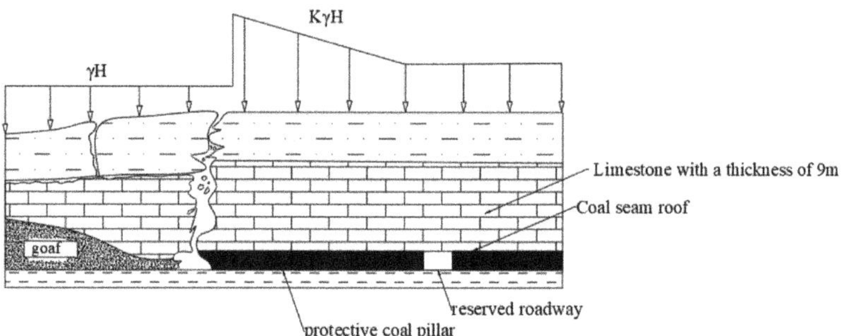

Figure 8. Schematic diagram of stress distribution after roof cutting of reserved roadway.

5.1.2. The Main Technical Parameters of Cutting Top Pressure Relief

According to the actual situation of the site and the workload situation, the end groove method is selected to weaken the top plate of the working face in order to reduce the initial pressure step distance of the working face. The average height of the working face is 1.7 m, so the effective blasting topping depth, H, can be calculated [37] using Equation (2) as 5.7 m, and 6 m is taken.

$$H = \frac{M}{K_P - 1} \tag{2}$$

M—height of mining; the value is 1.7 m.

K_p—volume expansion coefficient of rock after crushing; the value is 1.3.

Before the installation of the working face equipment, the roof is drilled 1 m away from the side of the old pond in the open-off cut. A total of 10 groups of blastholes are arranged in the 180 m long open-off cut: A, B, C, D, and E, and a, b, c, d, and e. Each group of A and a has two blastholes. B, C, and D and b, c, and d each have three holes. Both groups of E and e have four holes.

The specific parameters of A1 and a1 and A2 and a2 are as follows: the lengths of the blast holes are 14 m and 16 m, respectively, the elevation angles are 30° and 27°, respectively; the charge lengths are 6 m and 8 m, respectively; and the charge quantities are 15 kg and 20 kg, respectively.

The specific parameters of B1, b1; B2, b2; and B3, b3 are as follows: the hole lengths are 14 m, 14 m, and 16 m, respectively; the elevation angles are 32°, 30°, and 27°, respectively; the charge lengths are 7 m, 6 m, and 8 m, respectively; and the charge weights are 17.5 kg, 15 kg, and 20 kg, respectively.

The specific parameters of C1, c1; C2, c2; and C3, c3 are as follows: the hole lengths are 14 m, 14 m, and 16 m, respectively; the elevation angles are 32°, 30°, and 27°, respectively; the charge lengths are 7 m, 6 m, and 8 m, respectively; and the charge weights are 17.5 kg, 15 kg, and 20 kg, respectively.

The specific parameters of D1, d1; D2, d2; and D3, d3 are as follows: the hole lengths are 14 m, 14 m, and 16 m, respectively; the elevation angles are 32°, 30°, and 27°, respectively; the charge lengths are 7 m, 6 m, and 6 m, respectively; and the charge weights are 17.5 kg, 15 kg, and 20 kg, respectively.

The specific parameters of E1, e1; E2, e2; E3, e3; and E4, e4 are as follows: the lengths of the blast hole are 14 m, 14 m, 16 m, and 10 m, respectively; the elevation angles are 32°, 30°, 27°, and 30°, respectively; the charge lengths are 7 m, 6 m, 8 m, and 4 m, respectively; and the charge quantities are 17.5 kg, 15 kg, 20 kg, and 9.6 kg, respectively.

The remaining part of the blasthole is sealed with gun mud. Group A is symmetrical with group a, group B is symmetrical with group b, etc. The arrangement of the blasthole is shown in Figure 9.

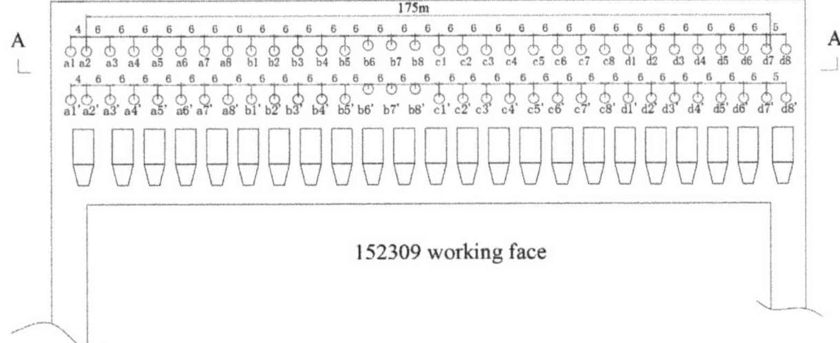

Figure 9. Blast hole layout plan.

5.2. 1523092 Roadway Roof Weakening Technology

5.2.1. Principle of Advanced Deep Hole Pre-Splitting Blasting Technology

Deep hole blasting can pre-split the thick and hard roof above the coal seam in advance, reduce the length of the main roof cantilever beam and the area of the suspended roof, and destroy the energy storage conditions of the roof. Deep hole blasting can loosen and pre-split the roof of the goaf in advance, release the roof pressure along the goaf side of the roadway, and effectively reduce the stress concentration degree of lateral pressure on the side of the protective coal pillar.

5.2.2. Main Technical Parameters of Advanced Deep Hole Pre-Splitting Blasting

A borehole with a length of 15 m and an elevation angle of 45° is arranged every 5 m from the cutting hole to the side of the coal pillar along the 1523092 roadway, with a horizontal angle of 0°, a charge length of 5 m, and a charge amount of 15 kg, and the remaining part of the borehole is filled with gun mud. Between the cut-through and the mining of the working face, 10 boreholes are arranged within 50 m of the cut-out to implement pre-splitting blasting. The specific blasthole parameters are shown in Table 2, and the blasthole parameters can be adjusted appropriately according to the on-site blasting effect. During the blasting, two rows of single pillars are arranged every 1 m from the current working face at 0.6 m from the two sides of the blasting roadway to be cracked, which is used to strengthen the advance support after roof cutting and pre-splitting and ensure the stability of the 1523092 roadway during pre-splitting blasting. With the mining of the working face, such a reciprocating cycle always maintains a 50 m advance support during blasting. The spatial relationship between the advance support and the roof cutting hole is shown in Figure 10.

Table 2. Pre-splitting blasting hole parameter table.

Hole Length (m)	Angle of Elevation (°)	Horizontal Angle (°)	Aperture (mm)	Charge Length (m)	Powder Charge (Kg)	Sealing Length (m)
15	45	0	75	5	15	10

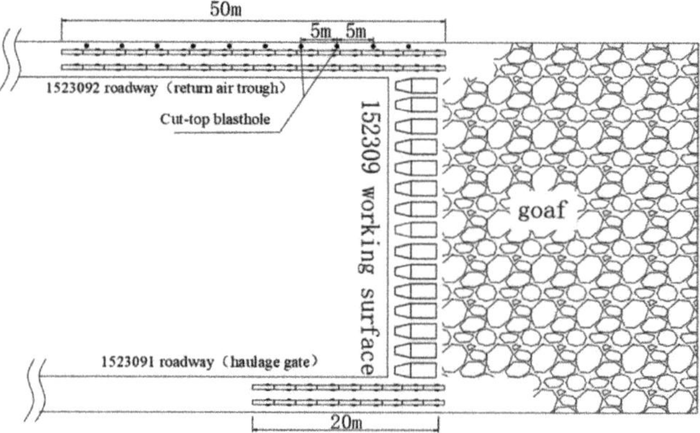

Figure 10. The spatial relationship diagram of advance support and roof cutting blasthole.

5.3. Grouting Anchor Cable Support Technology

5.3.1. Supporting Principle of Grouting Anchor Cable

The principle of grouting anchor cable reinforcement is to combine the supporting effect of the anchor cable with the effect of grouting reinforcement and work together on the surrounding rock of the roadway. The grouting of organic or inorganic slurry using a grouting anchor cable can not only fundamentally ensure the reliability of the anchor cable, but can also penetrate a large range of coal and rock mass around the drilling hole, produce a bonding and curing effect on the loose coal and rock mass, significantly improve its integrity, improve the self-supporting ability of the coal and rock mass, and greatly improve the supporting effect of the roadway. The structure diagram of the grouting anchor cable is shown in Figure 11.

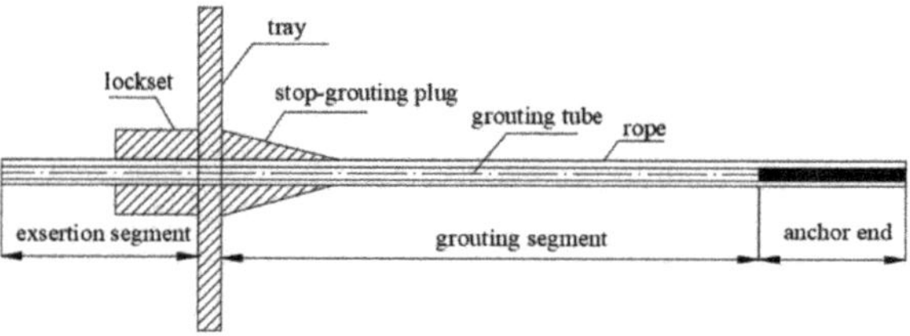

Figure 11. Structure diagram of grouting anchor cable.

5.3.2. Layout Scheme and Main Technical Parameters of Grouting Anchor Cable

On the basis of the original support of the 1523103 roadway, the grouting anchor cable is added to the coal wall from the two sides of the reserved roadway parallel to the floor to strengthen the control of the surrounding rock—that is, the hollow grouting anchor cable with dimensions of Φ22 mm $\times$ 5300 mm is added to the three-hole staggered type between the existing rows of anchor bolts and the upper and lower rows of the anchor cables are added with Φ14 mm trapezoidal steel strips, respectively. The grouting anchor cable layout is shown in Figure 12.

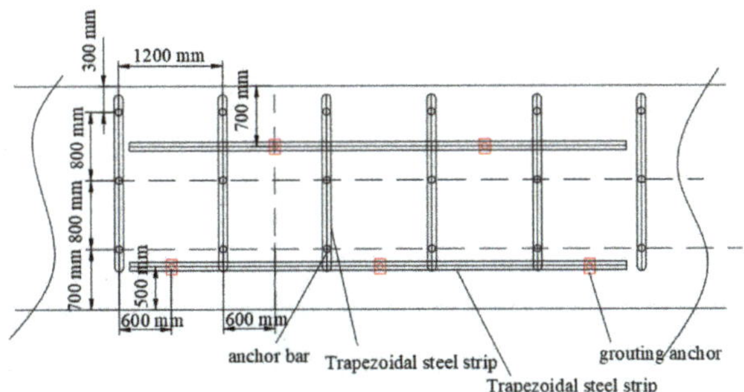

Figure 12. Arrangement diagram of grouting anchor cable.

According to the principle of the anchor cable length, strength, and key parts, the anchor cable row spacing is set to 1.2 m, the nominal diameter of the steel wire is 6.0 mm, the installation aperture is Φ32 mm, the strength is 1760 MPa, the breaking force is $\geq$420 KN, the resin anchorage length is 1000~1500 mm, the inner diameter of the hollow grouting pipe is Φ7.5 mm and the outer diameter is Φ10 mm, and the grouting pressure is 5.0~7.0 MPa. The new support scheme is shown in Figure 13.

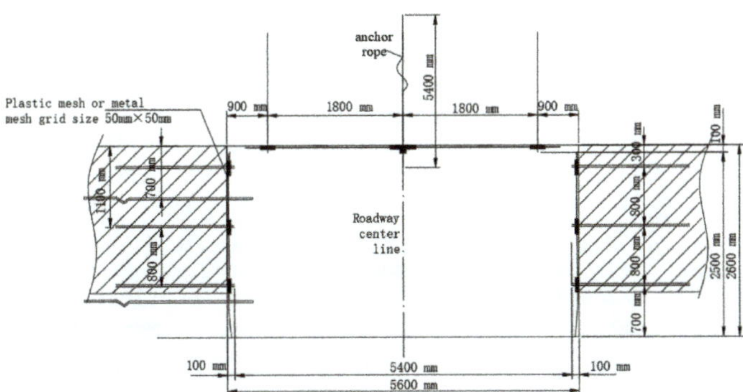

Figure 13. New support scheme diagram.

6. Analysis of Monitoring Results of Surrounding Rock Control in Reserved Roadway

6.1. Analysis of Monitoring Results of Lateral Support Pressure of Coal Pillar

Two survey areas were arranged outward along the contact lane at the open-off cut of the 1523103 roadway, and three boreholes were arranged in each survey area. In the #1 survey area, the #6 measuring point was 785 m away from the open-off cut, the #5 measuring point was 5 m away from the #6 measuring point, the #4 measuring point was 5 m away from the #5 measuring point, and the drilling installation depths were 5 m, 9.5 m, and 7 m, respectively. In the #3 survey area, the #4 measuring point was 900 m away from the open-off cut, the #2 measuring point was 5 m away from the #3 measuring point, the #1 measuring point was 5 m away from the #2 measuring point, and the drilling installation depths were 5 m, 6.5 m, and 8 m, respectively. Each borehole was 1 m higher than the floor and parallel to the floor of the coal pillar. The layout of the borehole stress meter is shown in Figure 14.

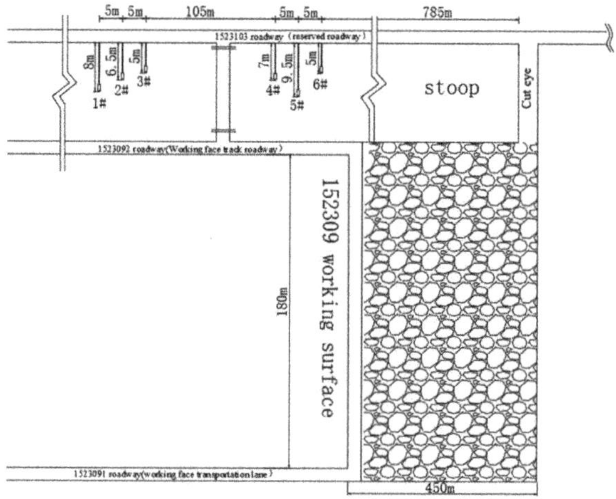

Figure 14. Borehole stress meter layout diagram.

According to the above design, six stress meters were arranged in the 1523103 roadway, and the placement depth of the stress meter was 7 m. Among them, the #5 stress gauge was destroyed by underground workers, and the #2 and #4 stress gauges had no pressure data after liquid injection, which was regarded as invalid, so there were three effective stress gauges. The stress curve is shown in Figure 15.

It can be seen in Figure 15 that there is no obvious change in the pressure value of the #1 borehole stress meter 50 m outside of the working face, and there is a decreasing trend between 9 m and 50 m in front of the working face. When the 1# drilling stress gauge is located 9 m in front of the working face, the stress value increases rapidly. When it is located 78 m behind the working face, the data basically do not increase. When the #1 borehole stress meter is located 9 m in front of the working face and 16 m behind the working face, its increment reaches 47% of the initial pressure. When the #1 borehole stress meter is located 16 m behind the working face, its increment only reaches 5% of the initial pressure. When the #1 borehole stress meter is located 78 m behind the working face, its value basically does not change. As can be seen from Figure 7a, the maximum stress at the drilling position before the top cutting is about 16 MPa; from Figure 15a, it can be seen that the maximum stress at the drilling position after the top cutting is only 4.2 MPa, which is about 73% lower than that without the top cutting.

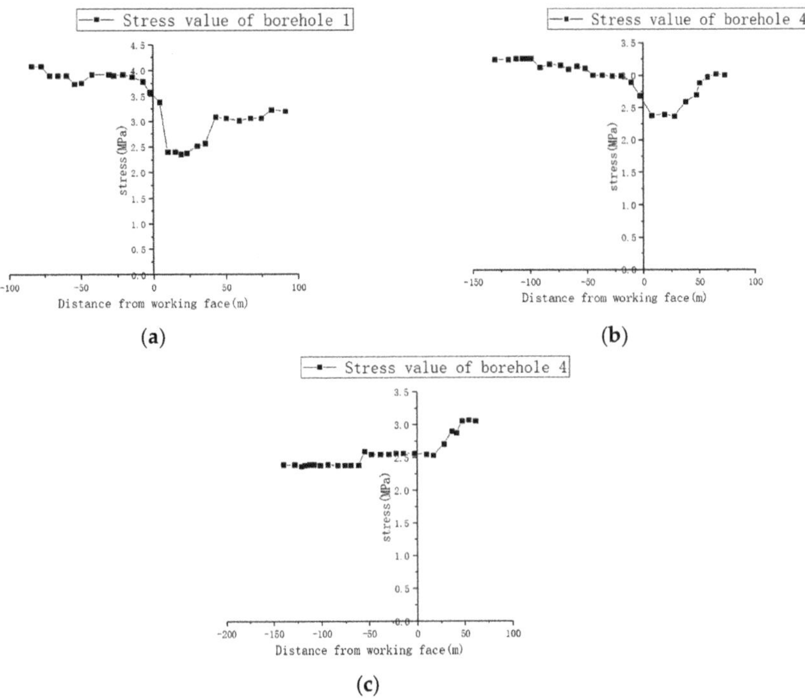

Figure 15. Stress curve. (**a**) Pressure curve of #1 borehole stress meter. (**b**) Pressure curve of #4 borehole stress meter. (**c**) Pressure curve of #6 borehole stress meter.

There is no obvious change in the pressure value of the #4 borehole stress meter 50 m outside of the working face, and there is a decreasing trend between 6 m and 40 m in front of the working face. When it is located in front of the working face at 6 m, the rate of increase is obvious. When it is located 90 m behind the working face, the data basically do not increase. When the #4 borehole stress meter is located in front of the working face at 6 m and 20 m behind the working face, the increment reaches 20% of the initial pressure. When the #4 borehole stress meter is located 20 m behind the working face, the increment only reaches 8% of the initial pressure. When the #4 borehole stress meter is located 90 m behind the working face, its value is basically unchanged. As can be seen from Figure 7a, the maximum stress at the drilling position before the top cutting is about 16 MPa; from Figure 15a, it can be seen that the maximum stress at the drilling position after the top cutting is only 3.3 MPa, which is about 79% lower than that without the top cutting.

The pressure value of the #6 borehole stress meter gradually decreases outside the working face at 26 m. When it is located at the working face of 26 m, the pressure meter value begins to stabilize and then basically does not change, indicating that the 1523103 roadway is 780 m away from the cut. The mining influence of the working face is very small, and there is no obvious significant influence range in this section. As can be seen from Figure 7a, the maximum stress at the drilling position before the top cutting is about 16 MPa; from Figure 15a, it can be seen that the maximum stress at the drilling position after the top cutting is only 3.2 MPa, which is about 80% lower than that without the top cutting.

Compared with before and after the top cutting of the working face, the lateral support pressure of the coal pillar is reduced by at least 73% and up to 80%, and the effect of roof cutting and pressure relief is obvious. In general, the influence range of the 1523103 roadway affected by mining is concentrated between 9 m in front of the working face and 90 m behind the working face, and the significant influence range is concentrated between

9 m in front of the working face and 20 m behind the working face. Even so, the relative value of the lateral coal wall support pressure of the reserved roadway is still very small, and the maximum is 4.2 MPa, that is, the mining of the working face has no obvious influence on the stability of the reserved roadway.

6.2. Analysis of Surface Displacement Monitoring Results of Reserved Roadway

In the 1523103 reserved roadway, six measuring points (#1~#6) are arranged from the inside to the outside of the open-off cut position. The #1 measuring point is 785 m away from the open-off cut, the #2 measuring point is 790 m away from the open-off cut, the #3 measuring point is 795 m away from the open-off cut, the #4 measuring point is 800 m away from the open-off cut, the #5 measuring point is 900 m away from the open-off cut, and the #6 measuring point is 910 m away from the open-off cut. The positions of #1, #2, #3, #5, and #6 correspond to those of the #6, #5, #4, #3, and #1 stress meters, respectively.

The surface displacement of the 1523103 reserved roadway was observed using a steel ruler and measuring rod. According to the data obtained from measuring points #1~#6, the surface displacement deformation observation results of each measuring point were drawn, as shown in Figure 16.

In the diagram, it can be seen that the maximum shrinkage of the two sides of the #1 measuring point is 44 mm, the maximum shrinkage of the two sides of the #2 measuring point is 23 mm, the deformation of the two sides of the #3, #4, #5, and #6 measuring points is within 15 mm, and the deformation curve is gentle. In general, there is almost no deformation in the two sides of the roadway.

When the #1 measuring point is located 33 m in front of the working face, the convergence of the roof and floor begins to increase greatly. When the measuring point is located 108 m behind the working face, the convergence of the roof and floor gradually tends to be gentle. The maximum shrinkage of the roof and floor of the #1 measuring point is 51 mm. The change trend of the roof and floor convergence of the #2, #3, and #4 measuring points is similar to that of the #1 measuring point. The maximum shrinkage of the roof and floor of the #2 measuring point is 23 mm, the maximum shrinkage of the roof and floor of the #3 measuring point is 40 mm, and the maximum shrinkage of the roof and floor of the #4 measuring point is 49 mm. The roof-to-floor convergence of the #5 and #6 measuring points is large. When the #5 measuring point is located 24 m in front of the working face, the roof-to-floor convergence increases greatly, and the maximum shrinkage of the roof-to-floor convergence is 91 mm. When the #6 measuring point is located 34 m in front of the working face, the convergence of the roof and floor increases greatly, and the maximum shrinkage of the roof and floor is 101 mm.

Based on the above analysis, the deformation pattern of the 1523103 reserved tunnel during the period affected by the mining face is roughly 29.5 m away from the leading working face as the dividing line. The roadway beyond 29.5 m of the advanced working face is basically not affected by the mining of the working face, the two sides of the roadway and the roof and floor of the roadway have no obvious deformation, and the roadway basically maintains its original stable state. The roadway within 29.5 m of the leading working face began to deform under the pressure of the leading support of the working face, and the approaching amount of the roof and floor began to increase rapidly. Through observation and data analysis, the deformation of the reserved roadway is mainly bottom heave, and the effect of controlling the surrounding rock deformation of the reserved roadway is better after adopting the new surrounding rock control scheme, as shown in Figure 17.

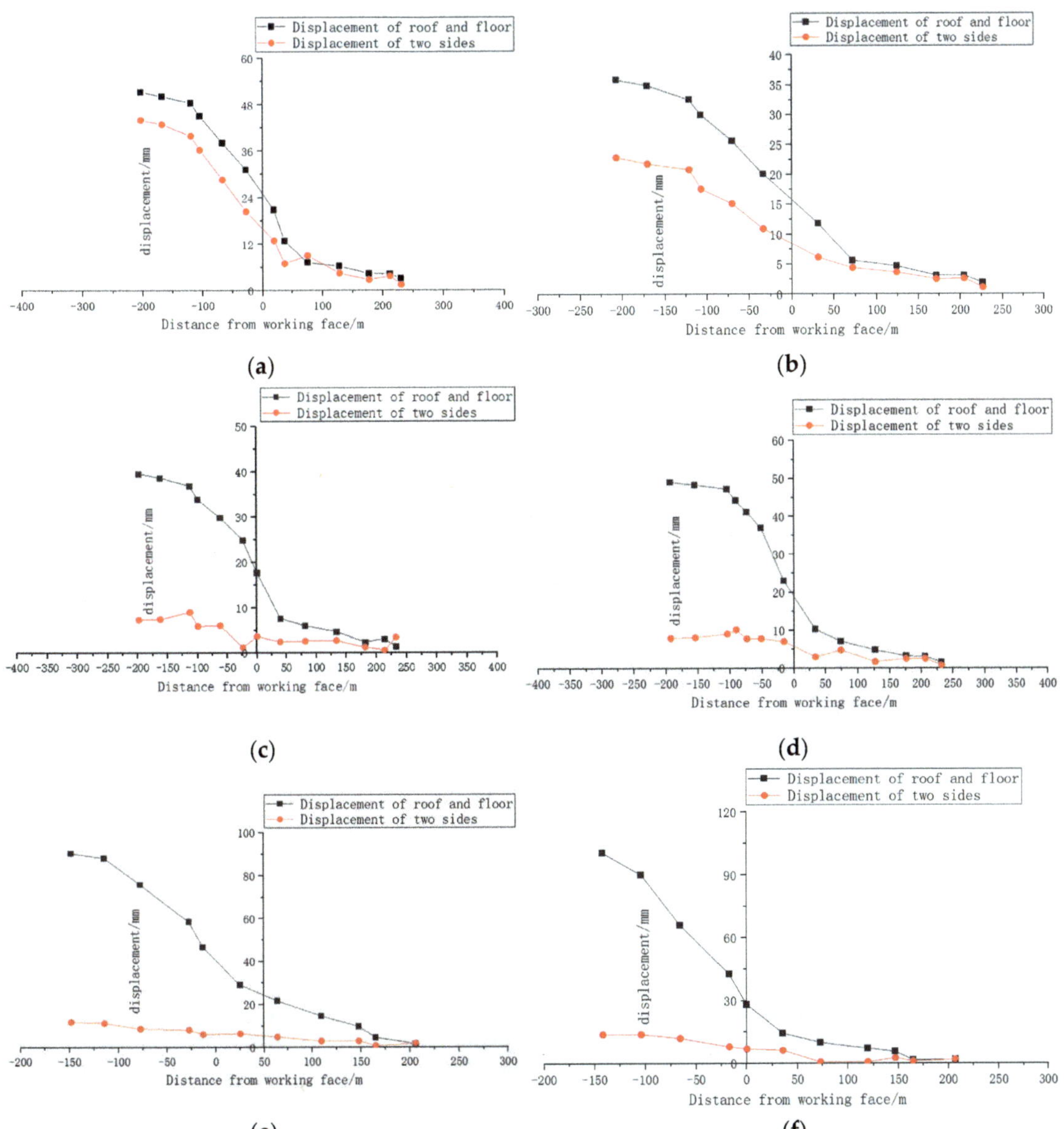

Figure 16. Surface displacement curve of measuring point. (**a**) Surface displacement curve of #1 measuring point. (**b**) Surface displacement curve of #2 measuring point. (**c**) Surface displacement curve of #3 measuring point. (**d**) Surface displacement curve of #4 measuring point. (**e**) Surface displacement curve of #5 measuring point. (**f**) Surface displacement curve of #6 measuring point.

Figure 17. Real picture of deformation control effect of reserved roadway surrounding rock.

7. Conclusions

The 1523103 roadway is greatly affected by the supporting pressure of the working face during mining. After the working face is excavated, the roof load is transferred to the coal pillars on both sides, causing fragments of the 1523103 reserved roadway. The floor of the 1523103 roadway is mudstone, and its own state and physical and chemical properties are easily affected by environmental factors such as water, pressure, temperature, etc. It easily swells when it encounters water, it shrinks after losing water, and it is prone to floor heaving. The roadway floor heave reduces the roadway section, restricts the coal mine transportation work, hinders the mine ventilation, consumes a lot of labor and material resources, seriously limits the output efficiency of the coal mine, and is not conducive to the sustainable development of the coal mine.

The cooperative control technology involving relieving pressure and improving the mechanical properties of the surrounding rock and high-strength support is adopted to control the large deformation of surrounding rock of the reserved roadway with a hard roof in the mine, reduce the roof pressure of the roadway, improve the stress state of the surrounding rock, increase the bearing capacity of the surrounding rock, and expand the bearing range of the rock mass on both sides of the roadway so as to reduce the destructive effect of both sides of the roadway and floor. After adopting the surrounding rock control scheme of roof cutting and pressure relief + bolt + grouting anchor cable support, the maximum bottom heave of the reserved roadway is reduced from 1.5 m to 0.1 m, and the maximum shrinkage of the cross section is reduced from 70% to 11%. The maximum deformation of the two sides of the roadway is 0.044 m, the deformation curve is smooth, and the surrounding rock control effect is good.

The successful treatment of the large deformation of the surrounding rock of the reserved roadway with a hard roof by cutting the top pressure + anchor rod + grouting anchor cable support reduces the number of protective coal pillars, improves the coal mining rate, ensures the safety of miners, prolongs the service life of coal mines, and is conducive to the sustainable utilization of non-renewable resources and the sustainable development of the coal industry.

Author Contributions: All the authors contributed to this paper. W.W. conceived of and designed the research; Y.H. and M.L. analyzed the data and wrote the manuscript; Z.L. and Z.Z. provided theoretical and methodological guidance in the research process; M.L. participated in revising the manuscript. All authors have read and agreed to the published version of the manuscript.

Funding: This research was funded by the Cultivation of National Major Achievements, grant number NSFRF230202; the Henan Province University Science and Technology Innovation Talent Support Program, grant number 23HASTIT011; and the Innovative Research Team of Henan Polytechnic University, grant number T2022-2.

Institutional Review Board Statement: Not applicable.

Informed Consent Statement: Not applicable.

Data Availability Statement: Not applicable.

Conflicts of Interest: The authors declare no conflict of interest.

References

1. Wang, Q.; He, M.; Wang, Y.; Liu, J.; Xue, H. Research progress and prospect of automatically formed roadway by roof cutting and pressure relief without pillars. *J. Min. Saf. Eng.* **2023**, *40*, 429–447.
2. Wang, Z.; Cai, X.; Liu, Z. Survival and Revival: Transition Path of the Chinese Construction Industry during the COVID-19 Pandemic. *Eng. Manag. J.* **2022**, 1–13. [CrossRef]
3. Wang, Z.; Liu, Z.; Liu, J. Innovation strategy or policy pressure? The motivations of BIM adoption in China's AEC enterprises. *J. Asian Arch. Build. Eng.* **2021**, *21*, 1578–1589. [CrossRef]
4. Wan, F. Application research on roof-cutting pressure relief blasting technology in gob-side entry retaining. *China Energy Environ. Prot.* **2018**, *40*, 4.
5. Ma, J.; Chen, J.; Chen, W.; Huang, L. A coupled thermal-elastic-plastic-damage model for concrete subjected to dynamic loading. *Int. J. Plast.* **2022**, *153*, 103279. [CrossRef]
6. Ma, J.; Ding, W.; Lin, Y.; Chen, W.; Huang, L. A fast and efficient particle packing generation algorithm with controllable gradation for discontinuous deformation analysis. *Geomech. Geophys. Geo-Energy Geo-Resour.* **2023**, *9*, 1–27. [CrossRef]
7. Ma, J.; Zhao, J.; Lin, Y.; Liang, J.; Chen, J.; Chen, W.; Huang, L. Study on Tamped Spherical Detonation-Induced Dynamic Responses of Rock and PMMA Through Mini-chemical Explosion Tests and a Four-Dimensional Lattice Spring Model. *Rock Mech. Rock Eng.* **2023**. [CrossRef]
8. He, M.; Wang, Y.; Yang, J.; Zhou, P.; Gao, Q.; Gao, Y. Comparative analysis on stress field distributions in roof cutting nonpillar mining method and conventional mining method. *J. China Coal Soc.* **2018**, *43*, 626–637.
9. He, M.; Chen, S.; Guo, Z.; Yang, J.; Gao, Y. Control of surrounding rock structure for gob-side entry retaining by cutting roof to release pressure and its engineering application. *J. China Univ. Min. Technol.* **2017**, *46*, 959–969.
10. Ma, X.; He, M.; Li, X.; Wang, E.; Hu, C.; Gao, R. Deformation mechanism and control measures of overlying strata with gob-side entry retaining formed by roof cutting and pressure releasing. *J. China Univ. Min. Technol.* **2019**, *48*, 474–483.
11. He, M.; Song, Z.; Wang, A.; Yang, H.; Qi, H.; Guo, Z. Theory of longwall mining by using roof cuting shortwall team and 110 method—The third mining science and technology reform. *Coal Sci. Technol. Mag.* **2017**, *1*, 1–9.
12. He, M.; Ma, X.; Yu, B. Analysis of strata behavior process characteristics of gob-side entry retaining with roof cutting and pressure releasing based on composite roof structure. *Shock. Vib.* **2019**, *2019*, 2380342. [CrossRef]
13. Zhu, Z.; Li, Y.; Chen, T.; Ruan, X.; Wang, G. Study on breaking law of hard thick layer roof and surrounding rock control. *Coal Technol.* **2023**, *42*, 25–29.
14. Chen, L.; Wang, X.; Huo, Y.; Ning, F.; Shi, J.; Di, X. Key parameters of retaining narrow coal pillar for roadway protection with deep hole roof cutting and pressure relief in hard roof. *Coal Eng.* **2023**, *55*, 13–19.
15. Wang, W.; Wu, Y.; Lu, X.; Zhang, G. Study on small coal pillar in gob-side entry driving and control technology of the surrounding rock in a high-stress roadway. *Front. Earth Sci.* **2023**, *10*, 1020866. [CrossRef]
16. Zhu, L.; Yao, Q.; Xu, Q.; Yu, L.; Qu, Q. Large Deformation Characteristics of Surrounding Rock and Support Technology of Shallow-Buried Soft Rock Roadway: A Case Study. *Appl. Sci.* **2022**, *12*, 687. [CrossRef]
17. Qin, D.; Wang, X.; Zhang, D.; Chen, X. Study on surrounding rock-bearing structure and associated control mechanism of deep soft rock roadway under dynamic pressure. *Sustainability* **2019**, *11*, 1892. [CrossRef]
18. Yang, H.; Zhang, N.; Han, C.; Sun, C.; Song, G.; Sun, Y.; Sun, K. Stability control of deep coal roadway under the pressure relief Effect of adjacent roadway with large deformation: A case study. *Sustainability* **2021**, *13*, 4412. [CrossRef]
19. Wang, F.; Wu, C.; Yao, Q.; Li, X.; Chen, S.; Li, Y.; Li, H.; Zhu, G. Instability mechanism and control method of surrounding rock of water-rich roadway roof. *Minerals* **2022**, *12*, 1587. [CrossRef]
20. Xie, S.; Li, H.; Chen, D.; Feng, S.; Ma, X.; Jiang, Z.; Cui, J. New technology of pressure relief control in soft coal roadways with deep, violent mining and large deformation: A key study. *Energies* **2022**, *15*, 9208. [CrossRef]
21. Xu, D.; Gao, M.; Yu, X. Dynamic Response Characteristics of Roadway Surrounding Rock and the Support System and Rock Burst Prevention Technology for Coal Mines. *Energies* **2022**, *15*, 8662. [CrossRef]
22. Majcherczyk, T.; Niedbalski, Z.; Małkowski, P.; Bednarek, Ł. Analysis of Yielding Steel Arch Support with Rock Bolts in Mine Roadways Stability Aspect. *Arch. Min. Sci.* **2014**, *59*, 641–654. [CrossRef]
23. Tu, H.; Tu, S.; Zhang, C.; Zhang, L.; Zhang, X. Characteristics of the Roof Behaviors and Mine Pressure Manifestations During the Mining of Steep Coal Seam. *Arch. Min. Sci.* **2017**, *62*, 871–891.
24. Guo, S.; Zhu, X.; Liu, X.; Duan, H. A Case Study of Optimization and Application of Soft-Rock Roadway Support in Xiaokang Coal Mine. *Adv. Civ. Eng.* **2021**, *2021*, 3731124. [CrossRef]
25. Shavarskyi, I.; Falshtynskyi, V.; Dychkovskyi, R.; Akimov, O.; Sala, D.; Buketov, V. Management of the longwall face advance on the stress-strain state of rock mass. *Min. Miner. Depos.* **2022**, *16*, 78–85. [CrossRef]
26. He, M.; Leal e Sousa, R.; Müller, A.; Vargas, E.; Ribeiro e Sousa, L.; Chen, X. Analysis of excessive deformations in tunnels for safety evaluation. *Tunn. Undergr. Space Technol.* **2015**, *45*, 190–202.
27. Wang, S.; Wang, J.; Dai, Y. Numerical analysis on movement and failure of coal and strees condition of support structures in mechanized top-coal caving in steep thick seam. *Chin. J. Rock Mech. Eng.* **2004**, *23*, 4531–4534.

28. Hashikawa, H.; Mao, P.; Sasaoka, T.; Hamanaka, A.; Shimada, H.; Batsaikhan, U.; Oya, J. Numerical simulation on pillar design for longwall mining under weak immediate roof and floor strata in indonesia. *Sustainability* **2022**, *14*, 16508. [CrossRef]
29. Yue, X.; Tu, M.; Li, Y.; Chang, G.; Li, C. Stability and Cementation of the Surrounding Rock in Roof-Cutting and Pressure-Relief Entry under Mining Influence. *Energies* **2022**, *15*, 951. [CrossRef]
30. Mondal, D.; Roy, P.N.S.; Kumar, M. Monitoring the strata behavior in the Destressed Zone of a shallow Indian longwall panel with hard sandstone cover using Mine-Microseismicity and Borehole Televiewer data. *Eng. Geol.* **2020**, *271*, 105593. [CrossRef]
31. Wang, F.; Zhang, C.; Zhang, X.; Song, Q. Overlying strata movement rules and safety mining technology for the shallow depth seam proximity beneath a room mining goaf. *Int. J. Min. Sci. Technol.* **2015**, *25*, 139–143. [CrossRef]
32. Ou, X.-F.; Ouyang, L.-X.; Xu, X.-X.; Wang, L. Case study on floor heave failure of highway tunnels in gently inclined coal seam. *Eng. Fail. Anal.* **2022**, *136*, 106224. [CrossRef]
33. Wang, J.; Guo, Z.; Yan, Y.; Pang, J.; Zhao, S. Floor heave in the west wing track haulage roadway of the Tingnan Coal Mine: Mechanism and control. *Int. J. Min. Sci. Technol.* **2012**, *22*, 295–299. [CrossRef]
34. He, W.; He, F.; Zhao, Y. Field and simulation study of the rational coal pillar width in extra-thick coal seams. *Energy Sci. Eng.* **2019**, *8*, 627–646. [CrossRef]
35. Liu, Q.; Xue, Y.; Ma, D.; Li, Q. Failure Characteristics of the Water-Resisting Coal Pillar under Stress-Seepage Coupling and Determination of Reasonable Coal Pillar Width. *Water* **2023**, *15*, 1002. [CrossRef]
36. Shuang-Chun, J.; Jia-chen, W.; Jian-ming, Z. Calculation and Application on Elastic-Plastic Coal Pillar Width of the Stope. *Procedia Eng.* **2011**, *26*, 1116–1124. [CrossRef]
37. Wang, Q.; Gao, H.; Jiang, B.; Li, S.; He, M.; Wang, D.; Lu, W.; Qin, Q.; Gao, S.; Yu, H. Research on reasonable coal pillar width of roadway driven along goaf in deep mine. *Arab. J. Geosci.* **2017**, *10*, 466. [CrossRef]

 *sustainability*

Article

Experimental Study of the Mechanical and Acoustic Emission Characteristics of Sandstone by Using High-Temperature Water-Cooling Cycles

Wen Wang [1,*], Lei Hong [1,*], Xuewen Cao [2], Xiaowei Lu [3], Fan Wang [1], Tong Zhang [4] and Weibing Zhu [5]

1   School of Energy Science and Engineering, Henan Polytechnic University, Jiaozuo 454003, China
2   School of Resources and Civil Engineering, Northeastern University, Shenyang 110819, China
3   School of Safety Engineering, China University of Mining and Technology, Xuzhou 221116, China
4   State Key Laboratory of Mining Response and Disaster Prevention and Control in Deep Coal Mines, School of Energy and Safety, Anhui University of Science and Technology, Huainan 232001, China
5   School of Mining Engineering, China University of Mining and Technology, Xuzhou 221116, China
*   Correspondence: wangwen2006@hpu.edu.cn (W.W.); 212102020005@home.hpu.edu.cn (L.H.)

Abstract: In order to study the physical and mechanical properties of sandstone under high-temperature water-cooling cycling conditions, an RMT-150B electrohydraulic servo rock testing system and a DS-5 acoustic emission detection and analysis system were used to conduct uniaxial compression acoustic emission tests on sandstone after high-temperature water-cooling cycles. The deformation, strength, and acoustic emission characteristics of sandstone were analyzed under different temperatures and cycle times. The results show that the high-temperature water-cooling effect caused changes in the physical properties of sandstone. The volumetric expansion rate of the rock samples first decreased, then increased in temperature, and the strength first increased, then decreased, whereas the number of cycles had less of an impact on the physical properties. At 200 °C, with increased cycle number, the elastic modulus increased by 20.1%, and the compressive strength increased from 63.9 MPa to 71.46 MPa. At 300–600 °C, the elastic modulus and compressive strength of sandstone gradually decreased with increases in the temperature and cycle number, with reductions of 6.04%, 7.24%, 28.7%, 35.57%, 17.6%, 18.2%, 20.4%, and 60.5%, respectively. With increased temperature and cycle times, the acoustic emission ringing counts increased, ringing counts and cumulative energy appeared earlier, the rock samples entered elastic deformation earlier, the yield stage length increased, and the samples showed a tendency to transition from brittle to ductile damage.

Keywords: rock mechanics; high-temperature water-cooling cycle; uniaxial compression; acoustic emission characteristics

Citation: Wang, W.; Hong, L.; Cao, X.; Lu, X.; Wang, F.; Zhang, T.; Zhu, W. Experimental Study of the Mechanical and Acoustic Emission Characteristics of Sandstone by Using High-Temperature Water-Cooling Cycles. *Sustainability* **2023**, *15*, 13358. https://doi.org/10.3390/su151813358

Academic Editor: Jianjun Ma

Received: 3 August 2023
Revised: 4 September 2023
Accepted: 5 September 2023
Published: 6 September 2023

1. Introduction

The rapid development of the global economy is fueled by the substantial demand for energy. The intensity of exploitation of traditional non-renewable energy sources increases year after year, resulting in insufficient reserves of shallow energy and challenges and risks associated with extracting deeper resources [1]. Geothermal energy from hot rock sources, which is renewable, abundant, and clean, has attracted attention worldwide. With advancements in research and technology used to harness geothermal energy from hot, dry rocks, the proportion of new energy in the overall energy structure continues to increase annually.

Enhanced geothermal systems (EGS) have been developed in many countries around the world to increase the utilization of geothermal energy [2–6]. EGS are geothermal reservoirs that enable the economic utilization of low-permeability conductive rocks by creating fluid connectivity in the initially low-permeability rocks through hydraulic, thermal, or chemical stimulation [7]. Unlike the hot, dry rock (HDR) systems, EGS not only have a high-temperature crust, but are also characterized by primary permeability, which needs to be enhanced before they can be developed [8]. While EGS improve the efficiency of geothermal energy extraction, the temperature of the rocks in the deep reservoir drops sharply when they are cooled by water; then, the surrounding rock heats them, and internal fractures in the rocks gradually develop under repeated hot and cold effects, causing the mechanical properties of the rocks to gradually weaken, resulting in decreased rock strength and possible secondary disasters, such as well wall collapse and earthquakes, which affect production safety [9–11].

Limited research has been conducted on sedimentary basin-type hot dry rock resources, which are primarily sandstone. Investigation of the physical and mechanical changes and damage evolution mechanisms of sandstone under high-temperature water-cooling cycles is important. Therefore, through the static mechanical testing of sandstone subjected to high-temperature water-cooling cycles, by comparing and analyzing the results, in this study, we aim to elucidate the physical and mechanical alterations and damage mechanisms in sandstone. The findings will provide theoretical guidance for the development of hot dry rock geothermal energy resources.

Previous studies have examined the changes in high-temperature rock characteristics. Xu Xichang et al. [12] found that mechanical parameters of granite decreased to varying extents after high-temperature heating, with physical properties exhibiting greater changes than mechanical properties. Su Chengdong et al. [13–16] subjected three sandstones to high temperatures and observed minimal effects on physical properties below 400 °C but significant impacts on the burn-off rate, bulk expansion rate, and longitudinal wave velocity above 400 °C. Their uniaxial compression tests on high-temperature sandstones indicated 600 °C as the approximate threshold for transition from strong to weak behavior. Zhang Z X, Araújo R G S, and Heueckel T et al. [17–19] obtained extensive results on the deformation and damage mechanisms of high-temperature rocks. Through Brazilian splitting tests of granite under different cooling conditions, Deng Longchuan et al. [20] identified thermal-stress-induced microcracking as the primary mechanism for rapid cooling damage in granite. Hou Di et al. [21] performed triaxial compression tests on thermally cycled marble to examine the effects of high-temperature cycling on rock strength and damage characteristics. X. G. Zhao et al. [22–24] observed a transition from brittle to plastic damage modes in rocks with increasing temperature. Zhu X et al. [25] identified 100 °C and 500 °C as threshold temperatures for changes in longitudinal wave velocity and mechanical properties, respectively, by testing sandstones containing pre-existing fractures at different temperatures.

Different cooling methods have varying effects on the properties of high-temperature rocks. For example, Kumari et al. [26,27] heated Australian Strathbogie granite to high temperatures and studied its mechanical and fracture behaviors under natural and water-cooling conditions using uniaxial compression and acoustic emission. They found that thermal damage occurred in granite cooled from above 400 °C, with water cooling exhibiting a greater impact than natural cooling. Other studies on sandstone showed that high temperatures increased porosity and weakened mechanical properties. Yu Li et al. [28] performed uniaxial compression and ultrasonic tests on thermally cycled granite to examine the resulting damage mechanisms and evolution of physical and mechanical properties. Li Yabo et al. [29] identified relationships between the physical and mechanical properties of granite with heating temperature and the number of thermal cycles. Through the mechanical testing of thermally cycled granite, Wang Chun et al. [10] proposed the use of maximum tensile strain as an indicator of damage initiation. Han Guangsheng et al. [30] observed a transition from brittle to plastic behavior when cooling sandstones with water. Analyses of microstructures and acoustic emissions during Brazilian splitting tests led Wu Shunchuan et al. [31] to identify 400–600 °C as the damage threshold range for granite.

In summary, domestic and international scholars have achieved some initial research results in the field of high-temperature rock mechanics. However, previous studies have primarily focused on mechanical testing and thermal–hydraulic coupling behaviors of rocks subjected to high-temperature heating and multiple high-temperature water-cooling cycles. A limited amount of research has been conducted on laminated hydrothermal reservoirs in sedimentary basins, which are chiefly comprised of sandstone. The damage evolution mechanism of sandstone under high-temperature water-cooling cycles is not yet well understood. Fundamental theoretical research can provide useful insights into relevant engineering challenges in geothermal resource development. Therefore, considering these research gaps and needs, in the present work, we investigate the physical and mechanical characteristic changes of sandstone under multiple high-temperature water-cooling cycles, as well as the associated damage mechanisms.

2. Materials and Methods

2.1. Specimen Preparation

Fine-grained sandstone was selected as the object of study. Intact sandstone blocks with no joints or visible cracks were chosen. Following the GB/T 23561 [32] series of standards issued for rock mechanics testing, the blocks were cut and polished to create 60 cylindrical specimens 50 mm in diameter and 100 mm in height. Because the initial damage and physical–mechanical properties present some variability between rock samples, the longitudinal wave velocity was measured using a ZT801 rock wave velocity meter prior to testing. The distribution range was 2967–3236 m/s, with an average velocity of 3124 m/s. Specimens with similar wave velocities were grouped together for testing. Some example specimens are shown in Figure 1.

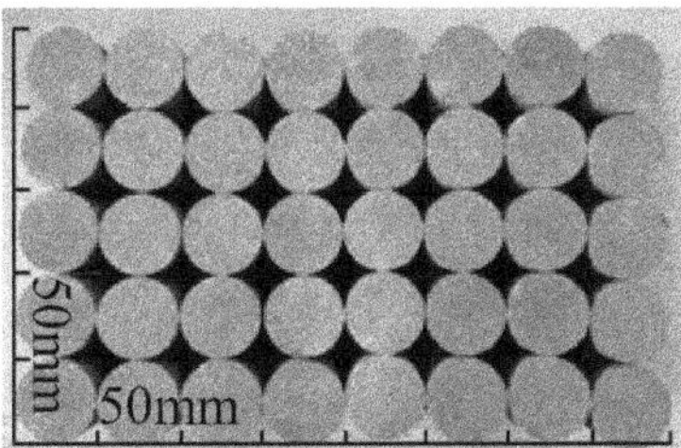

Figure 1. Part of the specimen diagram.

2.2. High-Temperature Water-Cooling Cycle Treatment

The rock samples were heated in a KSW-5D-12 high-temperature chamber furnace at 200 °C, 300 °C, 400 °C, 500 °C, and 600 °C. Groups of specimens were placed in a furnace and heated at a rate of approximately 5 °C/min to the target temperature. Upon reaching the set temperature, the specimens were held at a constant temperature for 4 h. Using crucible tongs, the samples were then weighed and immersed in ambient-temperature water for 8 h to cool, completing one high-temperature water-cooling cycle. After water cooling, the samples were dried for 24 h and tested for wave velocity. At each temperature level, the samples underwent 1, 3, 5, and 7 cycles, with 3 control specimens for each cycling condition, considering the variability between tests, as shown in Figure 2.

After high-temperature heating, color changes were observed in the rock samples at different temperatures. These color changes were attributed to oxidation reactions of iron present in the rocks, producing iron oxide (red) and ferrous oxide (black) at high temperatures. Smoke traces appeared on the sample surfaces during longitudinal wave velocity measurements, likely due to oxidation of coupling agents at elevated temperatures. Fracturing of some samples after cooling can be explained by thermal stresses generated by the sudden temperature change when hot samples were immersed in water. As the heating temperature increased, the thermal stress exceeded the tensile strength of the rocks, resulting in tensile fracture and cracking, as shown in Figure 3.

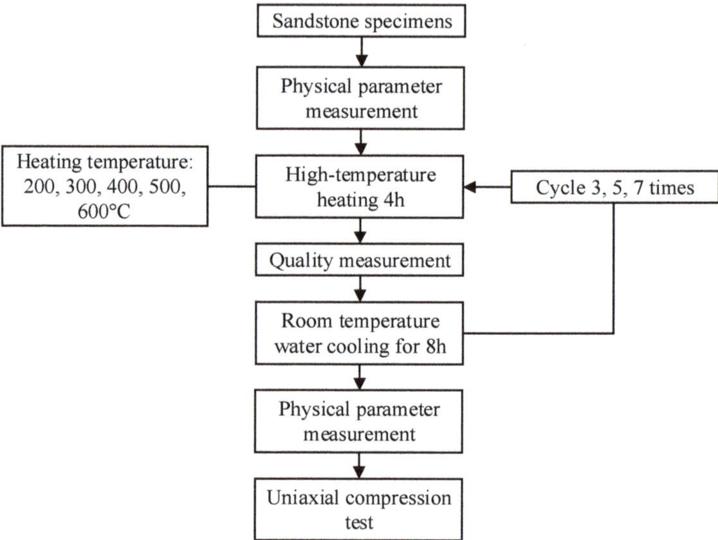

Figure 2. Flow chart of the experiment.

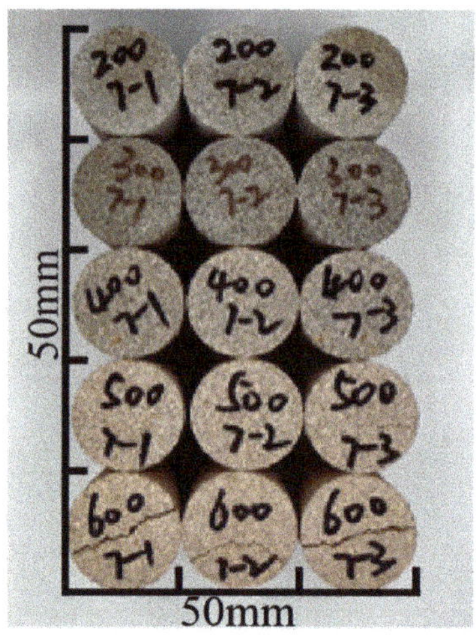

Figure 3. View of sandstone high-temperature water cooling.

2.3. Test Method

Uniaxial compression testing of sandstone was conducted using an RMT-150B electrohydraulic servo rock testing system (Figure 4). Axial load was measured using a 100 KN force transducer with 1.0×10^{-3} KN accuracy. Axial compression deformation was measured with a 5.0 mm displacement transducer. Testing was performed under a displacement control at a rate of 0.002 mm/s. Acoustic emission data were monitored simultaneously using a DS-5 8-channel acoustic emission detection and analysis system. Acoustic emission sensors were attached on both sides of the rock samples using a coupling agent and tape. The sampling frequency was set to 3 MHz with a 50 dB threshold and 40 dB amplification. During testing, acoustic emission amplitude, counts, and energy were recorded in real time.

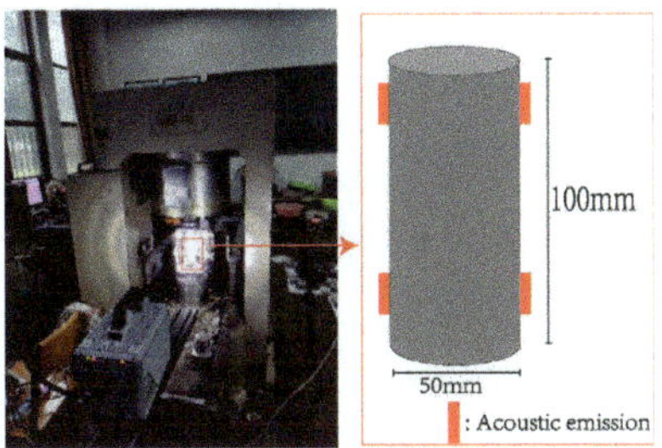

Figure 4. RMT-150B electrohydraulic servo rock test system.

3. Results

3.1. Physical Changes in Sandstone under High-Temperature Water-Cooling Cycles

During the high-temperature water-cooling cycles, some mineral components volatilize and decompose due to differing heat resistances, creating microporosity within the rock samples as the temperature increases. Repeated heating and cooling induces expansion and contraction of the various sandstone particles and pore minerals, which have different thermal expansion coefficients. This thermal cycling inevitably impacts the physical properties of the rock samples. High-temperature water-cooling cycle tests were conducted on the sandstone to obtain bulk expansion rate, mass change, and longitudinal wave velocity data. The results for bulk expansion rate, mass difference, and longitudinal wave velocity are shown in Table 1.

Table 1. Dynamically loaded sandstone sample body expansion rate change table.

Heating Temperature	Natural State		Number of Cycles	After High-Temperature Water-Cooling Cycle		Body Expansion Rate/%	Mass Difference (g)	Longitudinal Wave Speed (m/s)
	Diameter (mm)	High (mm)		Diameter (mm)	High (mm)			
200	49.95	50.34	1	49.98	50.32	0.08	0.154	3021.11
	49.9	50.49	3	49.9	50.37	−0.23	0.152	3210.78
	50.03	50.65	5	50.09	50.18	−0.69	0.162	3138.00
	50.3	50.3	7	50.01	50.34	−1.07	0.175	3177.78
300	49.84	50.16	1	49.77	50.41	0.21	0.161	2868.94
	50.04	50.24	3	50.1	50.27	0.29	0.217	2956.27
	49.89	50.44	5	49.92	50.54	0.31	0.247	2840.87
	49.9	50.31	7	49.9	50.55	0.47	0.259	2832.47
400	49.92	50.31	1	49.8	50.78	0.44	0.385	2764.33
	49.74	50.38	3	50.17	50.34	1.65	0.472	2361.20
	49.91	50.43	5	50.19	50.84	1.94	0.649	2309.07
	50.06	50.49	7	50.4	50.97	2.32	0.757	2329.33
500	49.71	50.65	1	49.95	50.37	0.40	0.385	2341.33
	49.9	50.32	3	50.21	50.57	1.74	0.472	1899.87
	49.85	50.21	5	50.08	50.85	2.21	0.649	1899.87
	49.72	50.33	7	49.9	51.15	2.36	0.758	1798.13
600	49.75	50.4	1	50.31	51.53	4.55	1.018	1369.60
	49.95	50.07	3	50.63	51.45	5.57	1.209	1126.60
	49.77	50.3	5	50.47	52.18	6.67	1.332	——
	49.69	50.22	7	51.01	51.01	7.04	1.787	——

As shown in Figure 5, the bulk expansion rate of the rock samples exhibited distinct trends with increasing heating temperature. After repeated high-temperature water-cooling cycles, the 200 °C samples showed a volume reduction compared to their original states, with a minimum expansion rate of −0.72%. This volume decrease at 200 °C can be explained by mineral decomposition and volatilization under heating, creating pores within the samples. The sandstone particles and minerals also thermally expand from generated thermal stresses. Rapid contraction then occurs during water cooling, compressing the pores formed at high temperatures and causing an overall volume reduction.

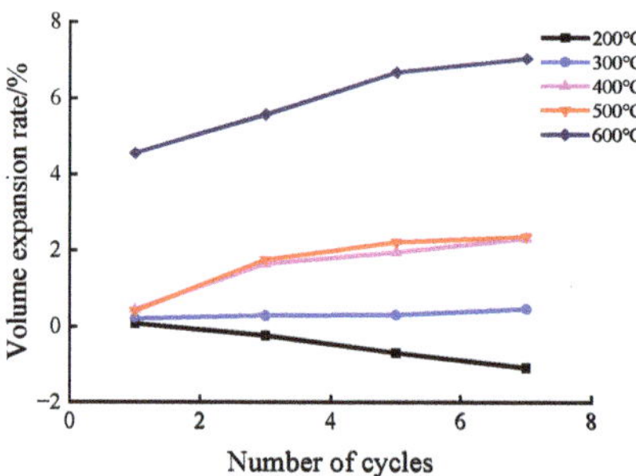

Figure 5. High-temperature water-cooling cycle sandstone body expansion rate.

At temperatures above 200 °C, the bulk expansion rate became greater than 0%, indicating that when the heating temperature exceeds a threshold, thermal stresses are generated within the rock samples. If the thermal stress exceeds the cohesive strength of the rock, microfractures and plastic deformation can occur. Moreover, the contraction effect during water cooling is reduced. As shown in Figure 5, the bulk expansion rate of sandstone increased with increased temperatures. The 200 °C samples had a negative expansion rate, whereas at 300 °C, the rate was close to 0%. Above 300 °C, the expansion rate escalated significantly. Additionally, the overall expansion rate tended to rise with more cycles. This is attributed to cumulative damage and increasing fracture development in the samples from repeated thermal cycling, leading to a gradual increase in bulk expansion.

Notably, all temperature samples exhibited a significant decrease in bulk expansion after the third cycle. The 600 °C samples produced approximately 1 mm of macroscopic crack damage after five cycles and were excluded from this analysis. The observed trend suggests that as the fissures become fully developed, the high-temperature heated rocks begin to expand into the existing cracks. However, with no further temperature increase and a constant thermal expansion coefficient, the bulk expansion rate gradually decreases. Additionally, the increasing high-temperature water-cooling cycles promote gradual fissure propagation towards the center of the rock samples.

The mass difference after each cycle is the main indicator of sandstone damage. As shown in Figure 6, the mass loss rates after the first cycle at increasing temperatures were 0.162%, 0.077%, 0.351%, 0.807%, and 1.527%, accounting for about 20.792% of the total mass loss. This significant initial mass loss is attributed to the decomposition of most mineral components and the vaporization of pore water at high temperatures during the first cycle. Additionally, the sharp contraction of sandstone particles during water cooling led to tensile failure and spalling of surface minerals with weaker cohesive strength.

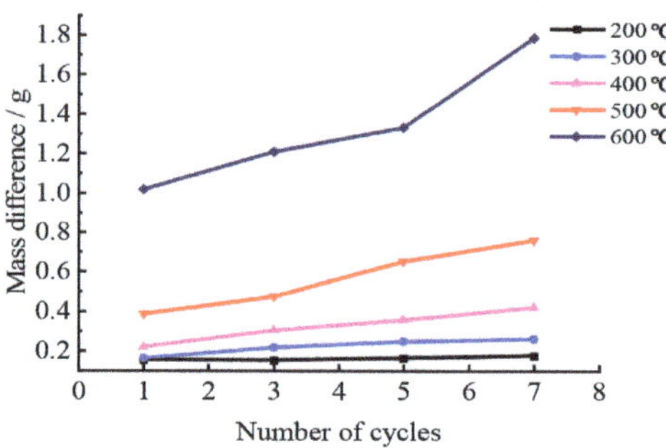

Figure 6. Poor-quality, high-temperature water-cooled recirculating sandstone.

After seven cycles, the mass loss rates at 200–600 °C were 0.238%, 0.535%, 0.582%, 1.083%, and 1.546%, respectively. The differences in mass loss between temperature levels were 0.238%, 0.297%, 0.047%, 0.501%, and 0.463% for the first and seventh cycles. However, the 500 °C samples showed increased mass loss in the fifth cycle due to macroscopic cracking and particle ejection. The results demonstrate that temperature has a close relationship with rock damage, with a threshold temperature that induces significant damage.

Based on the analysis of bulk expansion rate and mass loss, longitudinal wave velocities were measured before and after high-temperature water-cooling cycling. As shown in Figure 7, the wave velocities decreased by 4.6%, 9.4%, 12.71%, 26.06%, and 56.75% after the first cycle at each temperature, exhibiting similar patterns with additional cycles. Notably, the 200 °C samples showed increased wave velocity after cycling, aligning with the reduced bulk expansion rate and indicating a decrease in internal defects. The extent of velocity reduction escalated with higher temperatures, demonstrating more severe damage. The results indicate that temperature has a greater influence than the number of cycles on sandstone damage, with higher temperatures causing more deterioration.

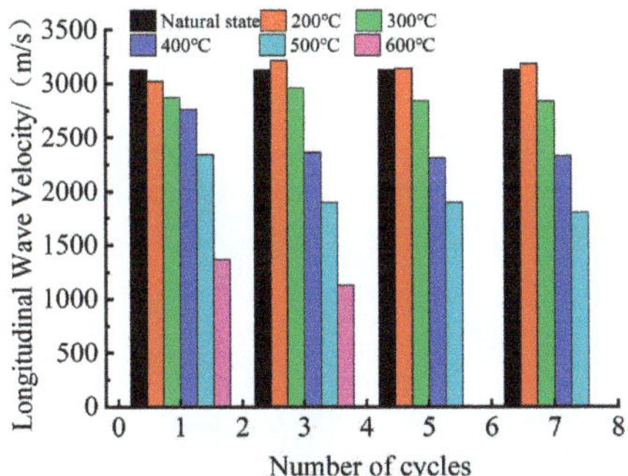

Figure 7. Longitudinal wave velocity in high-temperature water-cooled circulating sandstone.

3.2. Analysis of the Compressive Strength of Sandstone under the Action of High-Temperature Water-Cooling Cycles

The physical and mechanical properties of the sandstone specimens likely changed after high-temperature water-cooling cycling. The stress–strain curves under uniaxial compression for specimens subjected to the same number of cycles at different temperatures are shown in Figure 8. The curves exhibit four stages: compaction, elasticity, yielding, and damage.

The stress–strain curves after one cycle at each temperature are shown in Figure 8a. The 200–500 °C specimens display substantial overlap in the compaction and elastic phases, with a shortened compaction stage. The 200 and 300 °C specimens show stepped post-peak stress reduction corresponding to progressive crack growth. Alternatively, the 400 and 500 °C specimens exhibit brittle behavior with an instantaneous stress drop. The 600 °C curve has an extended compaction phase but a shortened elastic region. The peak stress is significantly lower, with an obvious yield plateau after peak stress, and the post-peak curve demonstrates ductile characteristics.

Figure 8d shows the stress–strain curves after seven high-temperature water-cooling cycles. Compression can be divided into compaction, elastic, yield, and damage stages. Microfractures produced by thermal cycling are compressed initially, lengthening the compaction phase as the elastic modulus decreases, reflecting reduced deformation resistance. In the elastic phase, stress–strain follows a linear relationship. At ~90% of peak strength, the sample enters the yield phase.

Figure 8d shows that peak strengths after seven cycles do not gradually decline with temperature but exhibit fluctuations for the 300, 400, and 500 °C samples. The 300 and 500 °C samples have two strength peaks, whereas the 400 °C sample displays a strength trough between peaks.

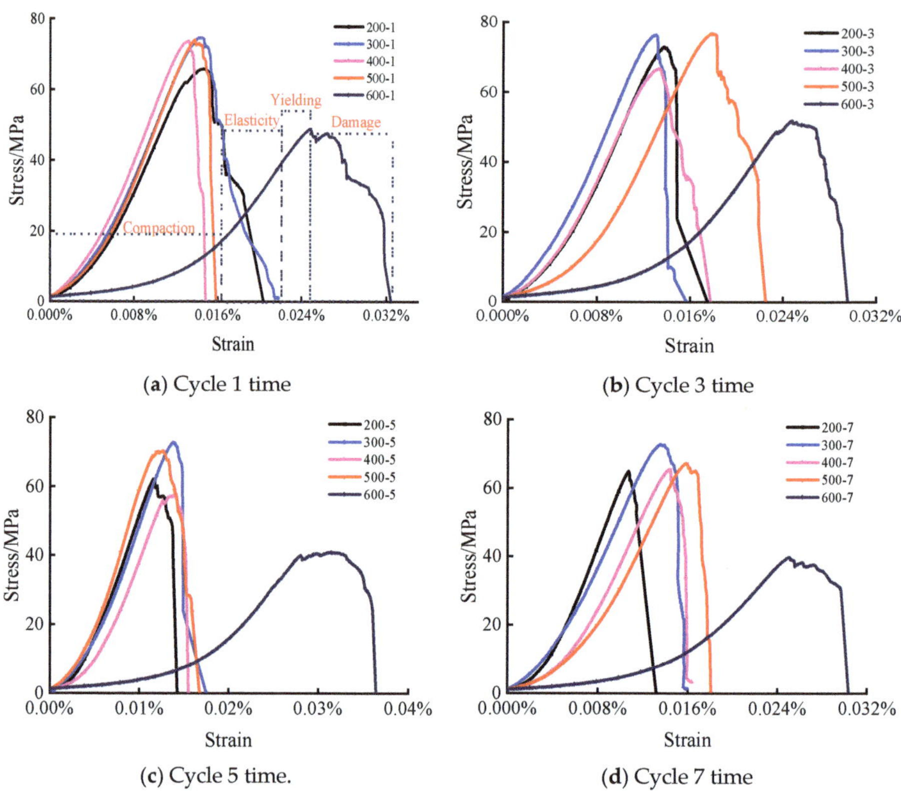

(**a**) Cycle 1 time

(**b**) Cycle 3 time

(**c**) Cycle 5 time.

(**d**) Cycle 7 time

Figure 8. Stress–strain curves of sandstone samples under the same cycle and different temperature conditions.

Figure 9 shows the elastic modulus for sandstone specimens at each temperature and cycle number. The modulus exhibited two distinct trends with cycling: at 200 °C, it increased by 4.06% from one to seven cycles; from 300 to 600 °C, it decreased with cycles, by 6.6% at 300 °C, 7.43% at 400 °C, 26.67% at 500 °C, and 35.56% at 600 °C.

The relationship between modulus and cycles aligns with the pattern of longitudinal wave velocity changes, indicating that at 200 °C, strength increased from thermal cycling, likely due to internal pore shrinkage. A threshold temperature appears to governs the cyclic effect on modulus. Below this temperature, the internal structure densifies with cycling, increasing or maintaining modulus. Above this threshold, the modulus rapidly decreases with more cycles as internal damage accumulates.

Figure 10 shows the compressive strength versus the number of cycles for each temperature. At 200 °C, the strength gradually increased from 63.9 MPa after one cycle to 71.46 MPa after seven cycles, corresponding to a 12.2% increase. From 300 to 600 °C, the strength decreased with more cycles. At 300 °C, the strength reduced from 77.27 MPa to 71.99 MPa, corresponding to a 17.6% decrease. At 400 °C, the strength decreased from 72.1 MPa to 66.63 MPa, corresponding to an 18.2% reduction. At 500 °C, the strength declined from 74.05 MPa to 67.92 MPa, corresponding to a 20.4% decrease. At 600 °C, the strength dropped from 56.76 MPa to 38.59 MPa, corresponding to a substantial 60.5% reduction. The data show that compressive strength reductions grew progressively more severe at higher temperatures as the number of cycles increased.

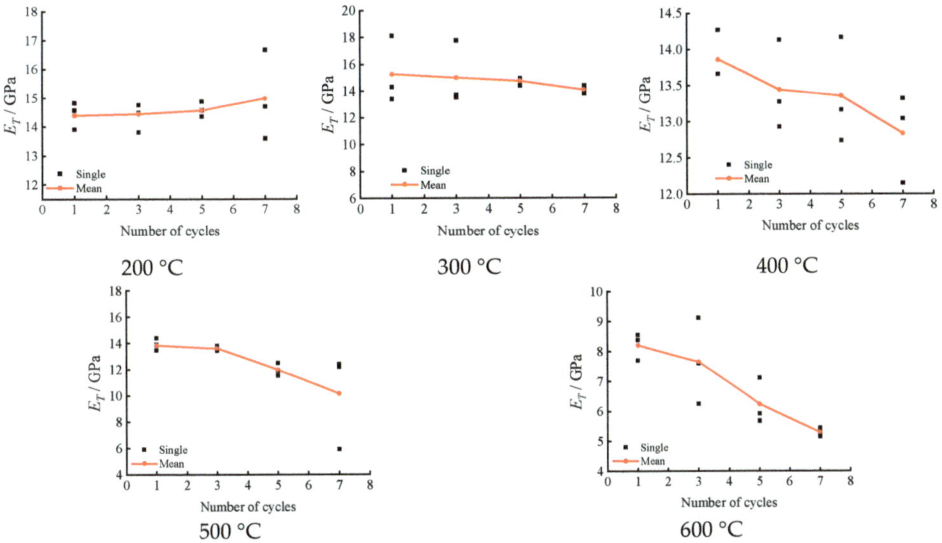

Figure 9. Modulus of elasticity as a function of the number of cycles.

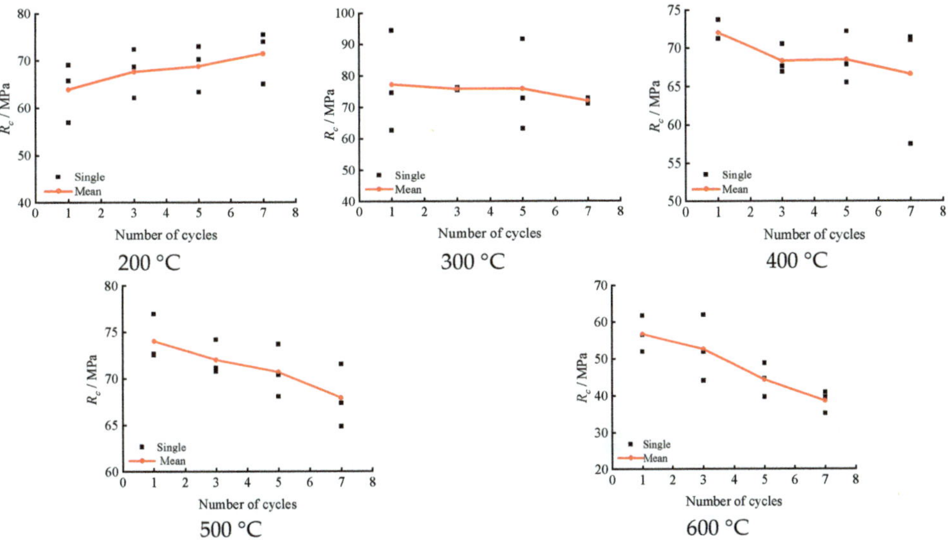

Figure 10. Peak stress versus cycle number variation.

3.3. Acoustic Emission Characteristics of Sandstone under a High-Temperature Water-Cooling Cycle

Acoustic emission data reflect crack development and damage accumulation in rocks under load. In this study, the acoustic emission characteristics of sandstone during uniaxial compression after high-temperature water-cooling cycles were analyzed based on ring counts and energy. The ring count indicates the frequency of acoustic emissions per unit time, with amplitudes exceeding a threshold corresponding to deformation and damage events in the sample. Figure 10 shows the load–time curves and acoustic emission profiles for sandstone uniaxial compression tests. Owing to space constraints, the acoustic emission features are analyzed for 400 °C samples with different cycle numbers and samples with seven cycles at various temperatures.

3.3.1. Different Cycles at the Same Temperature

1. Characteristics of the change in ringdown count

The relationship between the ringdown count of rock samples with loading and cycling can be described in two parts: when the number of cycles is less than five and when the number of cycles is five or more.

When the number of cycles is less than five, the acoustic emission ringdown behavior during loading can be divided into four stages: calm period, growth period, steep increase period, and decay period.

Calm Period: As depicted in Figure 11a,b, ringdown counts initially emerge later in the loading process, with the first occurrence at 75 s and 20 counts for one cycle and 120 s and 25 counts for three cycles. The low ringdown rate and long intervals between events signify an energy accumulation phase early on during which internal pores and cracks are closed under pressure without fracture growth, producing minimal acoustic emissions.

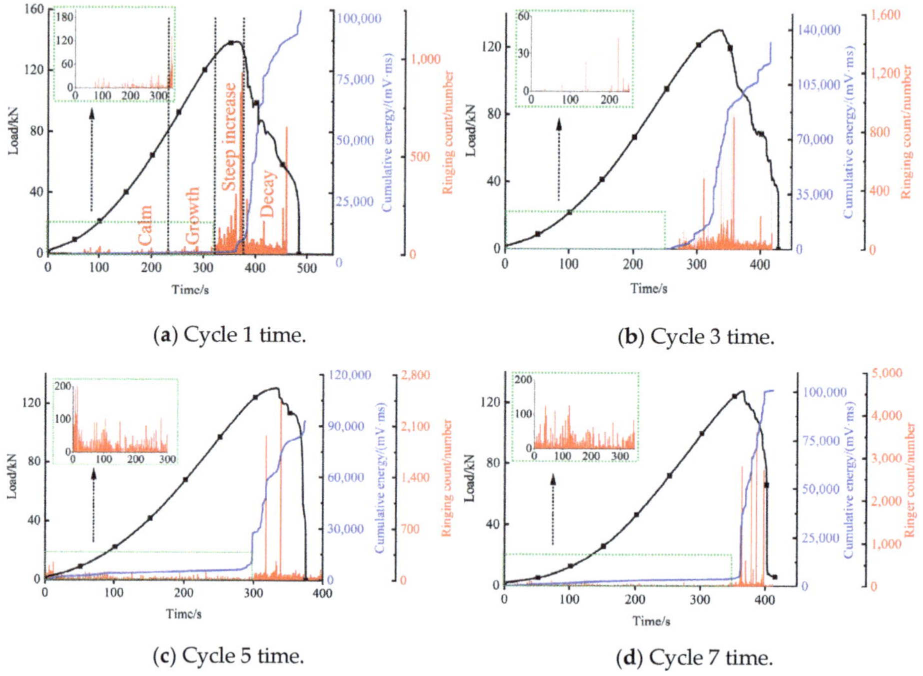

(**a**) Cycle 1 time.

(**b**) Cycle 3 time.

(**c**) Cycle 5 time.

(**d**) Cycle 7 time.

Figure 11. Acoustic emission characteristics of the 400 °C rock sample.

Growth Period: For one cycle, the loading time from 220 to 330 s and for three cycles from 250 to 300 s exhibit continuous ringdown occurrence without increased rates, indicating persistent but limited fracture development as stored energy begins overcoming local cohesive strength.

Steep Increase Period: At 330–370 s for one cycle and 300–350 s for three cycles, corresponding to the transition from yielding to failure, ringdown rates escalate rapidly to a peak far exceeding previous levels. As the load increases, abundant internal microcracks coalesce into dominant fractures, prompting the peak rate.

Decay Period: Post failure with further loading, ringdown counts rapidly decrease, likely indicating a shift from substantial fracture damage to distributed microcracking.

With five or more cycles, ringdown counts persist from 0 s until the yield point, but at lower magnitudes, signifying an extended calm period before ringdown growth begins, indicating that the rock samples undergo concurrent fracture compression and densification. Some fractures initiate small-scale development, suggesting easier crack growth with more cycles as internal damage accumulates, further implying that as loading continues, external forces progressively extend fractures, forming larger cracks and amplifying the scale of fracture propagation.

2. Characteristics of cumulative energy change

At a given temperature, cumulative acoustic emission energy trends can be examined. With more cycles, the emergence of cumulative energy advances, and the duration of rapid energy growth decreases. For instance, at one cycle, cumulative energy emerges around 24 s and rises slowly until ~260 s, then escalates rapidly until failure at 430 s. The time from emergence to rapid growth is ~236 s, with 164 s of fast accumulation. At five cycles, cumulative energy appears earlier, around 0 s, slowly increasing until ~280 s before rising rapidly to failure at 375 s. The time to accelerated growth is ~280 s, with just 95 s of rapid escalation. Such a shrinking rapid growth window indicates that the time between yielding and failure decreases with more cycles because crack expansion occurs sooner in loadings with more cycles, hastening fracture development and reducing the sample destruction time.

3.3.2. Same Number of Cycles at Different Temperatures

3. Characteristics of the change in ringdown count

Figure 12 depicts the time–load curves, acoustic emission ringdown trends, and cumulative acoustic emission energy for rock samples subjected to the same number of cycles at varying temperatures. The acoustic emission ringdown behavior during loading can be characterized in two temperature ranges: (1) below 400 °C and (2) at and above 400 °C.

As depicted in Figure 12, the 200 °C and 300 °C samples exhibited low initial ringdown counts during early loading, with sparse occurrence. The peak ringdown reached 126, with intervals around 40 s between some events, signifying gradual compaction of internal fractures with limited propagation, as the rock accumulates energy and minimal release. The rise in ringdowns from 278 s to 327 s indicates new crack development in the samples. The sudden spike to 6542 at 328 s suggests abundant fracture creation and initial damage from dominant cracks, although load-bearing capacity remains. The sharp ringdown drop follows from the compaction of cracks. The continued counts may arise from frictional acoustic emissions between compacting fractures. The spike at 356 s signifies the penetration of the main fracture and loss of load capacity. The 300 °C sample exhibited similar behavior to that of the 200 °C sample.

In the 400 °C, 500 °C, and 600 °C samples, ringdown counts consistently remained below 250 during early loading. Such a prolonged low-level period signifies small-scale crack initiation and energy accumulation, lasting 364 s at 400 °C, 410 s at 500 °C, and 726 s at 600 °C. The extended duration indicates that microcracking emerges earlier with rising temperature at a fixed cycle number. Figure 12 depicts the acoustic emission profile for the 400 °C sample, showing the peak ringdown changes for each temperature under seven cycles. The counts decreased from 6542 to 1384 from the lowest to highest temperature, corresponding to a 78.85% reduction, which is attributed to the escalating internal damage at higher temperatures, as evidenced by sustained ringdown generation during loading as microcracks exceeded the threshold. Sudden multiorder leaps in ringdowns signify the coalescence and propagation of dominant fractures, damaging the specimen. The decline in peak ringdowns with temperature points to a transition from brittle to ductile damage modes, demonstrating that temperature exerts a greater impact than cycling on rock damage.

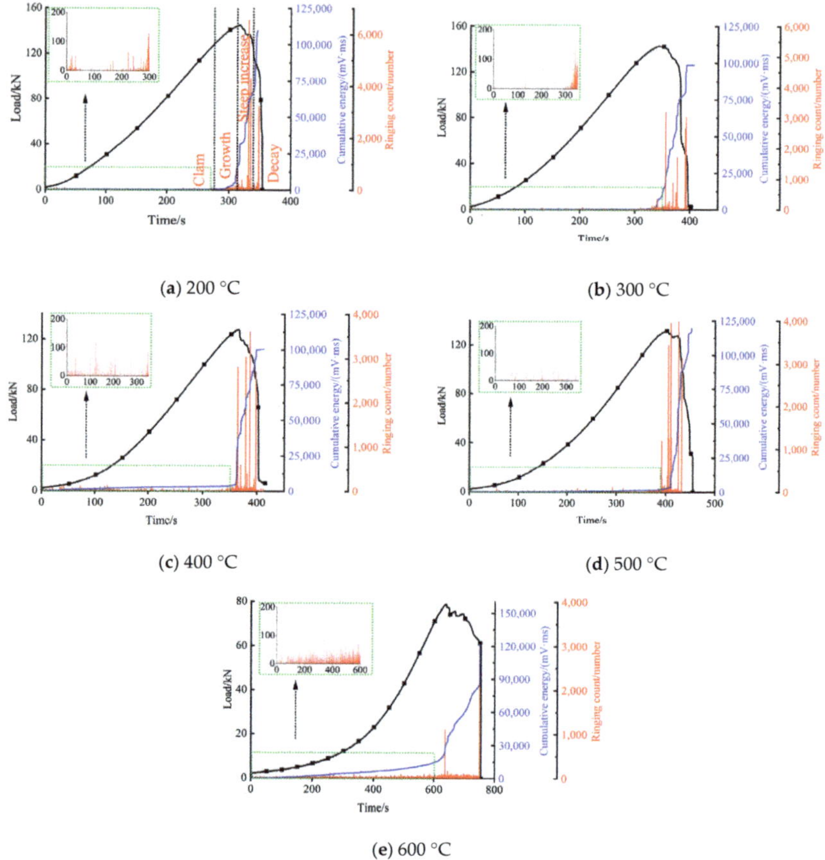

(**a**) 200 °C

(**b**) 300 °C

(**c**) 400 °C

(**d**) 500 °C

(**e**) 600 °C

Figure 12. Acoustic emission characteristics of seven cycles of rock samples at different temperatures.

4. Cumulative energy change characteristics

At temperatures below 400 °C, the cumulative energy change is minimal, indicating slow crack growth with gradual energy accumulation during loading. However, owing to limited crack development, the acoustic emission cumulative energy remains low. The 200 °C sample shows a rapid escalation in cumulative energy at 287 s, signifying intensified cracking. The 300 °C sample exhibited similar behavior.

Above 400 °C, the cumulative acoustic emission energy trends differ from those at 200 °C and 300 °C. Gradual increases emerge early in loading, with greater rises at higher temperatures, suggesting that cracking is initiated at loading onset, consuming internally stored energy. With elevated heating temperatures, fracture development becomes more active early in the loading process, revealing increasing internal damage and more severe degradation at higher temperatures.

3.4. Characteristic Analysis of Crack Growth and Failure Modes

As depicted in Figure 13, the 200 °C sandstone specimens displayed pronounced brittle behavior after one, three, and five high-temperature water-cooling cycles. The samples failed predominantly through splitting, producing numerous fragments. Upon failure, substantial elastic energy release occurred, ejecting large rock pieces. After seven cycles, damage initiation occurred, still exhibiting brittle cleavage, with major crack growth at an angle to the maximum principal stress direction.

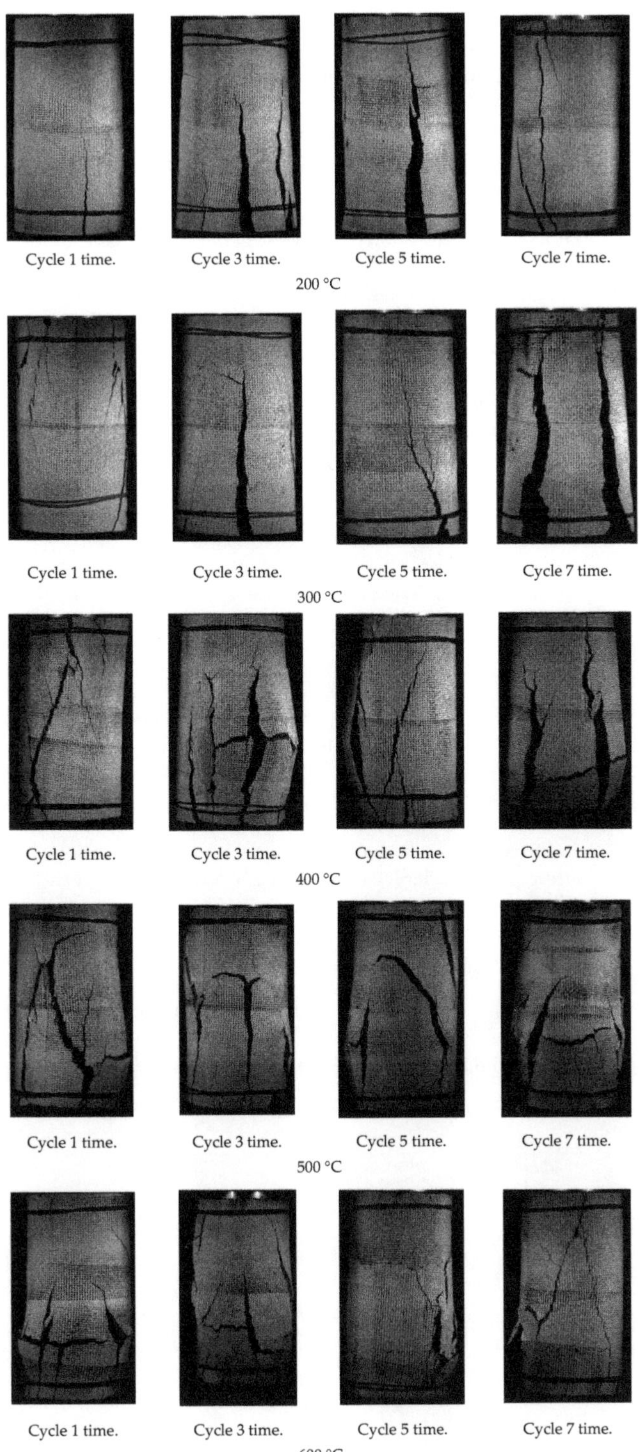

Cycle 1 time.   Cycle 3 time.   Cycle 5 time.   Cycle 7 time.

200 °C

Cycle 1 time.   Cycle 3 time.   Cycle 5 time.   Cycle 7 time.

300 °C

Cycle 1 time.   Cycle 3 time.   Cycle 5 time.   Cycle 7 time.

400 °C

Cycle 1 time.   Cycle 3 time.   Cycle 5 time.   Cycle 7 time.

500 °C

Cycle 1 time.   Cycle 3 time.   Cycle 5 time.   Cycle 7 time.

600 °C

Figure 13. Sandstone destruction patterns following high-temperature water-cooling cycles.

At 300 °C, significant brittleness occurred after one and three cycles, with predominant cleavage failure producing many fragments. Large energy release and block ejection occurred upon failure, with cracks following the maximum principal stress direction and gradually increased elongation. After three, five, and seven cycles, brittle failure persisted but with smaller damaged blocks and larger fractures.

The 400 °C specimens still showed primarily brittle damage after one cycle but with reduced block sizes and surface spalling. After three, five, and seven cycles, the damage transformed from brittle to ductile, with initial mid-specimen flaking during uniaxial loading. More particles were generated during failure, with evident shear slip and fewer blocks.

At 500 °C and 600 °C after one, three, five, and seven cycles, the damage shifted from brittle to ductile. Flaking initiated mid-specimen during loading, and shear slip escalated during failure. At 600 °C, tensile splitting occurred initially after one cycle, with more particles and fewer blocks after failure. After three, five, and seven cycles, flaking became granular, with a strong hierarchical texture on the damaged surface and abrasive damage in the middle of the specimen.

4. Discussion

To clearly characterize the damage to sandstone after high-temperature water-cooling cycles, the damage variable (D) was calculated for the sandstone specimens based on the modulus of elasticity determined in uniaxial compression tests under static loading after thermal cycling. The damage variable (D) was computed as (1):

$$D = 1 - \left(\frac{E_N}{E_0} \right) \tag{1}$$

where E_0 is the modulus of elasticity of the sandstone in its natural state averaged around 14 GPa under uniaxial compression, and E_N is the modulus of elasticity of the sandstone specimen after N cycles. Using the average damage variables of the thermally cycled sandstone specimens at different temperatures and cycle numbers, the data were fitted to obtain empirical formulas and plots describing the relationships between damage variables and temperature, as well as cycles. The fitting equations are shown in (2) and (3), and the fitting plots are depicted in Figures 14 and 15. However, at 200 °C, the sandstone specimens exhibited strengthening with more cycles, so damage variables at 200 °C were not included in the fitting analysis.

Fitting equation for damage variables with number of cycles:

$$D = 0.1891N + 0.3853, \; R^2 = 0.92671 \tag{2}$$

Fitting equation for damage variables with temperature (T):

$$D = 2.09509T^3 - 1.58772T^2 + 0.00272T - 0.04575, \; R^2 = 0.92197 \tag{3}$$

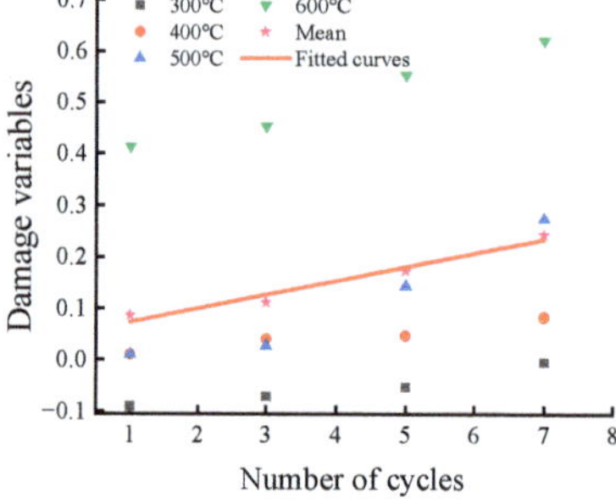

Figure 14. Relationship between damage variable and cycle times.

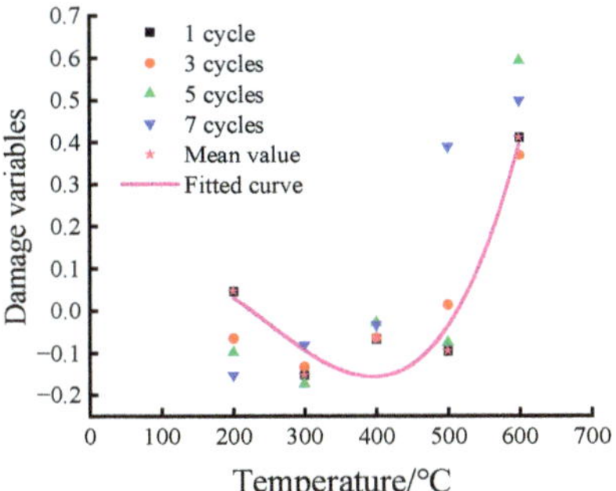

Figure 15. Relationship between damage variable and temperature.

The changes in the physical and mechanical properties of sandstone specimens after undergoing high-temperature water-quenching cycles can be attributed to two fundamental factors. First, the elevated temperatures cause alterations in the composition of the sandstone specimens, which consequently lead to variations in their mechanical properties. Secondly, the differential expansion rates of internal air, quartz, and other mineral components within the sandstone specimens generate tensile stresses between minerals due to repeated thermal cycles. When the tensile stress exceeds the thermal stress, cracks form, resulting in a reduction in the specimens' physical properties and strength. Moreover, water seeping into the specimens dissolves specific mineral components, weakening the bonding between mineral particles and facilitating the propagation and development of cracks [28].

As depicted in Figure 14, the sandstone damage variable gradually increases with the number of cycles, although the rise is small because under high temperatures, crystalline water and some minerals within the specimen vaporize, dissolve, and create internal pores. When the hot sample is immersed in water for cooling, considerable temperature differences generate thermal stresses. The tensile forces between crystals can produce microcracks. Repeated thermal cycling causes separation between crystals, forming fissures. Once cracks form, water ingress promotes further propagation under hydraulic and tensile stresses. Therefore, more cycles lead to increased crack development and more severe cumulative damage in the sandstone.

From the perspective of changes in mineral composition, Su Chengdong et al. [15] conducted X-ray diffraction analysis on high-temperature sandstone. They discovered that when the temperature is below 400 °C, kaolin disappears, and the overall mineral content decreases. A new substance called calcium feldspar emerges as the primary constituent of the sandstone. With increasing temperature, the quartz content in the sandstone increases, resulting in an apparent increase in its strength. However, at 600 °C, the strength of the specimen decreases due to a reduction in quartz content.

As shown in Figure 15, at a fixed number of cycles, the sandstone damage variable exhibits two distinct trends with temperature. Below 400 °C, it is hypothesized that the specimen experiences increased internal porosity after high-temperature exposure. Rapid water cooling then causes shrinkage as a result of thermal contraction, densifying the internal structure and causing apparent strengthening. Above 400 °C, internal damage gradually rises with increasing temperature, as reflected by escalating damage variables. This is attributed to heightened porosity from escaping crystalline water and mineral decomposition at high temperatures, culminating in damage growth. Between 500 and 600 °C, the quartz phase change severely weakens the internal structure, further increasing damage. Aligning with the physical property changes presented earlier, this phenomenon suggests that at a constant cycle number, the sandstone structure initially strengthens, then weakens with rising temperature, as evidenced by the damage variable analysis.

5. Conclusions

Analysis of the deformation, strength, and acoustic emission characteristics of sandstone after high-temperature water-cooling cycling leads to the following conclusions:

(1) With rising temperature, the compression–densification and yield stages of the stress–strain curve were prolonged, the elastic stage shortened, and the post-peak damage stage transitioned from sudden to gradual strength loss. Peak stress decreased and strain increased, indicating enhanced sandstone ductility.

(2) At 200 °C, the elastic modulus exhibited an increasing trend, rising by 20.1%. From 300 to 600 °C, the modulus showed declining trends, reducing by 6.04%, 7.24%, 28.7%, and 35.57%, respectively. At one and three cycles, the modulus rose, then fell in waves as the temperature increased from 200 °C to 600 °C. At five and seven cycles, the modulus declined overall.

(3) Compressive strength increased from 63.9 MPa after one cycle, and to 71.46 MPa after seven cycles at 200 °C. From 300 to 600 °C, strength gradually decreased by 17.6%, 18.2%, 20.4%, and 60.5%, respectively. With rising temperature, strength displayed a pattern of initial increase, subsequent decrease, another increase, and an eventual sharp reduction.

(4) At <5 cycles, acoustic emission accumulated energy was initially low but spiked at yield. At >5 cycles, energy rose substantially from early loading.

(5) Fitting of damage variables showed the 200 °C sample experienced a negative increase, the 300 °C sample experienced negligible change, and the 400–600 °C samples underwent increased damage with rising temperature, signifying sandstone strengthening below 400 °C but escalating degradation above 400 °C.

Author Contributions: Conceptualization, W.W., L.H. and X.C.; methodology, X.C.; formal analysis, L.H. and X.C.; investigation, F.W. and X.L.; resources, T.Z. and W.Z.; data curation, L.H.; writing—original draft preparation, W.W. and L.H.; writing—review and editing, W.W. and L.H.; funding acquisition, W.W. All authors have read and agreed to the published version of the manuscript.

Funding: This study was supported by the Program for National Major Achievement Cultivation, Theory and Key Technology of Natural Gas and Coal Resources Cooperative Mining, NSFRF230202; Theory and Technology of Natural Gas-Coal-Uranium Mining Synergy; Research on the Theory and Key Technology of Coordinated Natural Gas and Coal Mining, 23HASTIT011; Research on the Mechanism of Coordinated Coal and Natural Gas Exploitation and Disaster Warning in Ordos Basin, T2022-2; National Natural Science Foundation of China funded project (52174109).

Data Availability Statement: Due to the nature of this research, participants of this study did not agree for their data to be shared publicly, so supporting data is not available.

Conflicts of Interest: The authors declare no conflict of interest.

References

1. Xie, H.; Gao, F.; Ju, Y. Research and exploration on the mechanics of deep rock masses. *J. Rock Mech. Eng.* **2015**, *34*, 2161–2178.
2. Boretti, A. Enhanced geothermal systems (EGS) a key component of a renewable energy-only grid. *Arab. J. Geosci.* **2022**, *15*, 72. [CrossRef]
3. Fang-chao, K.A.N.G.; Chun-an, T.A.N.G.; Ying-chun, L.I.; Tian-jiao, L.I.; Jin-long, M.E.N. Challenges and opportunities of enhanced geothermal systems: A review. *Chin. J. Eng.* **2022**, *44*, 1767–1777.
4. Kang, F.C.; Li, Y.C.; Tang, C.A. Grain size heterogeneity controls strengthening to weakening of granite over high-temperature treatment. *Int. J. Rock. Mech. Min. Sci.* **2021**, *145*, 104848. [CrossRef]
5. Kang, F.C.; Tang, C.A. Overview of enhanced geothermal system (EGS) based on excavation in China. *Earth Sci. Front.* **2020**, *27*, 185.
6. Cai, M.F.; Dor, J.; Chen, X.S. Development strategy for Co-mining of the deep mineral and geothermal resources. *Strat. Study CAE* **2021**, *23*, 43. [CrossRef]
7. Liu, X.; Falcone, G.; Alimonti, C. A systematic study of harnessing low-temperature geothermal energy from oil and gas reservoirs. *Energy* **2018**, *142*, 346–355, (In Chinese with English Abstract). [CrossRef]
8. Ma, B.; Jia, L.; Yu, Y.; Wang, H. Current status and prospects of geothermal energy development and utilization in the world. *Geol. China* **2021**, *48*, 1734–1747.

9. Zhou, C.; Wan, Z.; Zhang, Y.; Gu, B. Experimental study on hydraulic fracturing of granite under thermal shock. *Geothermics* **2018**, *71*, 146–155. [CrossRef]
10. Wang, C.; Wang, H.; Xiong, Z.; Wang, C.; Cheng, L.; Zhan, S. Experimental study on the mechanical characteristics of radial compression of circular granite under warm and wet cyclic conditions. *J. Rock Mech. Eng.* **2020**, *39*, 3260–3270.
11. Wang, W.; Li, H.; Yuan, R.; Gu, H.; Wang, H.; Li, H. Mechanical characteristics and fine mechanics analysis of water-bearing coal samples loaded by combined dynamic and static loading. *J. Coal* **2016**, *41*, 611–617.
12. Xu, X.; Liu, Q. Preliminary study on the basic mechanical properties of granite under high temperature. *J. Geotech. Eng.* **2000**, *22*, 4.
13. Su, C.; Guo, W.; Li, X. Experimental study on the mechanical effect of coarse sandstone after high temperature action. *J. Rock Mech. Eng.* **2008**, *6*, 1162–1170.
14. Su, C.; Wei, S.; Qin, B.; Yang, Y. Analysis of the effect of high temperature action on the characteristics of physical parameters of three sandstones. *J. Undergr. Space Eng.* **2018**, *14*, 341–348.
15. Su, C.; Wei, S.; Qin, B.; Yang, Y. Experimental study on the mechanism of high temperature effect on mechanical properties of fine sandstone. *Geotechnics* **2017**, *38*, 623–630.
16. Ma, J.; Wei, S.; Su, C. Effect of high temperature on the physical and mechanical properties of weakly cohesive medium sandstone. *J. Min. Saf. Eng.* **2017**, *34*, 155–162.
17. Zhang, Z.X.; Yu, J.; Kou, S.Q.; Lindqvist, P.A. Effects of high temperatures on dynamic rock fracture. *Int. J. Rock Mech. Min. Sci.* **2001**, *38*, 211–225. [CrossRef]
18. Araújo, R.G.S.; Sousa, J.L.A.O.; Bloch, M. Experimental investigation on the influence of temperature on the mechanical properties of reservoir rocks. *Int. J. Rock Mech. Min. Sci.* **1997**, *34*, 459–466. [CrossRef]
19. Hueckel, T.; Peano, A.; Pellegrini, R. A constitutive law for thermo—Plastic behavior of rocks: An analogy with clays. *Surv. Geophys.* **1994**, *15*, 643–671. [CrossRef]
20. Deng, L.; Li, X.; Wu, Y.; Xu, Z.; Huang, Z.; Jiang, C.; Yuan, D. Effect of different cooling methods on the mechanical damage characteristics of granite. *J. Coal* **2021**, *46*, 13.
21. Hou, D.; Peng, J.; Zeng, S.C. Mechanical properties of triaxial compression of marble under high temperature thermal cycling. *Nonferrous Met. (Min. Sect.)* **2021**, *73*, 111–116, 129.
22. Zhao, X.G.; Zhao, Z.; Guo, Z.; Cai, M.; Li, X.; Li, P.F.; Chen, L.; Wang, J. Influence of thermal treatmenton the thermal conductivity of beishan granite. *Rock Mech. Rock Eng.* **2018**, *51*, 2055–2074. [CrossRef]
23. Zhao, X.G.; Xu, H.R.; Zhao, Z.; Guo, Z.; Cai, M.; Wang, J. Thermal conductivity ofthermally damaged beishan granite under uniaxial compression. *Int. J. Rock Mech. Min. Sci.* **2019**, *115*, 121–136. [CrossRef]
24. Zhu, D.; Jing, H.; Yin, Q.; Han, G. Experimental study on the damage of Granite by acoustic emission after cyclic heating and cooling with circulating water. *Processes* **2018**, *6*, 101. [CrossRef]
25. Zhu, X.; Xu, Q.; Zhou, J.; Tang, M. Experimental study of infrasonic signal generation during rock fracture under uniaxial compression. *Int. J. Rock Mech. Min. Sci.* **2013**, *60*, 37–46. [CrossRef]
26. Kumari, W.G.P.; Ranjith, P.G.; Perera, M.S.A.; Chen, B.K.; Abdulagatov, I.M. Temperature-dependent mechanical behaviour of Australian Strathbogie granite with different cooling treatments. *Eng. Geol.* **2017**, *229*, 31–44. [CrossRef]
27. Kumari, W.G.P.; Ranjith, P.G.; Perera, M.S.A.; Chen, B.K. Experimental investigation of quenching effect on mechanical, microstructural and flow characteristics of reservoir rocks: Thermal stimulation method for geothermal energy extraction. *J. Pet. Sci. Eng.* **2018**, *162*, 419–433. [CrossRef]
28. Yu, L.; Peng, H.; Li, G.; Zhang, Y.; Han, Z.H.; Zhu, H.Z. Experimental study of granite under high temperature-water cooling cycle. *Geotechnics* **2021**, *42*, 1025–1035.
29. Li, Y.; Zhai, Y.; Zhang, E.; Peng, Z.P.; Yang, Q.I.Y. Effect of high temperature heating-water cooling cycle on physical and mechanical properties of granite. *China High-Tech* **2021**, *79*, 67–70.
30. Han, G.; Jing, H.; Su, H.; Yin, Q.; Wu, J.; Gao, Y. Study on the mechanical behavior of sandstone in high temperature condition after water cooling. *J. China Univ. Min. Technol.* **2020**, *49*, 69–75.
31. Wu, S.; Guo, P.; Zhang, S.; Zhang, G.; Jiang, R. Study on thermal damage of granite based on Brazilian splitting test. *J. Rock Mech. Eng.* **2018**, *37*, 3805–3816.
32. *GB/T 23561.1-2009*; Methods for Determining the Physical and Mechanical Properties of Coal and Rock. China National Coal Association: Beijing, China, 2009.

 *sustainability*

Article

Damage Mode and Energy Consumption Characteristics of Paper-Sludge-Doped Magnesium Chloride Cement Composites

Shuren Wang [1,2], Zhixiang Wang [1], Jian Gong [1,*] and Qianqian Liu [1]

[1] International Joint Research Laboratory of Henan Province for Underground Space Development and Disaster Prevention, Henan Polytechnic University, Jiaozuo 454003, China; w_sr88@163.com (S.W.); wangzhixiang998@163.com (Z.W.); lqq0635@163.com (Q.L.)
[2] Collaborative Innovation Center of Coal Work Safety, Henan Polytechnic University, Jiaozuo 454003, China
* Correspondence: gongjian@hpu.edu.cn

Abstract: To reduce the pollution caused by paper sludge in the environment and overcome issues of poor water resistance and brittleness in Magnesium Oxychloride Cement (MOC), the MOC was modified by adding different dosages of paper sludge. The mechanical properties and damage modes of composite MOC materials containing paper sludge were studied by uniaxial compression tests. Under cyclic loading conditions, the damage progression of MOC composites was characterised using the tensile-shear conversion factor (Tsc) and by monitoring the energy parameters (elastic strain energy, plastic strain energy and dissipation energy). The results show that the average peak stress drop of MOC composites gradually increases with the increase in paper sludge dosage. Under uniaxial compression conditions, the Tsc of the MOC composites decreases from 0.99 to 0.44, and the damage mode is transitioned from brittle tensile damage to tensile-shear damage, X-shaped conjugate surface shear damage and finally to pure shear damage with an increase in doping. During cyclic loading conditions, the brittleness of MOC composites gradually decreases with an increase in paper sludge doping, which verifies the effect of paper sludge on the mechanical properties of MOC materials and the change in damage modes from the perspective of energy dissipation.

Keywords: paper sludge; magnesium chlorox cement; tensile-shear conversion factor; strain energy; damage modes

Citation: Wang, S.; Wang, Z.; Gong, J.; Liu, Q. Damage Mode and Energy Consumption Characteristics of Paper-Sludge-Doped Magnesium Chloride Cement Composites. *Sustainability* **2023**, *15*, 13051. https://doi.org/10.3390/su151713051

Academic Editor: Hosam Saleh

Received: 5 August 2023
Revised: 18 August 2023
Accepted: 23 August 2023
Published: 30 August 2023

1. Introduction

In recent years, tens of thousands of tonnes of general industrial solid waste have been dumped and discarded in large and medium-sized cities in China. As the terminal waste of paper-making enterprises, paper sludge is mainly managed through sanitary landfilling within the industry. Due to the various treatment methods, the landfill will contain trace amounts of lead, copper, nickel, chromium and other heavy metals which can lead to secondary soil pollution. Moreover, landfill sludge occupies a large amount of agricultural land, which is not conducive to China's adherence to the concept of green and sustainable development. As a result, China began to gradually implement refined and systematic treatments for paper sludge.

Some researchers have conducted related studies about paper sludge. For example, Chen et al. [1] compared the composition and properties of different paper sludges and found that the incorporation of paper sludge significantly reduced its thermal conductivity. Tofani et al. [2] evaluated the thermal stability of the incinerated paper sludge by infrared analysis. Aiken et al. [3] revealed the crystalline structure, microscopic morphology and physical phase composition of each Magnesium Oxychloride Cement (MOC) building board and explored the feasibility of MOC building boards to replace traditional thin sheet materials, such as plywood, gypsum plasterboard and fibre cement board. Wang et al. [4] investigated the effects of fly ash and polyethylene fibre incorporation on the flow, tensile properties and compressive properties of MOC. Ye et al. [5] prepared corn starch/poly(sodium acrylate) (PAAS) MOC composites with higher compressive strength and water resistance. Gong et al. [6] investigated the effect of paper sludge with different dosages, particle sizes and water contents on the strength and microscopic properties of magnesium chloroxygenated cement by means of compressive strength test, water absorption test, SEM and XRD. Chen et al. [7] investigated the effect of phosphoric acid and tartaric acid addition on the water resistance of MOC cement pastes. Devi et al. [8] used waste paper sludge ash to replace M25 grade concrete at different percentages (2.5%, 5.0% and 7.5%), and they found that a 5% substitution rate yielded superior mechanical properties. Ingale et al. [9] comprehensively tested the mechanical and durability properties of the composites, and they finally achieved the best overall performance of the material at a 5–10% substitution rate. Gomes et al. [10] prepared magnesium chloroxylate fibre cement boards with different molar ratios, and they found that rice husk silica (RHS) increased the modulus of rupture and toughness of the MOFC and improved the durability of the material through the filler effect. He et al. [11] prepared straw/sawdust–magnesium chloroxylate cement composites (SMOCC) from straw, sawdust and MOC using a compression moulding process, and they found that increasing the compression pressure can effectively reduce the porosity to redistribute pore sizes, which resulted in an increase in compressive strength and flexural strength. Singh et al. [12] found that the calcination temperature, the bohemian of $MgCl_2$ solution, and the fineness of the filler were the key factors controlling the development of MOC strength.

Zhao et al. [13] used the phase inversion method to modify a novel polyvinyl chloride (PVC) membrane with graphene oxide (GO) to improve its hydrophilicity and mechanical properties. Yu et al. [14] found that the prepared high-performance MOC possessed high early strength, high ductility and good water resistance. Xu et al. [15] revealed that the changes in the microscopic properties of MOC cementitious composites induced by high-temperature curing are the macro-mechanical changes significantly. Liu et al. [16] conducted uniaxial compression cyclic loading and unloading tests at different rates and micro-morphological analysis of crack surfaces, showing that with an increase in loading and unloading rates, the energy evolution eventually changed from being dominated by the damping energy to being dominated by the damage energy. Rapid loading and unloading rates were the main factors contributing to the larger dissipation and damage energy in the rock. Meng et al. [17] and Liu et al. [18] investigated the effect of loading and unloading rates on the energy evolution and damage characteristics of rocks under cyclic loading using four rock samples, and they found that the dissipated energy increased nonlinearly with an increase in loading and unloading rates for all of them. Wang et al. [19] used the loading and unloading response ratio (LURR) to evaluate the cumulative damage of porous coal under cyclic loading. Wang et al. [20] proposed a micromechanical model of LWAC using the nearest surface distribution function, generalised self-consistent scheme and two-phase spherical model to analyse the damage process of lightweight shale vitrified concrete. Gao et al. [21] described the damage mechanism of tectonic coal from the perspective of energy distribution by introducing the concepts of crushing and friction energy. Xu et al. [22] analysed the energy, deformation and crack development mechanism of rubber-cemented mortar under three different cyclic loading and unloading modes, failure mode and crack development mechanism. Song et al. [23] demonstrated that material energy dissipation and damage evolution were related to the stress path. Lin et al. [24] found that the damage variables reached a stable growth stage with a gradual increase in loading and unloading cycles under different circumferential pressures by conducting triaxial unloading tests on salt rock specimens.

The above researchers mainly worked on the treatment of paper sludge, the improvement of water resistance of MOC composites and the study of solid waste fillers. They found that different substitution rates, molar ratios, filler finenesses and pressing processes affect the micro-morphology and mechanical properties of MOC composites. In order to improve the water resistance of MOC composites, some researchers have used a range of modifiers, such as phosphoric acid, citric acid, tartaric acid, phenyl propylene emulsions and solid waste fillers, to study their water resistance. Meanwhile, regarding the energy consumption of rock-based materials, other researchers have conducted in-depth studies on the energy loss of different types of rocks under cyclic loading. However, there is no substantial solution for the defects of MOC composites, such as brittleness and susceptibility to cracking.

In this study, the brittleness of MOC cementitious materials was improved by using components such as lignin fibres derived from industrial solid waste paper sludge as well as using tiny particle size fillers, analysing the changes in damage modes and characterising the different mechanical properties embodied in each loading stage, and describing the damage mechanism from the perspective of energy dissipation by combining the elastic strain energy, the plastic strain energy and the dissipation energy. Under the cyclic loading conditions, the energy parameters, such as stage elastic strain energy, stage plastic strain energy and stage dissipation energy, were analysed, and a new variable parameter, i.e., the tensile-shear conversion factor (Tsc), was introduced to further investigate the damage evolution of MOC composites. Compared with other related articles, further explorations in green utilisation of solid waste and material modification were made.

2. Materials and Methods

2.1. Material Preparation

Magnesium oxide with 65% activity was manufactured in Dashiqiao City, Liaoning Province, China. Magnesium chloride with 46.4% purity was obtained from Tianjin Tanggu Jinlun Salt Chemical Co., Tianjin, China. Paper sludge (PS) was obtained from Xinmi Paper Mill, Xinmi, China. The paper sludge (PS) was selected from a paper mill waste in Xinmi City, China. The basic physical and chemical properties of the raw materials are given in Tables 1 and 2. Data used in Tables 1 and 2 were from manufacturers. The results of the XRD spectra of the paper sludge used in the experiments are shown in Figure 1. The blue line represents the diffraction pattern of XRD and the red line represents the standard card. XRD testing is conducted to analyse whether the composition of the paper sludge has an effect on the generation of phases 3 and 5 within the MOC material.

Table 1. Chemical composition of magnesium oxide.

Component	MgO	IL	SiO_2	CaO	Al_2O_3	Fe_2O_3	Activated MgO
Content (%)	86.04	5.50	5.33	2.02	0.50	0.61	63.50

Table 2. Chemical compositions of $MgCl_2 \cdot 6H_2O$.

Component	$MgCl_2$	Ca^{2+}	SO_4^{2-}	Water-Insoluble Substance	Pb	As	NH^{4+}
Content (%)	46.9	0.09	0.12	0.01	1.0	0.5	37

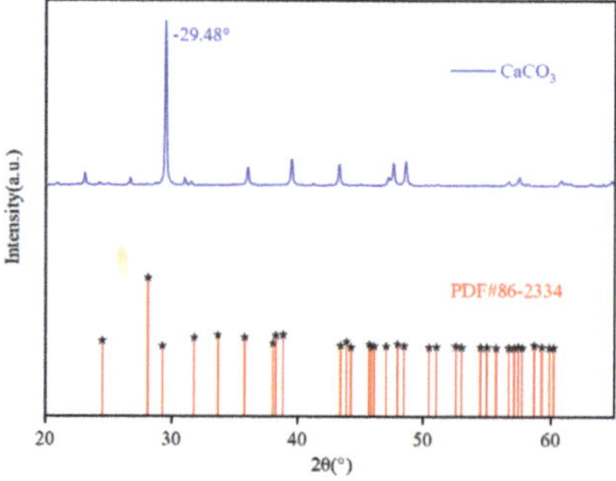

Figure 1. Diffraction pattern of paper sludge and schematic diagram of a standard card.

The paper sludge used in the test was first dried to remove excess moisture. It was processed into fragments using an overhead crusher. It was ground into fine particles and small fragments using a ball mill. Finally, it is completely pulverised into powder using a small dry mill with a diameter of 15 cm and a high-speed crusher. The crushed sludge powder was then sieved through a 200 mesh square hole sieve to obtain the 75 μm sludge powder required for the test. The whole process is shown in Figure 2.

(a)   (b)   (c)   (d)   (e)

Figure 2. Materials and equipment required for preparation of experimental materials: (**a**) initial lumpy paper sludge, (**b**) ball mill equipment, (**c**) high-speed pulveriser, (**d**) 200 mesh sieve and (**e**) 75-micron powder.

A molar ratio of $MgO:MgCl_2:H_2O = 7:1:15$ was used for the experiments. $MgCl_2 \cdot 6H_2O$ crystals were mixed with water and stirred thoroughly until the liquid in the container was free of visible granular crystals. It was allowed to stand for 24 h at room temperature around 20 °C and slowly poured into a measuring cylinder along with the mixture, ensuring that the temperature of the test liquid remained around 20 °C. The Baume degree of the measured agent should be observed using the scale-level intersection angle of the liquid surface. If the measured temperature significantly deviates from 20 °C, correct the measured value accordingly.

2.2. Experiment Method

As shown in Figure 3, this experiment considers paper sludge doping as the main variable. Firstly, 0%, 20%, 40%, 60% and 80% of paper sludge and magnesium oxide were weighed using an electronic balance. The dosage of paper sludge depended on the required strength, and when the dosage reached 80%, the strength decreased to a greater extent and the fluidity was poor; as a result, it was necessary to add a water-reducing agent. The practicality was poor, so higher dosage tests were not carried out. The dosage of paper sludge was measured based on the mass of magnesium oxide using the internal mixing method. The vertical mortar-concrete mixer manufactured in Cangzhou, China, was used for the experiment. Mixing was performed at a low speed for 120 s to ensure thorough mixing of the dry materials. After mixing for about 180 s, the concrete was quickly poured into a cylindrical iron mould with a diameter of 50 mm and a height of 100 mm. The moulds were removed and numbered after 28 d in a standard curing room. Then, the dry density test and static mechanical property test were carried out. Two strain gauges were affixed to the middle of each specimen's side positioned symmetrically in both the transverse and longitudinal directions, and the stress–strain conditions were observed using the static strain gauges. In the multistage cyclic loading experiment, 20 kN was used as the loading increment interval. The loading rate and unloading rate were both 1 kN/s.

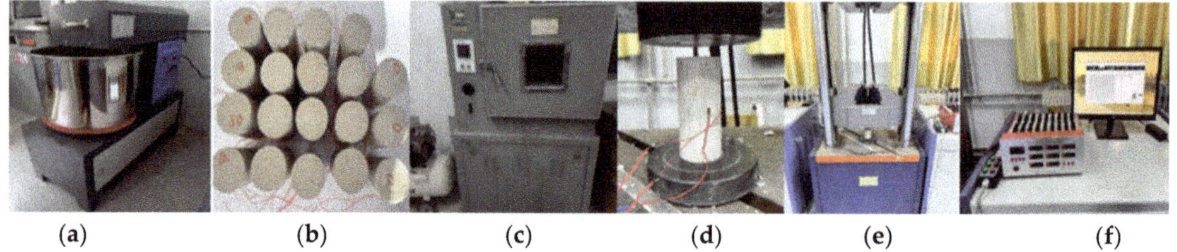

(a)   (b)   (c)   (d)   (e)   (f)

Figure 3. Materials and equipment required for preparation of experimental materials: (**a**) mortar mixer, (**b**) MOC composite specimens, (**c**) drying equipment, (**d**) pressure test, (**e**) press machine and (**f**) static strain gauge.

3. Results Analysis and Discussion

3.1. Variation Characteristics of Mechanical Parameters of MOC Composites

As shown in Figure 4, the initial state strength of the unblended paper sludge MOC material is high. With the increase in pressure, the primary pores and cracks within the MOC cement gradually healed. During the stage of compression and densification of the pore cracks, the modulus of elasticity of the material gradually tends to stabilise. The curvature of the curve presented by the material after compaction first becomes larger and starts to concave, as evident in the stress–strain curve of the 0%-doped MOC material. This stage has a greater effect on fractured rocks and a much smaller effect on materials with lower fracture and pore content and brittle materials. After the healing of the pores and fissures, the material gradually approaches a nearly straight elastic phase. At this stage, the pore cracks in the paper-mud composite MOC cementitious materials did not show significant compaction. This suggests that paper mud can indeed play a role in inhibiting the generation of pores and cracks.

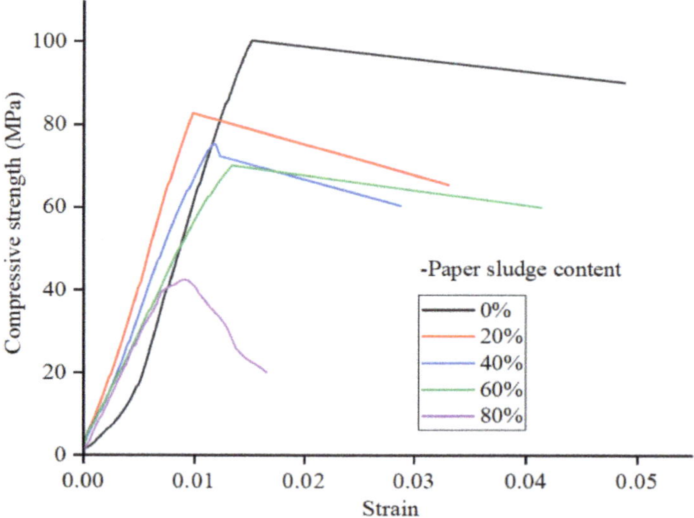

Figure 4. Effect of paper sludge content on stress and strain of MOC.

In the elastic deformation phase, the stress–strain curve shows that the modulus of elasticity generally decreases with an increase in paper sludge dosage. When the dosage exceeds 20%, it will have an effect on the strength system generated by phases 3 and 5 within the composite MOC matrix. Excessive dosage produces a dilution effect. Figure 5 shows that the dry density of the MOC material increases with an increase in paper sludge dosage, but the decrease in compressive strength increases gradually, which confirms the above observation.

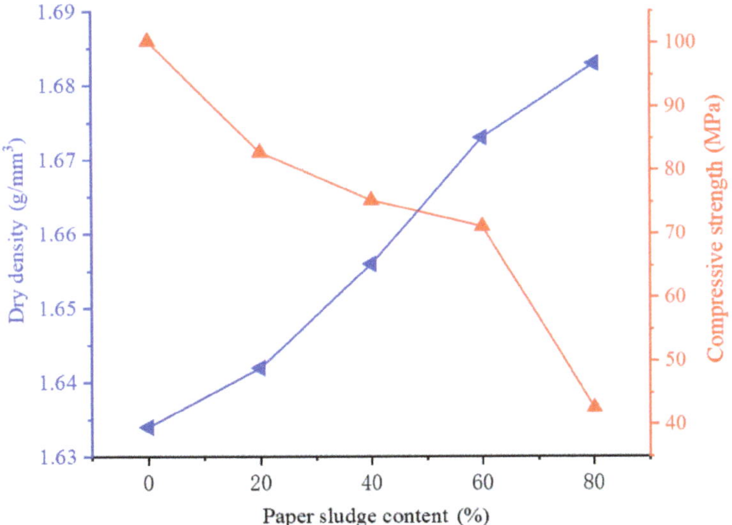

Figure 5. Variation curves of compressive strength and dry density of MOC.

With further increases in pressure, the material progressed from the microelastic cleavage stable development stage to the elastic deformation limit stage. The characteristics of brittle materials began to manifest. The slope of the stress–strain curve of the specimen with uncompounded paper sludge still shows a small rate of change. It indicates that the elastic modulus of MOC material is basically stable in this process. In the case of composite paper sludge in the MOC material, it gradually shows a yield stage that is inconsistent with brittle materials. The 3D mapping surface plots of Poisson's ratio, elastic modulus and compressive strength of MOC cement are shown in Figure 6. Figure 6 reveals the correlation between the elastic modulus as a response variable and the other two variables. The elastic modulus in the stress–strain curve can be regarded as the slope of the cut line at the instantaneous damage state point. The deterioration process of elastic modulus is actually a slow decrease in the slope of the cut line. Accompanied by the change in elastic modulus with step-by-step loading and unloading, the unloading modulus of the rock can also to some extent match the characterisation of the instantaneous elastic modulus of the damaged rock material under uniaxial compression. The Poisson's ratio decreases to some extent from 0.27 to 0.22. The elastic phase also starts to end progressively earlier with the increase in dosage, and the later curve explains the earlier and earlier nature of the plastic material at the time of material damage under a high dosage of paper sludge. This stage produces a clear difference with the incorporation of paper sludge from the 0%-doped matrix. The brittleness of the material gradually decreases as the matrix reaches a 60% dosage, starting to produce significant plastic deformation. The plastic deformation is most pronounced at 80% doping.

The final breaking strength of MOC material is 100.35 MPa. The increase in dosage of paper sludge can help fill the pores within MOC cement. The pre-compaction stage of the material is shortened, and the material can enter the elastic stage to some extent faster when compressed, but the strength of the material decreases. The strength of the specimen decreased by 17.44% at 20% dosage. At 40% dosage, the strength decreased by 24.94%. At 60% dosage it decreased by 29.01%, and at 80% dosage strength decreased by 57.51%.

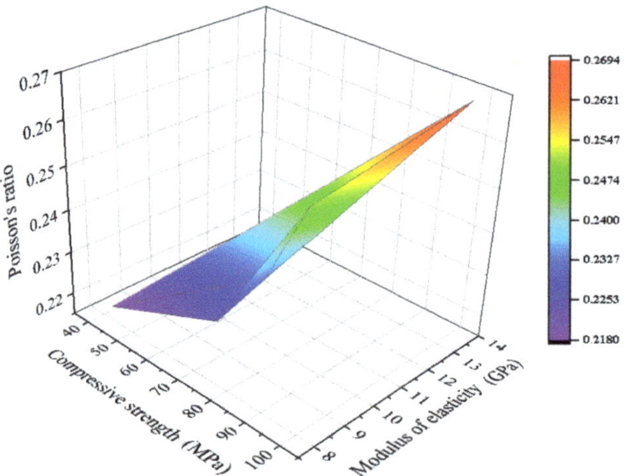

Figure 6. 3D diagram of Poisson's ratio, elastic modulus and compressive strength of magnesium oxychloride cement.

3.2. Transformation and Characterisation Methods for MOC Composite Damage Modes

The specimen has no obvious plastic changes in the preloading period. When reaching the destructive strength, the specimen shows complete destruction with an obvious popping sound due to the destruction. The edge of the specimen fragments splashes. The middle damage section is close to 90° showing a considerable number of rupture surfaces, such as Figures 7a and 8a. There are multiple rupture surfaces in the middle running axially through the entire specimen. The rupture surfaces are smooth and flat. The middle part of the specimen mainly experiences tensile damage caused by the specimen exceeding the ultimate tensile strength of the material. However, a small amount of less pronounced shear damage still exists at its edges. Specimens doped with 0% paper sludge, under pressure, even hide the shear damage surface to some extent, which shows obvious tensile damage with the fracture surface of brittle material parallel to the axial direction. Although minor shear damage is still present on the lateral side of the specimen, it accounts for a relatively small amount compared to the main damage. What is presented in the figure belongs to the typical tensile damage of rocks. Due to the greater brittleness as well as the higher strength of the magnesium chlorite cementitious material, the final damage generates more energy. The specimen is damaged by peeling off from the outer wall in a split second. The damage depends on the material being subjected to tensile stresses exceeding the maximum stress limit due to the "Poisson effect".

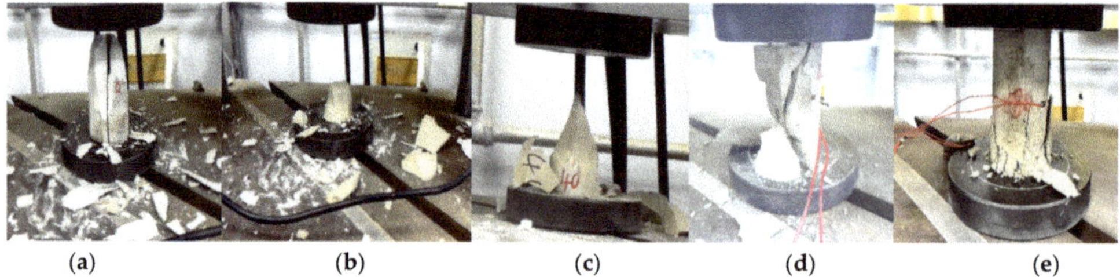

(a)   (b)   (c)   (d)   (e)

Figure 7. Failure modes of MOC with different dosages of paper sludge under normal pressure: (**a**) actual damage diagram for 0% content MOC composites, (**b**) actual damage diagram for 20% content MOC composites, (**c**) actual damage diagram for 40% content MOC composites, (**d**) actual damage diagram for 60% content MOC composites and (**e**) actual damage diagram for 80% content MOC composites.

(a)   (b)   (c)   (d)   (e)

Figure 8. Failure cross-sections and failure modes of MOC under normal pressure: (**a**) brittle tensile damage, (**b**) tensile-shear damage-1, (**c**) tensile-shear damage-2, (**d**) X-shaped conjugate surface shear damage and (**e**) diagonal pure shear damage.

The damage pattern of magnesium chloroxylate cement specimens in the presence of 20%-doped paper sludge was altered. Compared to the 0%-doped specimens that incurred damage upon reaching the tensile stress limit, the 20%-doped edges still showed axial tensile damage, corresponding to Figures 7b and 8b. The shear fracture surface inside the specimen close the end shows preliminary shear damage. However, the slip surface of shear damage is less obvious. The centre is still showing tensile damage, which indicates that the damage pattern of the material is gradually changing with the change in doping.

The axial compression of the composite MOC specimens with 40% doping resulted in significant changes in the damage slip surface. At the end of the specimen, there appeared a localised conical surface as shown in Figures 7c and 8c. The angle between the damage section and the horizontal plane also changed to different degrees. This explains that the damage mode of the specimen is gradually changing. The ratio of tensile damage and shear damage is converted to some extent. The tensile damage is gradually converted to shear damage.

The strength of the material decreased as the paper sludge doping increased. Unlike the previous damage pattern, the specimens at 60% dosage no longer showed high-energy specimen bursting. The sides still show tensile damage as shown in Figures 7d and 8d. However, the fragmentation of the sidewalls is no longer towards the fragmentation demonstrated at small dosages. The parts resulting from tensile damage tend to be increasingly monolithic. The X-shaped conjugate shear damage occurs within the composite MOC material, and the localised conical surfaces appear at the upper and lower ends. The two cone-shaped vertices are almost connected. The tensile damage on the sides no longer appears fragmented, but increasingly shows a trend towards wholeness.

Finally, the specimen at 80% paper sludge doping showed obvious plasticity changes in the early stage. The plasticity showed a significant enhancement. The damage is manifested in the specimen extending from the end of the diagonal shear surface through the specimen, as shown in Figures 7e and 8e. The damage surface is rougher at this point, which is different from the smoother tensile damage surface presented by the 0%-doping specimen.

The tensile-shear damage of rock is a form of combined rupture based on the nature of the material, which is subjected to normal tensile stress perpendicular to the axial compression direction and also damage from shear force horizontal to the rupture surface at the same time. This phenomenon is mainly explained by the third strength theory proposed by Coulomb. In rock mechanics, this theory is often applied to the study of various engineering problems related to the description of rocks damaged by shear. Coulomb's law states that when we study the shear damage of rocks, we address two main aspects. One is the bond between rock and soil particles, and the other part is the internal friction which is positively related to the positive stress. The angle of internal friction is between different levels within the rock. The maximum shear stress at which a rock resists damage by shear is called shear strength. The shear strength of rock, like that of soil, is also composed of two parts, cohesion and internal friction resistance, both of which are larger than that of soil, which is related to the fact that rock has a strong connection.

By collecting and reassembling the specimen fragments at different doping levels in each group and analysing the damage cross-section, the internal tensile strength of MOC composites was simulated by using revit modelling software. Combined with the actual cross-sectional damage morphology of MOC composites, the revit modelling software was used to simulate the internal perspective of tensile-shear damage within MOC composites at five doping levels: 0%, 20%, 40%, 60% and 80% (Figure 9I–V). The characteristics of the damage pattern changing with the doping amount and the energy dissipation in the whole process are combined to jointly corroborate the damage evolution law of MOC composites.

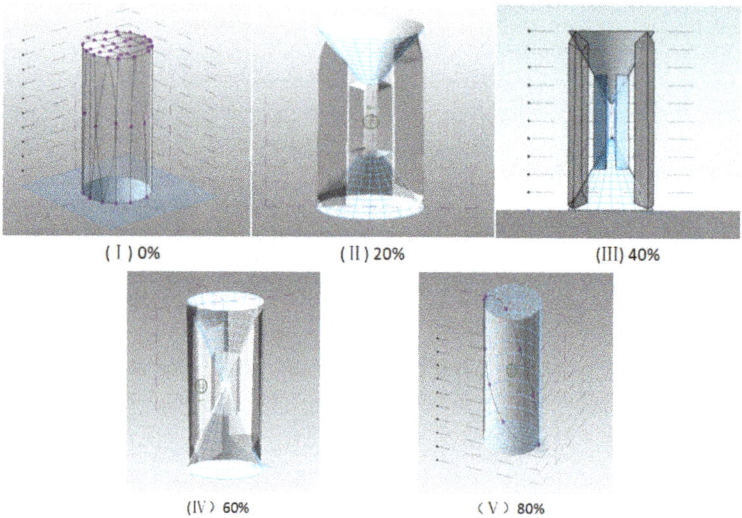

Figure 9. Perspective view of damage modes of MOC with different dosages of paper sludge.

The following equations were introduced to calculate the actual various mechanical damage parameters of MOC:

$$\varepsilon_{max1} = \mu\varepsilon_{max2} \tag{1}$$

where ε_{max1} is the transverse ultimate strains corresponding to MOC composites. ε_{max2} is the longitudinal ultimate strains corresponding to MOC composites. μ is the Poisson's ratio corresponding to MOC composites.

$$\sigma_a = E\varepsilon_{max1} \tag{2}$$

where E is the modulus of elasticity of the material at different dosages. σ_a is the ultimate tensile stress of the material at different dosages.

$$\alpha = \frac{\pi}{4} + \frac{\varphi}{2} \tag{3}$$

where α is the angle between the surface of destruction of the material and the horizontal plane. φ is the angle of internal friction of the material.

$$\sigma_n = \sigma_x\cos^2\alpha + \sigma_y\sin^2\alpha - 2\tau_{yx}\sin\alpha\cos\alpha \tag{4}$$

$$\tau_n = (\sigma_x - \sigma_y)\cos\alpha\sin\alpha + \tau_{xy}\left(\cos^2\alpha - \sin^2\alpha\right) \tag{5}$$

where σ_x is the longitudinal stresses received by MOC composites. σ_y is the transverse stresses on MOC composites. σ_n is the positive stress perpendicular to the damage surface of the material.

$$C = \frac{\sigma_n(1 - \sin\varphi)}{2\cos\varphi} \tag{6}$$

where C is the internal cohesion of the material.

$$\tau = C + \sigma_n\tan\varphi \tag{7}$$

where τ is the actual shear force on the damage surface.

The mechanical parameters of the composite MOC material calculated from the above equations are listed in Table 3.

Table 3. Mechanical parameters of MOC composites.

Paper Sludge Content (%)	0%	20%	40%	60%	80%
σ_a (MPa)	29.26	26.00	25.24	20.01	10.20
α (°)	/	65.5	58.0	61.0	63.4
Φ (°)	/	41.1	26.0	31.9	36.9
C (MPa)	/	28.38	42.30	33.57	17.06
τ (MPa)	/	36.87	52.62	44.018	23.45
Tsc	0.99	0.669	0.480	0.455	0.435

The characteristics of the different damage modes of MOC were analysed by introducing the tensile-shear conversion factor (Tsc) as shown in Figure 10. Based on the damage modes, it is broadly classified into three stages. In the initial damage stage of MOC composites, the change from tensile to shear conversion rate exhibited by the tensile-shear conversion factor from the first stage to the second stage is high. This stage corresponds to a high variation in material compressive strength from 100.35 MPa to 82.85 MPa. The specimen undergoes the compaction of the natural pores within the specimen and the gradual healing of the cracks leading to the generation of nascent cracks. The material's tensile-shear conversion factor starts to accumulate at this stage of the damage. The Tsc index decreases from 0.99 to 0.66. The second stage of damage is persistent uniform damage. The fracture compaction process reduces the energy lost in the process due to the addition of a certain amount of paper sludge to fill the internal porosity.

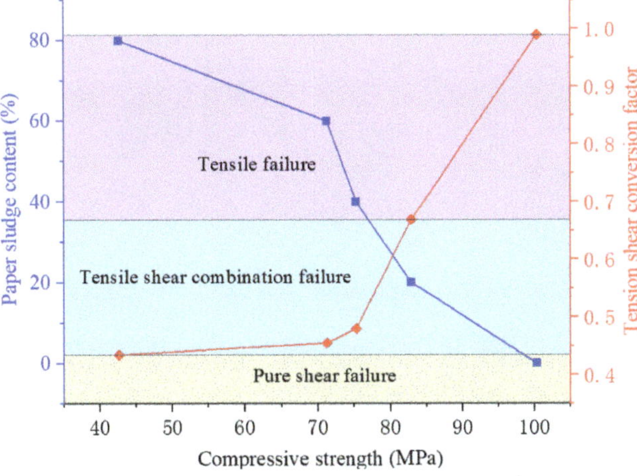

Figure 10. Variation curves of compressive strength, tensile-shear conversion factor and paper sludge content.

The Tsc index decreases from 0.67 to 0.46, indicating that the damage of the 20%, 40% and 60% MOC composites develops in a slow incremental progression. The change in slope is relatively small. The compressive strength decreases from 82.84 MPa to 71.19 MPa, indicating that the compressive damage mode of MOC materials in the 20–60% doping interval is mainly tensile and shear combined damage. However, the influence ratios are not the same when leading to the damage. The material damage accumulates faster when the Tsc index is from 0.455 to 0.435. The deformation during loading is more obvious compared to other dosages. The final compressive strength decreased from 71.19 MPa to 42.63 MPa. The decrease in compressive strength at this stage was relatively large, which is due to the substantial reduction in paper sludge within the strength structure of the 3-phase and 5-phase proportions, resulting in the more obvious "dilution effect" on the overall performance. At this time, the specimen will not show violent brittle damage. In the case of the 80%-doped MOC specimens, the crack expansion rate is slower. The mechanical properties of MOC materials exhibited in these three stages are very different, which is useful for future experiments in the modification of MOC cement.

3.3. Cyclic Loading and Unloading Experimental Methods and the Curves for MOC Composites

A universal testing machine was used to test the mechanical performance of MOC cementitious materials under cyclic loading with different paper sludge dosages. The ultimate stresses of MOC materials with different dosages under actual uniaxial loading were selected. Finally, a loading interval of 20 kN was adopted as the standard for adjusting cyclic loading. After the loading and unloading applications were debugged and stabilised, the standard specimens were placed on the experimental bench to start cyclic loading.

The mechanical properties of MOC materials subjected to cyclic loading were significantly changed by changing the dosage of paper sludge. The total energy input, elastic strain energy, plastic strain energy, dissipation energy, residual deformation and other parameters measuring the material damage during the loading process also showed obvious regularity. The maximum loading times and cyclic loading paths corresponding to uniaxial graded loading are shown in Figure 11a,b.

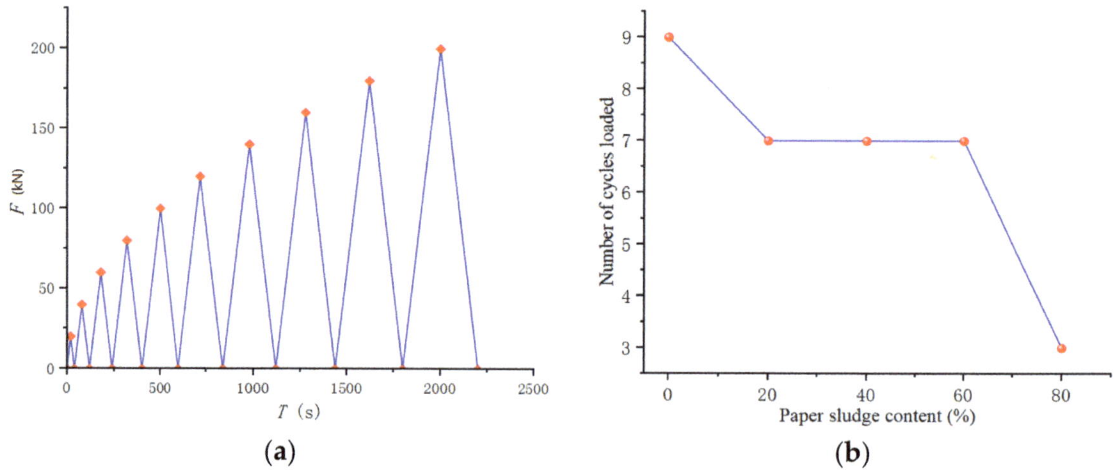

Figure 11. Cyclic loading path and paper sludge dosage versus the maximum number of loads to the specimen: (**a**) cyclic loading path and (**b**) paper sludge dosage versus the maximum number of loads.

3.4. Energy Evolution and Damage Characterisation Based on MOC Composites

The deformation, damage and destruction processes of MOC composites are accompanied by complex energy changes. The total energy input and loss of the system can serve as important factors for measuring and responding to the internal accumulation of material damage. The process includes the healing of the primary cracks of the original MOC composites, the generation of nascent cracks, the stage of the nascent cracks development and the eventual damage caused by the original MOC composites. According to the law of energy conservation, the energy input from the external load is referred to as the total energy W_1, and the total energy exerted by the whole test system on the MOC material under cyclic loading has the following energy relationship:

$$W_1 = E_d + E_e \tag{8}$$

$$E_d = E_p + E_1 \tag{9}$$

$$W_1 = \int_{\epsilon_0}^{\epsilon_2} \sigma_x \mathrm{d}\epsilon \tag{10}$$

$$E_d = \int_{\epsilon_1}^{\epsilon_2} \sigma_y \mathrm{d}\epsilon \tag{11}$$

Figure 12a represents the loading and unloading situation in a certain stage. σ_x is the instantaneous stress corresponding to the loading curve under the cyclic loading and unloading stages. σ_y is the instantaneous stress corresponding to the unloading curve under the cyclic loading and unloading stages. W_1 is the total energy inputted by the whole test system in this stage as shown in Figure 12b, which is obtained by integrating the loading curve and the area enclosed by the transverse strain in this stage. Ee is the elastic strain energy stored within the specimen, which denotes the strain energy that is present within the specimen and can be released, as shown in Figure 12c. It is obtained by integrating over the unloading curve and the area enclosed by the transverse strain at this stage. E_d is the total energy loss generated during cyclic loading and unloading of the platform, which is obtained by subtracting the elastic strain energy from the integral over the region of the total energy input to the platform. E_p is the plastic strain energy used for plastic deformation during loading and unloading, as shown in Figure 12d, which is obtained by subtracting the total input energy for the next stage of loading from the total energy dissipated during that stage of loading and unloading. E_1 is the actual energy dissipated during loading and unloading, as shown in Figure 12e. The plastic strain energy resulting from plastic strain is subtracted from the total dissipated energy.

In the unloading process, part of the deformation will be recovered as elastic deformation. The energy accumulated at this stage is the elastic strain energy. On the other hand, the remaining unrecoverable deformation is the residual deformation of the composite MOC material during the process. This is caused by the closure, slip and misalignment of the structural surfaces of the MOC material during the axial compression. Dissipated energy includes energy in the form of heat during cyclic loading and unloading and energy consumed by internal damage destruction. The elastic strain energy at each stage of loading increases with the number of loading cycles. This results in more thermal energy being lost during the process, and more energy is consumed due to internal damage. The cyclic loading curve produces less energy consumption due to the inelastic deformation of the structural surfaces compared to a typical rock. The overall curve is still consistent with the characteristics of brittle materials under pressure. There is no significant plastic deformation observed during the loading and unloading phases.

With the increase in the number of cyclic loading, the elastic strain energy, plastic strain energy and dissipated energy all showed a gradual increase in the MOC composites with 0% doping. At a paper sludge dosage of 40%, the average plastic strain energy per cycle in the unloading stage of the curve starts to increase compared to the average plastic strain energy per cycle at 0% and 20%. As the paper sludge dosage increases, the peak value of the envelope decreases. The residual strain also increases with an increase in plastic strain energy. The overall brittleness of the specimens was reduced and the plasticity was improved to some extent.

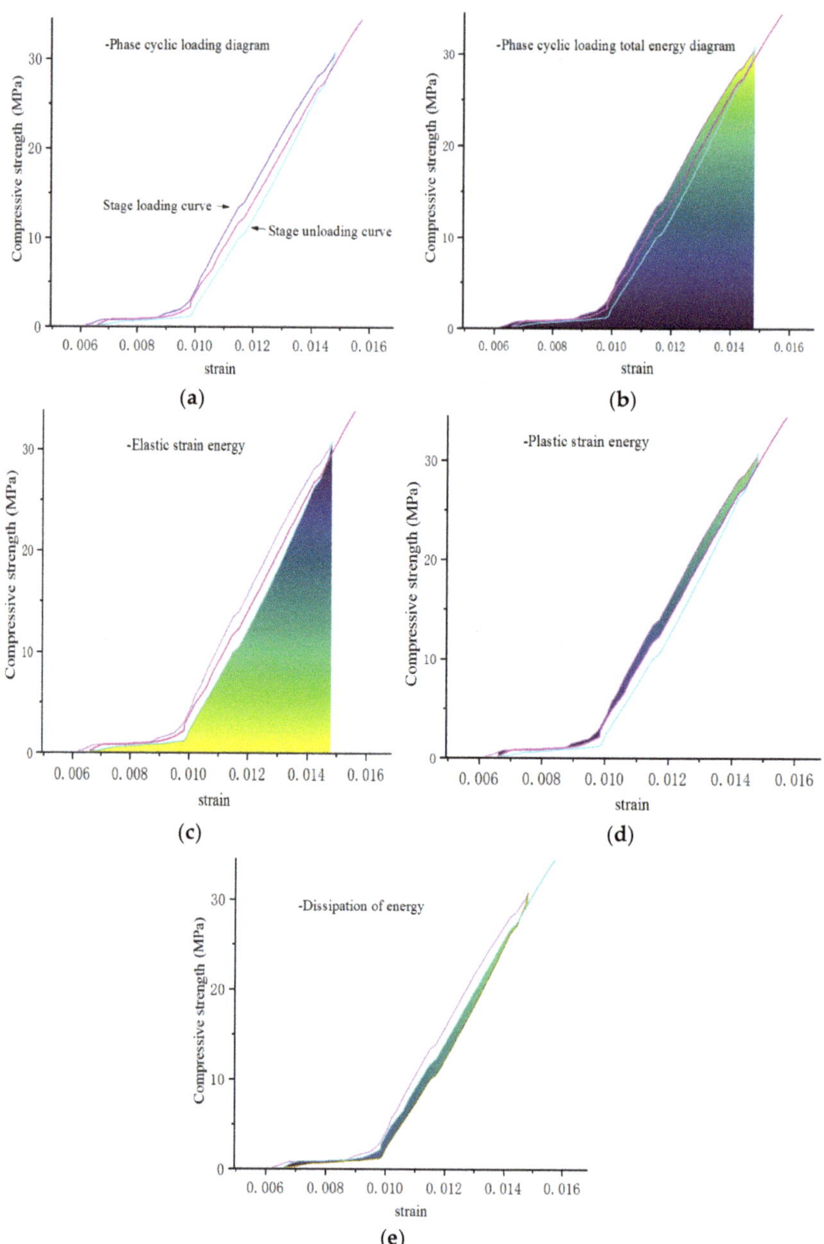

Figure 12. Cyclic loading path and different energy diagram under cyclic loading: (**a**) cyclic loading path, (**b**) total energy under cyclic loading, (**c**) elastic strain energy under cyclic loading, (**d**) plastic strain energy under cyclic loading and (**e**) dissipated energy under cyclic loading.

As shown in Table 4 and Figure 13, most of the energy dissipation generated at each level of the loading stage in MOC is attributed to plastic deformation with a small amount of heat energy dissipation. As the plastic deformation and residual strain increase in each stage, the cumulative damage reflected in the MOC composites is larger. Different average dissipation energies result in different levels of damage, and as a result, different damage modes occur. Before reaching the ultimate stress yield point of the material, the material passes through a compaction phase. The damage caused by this stage is mainly due to the primary state of the internal micro-cracks resulting from repeated compaction and crack healing damage. This damage is more obvious when the paper sludge is not mixed. The internal microporosity is higher than that of the paper sludge, which is the absolute damage in the early stage. As the cyclic loading test force increases, new cracks begin to appear in the MOC material. This marks the beginning of the accumulation of damage in the second stage of the material. Starting from the first loading, the average elastic strain energy of the loading stage gradually decreased. The properties of the MOC material were analysed from the perspective of energy loss under cyclic loading and from the mechanical point of view through Tsc, and the changes in the damage modes of paper sludge with different dosages were confirmed.

Table 4. Energy of MOC composites under cyclic loading at different dosages.

Paper Sludge Content (%)	Total Energy (MJ·m^{-3})	Elastic Strain Energy (MJ·m^{-3})	Plastic Strain Energy (MJ·m^{-3})	Dissipative Energy (MJ·m^{-3})	Residual Strain
0	1.635	1.471	0.091	0.073	0.0025
20	0.899	0.823	0.051	0.025	0.0035
40	0.679	0.867	0.085	0.044	0.0125
60	0.999	0.778	0.166	0.072	0.0078
80	0.056	0.043	0.006	0.015	0.0066

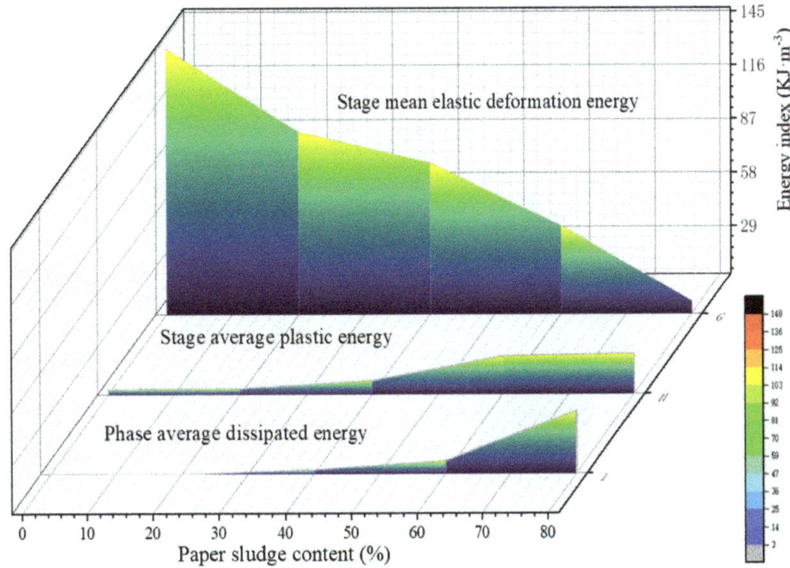

Figure 13. Average energy of each stage under cyclic loading of MOC.

4. Conclusions

The paper sludge was used to change the brittleness of MOC composites by varying its dosage. The uniaxial compression experiments and cyclic loading mechanical experiments were carried out. The Tsc was introduced to further analyse the damage mechanism from both the mechanical and the energy loss perspectives. The mechanical properties and damage mechanisms of MOC composites were studied with the increase in paper sludge dosage. The following conclusions were drawn:

(1) The modulus of elasticity of MOC gradually decreases from 13.30 GPa to 8.04 GPa with the increase in paper clay doping. Poisson's ratio also decreases to a certain extent within a small range of 0.27–0.22, which indicates that the proportion of transverse and longitudinal strains generated by uniaxial compression of MOC decreases at different doping levels. MOC can be reasonably substituted for conventional silicate cement without affecting the actual applied strength. This not only improves its mechanical properties but also reduces the cost of disposal of this industrial solid waste and improves the environment.

(2) In the process of increasing the doping level from 20% to 60%, the damage mode gradually shifts from tensile damage to tensile-shear conjugate damage mode. The proportion of tensile damage in the damage process is gradually decreasing. Subsequently, the tensile-shear combined damage is over to X-shaped conjugate shear damage and finally becomes pure shear damage. To characterise the changes in damage patterns, a tensile-shear conversion factor (Tsc) was introduced, which is of significance for the subsequent study of the damage mechanism of MOC materials.

(3) With the increase in sludge dosage, the elastic strain energy under the loading and unloading stages gradually decreases, and the plastic strain energy, dissipation energy and residual strain increase to different degrees. Since different MOC materials can withstand different numbers of loading and unloading, it is suggested that the elastic strain energy, plastic strain energy and dissipation energy at each stage of the loading and unloading process should be used to reveal the ratio of the energy of a particular loading stage to the overall damage process.

Based on the modification of MOC composites using different amounts of paper sludge, the sludge after high-temperature calcination has the characteristics of volcanic ash. The next stage is expected to carry out more in-depth research on MOC materials by calcining paper sludge under different temperature gradients and adjusting the particle size of paper sludge. In the experiment, the acoustic emission sensors and other detection instruments will be used to further investigate the crack expansion pattern of MOC composites.

Author Contributions: Formal analysis, funding acquisition, methodology, supervision and writing—review and editing, S.W.; data analysis and writing—original draft, Z.W.; writing—review and editing, J.G.; data analysis, Q.L. All authors have read and agreed to the published version of the manuscript.

Funding: This work was supported by the Science and Technology Project of Henan Province (222102320262), the Key Project of Natural Science Foundation of Henan Province (232300421134) and First-Class Discipline Implementation of Safety Science and Engineering (AQ20230103), China.

Institutional Review Board Statement: Not applicable.

Informed Consent Statement: Not applicable.

Data Availability Statement: Data were curated by the authors and are available upon request.

Conflicts of Interest: The authors declare no conflict of interest.

References

1. Chen, M.; Li, L.; Zhao, P.; Wang, S.; Lu, L. Paper sludge functionalization for achieving fiber-reinforced and low thermal conductivity calcium silicate insulating materials. *J. Therm. Anal. Calorim.* **2019**, *136*, 493–503. [CrossRef]
2. Tofani, G.; de Nys, J.; Cornet, I.; Tavernier, S. Alternative filler recovery from paper waste stream. *Waste Biomass Valorization* **2021**, *12*, 503–514. [CrossRef]
3. Aiken, T.; Russell, M.; McPolin, D.; Bagnall, L. Magnesium oxychloride boards: Understanding a novel building material. *Mater. Struct.* **2020**, *53*, 118. [CrossRef]
4. Wang, Y.; Wei, L.; Yu, J.; Yu, K. Mechanical properties of high ductile magnesium oxychloride cement-based composites after water soaking. *Cement Concrete Comp.* **2019**, *248–258*, 0958–9465. [CrossRef]
5. Ye, Q.; Han, Y.; Zhang, S.; Gao, Q.; Zhang, W.; Chen, H.; Gong, S.; Shi, S.; Xia, C.; Li, J. Bioinspired and biomineralized magnesium oxychloride cement with enhanced compressive strength and water resistance. *J. Hazard. Mater.* **2020**, *383*, 121099. [CrossRef]
6. Gong, J.; Liu, Q.; Wang, S.; Wang, Z.; Li, C. Mechanical properties and microscopic mechanism of paper mill sludge-magnesium oxychloride cement composites. *DYNA* **2023**, *98*, 57–63. [CrossRef]
7. Chen, X.; Zhang, T.; Bi, W.; Cheeseman, C. Effect of tartaric acid and phosphoric acid on the water resistance of magnesium oxychloride (MOC) cement. *Constr. Build. Mater.* **2019**, *383*, 528–536. [CrossRef]

8.   Devi, P.; Sanchaya, M.; Harikaran, M.; Krishna, J.; Kaviyarasan, V.; Venkatesh, N. Effective utilization of waste paper sludge ash as a supplementary material for cement. *Mater. Today* **2023**, 599. [CrossRef]
9.   Ingale, S.; Nemade, P. Effect of paper sludge ash on properties of cement concrete: A review. *Mater. Today* **2023**, 492. [CrossRef]
10. Gomes, C.; Garry, A.; Freitas, E.; Bertoldo, C.; Siqueira, G. Effects of Rice Husk Silica on microstructure and mechanical properties of Magnesium-oxychloride Fiber Cement (MOFC). *Constr. Build. Mater.* **2020**, *241*, 118022. [CrossRef]
11. He, H.; Zhang, H.; Yang, J.; Fan, Z.; Chen, W. Effect of pressing pressure on the mechanical properties and water resistance of straw/sawdust-magnesium oxychloride cement composite. *Constr. Build. Mater.* **2023**, *383*, 131362. [CrossRef]
12. Singh, A.; Kumar, R.; Goel, P. Factors influencing strength of magnesium oxychloride cement. *Constr. Build. Mater.* **2021**, *303*, 124571. [CrossRef]
13. Zhao, Y.; Lu, J.; Liu, X.; Wang, Y.; Lin, J.; Peng, N.; Li, J.; Zhao, F. Performance enhancement of polyvinyl chloride ultrafiltration membrane modified with graphene oxide. *J. Colloid Interface Sci.* **2016**, *480*, 1–8. [CrossRef] [PubMed]
14. Yu, K.; Guo, Y.; Zhang, Y.; Soe, K. Magnesium oxychloride cement-based strain-hardening cementitious composite: Mechanical property and water resistance. *Constr. Build. Mater.* **2020**, *261*, 119970. [CrossRef]
15. Xu, B.; Ma, H.; Hu, C.; Yang, S.; Li, Z. Influence of curing regimes on mechanical properties of magnesium oxychloride cement-based composites. *Constr. Build. Mater.* **2016**, *102*, 613–619. [CrossRef]
16. Liu, J.; Lyu, X.; Liu, Y.; Zhang, P. Energy evolution and macro-micro failure mechanisms of frozen weakly cemented sandstone under uniaxialcyclic loading and unloading. *Cold Reg. Sci. Technol.* **2023**, *214*, 103947. [CrossRef]
17. Meng, Q.; Liu, J.; Pu, H.; Huang, B.; Zhang, Z.; Wu, J. Effects of cyclic loading and unloading rates on the energy evolution of rocks with different lithology. *Geomech. Energy Envir.* **2023**, *34*, 100455. [CrossRef]
18. Liu, Z.; Cao, P.; Zhao, Q.; Cao, R.; Wang, F. Deformation and damage properties of rock-like materials subjected to multi-level loading-unloading cycles. *J. Rock Mech. Geotech. Eng.* **2023**, *15*, 1768–1776. [CrossRef]
19. Wang, H.; Li, J. Mechanical Behavior Evolution and Damage Characterization of Coal under Different Cyclic Engineering Loading. *Geofluids* **2020**, *2020*, 8812188. [CrossRef]
20. Wang, S.; Zhao, J.; Wu, X.; Yang, J.; Wang, Q. Elastic properties and damage evolution analysis for lightweight shale ceramsite concrete. *Int. J. appl. Mech.* **2023**, *15*, 1950048. [CrossRef]
21. Gao, D.; Sang, S.; Liu, S.; Wu, J.; Geng, J.; Tao, W.; Sun, T. Experimental study on the deformation behaviour, energy evolution law and failuremechanism of tectonic coal subjected to cyclic loads. *In. J. Min. Sci. Technol.* **2022**, *32*, 1301–1313. [CrossRef]
22. Xu, Y.; Yang, R.; Chen, P.; Ge, J.; Liu, J.; Xie, H. Experimental study on energy and failure characteristics of rubber-cement composite short-column under cyclic loading. *Case Stud. Constr. Mat.* **2022**, *16*, e00885. [CrossRef]
23. Song, Z.; Frühwirt, T.; Konietzky, H. Characteristics of dissipated energy of concrete subjected to cyclic loading. *Constr. Build. Mater.* **2018**, *168*, 47–60. [CrossRef]
24. Lin, H.; Liu, J.; Yang, J.; Ran, L.; Ding, G.; Wu, Z.; Lyu, C.; Bian, Y. Analysis of damage characteristics and energy evolution of salt rockunder triaxial cyclic loading and unloading. *J. Energy Storage* **2022**, *56*, 106145. [CrossRef]

sustainability

MDPI

Article

Experimental Study on the Solidification of Uranium Tailings and Uranium Removal Based on MICP

Lin Hu [1,2], Zhijun Zhang [1,2], Lingling Wu [1,2,*], Qing Yu [1,2], Huaimiao Zheng [3], Yakun Tian [1,2] and Guicheng He [1,2]

[1] School of Resource & Environment and Safety Engineering, University of South China, Hengyang 421001, China; 343737982229@usc.edu.cn (L.H.); 130000148665@usc.edu.cn (Z.Z.); 2009000547@usc.edu.cn (Q.Y.); 2017000012@usc.edu.cn (Y.T.); 2005000509@usc.edu.cn (G.H.)
[2] Hunan Province Engineering Technology Research Center for Disaster Prediction and Control on Mining Geotechnical Engineering, Hengyang 421001, China
[3] School of Economics, Management and Law, University of South China, Hengyang 421001, China; 2004001109@usc.edu.cn
* Correspondence: wllshmily@foxmail.com

Abstract: The governance of uranium tailings aims to improve stability and reduce radionuclide uranium release. In order to achieve this goal, the uranium removal solution test and uranium tailings grouting test were successively carried out using microbially induced calcium carbonate precipitation (MICP) technology. The effect of MICP on the reinforcement of uranium tailings and the synchronous control of radionuclide uranium in the tailings were discussed. The solution test results show that *Sporosarcina pasteurii* could grow and reproduce rapidly in an acidic medium with an initial pH of 5. The uranium concentration decreased with the increase in MICP reaction time, and the removal efficiency reached 60.9% at 24 h. In the solidification test of tailings, the strength of tailings improved significantly after 12 days of reinforcement, with an increase in the cohesion of tailings by 2.937 times and an increased internal friction angle of 8.393°. The peak stress value of solidified tailings at the surrounding pressure of 50 kPa increased by 1.87 times, and the uranium concentration in the discharge fluid decreased by 76.91% compared to the blank group. This study provides valuable insights and references for safely disposing of uranium tailings.

Keywords: uranium tailings; solidification; uranium removal; microorganism; calcium carbonate

Citation: Hu, L.; Zhang, Z.; Wu, L.; Yu, Q.; Zheng, H.; Tian, Y.; He, G. Experimental Study on the Solidification of Uranium Tailings and Uranium Removal Based on MICP. *Sustainability* **2023**, *15*, 12387. https://doi.org/10.3390/su151612387

Academic Editors: Hong-Wei Yang, Shuren Wang and Chen Cao

Received: 20 July 2023
Revised: 9 August 2023
Accepted: 10 August 2023
Published: 15 August 2023

1. Introduction

In recent years, there has been a surge of interest in microbially induced calcite precipitation (MICP) as a highly effective and sustainable reinforcement technology, as evidenced by numerous domestic and international studies [1]. These studies demonstrate that MICP can induce the formation of crystals with exceptional cementing properties through the metabolic activities of microorganisms. When applied to geotechnical materials, this technology significantly boosted their strength [2,3]. MICP technology empowers an ecologically friendly approach to enhancement since it can improve the stiffness, bearing capacity, permeability, and liquefaction resistance of earth and rock. Song et al. [4] conducted several cycles of MICP via grouting on sandstone. This achieved remarkable results: a 229% increase in uniaxial compressive strength, a 179% increase in elastic modulus, and a 177% increase in brittleness index compared to the pre-grouting conditions. The overall mechanical properties and permeability of sandstone were shown to be primarily impacted by the amount of cemented minerals present, which, in turn, directly govern the microscopic distribution of $CaCO_3$, amplifying the efficacy of bio-cementation in sandstone. Banik et al. [5] explored the microstructures and mechanical characteristics standard sand specimens of after MICP reinforcement. The resulting stress–strain data and the strength gain observed under different pore volumes of cementing fluids provided insights into quantifying the strength of microbially reinforced sand and enhancing the low-strain shear modulus. Zhao et al. [6] carried out a solution and cyclic triaxial tests while regulating the concentration of NaCl and observed that the stiffness and cyclic resistance of the additive solids gradually decreased as the NaCl concentration increased. However, these values remained higher than those of unreinforced sand, and it was discovered that the decline in liquefaction resistance was caused by the conversion of calcium carbonate crystals from clusters to single crystals induced by the technology. Xiao et al. [7] leveraged the application of MICP-reinforced coral sand by acclimating microorganisms to various environments to improve their enzymatic activities, developing a novel three-stage reinforcement method. Experimental results confirm that this method considerably increased the strength of MICP treatment and brought about a 32% increase in induced calcium carbonate content. Additionally, suitable cementing solution concentrations proved advantageous for enhancing reinforcement strength.

In addition to studying the reinforcement effect of MICP on soils, considerable attention has been given to the application of MICP technology for the remediation of heavy metal pollution. Chen et al. [8] used bacillus *pasteurii* to immobilize lead, reducing the leaching of Pb^{2+} in water by 76.34%. Achal et al. [9] used a copper-resistant strain of *Kocuria flava* CR1 isolated from a mining area. The bacteria showed a high urease activity under culture medium conditions at an initial Cu^{2+} concentration of 1000 mg/L, pH value of 8.0, calcium source of $CaCl_2$, urea concentration of 2%, and a temperature of 30 °C. At that time, the Cu^{2+} removal rate was 97%. Li et al. [10] discovered a strain of *Sporosarcina pasteurii* UR31 in soil that is highly tolerant to zinc. When $ZnCl_2$ was introduced to the culture medium at an initial concentration of 2 g/L, this strain was able to remove as much as 99% of the zinc. Jalilvand et al. [11] isolated two rhizobacterial strains, *Stenotrophomonas rhizophila* A323 and *Variovorax boronicumulans* C113, from grassroots. The study found that these strains exhibited cadmium tolerance and produced urease. In aqueous solutions, these bacteria were discovered to transform more than 70% of soluble Cd into insoluble carbonate mineral forms. Moreover, the researchers utilized isolated strains of *Sporosarcina pasteurii* to eliminate Cd^{2+} and accomplished a removal rate of 97.15%. These studies show that MICP technology holds promising potential and has good prospects for the bioremediation of heavy metal pollution.

Uranium tailings contain a large amount of low-grade uranium ore, and the uranium in the tailings accounts for about 1~5% of the original ore. It contains 85% of the radioactivity found in raw ore [12]. Although uranium tailings have a low average grade of uranium, ranging from 0.005% to 0.03%, the total amount of uranium present must be taken into consideration and addressed. Additionally, the long-term storage of uranium tailings causes severe radioactive pollution to the surrounding environment and results in a significant waste of uranium resources [13,14]. Therefore, the current governance objective of uranium tailings (slag) ponds is to address the above environmental issues at the source by implementing reasonable and practical measures to enhance their stability and reduce the release of residual radionuclides [15].

The objective of this study is to enhance the stability of uranium tailings and prevent radioactive contamination. Through the utilization of MICP technology, the experimental microbial grouting reinforcement of uranium tailing sand was conducted to investigate the impact of MICP on uranium tailings reinforcement and radionuclide control. As opposed to the single focus on the reinforcement effect in most current studies, the synchronous control effect of pollutants is also considered in this study. Microbial grouting reinforcement is characterized by a low cost and short reinforcement time and is environmentally friendly. Compared with the inclusions of tailings formed via traditional chemical grouting, the biobased body is more convenient for grinding dissociation and recovery, which is conducive to the recovery and comprehensive utilization of effective resources in cemented tailings. This is important for the theoretical innovation and technical promotion of the safe disposal of uranium tailings.

2. Materials and Methods

2.1. MICP Aqueous Solution Tests

2.1.1. Purposes

To examine the effectiveness of mineralizing bacteria in sand grouting tests on uranium tailings, the activity and ability of the mineralizing bacteria to perform normal mineralization reactions were determined in an acidic and radioactive solution.

2.1.2. Materials

The test was carried out using *Sporosarcina pasteurii* as the dominant species, purchased from the American Type Culture Collection (Rockefeller, MD, USA), and numbered ATCC11859. The liquid medium consisted of 15 g/L casein peptone, 5 g/L soy peptone, 5 g/L NaCl, 20 g/L urea, and 1000 g/L distilled water. The solid medium was added to 20 g/L agar based on the liquid medium. The cement solution comprised 1 M urea solution and 1 M $CaCl_2$ solution with a volume ratio of 1:1.

2.1.3. Methods

In order for the subsequent mineralization reaction to be effective in uranium tailings, the acidity of the solution and the concentration of uranium must be equal to or stronger than that of uranium tailings. Therefore, in order to set the appropriate initial pH value and uranium concentration for the solution, the pH value and uranium concentration of the uranium tailings must be measured first. Uranium tailings samples from a local large uranium tailings pond in Hunan Province were soaked in deionized water for 48 h, the soaking solution was sampled, and a method was used to determine pH and uranium concentration [16]. The result was a pH of 5.3 and uranium concentration of 0.365 mg/L. With reference to this value, the initial pH of the solution was set to 5 and the initial concentration of uranium was set to 2 mg/L.

After the mineralization reaction was completed, the uranium concentration was determined using inductively coupled plasma—mass spectrometry (ICP-MS). When the concentration of uranium was measured using ICP-MS, the quality control methods were as follows: ① Because the pH of the solution after the mineralization reaction increases and becomes alkaline, nitric acid was used to dissolve the filtered solution. ② Before testing the sample, the machine was preheated for 30 min, so that the temperature of the instrument itself was stable, and the wavelength was unaffected by the temperature. ③ The calibration was first carried out with the standard liquid of uranium. ④ The argon gas flow rate was slowly adjusted to 15 L /min, and the auxiliary gas flow rate was 2 L /min.

When the concentration of the original bacterial solution reached its maximum, an appropriate amount of the bacterial solution was transferred to the culture above the medium for expansion. After expansion, 100 mL of the bacterial solution was poured into the reaction vessel. Simultaneously, 400 mL of cement solution (concentration specified in Section 2.1.2) was slowly injected through a peristaltic pump. Then, the reaction vessel was placed on a magnetic stirrer and stirred for 24 h. During the experiment, every 2 h, the bacterial solution was taken out to determine the pH, OD600 value, and uranium concentration, with a total of 12 determinations. After 24 h, the precipitate was taken for further testing.

The main detection instruments involved in this study are shown in Table 1.

Table 1. List of testing instruments.

Experimental Instrument	Company	Model
pH	Mettler Toledo, Shanghai, China	S220-K
OD600	Eppendorf, Hamburg, Germany	Biophotometer
ICP-MS	Agilent, Santa Clara, CA, USA	7700×
XRD	Hitachi, Tokyo, Japan	Regulus8100
SEM	Bruker, Karlsruhe, Germany	Bruker D8 ADVANCE
CT	Zeiss Xradia, Turingen, Germany	410Versa
Triaxial testing	Teco, Nanjing, China	TKA-TTS-IS

2.2. Grouting Tests on Uranium Tailings

2.2.1. Purposes

In order to evaluate the effectiveness of MICP technology in stabilizing uranium tailings and removing uranium, experiments were conducted on the uranium tailings using MICP technology, as per the findings presented in Section 2.1.

2.2.2. Materials

The experiment tailings were collected from a large uranium tailings pond in Hunan Province, China. The size of the uranium tailing particles is relatively uniform, mainly being translucent granular minerals with metallic luster and little debris. Impurities such as quartz can be seen. The hardness of the tailings can be felt by kneading them by hand. The relative density of the uranium tailings is 2.53, with a natural dry density of 1.446 g/cm^3 and a void ratio of 58.8%. In order to determine the primary parameters of particle size for uranium tailings, a series of particle analysis tests were conducted on the tailing sand. Based on the test results, the cumulative curve of particle size distribution of tailings is drawn, as shown in Figure 1. According to the curve, the relevant parameters of tailings particle size were determined, such as the restricted particle size d_{60} = 1.00 mm, effective particle size d_{10} = 0.16 mm, median particle size d_{30} = 0.5 mm, inhomogeneity coefficient C_U = 6.25 > 5, and curvature coefficient C_C = 0.15625 > 1. According to the Standard for Engineering Classification of Soil, the size distribution of uranium tailings is continuous and particle size distribution is good [17].

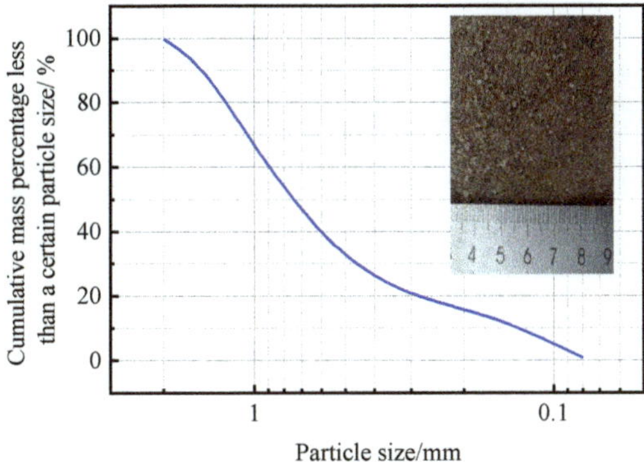

Figure 1. Cumulative curve of particle gradation of tailings.

The microorganisms, culture medium, and cement solution used for grouting are the same as those described in Section 2.2.1.

2.2.3. Test Devices and Methods

The tailing sand was filled and compacted into the grouting reinforcement device before the test. The grouting reinforcement device is shown in Figure 2. The thickness, width, and length of tailings are 30 cm, 40 cm, and 100 cm, respectively. The slope of tailings is 30°, and the volume of tailings is calculated as follows: 100 cm × 40 cm × 30 cm − 1/2 × 40 cm × 30 cm × 60 cm × cos30° = 88,800 cm^3. In tailings with a moisture content of approximately 10% and void ratio of 58.8%, the total amount of grout required was calculated as 88.8 × 48.8 = 43.33 L [18]. To make the calculation easier, the total grouting volume was fixed at 44 L, which is composed of 8.8 L of bacterial liquid, 17.6 L of CaCl$_2$ solution, and 17.6 L of urea solution. The reaction liquid was injected in a specific order, starting with bacterial solution, followed by urea solution, and CaCl$_2$ solution. The rate of grout injection was 20 mL/min, and the grouting was carried out once per day. After each grouting, the drain pipe valve was opened to collect residual liquid. This process was repeated for 12 days, and, at the end, samples were taken for testing and analysis.

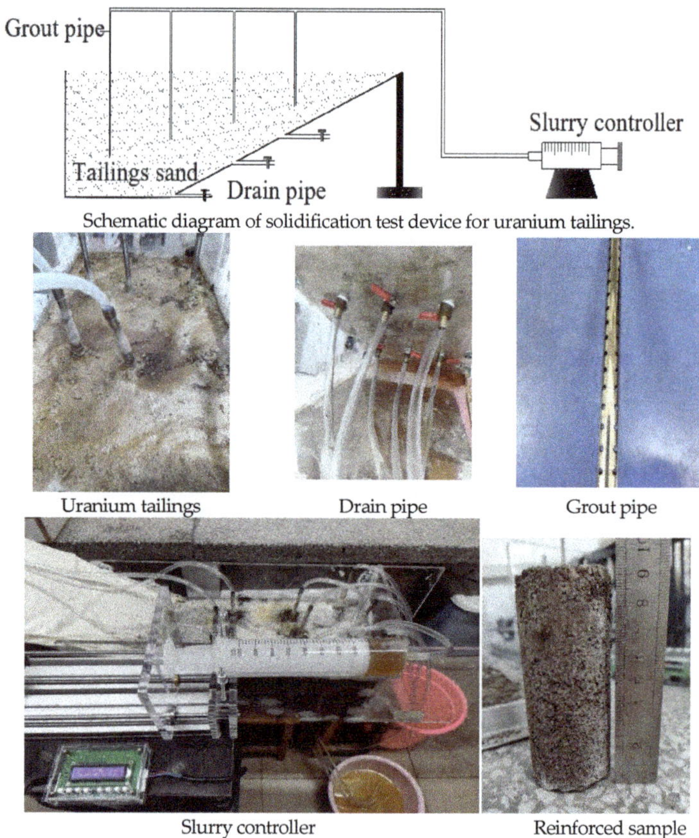

Figure 2. Tailing reinforcement device diagram.

3. Results and Analysis

3.1. Solution Test Results

3.1.1. Biochemical Characterization of Bacteria

The change in pH significantly impacts the physiological metabolic activities and nutrient absorption abilities of *Sporosarcina pasteurii*. Moreover, pH has a more significant impact on intracellular and extracellular urease activity produced during bacterial metabolism. Therefore, the biochemical characteristics of the bacteria are reflected by measuring the pH and the bacterial cell density OD600 value of the bacterial liquid. Every 2 h, the pH of the bacterial solution was measured, while the OD600 value was measured using a protein–nucleic acid analyzer. According to the test results, the growth and pH change curves of the bacteria were drawn and are shown in Figure 3.

Figure 3 shows that when the initial pH is 5 and the uranium mass concentration is 2 mg/L, the number of OD600 bacteria linearly increases within 4–8 h, indicating vigorous multiplication. However, with the gradual depletion of nutrients and accumulation of metabolic waste, the growth rate of OD600 bacteria decrease between 10 and 24 h. It can also be seen from Figure 3 that the pH value rapidly increases and stabilizes, eventually staying at approximately 10, Indicating that the bacteria can multiply in the medium with an initial pH of 5. The pH increased almost linearly within 0~6 h, and then stabilized from 6 to 8 h. The pH values showed the same trends as the OD600 values. The pH values varied within 0~6 h, mainly due to the rapid increase in the number of bacteria and urea hydrolysis rate caused by urease, which produced a considerable number of carbonate and ammonium ions. When the pH value reached around 10 at 6 h, it provided an alkaline environment for bacterial growth and reproduction. However, the ability of bacteria to multiply reached its maximum, and the nutrient resources in the culture medium were consumed at the later stages, resulting in the pH value gradually stabilizing and slowly decreasing.

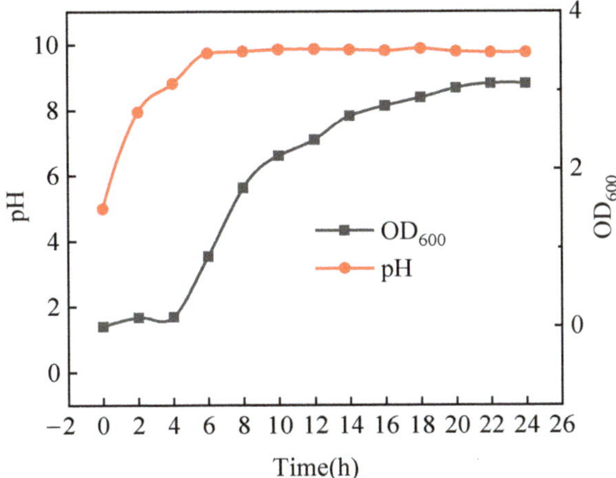

Figure 3. Growth curve and pH change curve of *Sporosarcina pasteurii*.

3.1.2. Changes in Uranium Concentration

The uranium concentration in the solution was measured every 6 h during the reaction, and the results are shown in Figure 4. According to Figure 4, the uranium concentration significantly decreases within the first 10 h and slightly increases after 12 h. This is because some uranium was adsorbed in the form of unstable combinations of anions and cations, resulting in a small amount of solidified uranium being reintroduced into the solution over time. When the reaction time was 24 h, the rate of uranium removal was 60.9%.

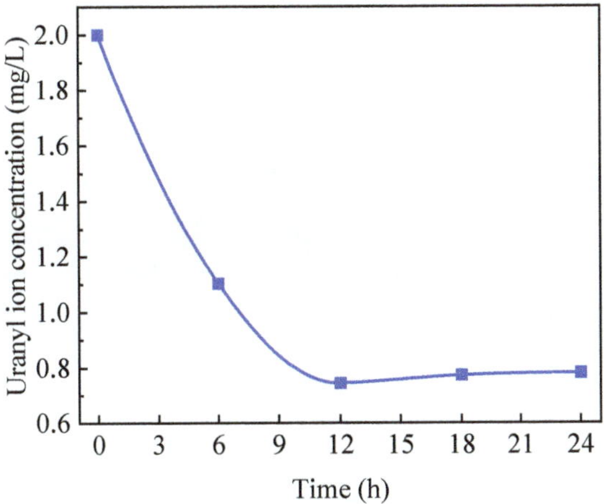

Figure 4. Change curve of uranium concentration in solution.

3.1.3. Mineralization Product Test Results
X-ray Diffraction (XRD)

After the reaction, the resulting crystal precipitate was washed, dried, and weighed, yielding a mass of 4.226 g. The crystalline products appeared slightly yellow and white under visual observation, with stacked and arranged crystals, large particle agglomeration and interparticle stacking connections, dense crystal–crystal stacking, rich pores, a uniform crystal texture, and good structure. The mineralized products were taken for XRD test analysis, and the results are shown in Figure 5. Figure 5 shows that the crystalline product is calcium carbonate crystals, which contain crystalline forms of calcite and vaterite.

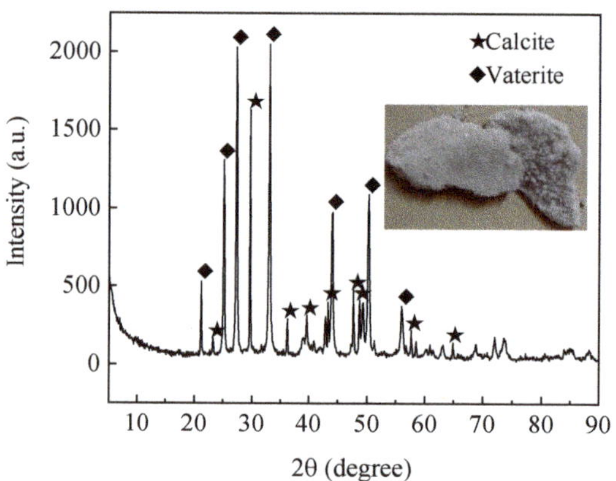

Figure 5. XRD test results of mineralized products.

Scanning Electron Microscopy (SEM)

SEM tests were carried out on calcium-carbonate-mineralized samples, and the results are shown in Figure 6. The scanned images from Figure 6 indicate that the main forms of calcium carbonate crystals are regular and uniformly shaped vaterite and calcite with various shapes (massive and granular) and uneven grain sizes, consistent with XRD results. The scanned images also show that the crystal surface is uneven and densely porous, and vaterite and calcite overlap with each other and stack in clusters, creating a structure conducive to stabilizing the crystals.

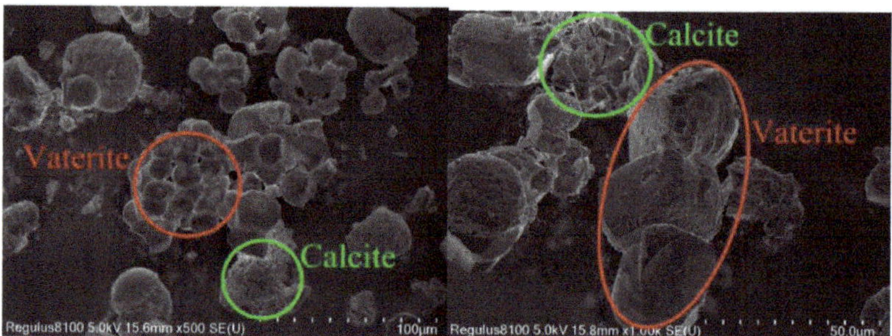

Figure 6. SEM image of calcium carbonate crystal.

3.2. Test Results Uranium Tailing Sand Grouting

3.2.1. The Triaxial Test Results

The strain-controlled triaxial apparatus based on the Mohr–Coulomb strength theory (as shown in Formula (1)) was utilized for the undrained shear test. Confining pressures of 50 kPa, 100 kPa, and 200 kPa were set, and the axial force was applied immediately after the specimen became stable, resulting in the tailings specimen's shear failure under undrained conditions. During the test, critical parameters, such as the shear strength index of tailings, were monitored and analyzed, and their cohesion and internal friction angle were calculated. The stress–strain curves of tailing samples strengthened by each group of tailings under the consolidated undrained triaxial shear test are depicted in Figure 7.

$$\tau = \sigma \cdot \tan\varphi + c \tag{1}$$

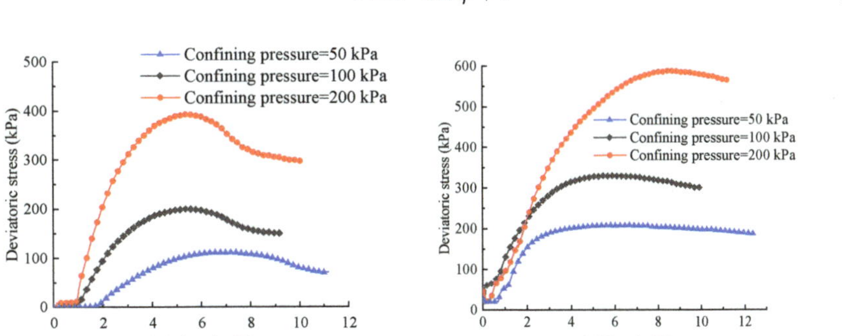

Figure 7. Stress–strain relation curve of tailing samples.

τ is shear strength; σ is normal stress; φ is the angle of internal friction; c is the cohesion.

The stress–strain curve reveals the changing trend in the stress–strain responses of the tailings samples under axial load, broadly categorized into four stages. In the initial loading stage, the pores within the tailing samples begin to compact under axial force, leading to small transverse deformations and a decrease in sample volume over time. This is the pore compaction stage. Moving into the second stage, the load increases and the deviator stress of the sample reaches its peak value, followed by entry into the third stage. In this stage, the deviator stress gradually decreases, and the specimen incurs some damage. Continued loading leads to shear failure along cracks within the sample. Entering the fourth stage, axial deformation increases while deviator stress decreases and gradually stabilizes. Upon reaching the set axial strain value, the specimen experiences damage, although it still displays a certain degree of strength due to mutual embedding and occlusion between particles.

Compared to the unstrengthened group, the deviatoric stress of the MICP grouting group was enhanced. The peak value increased by 1.87 times at a confining pressure of 50 kPa, the peak value increased by 1.65 times at a confining pressure of 100 kPa, and the peak value increased by 1.49 times at a confining pressure of 200 kPa.

The Mohr–Coulomb strength theory formula was used to process triaxial shear strength test data, obtaining the characteristic values of mechanical parameters of tailings samples under different confining pressures, as displayed in Table 2. It can be seen from Table 2 that under the same test conditions, the cohesion of unreinforced uranium tailings is 5.564 kPa and the angle of internal friction is 17.891°. By contrast, the cohesion of the reinforced tailing samples greatly improved. The cohesion of the tailing samples after 12 days of reinforcement increased by 2.937 times compared to the original reinforcement. The internal friction angle increased by 8.393°. This shows that after 12 days of cyclic reinforcement, the reinforcement of tailings had a certain effect.

Table 2. Mechanical characteristic parameters of the triaxial test.

Test Group	Confining Pressure (kPa)	Deviatoric Stress (kPa)	Cohesion (kPa)	Internal Friction Angle (°)
original tail sand	50 100 200	112.15928 200.29918 393.86521	5.564	17.891
Reinforcement (12 days)	50 100 200	199.62468 329.99297 588.21775	21.906	26.284

3.2.2. Comparative Analysis of the Micro-Morphology of Cement

To analyze the microstructure of samples of original tailings and reinforced tailings, a scanning electron microscope (SEM) was employed. From Figure 8, it can be seen that the original tailings have relatively flat and smooth surfaces, without obvious pores and cracks, and there are few surface attachments.

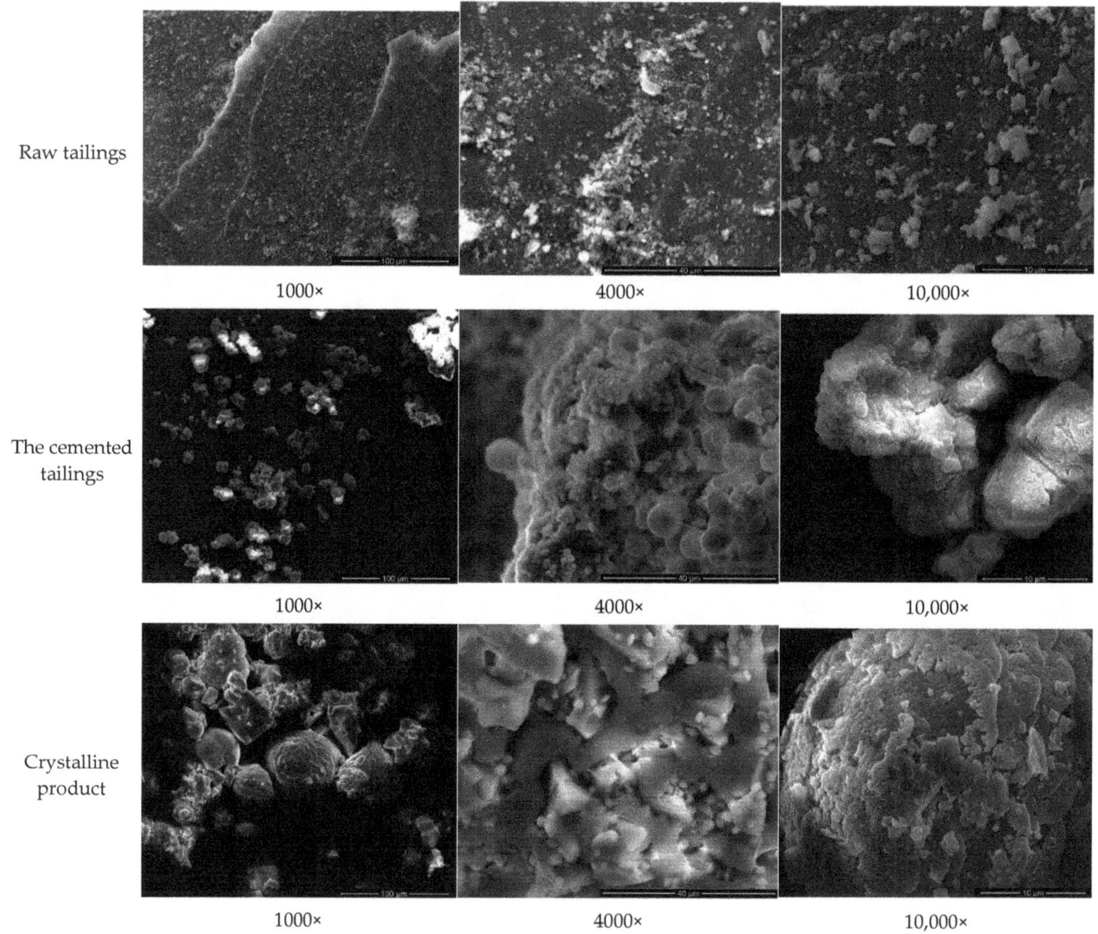

Figure 8. SEM images of the tailings before and after reinforcement.

The SEM image of the cemented sand shows that after cementation, many calcium carbonate crystals adhere to the surface of the tailings, and the crystals are stacked and arranged. Particle aggregation is clearly visible, uneven, and porous. From the 4000× image, it can be seen that calcium carbonate almost completely covers the tailing particles. This coverage and cementation form an intricate stacking pattern that is irregular in connection and stacking, which is beneficial for improving compactness and enhancing the consolidation effect.

It is clear from the SEM image of calcium carbonate crystal that the generated crystal is mainly composed of vaterite with regular and uniform shape and calcite with various shapes (massive and granular) and uneven grain size. Vaterite and calcite are embedded and distributed, stacked in clusters, and bonded together. The image results are consistent with the XRD detection results of mineralized products in the solution experiment.

3.2.3. XRD Detection

Since the SEM test only provides the surface morphology information of calcium carbonate crystals, it cannot be used as a sole basis for determining the crystal form. Therefore, an XRD test was conducted to obtain crystal parameters and characterize the substance structure. The representative samples were selected for XRD detection, and the results are shown in Figure 9. From Figure 9, it can be seen that the composition of the tailing sand particles after cementation is mainly calcite, vaterite, and quartz. The calcium carbonate crystal form mainly consists of calcite and vaterite, which is consistent with the SEM test results.

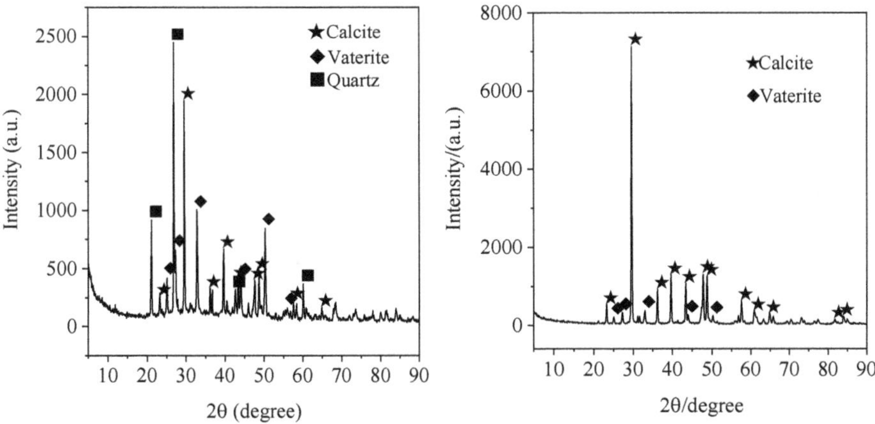

Figure 9. XRD detection diagram of tailings.

3.2.4. Quantitative Characterization of Pore Structure Based on CT Scanning

After being treated with microbial grouting, the uranium tailing samples were scanned using computed tomography (CT) technology. Three-dimensional imaging was utilized to extract information on porosity and particle characteristics, allowing for the analysis of microscopic processes and quantitative characterization of structural changes during the grouting process [19]. The scanned sample data were reconstructed in three dimensions using visualization software to obtain the 3D pore image of the tailing column, as shown in Figure 10. From Figure 10, it can be observed that the pore size and number of tailing samples after 8 days of grouting are significantly greater than those after 12 days of reinforcement. It can also be seen from the figure that the pore volume in the upper part of the sand column is smaller than that in the lower part.

To further quantify the change in porosity inside the tailing column, the porosity of each layer along the Z-axis direction (from bottom to top) of the column at different periods during the grouting process was statistically analyzed using Dragonfly software (Dragonfly 2021.1 Build 977), and the results are shown in Figure 11.

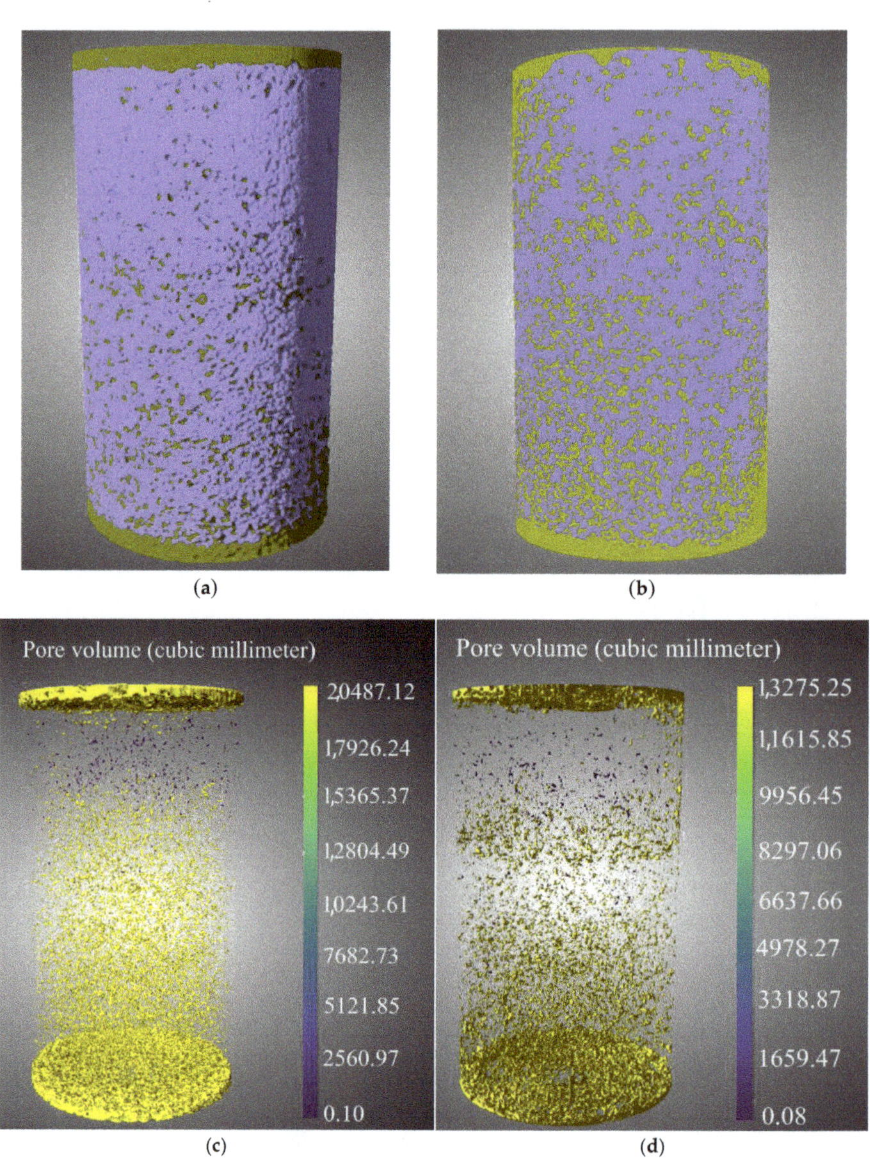

Figure 10. Three-dimensional pore structure of tailing samples. (**a**) Reinforced 8-day 3D re-composition of the tailing column. (**b**) Reinforced 12-day 3D re-composition of the tailing column. (**c**) Reinforced pore volume distribution of the tailing column for 8 days. (**d**) Reinforced pore volume distribution of the tailings column for 12 days. Description: In subfigures (**a**,**b**), purple represents solid particles and yellow represents pores.

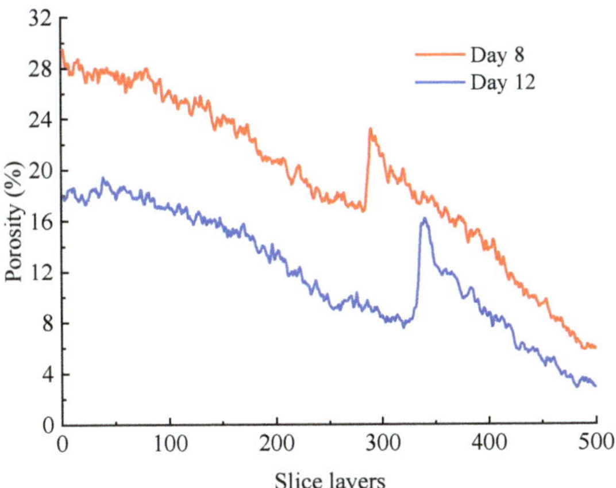

Figure 11. Porosity statistics of tailings layer-by-layer at each grouting time.

According to the statistical analysis shown in Figure 11, The variation interval of layer-by-layer porosity for 8 days of grouting ranged from 4.94% to 29.59%, with an average value of 18.85%. Similarly, after 12 days of grouting, the porosity ranged from 2.51% to 19.477%, with an average value of 11.75%. It can be observed from Figures 10 and 11 that the porosity of the tailing column within the same region increases and then decreases with grouting time, indicating that microbial grouting reinforcement significantly affects porosity. Furthermore, the porosity gradually decreases from the bottom to the top of the tailing column with the increase in the number of Z-axis slices, mainly due to slurry being injected from the upper part of the column and penetrating downward layer-by-layer, resulting in the higher accumulation of calcium carbonate precipitates in the upper part than in the lower part. This is consistent with the previous results shown in Figure 10.

3.2.5. Determination of Changes in Uranium Concentration

To investigate the effect of microorganisms on the immobilization of radioactive uranium in uranium tailings during microbial mineralization, the discharge solution was taken out to measure the uranium concentration every two days. The results are shown in Figure 12.

According to Figure 12, the uranium concentration in the discharge fluid showed a trend of increasing and then decreasing with increasing grouting time. In contrast, the uranium concentration in the discharge fluid rapidly increased and then gradually stabilized for the blank group, which was only injected with pure water. The uranium concentration in the microbial grouting group was slightly lower than that in the blank group: the maximum uranium concentration was 2.405 mg/L. In contrast, the maximum uranium concentration of the microbial grouting group was 2.021 mg/L. Furthermore, it can also be seen from the figure that the uranium concentration in the discharge fluid reached the highest value at 6 days and then started to significantly decrease. Compared to the blank group, a peak uranium concentration was observed in the discharge fluid on the 8th day and remained high. The uranium concentration in the discharge fluid was 0.533 mg/L after 12 days of grouting, while for the blank group, it was 2.308 mg/L, which is 76.91% less than that of the blank group. Therefore, microbial grouting reinforcement of uranium tailings can effectively lock a portion of the uranium, thereby reducing the uranium concentration in the discharge fluid.

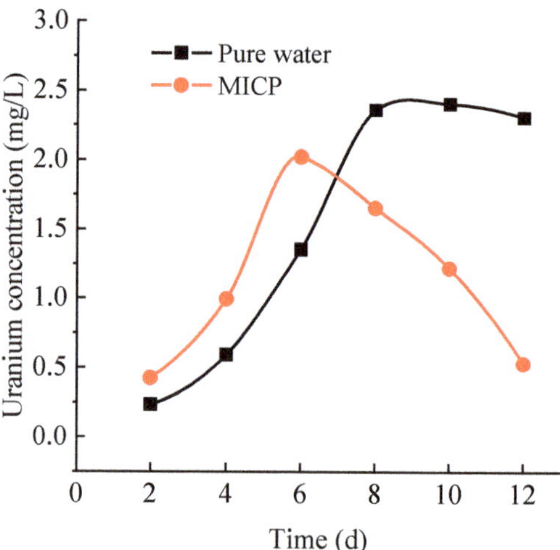

Figure 12. Change curve of uranium concentration.

The reason for the change in uranium concentration in the microbial grouting group was analyzed as follows. At the early stage of grouting, the bacterial population is small, resulting in relatively minimal adsorption and fixation of uranium. Meanwhile, the slurry absorbs the uranium in uranium tailing sand. The higher the uranium concentration in the discharge fluid, the longer the soaking time, so the concentration curve shows an upward trend at this stage. With the increased number of grouting cycles, microorganisms multiply rapidly and generate calcium carbonate precipitates that continuously accumulate, and some uranium is adsorbed and fixed in calcium carbonate precipitates. The calcium carbonate also fills in the tailing sand pores, decreasing porosity and hindering the discharge of liquid containing uranium. As a result, the uranium that can be flushed out by the slurry ultimately decreases, and the curve also shows a decreasing trend.

4. Discussion

This study demonstrates that MICP has a reinforcing effect on uranium tailings, but further investigations are required to control the uranium content in the tailings. The experiments were conducted under environmental conditions with a pH of 5, which may not apply to tailing ponds with extremely acidic characteristics or tailings located deep within the pond. Therefore, the suitability of MICP-reinforced uranium tailings was studied, mainly due to the significantly reduced microbial activity in highly acidic environments. Furthermore, the uranium concentration in uranium tailings decreased, which may be due to the uranium co-precipitated with calcium carbonate in solid crystals or locked in the tailings particles. The uranium was removed via microbial adsorption, and thus further research of the mechanisms of uranium removal and immobilization is required.

5. Conclusions

In this study, solution tests and uranium tailing grouting tests were carried out to systematically investigate the reinforcing effect of MICP on uranium tailing and the control of radionuclides. Based on the results, the following conclusions can be drawn.

The *Sporosarcina pasteurii* grew rapidly, and the bacterial number showed a vigorous linear increase and reproduction under environmental conditions with a pH of 5 and uranium concentration of 2 mg/L. After cementing the solution, the mineralization reactions could be carried out as usual, inducing calcium carbonate precipitation.

The uranium concentration in the solution was effectively reduced during microbially induced calcium carbonate precipitation. The uranium concentration was reduced most rapidly within the first 10 h, and the efficiency of the removal of elemental uranium was 60.9% at 24 h.

The mechanical strength of uranium tailing sand was improved via microbial grouting for 12 days. Compared with the control group, the cohesion of solidified tailing sand increased by 2.937 times, and the internal friction angle increased by 8.393°. Furthermore, the curve characteristics of stress–strain and ordinary consolidated clay curves were alike, and the peak value of stress increased by 1.87 times at the surrounding pressure of 50 kPa.

The CT scan results show that the tailing sand pores are significantly affected by microbial grouting reinforcement. With a higher number of grouting days, the porosity gradually decreased, and the upper porosity was smaller than the lower.

The uranium concentration of the discharge fluid showed a trend of increasing and then decreasing with the higher number of grouting days. After 12 days, compared to the control group, the uranium discharge concentration decreased by 76.91%.

Author Contributions: Conceptualization, L.H. and H.Z.; Data curation, Y.T.; Funding acquisition, Z.Z. and L.W.; Investigation, Y.T.; Methodology, L.W.; Project administration, L.W.; Software, G.H.; Supervision, Z.Z.; Validation, Z.Z. and Q.Y.; Visualization, Q.Y. and G.H.; Writing—original draft, L.H. and H.Z.; Writing—review and editing, L.H. All authors have read and agreed to the published version of the manuscript.

Funding: This research was funded by the Research Foundation of Education Bureau of Hunan Province (grant numbers: 22B0410, 22A0291, 20B496), the National Natural Science Foundation of China (grant numbers: 52274167, 52274127), the Natural Science Foundation of Hunan Province (grant numbers: 2022JJ40374, 2023JJ30516, 2021JJ30580), the Hengyang City Science and Technology Program Project Funding (grant numbers: 202150063769), and the Hunan Province's technology research project "Revealing the List and Taking Command" (grant numbers: 2021SK1050).

Informed Consent Statement: Not applicable.

Data Availability Statement: The data presented in this study are available on request from the corresponding author.

Conflicts of Interest: The authors declare no conflict of interest.

References

1. Tang, C.S.; Yin, L.-Y.; Jiang, N.-J.; Zhu, C.; Shi, B. Factors affecting the performance of microbial-induced carbonate precipitation (MICP) treated soil: A review. *Environ. Earth Sci.* **2020**, *79*, 94. [CrossRef]
2. Lv, C.; Tang, C.S.; Zhang, J.Z.; Pan, X.H.; Liu, H. Effects of calcium sources and magnesium ions on the mechanical behavior of MICP-treated calcareous sand: Experimental evidence and precipitated crystal insights. *Acta Geotech.* **2022**, *18*, 2703–2717. [CrossRef]
3. Wang, Y.K.; Wang, G.; Zhong, Y.H.; Shao, J.G.; Zhao, J.C.; Li, D.B. Comparison of different treatment methods on macro-micro characteristics of Yellow River silt solidified by MICP technology. *Mar. Georesources Geotechnol.* **2022**, *41*, 425–435. [CrossRef]
4. Song, C.; Elsworth, D.; Jia, Y.; Lin, J. Permeable rock matrix sealed with microbially-induced calcium carbonate precipitation: Evolutions of mechanical behaviors and associated microstructure. *Eng. Geol.* **2022**, *304*, 106697. [CrossRef]
5. Banik, N.; Sarkar, R.; Uddin, M.E. Assessment of strength and low-strain shear modulus of bio-cemented sand considering MICP treatment. *Environ. Earth Sci.* **2023**, *82*, 1–20. [CrossRef]
6. Zhao, J.; Shan, Y.; Tong, H.; Yuan, J.; Liu, J. Study on calcareous sand treated by MICP in different NaCl concentrations. *Eur. J. Environ. Civ. Eng.* **2022**, *27*, 3137–3156. [CrossRef]
7. Xiao, R.; Liang, B.Y.; Wu, F.; Huang, L.C.; Lai, Z.S. Biocementation of coral sand under seawater environment and an improved three-stage biogrouting approach. *Constr. Build. Mater.* **2023**, *362*, 129758. [CrossRef]
8. Chen, M.; Li, Y.; Jiang, X.; Zhao, D.; Liu, X.; Zhou, J.; He, Z.; Zheng, C.; Pan, X. Study on soil physical structure after the bioremediation of Pb pollution using microbial-induced carbonate precipitation methodology. *J. Hazard. Mater.* **2021**, *411*, 125103. [CrossRef] [PubMed]
9. Achal, V.; Pan, X.L.; Zhang, D.Y. Remediation of copper-con taminated soil by Kocuria flava CR1, based on microbially induced calcite precipitation. *Ecol. Eng.* **2011**, *37*, 1601–1605. [CrossRef]
10. Li, M.; Cheng, X.H.; Guo, H.X. Heavy metal removal by biomin-eralization of urease producing bacteria isolated from soil. *Int. Biodeterior. Biodegrad.* **2013**, *76*, 81–85. [CrossRef]
11. Jalilvand, N.; Akhgar, A.; Alikhani, H.A.; Rahmani, H.A. Removal of heavy metals zinc, lead, and cadmium by biomineralization of urease-producing bacteria isolated from Iranian Mine cal careous soils. *J. Soil Sci. Plant Nutr.* **2020**, *20*, 206–219. [CrossRef]

12. Waggitt, P. *A Review of Worldwide Practices for Disposal of Uranium Mill Tailings*; Australian Government Publishing Service: Canberra, Australia, 1994.
13. Ting, L.; Qiaofeng, L. Uranium Resources Present Situation and Development Trend of Nuclear Power Around World. *Mod. Min.* **2017**, *4*, 98–103. (In Chinese)
14. Li, G.; Liu, Y.; Zhao, G.; Zang, X. Analysis and Application of Environmental Stability of Decommissioning Uranium Tailings Pond. *Ind. Saf. Environ. Prot.* **2016**, *42*, 68–71. (In Chinese)
15. Lui, R.; Chen, J.; Li, J.; Wang, X.; Huang, M. Study on Countermeasures of Supervision and Radiation Safety of Uranium Mine and Mill in North China. *Uranium Min. Metall.* **2017**, *36*, 151–154.
16. Howe, S.E.; Davidson, C.M.; McCartney, M. Determination of uranium concentration and isotopic composition by means of ICP-MS in sequential extracts of sediment from the vicinity of a uranium enrichment plant. *J. Anal. At. Spectrom.* **2002**, *17*, 497–501. [CrossRef]
17. ASTM. *Standard Classification of Soils for Engineering Purposes (Unified Soil Classification System)*; American Society for Testing and Materials: West Conshohocken, PA, USA, 2006.
18. Zhang, Z.-J.; Tong, K.-W.; Hu, L.; Yu, Q.; Wu, L.-L. Experimental study on solidification of tailings by MICP under the regulation of organic matrix. *Constr. Build. Mater.* **2020**, *265*, 120303. [CrossRef]
19. Saxena, N.; Hofmann, R.; Alpak, F.O.; Dietderich, J.; Hunter, S.; Day-Stirrat, R.J. Effect of image segmentation & voxel size on micro-CT computed effective transport & elastic properties. *Mar. Pet. Geol.* **2017**, *86*, 972–990.

 *sustainability*

Article

Geological Strength Index Relationships with the Q-System and Q-Slope

Samad Narimani [1], Seyed Morteza Davarpanah [1], Neil Bar [2], Ákos Török [1] and Balázs Vásárhelyi [1,*]

1   Department Engineering, Geology & Geotechnics, Faculty of Civil Engineering, Budapest University of Technology and Economics, 1111 Budapest, Hungary
2   Gecko Geotechnics LLC, Kingstown 1471, Saint Vincent and the Grenadines
*   Correspondence: vasarhelyi.balazs@emk.bme.hu

Abstract: The Q-system and Q-slope are empirical methods developed for classifying and assessing rock masses for tunneling, underground mining, and rock slope engineering. Both methods have been used extensively to guide appropriate ground support design for underground excavations and stable angles for rock slopes. Using datasets obtained from igneous, sedimentary, and metamorphic rock slopes from various regions worldwide, this research investigates different relationships between the geological strength index (GSI) and the Q-system and Q-slope. It also presents relationships between chart-derived GSI with GSI estimations from RMR89 and Q' during drill core logging or traverse mapping. Statistical analysis was used to assess the reliability of the suggested correlations to determine the validity of the produced equations. The research demonstrated that the proposed equations provide appropriate values for the root mean squared error value (RMSE), the mean absolute percentage error (MAPE), the mean absolute error (MAE), and the coefficient of determination (R-squared). These relationships provide appropriate regression coefficients, and it was identified that correlations were stronger when considering metamorphic rocks rather than other rocks. Moreover, considering all rock types together, achieved correlations are remarkable.

Keywords: rock mass classification; Q-system; Q-slope; geological strength index (GSI)

Citation: Narimani, S.; Davarpanah, S.M.; Bar, N.; Török, Á.; Vásárhelyi, B. Geological Strength Index Relationships with the Q-System and Q-Slope. *Sustainability* **2023**, *15*, 11233. https://doi.org/10.3390/su151411233

Academic Editors: Hong-Wei Yang, Shuren Wang and Chen Cao

Received: 16 May 2023
Revised: 13 July 2023
Accepted: 17 July 2023
Published: 19 July 2023

1. Introduction

Rock masses can be described as a complex combination of intact rock material separated by geological discontinuities, including joints, bedding planes, veins, shears, and faults. It is practically impossible to identify and characterize every single intact block and discontinuity in a rock mass with respect to the engineering scale of projects (e.g., in tunnels, slopes, and mines).

Rock mass classification systems are a process of grouping or classifying a rock mass based on defined relationships [1] and assigning it a unique description (and number) based on similar geomechanical properties or characteristics so that its behavior may be predicted. Rock mass classifications and design charts are particularly useful for providing:

- Assessment of ground conditions by converting engineering geological descriptions to "numbers" which can be used for engineering purposes;
- Fast prediction of underground excavation and slope performance;
- Guidance on support requirements and stable slope geometry.

The boundaries of the structural zones are typically defined by a significant structural element, such as a fault or a change in the type of rock. Within the same rock type, there may be instances where significant variations in discontinuity spacing or features call for the partitioning of the rock mass into a number of tiny structural sections.

Empirical methods, including rock mass classifications, are most effective when the geometry, geology, hydrogeology, and geomechanical characteristics of the engineering problem (underground excavations, slopes, etc.) under investigation are similar to the known performance of precedent engineering problems [2]. An excavation will be usable for a certain amount of time without support; after that, major caving and failure may happen [3]. The only basis for empirical design is experience or observation; it is not based on any theory or method of science. Its use in engineering design focuses on comparing the outcomes of prior trials to project future behavior based on the most crucial components of the design.

The design process using rock mass classification (Q-slope for rock slopes) and calculation of the rock mass behavior (using the geological strength index and the Hoek–Brown failure criterion) is shown in Figure 1 [4]. As shown, to design the slope's angle, rock mass classification must be provided by using the Q-slope method, which needs field investigations, and describing block size, roughness, and project factors. In terms of rock mass characterization, the well-known method is the geological strength index introduced by Hoek–Brown. The necessary parameters to analyze the stability of slope in this method are friction, cohesion and uniaxial compressive strength of intact rock.

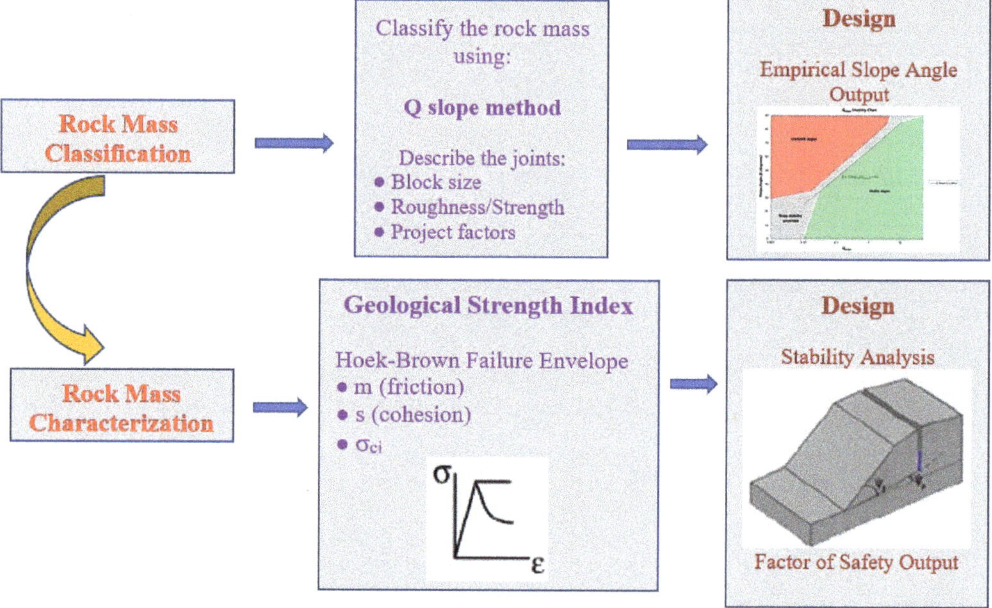

Figure 1. Rock mass characterization vs. Classification for rock slopes [4].

In the engineering design of rock slopes, it is assumed that all four input parameters (GSI, m, s, and σ_{ci}) may be represented by normal distributions. The standard deviations given to these four distributions are based on geotechnical program experience for significant civil and mining projects where sufficient funding is available for high-quality studies [5].

Rock mass behavior is mainly characterized by strength and deformation modulus of rock mass. In order to provide a good estimation of rock mass behavior, geological strength index plays an important role. Providing accurate values, the error in designing rock slope stability would be minimized and lead to realistic safety factor, as discussed by Ván and Vásárhelyi [6]. They analyzed the sensitivity of GSI-based equations and discovered that relationships are extremely reliant on the input parameters and that changing one parameter by 5% could significantly impact the outcomes. The well-known practical GSI chart provided by Hoek–Brown and non-linear failure criteria can be used for further design as shown in Figure 2.

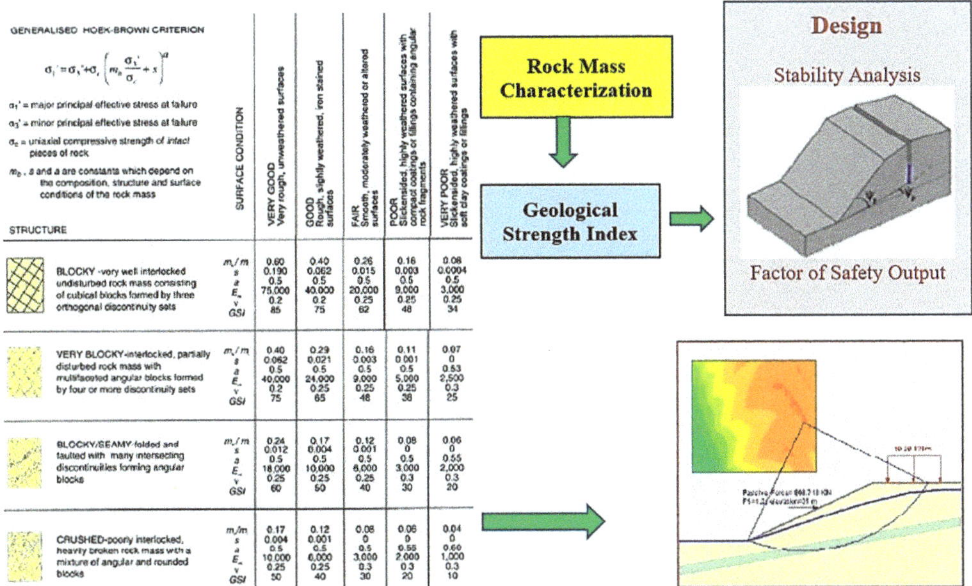

Figure 2. Estimation of constants m_b/m_i, s, a, deformation modulus E, and the Poisson's ratio (ν) for the generalized Hoek–Brown failure criterion based upon rock mass structure and discontinuity surface conditions used for the design stage [4].

Furthermore, several researchers established, improved, and updated several rock mass rating methods for tunnel support design. These classification systems also have applications in diverse engineering projects including rock slopes [7–12].

Santos et al. [13] showed using machine learning techniques that it is possible to achieve an accurate RMR classification system using only factors that are directly connected to rock mass quality rather than all RMR variables.

2. Rock Mass Classification System

The most commonly used rock mass classifications include rock mass rating $(RMR)_{89}$ [1], rock mass quality (Q-system) [14] and geological strength index (GSI) [15]. Since its relatively recent introduction, the Q-slope method for rock slope engineering [16] has also become commonly used for slope stability appraisals. With the use of Q-slope, engineering geologists and rock engineers can evaluate the stability of rock slopes that have been excavated in the field and possibly change the slope angles when the condition of the rock mass becomes clear during construction. The correlation between the geological strength index (GSI) and the rock mass rate was also recently analyzed by Somodi et al. [17].

RMR$_{89}$ and Q are empirical methods that directly guide underground excavation support requirements. Similarly, Q-slope provides direct guidance for long-term stable slope angles. These methods have been used by visually assessing tunnel and slope faces, as well as for the characterization of drill core. The GSI system on the other hand, is an input parameter to the Hoek–Brown failure criterion [18].

Hoek and Brown [19] expanded their intact rock failure criterion for applicability to homogeneous and isotropic rock masses under the assumption that the same type of non-linear envelopes continued to be valid. Rock mass rating, RMR$_{76}$, and modified rock mass quality, Q', were initially utilized to downgrade intact rock to rock mass strength in the Hoek–Brown failure criterion [20].

2.1. Geological Strength Index (GSI)

Unless a rock mass failure criterion could be connected to a geological description that engineering geologists could make quickly, it would be of no practical use. The geological strength index (GSI) is a system for classifying rocks that was created in the field of rock engineering to address the requirement for accurate estimation of the attributes of rocks for the design of engineering projects [15]. The foundation of the GSI system is an in-depth engineering geology description of the rock mass encountered in engineering projects. Structure and the condition of discontinuities in the rock mass, which can be determined through visual inspection of the rock mass exposed in outcrops, are two essential criteria that determine the value of the GSI [21]. The basic GSI chart's use entails some subjectivity because there are not any quantifiable or more representative metrics, linked interval limits, or ratings for describing the rock structure and surface conditions of the discontinuities. In order to make the method easier to use, particularly for new engineers, numerous publications [18,22,23] presented quantitative GSI charts to quantify the estimation of GSI. In circumstances when there are only one or two sets of discontinuities, the GSI should not be utilized. Instead, it should only be employed when the rock mass is intact or heavily joined. When working with blocky rock masses that have minimum anisotropy, extreme caution should be used. However, it is occasionally reasonable to assign the GSI to the entire rock mass and include the single discontinuity as a layer element in the numerical model for a rock mass that has a single well-defined discontinuity [20]. According to the definition of GSI and subsequent application, it is independent of the water inflow and the in situ stress state. As mentioned, GSI has limits, as do any categorization systems, including the differences in the results of individuals. Others have suggested that the generic GSI may not appropriately reflect the size of the rock mass problem because its parameters, particularly RQD and J_{Cond}, are scale-dependent [24,25].

The two main geological elements behind GSI were the surface conditions of discontinuities and their shear strength, as well as the rock mass structure, which represented the degree of fracture [26]. GSI subsequently replaced RMR in the Hoek–Brown failure criterion. To encourage early adoption, Hoek et al. [15,27] provided the following relationships for estimating GSI, termed GSI_{calc} in this paper from RMR and Q':

$$GSI_{calc} = RMR_{89} - 5 \tag{1}$$

for $RMR_{89} \geq 23$

$$GSI_{calc} = 9\ln Q' + 44 \tag{2}$$

for $RMR_{89} < 23$

GSI was developed to assess tunnel and slope faces using charts (Figure 3) and cannot be estimated directly from the characterization of the drill core. Equations (1) and (2) also enable the estimation of GSI_{calc} from the drill core.

Hoek et al. [19] suggested quantifying the GSI chart as a function of RQD and the joint condition parameters presented by either [1]: $JCond_{89}$ from RMR_{89} or Barton et al. [14]; J_r and J_a from Q and Q-slope, on the premise that professionals with less expertise and who are "less at ease with the qualitative descriptions" may utilize this new GSI chart. The suggested equations included [18]:

$$GSI_{2013} = 1.5 JCond_{89} + RQD/2 \tag{3}$$

$$GSI2013 = \frac{52 J_r/J_a}{\left(1 + \frac{J_r}{J_a}\right)} + RQD/2 \tag{4}$$

Vásárhelyi et al. [28] and Somodi et al. [29] discovered that visual observation by an expert engineering geologist is the best technique to estimate GSI, compared to the computational and estimation methods examined. Moreover, Deák et al. [30] and Somodi et al. [31] analyzed the relationship between the parameters of the Q-system and the GSI value and the equations are suggested based on the empirical results of different rock engineering projects in Hungary and Australia.

Bertuzzi et al. [32] observed a decent connection between the GSI values derived by the chart (GSI_{chart}) and the GSI quantified (GSI_{2013}) using the Hoek et al. [18] technique (Figure 3). Santa et al. [33] compared the differences between the latest edition of GSI_{2013} and the prior version (GSI_{98}) in an anisotropic media. They discovered that the 2013 version [18] allocated lower GSI ratings to rock masses with lesser quality and higher values to rock masses with greater quality. Winn et al. [34] revealed that the three alternative techniques proposed by [22,23] for a sedimentary rock mass in a Singapore dive, a variety of outcomes comparable to the field-assessed qualitative GSI, showing their good performance. However, Winn et al. [34] established a new GSI relationship by replacing RQD/2 with RQD/3 in the calculation of GSI_{2013}, which provided substantially higher GSI values that did not correspond to the qualitative field values.

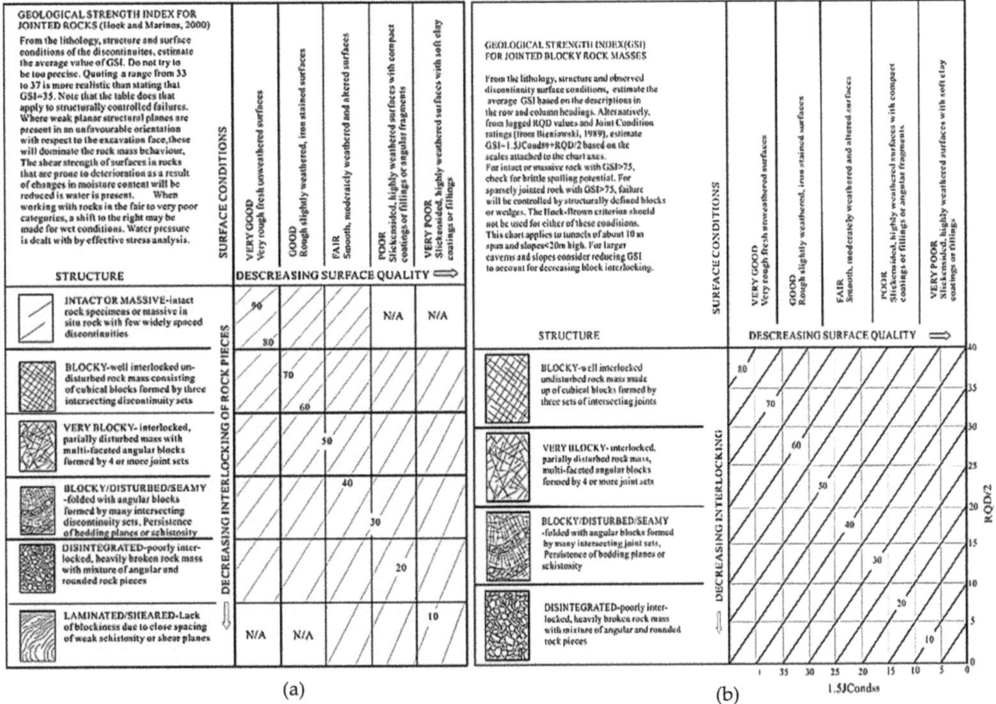

Figure 3. (**a**) GSI chart for jointed rocks [35]. (**b**) Quantification of GSI through joint conditions and RQD [18].

Yang and Elmo [36] argue against the quantification paradigm for GSI determination without taking into account the constraints of the idea that GSI quantification approaches may convert subjectivity into objectivity since the parameters under consideration are not quantitative measurements.

2.2. The Q-Slope Method

Barton and Bar [37] are credited with developing the Q-slope, an empirical engineering method for assessing the stability of rock slopes that allows for immediate access to stability with few presumptions. It is descended from the Q-system, which has been used globally for over 40 years to characterize rock exposures, drill core, and tunnels that are under construction [14,38]. Q-slope allows engineering geologists and rock engineers to analyze the stability of excavated rock slopes in the field and make prospective slope angle adjustments as rock mass conditions become obvious during construction [37,39]. The Q_{slope} used Q system components to slope stability assessments are calculated using the expression [16]:

$$Q_{slope} = \frac{RQD}{J_n} \times \left(\frac{J_r}{J_a}\right)_O \times \frac{J_{wice}}{SRF_{slope}} \tag{5}$$

RQD, J_n, J_r, and J_a are the same four parameters used in Q-system. On the other hand, the frictional resistance pair Jr and Ja as necessary, each side of a wedge that may be unstable will receive this treatment. If suitable, simple orientation variables such as $(J_r/J_a)_O$ give an estimate of the reduction in overall whole-wedge frictional resistance. The term J_w, which has now been modified to J_{wice}, refers to a broader range of environmental conditions ideal for rock slopes exposed to the elements for a long time. Extremes of erosive rains and ice wedging, which can occur periodically at opposing extremities of the rock-type and regional spectrum, are examples of these situations. Slope-relevant SRF classifications are also available in slope surface conditions, stress–strength ratios and major discontinuities such as faults, weakness zones or joint swarms [37].

Q′ and Q-slope′ are estimated using Equations (6) and (7), which merely remove the water, environment, stress, and strength reduction factors from the full Q-system and Q-slope equations:

$$Q' = \left(\frac{RQD}{J_n}\right) \times \left(\frac{J_r}{J_a}\right) \tag{6}$$

$$Q\text{-slope}' = \left(\frac{RQD}{J_n}\right) \times \left(\frac{J_r}{J_a}\right)_A \times \left(\frac{J_r}{J_a}\right)_{B \text{ where applicable}} \tag{7}$$

where RQD = rock quality designation (%).

- J_n = joint set number.
- J_r = joint roughness number.
- J_a = joint alteration number.

Q-system and Q-slope ratings for RQD, J_n, J_r, and J_a are described by Barton et al. [14] and Bar and Barton [16]. These parameters are commensurate with the primary factors in GSI: rock mass structure and discontinuity surface conditions.

The main distinction between Q and Q-slope is that Q uses J_r/J_a from the collection of discontinuities with the least favorable discontinuities. In contrast, J_r/J_a for both sides (A and B) of the wedge can be considered necessary when wedges are encountered while utilizing Q-slope. In order to predict various parameters from others in material sciences and engineering properties of various rock types from their petrographic characteristics in engineering geology and rock mechanics, several researchers have attempted to construct various soft computing models [40–43]. Recent research has examined the relationships between the Q-system, Q-slope, and GSI as well as for other rocks by employing statistical analyses and various soft computing techniques such as root mean squared error value (RMSE), the mean absolute percentage error (MAPE), mean absolute error (MAE), and coefficient of determination (R-squared).

3. Study Area

The current study reviews over 192 case records collected from across 11 countries on five continents to investigate the relationship between the GSI obtained using the popular chart methods in more depth. (GSI_{chart}) and its correlation with Q′ and Q-slope′.

The case records were obtained from rock slope faces that were assessed during the Q-slope method development [16] from a range of mining commodities, including gold, copper, iron ore, coal, and diamonds, and civil engineering applications, including roads, highways, and quarries. GSI was estimated for these case studies using GSI chart for jointed rocks (GSI_{chart}). Figure 4 graphically illustrates the results from a selection of case studies. The case records include over 20 different rock types in vastly different engineering geological, environmental, and climatic settings:

- Igneous rocks include andesite, basalt, diorite, granite, kimberlite, monzodiorite, rhyolite, and tuff.
- Sedimentary rocks include chert, greywacke, limestone, mudstone, siltstone, sandstone, and banded iron formation.
- Metamorphic rocks include marble, metasandstone, phyllite, quartzite, schist, and shale.

Figure 5 presents a selection of photographs of rock slopes hosted within vastly different rock masses and their respective GSI, Q-slope, and Q-slope' ratings for context:

- Case records A and B are both granites. A is massive whilst B is blocky.
- Case records C, D, and E are all siltstones, which are blocky, seamy, and very blocky, respectively. Joint surface quality reduces from very good to good in C and E to fair in F.
- Case record F is a sheared mudstone comprising claystones and siltstones with slickensided, graphitic infilling along bedding planes and bedding shears.
- Case records B and C have the same GSI, but vastly different Q-slope values, due to the orientation of the discontinuities.
- Case records D and E have different GSI and Q-slope' values, but very similar Q-slope values. Both are stable slopes with bench face angles of approximately 65°.

Data analysis was conducted using SPSS23 software to outline the relations between parameters and describe the frequency distribution of GSI values [44]. Three rock types were picked and tested independently, and reliability analyses were also made. Figure 6 indicates the frequency distribution of GSI values, and Table 1 indicates the statistical analysis of the data for all the rocks and each type of rocks separately. According to Table 1, GSI values range between 20 and 90 for all the rock types, indicating that our datasets include very good through very poor rock mass.

Table 1. Statistical analysis of measured GSI values.

	Number of Estimates	Median	Minimum	Maximum
Igneous	26	65	27	89
Sedimentary	139	55	20	85
Metamorphic	27	60	30	90
Total	192	55	20	90

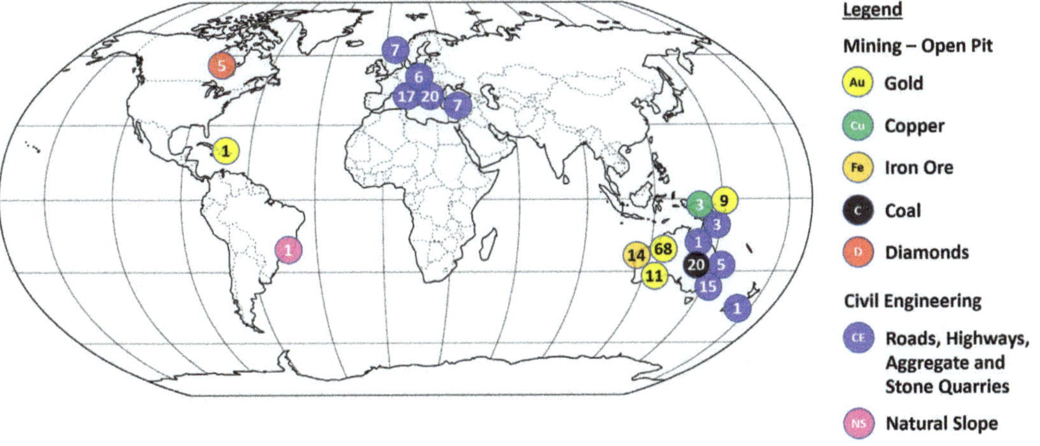

Figure 4. Case records: geographic location and industry and commodity. (The numbers indicate the locations of sampling).

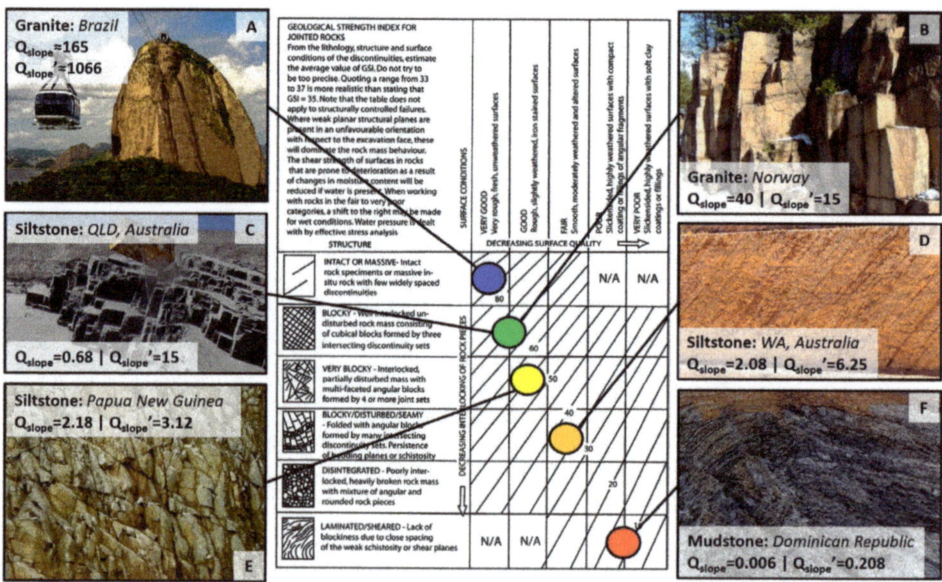

Figure 5. Selection of case studies comparing GSI, Q-slope, and Q-slope'.

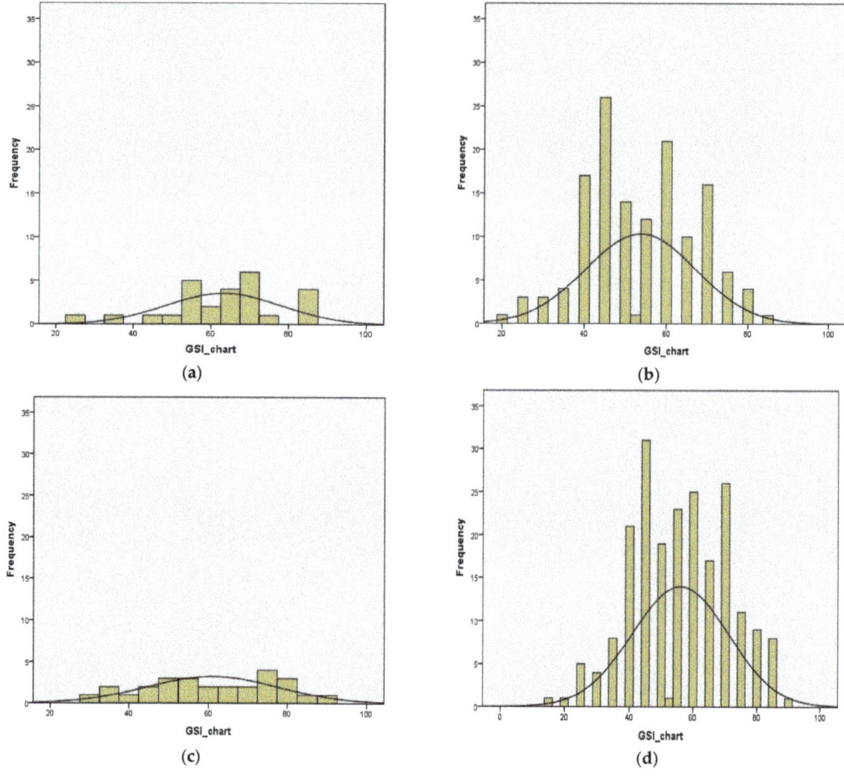

Figure 6. Histogram of measured GSI for (**a**) igneous rocks, (**b**) sedimentary rocks, (**c**) metamorphic rocks, and (**d**) all the rock types.

4. Derived Relationships between the Different Methods

4.1. Correlation between GSI and Q′

The Q′ values were calculated by using Equation (6) and correlated to the GSI_{chart}. These results are summarized in Figure 7. Figure 7a shows the correlation for three types of rocks (igneous, sedimentary, and metamorphic) separately, and Figure 7b shows the correlation for all rocks together. The more reliable correlation relates to metamorphic rocks ($R^2 = 79\%$). Also, for all cases, the best fitting is in logarithmic. These relationships have similarities with Equation (2).

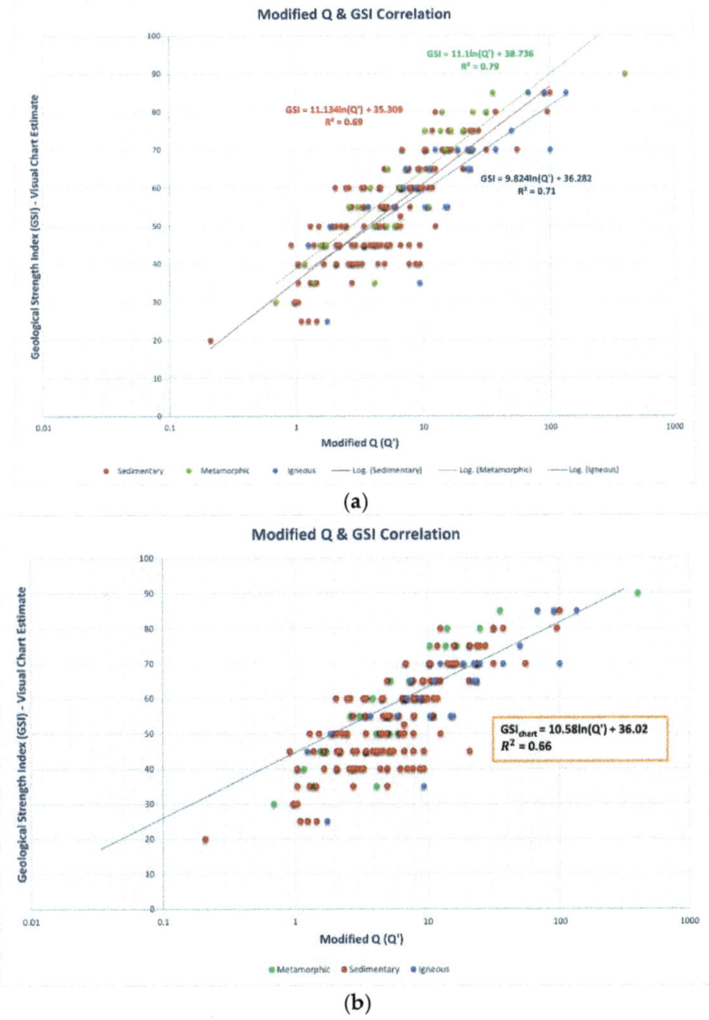

Figure 7. Relationships between GSI_{chart} and Q′: (**a**) separated for igneous, sedimentary, metamorphic rocks, (**b**) for all rocks.

4.2. Correlation between GSI and Q-Slope′

In addition, using Equation (7), the Q-slope′ values were calculated and correlated to the GSI_{chart}. Figure 8 summarizes these findings. Figure 8a shows the correlation for three different types of rocks (igneous, sedimentary, metamorphic), while Figure 8b shows the correlation for all rocks combined. The more reliable correlation is with metamorphic rocks ($R^2 = 63\%$). Also, in all cases, the best fitting is logarithmic. These relationships also have similarities with Equation (2).

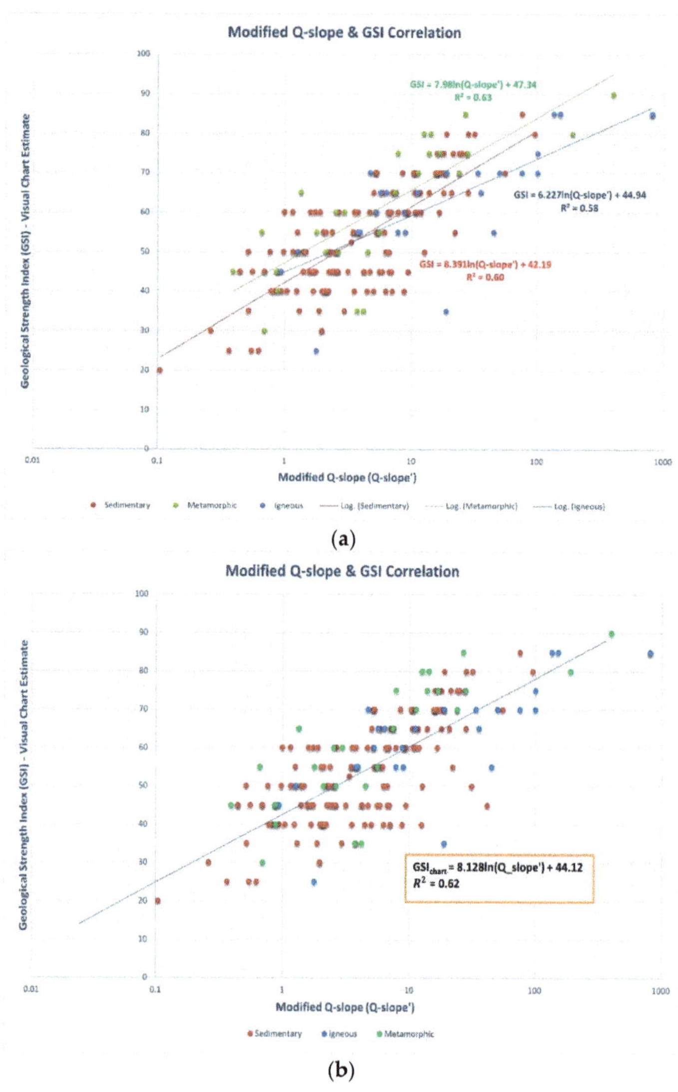

Figure 8. Relationships between GSI_{chart} and Q-slope': (**a**) separated for igneous, sedimentary, metamorphic rocks, (**b**) for all rocks.

4.3. Correlation between GSI_{chart} and GSI_{2013}

Furthermore, Equation (3) was used to produce the GSI_{2013} values, which were then compared with GSI_{chart}. Figure 9 summarizes the findings. Figure 9a depicts the correlation for three different types of rocks (igneous, sedimentary, metamorphic), while Figure 9b depicts the correlation for all rocks. Metamorphic rocks have a more reliable correlation ($R^2 = 88\%$). In addition, the optimal fitting is linear in all circumstances.

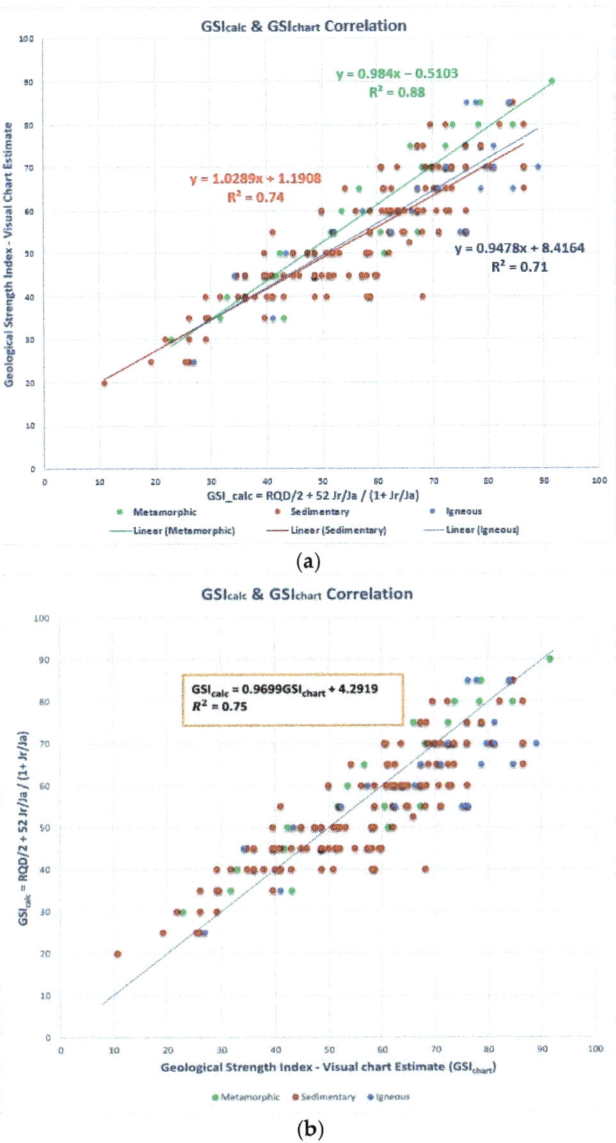

(a)

(b)

Figure 9. Relationships between GSI_{chart} and GSI_{2013}: (**a**) separated for igneous, sedimentary, metamorphic rocks, (**b**) for all rocks.

5. Key Findings

Several different relationships between GSI (GSI_{chart}) and Q′ and Q-slope′ have been identified using over 200 new case records. The relationship between the GSI value and the Q value is:

$$GSI = A \ln(Q) + B, \tag{8}$$

where A and B are material constants, depending on the rock type [30].

- For Q′: A and B values range from 9.82 to 11.13 and from 35.31 to 38.74, respectively.

- For Q-slope': A and B values range from 6.23 to 8.39 and from 42.19 to 47.34, respectively.

All derived relationships are summarized in Table 2.

Table 2. Empirical equations derived in this study.

Types of Correlation	Equation	R^2
GSI_{chart} & Q' for Metamorphic rocks	$GSI = 11.10\ln(Q') + 38.74$	0.79
GSI_{chart} & Q' for Sedimentary rocks	$GSI = 11.13\ln(Q') + 35.31$	0.69
GSI_{chart} & Q' for Igneous rocks	$GSI = 9.82\ln(Q') + 36.28$	0.71
GSI_{chart} & Q' for all rocks	$GSI = 10.58\ln(Q') + 36.02$	0.66
GSI_{chart} & Q-slope' for Metamorphic rocks	$GSI = 7.98\ln(Q\text{-slope}') + 47.34$	0.63
GSI_{chart} & Q-slope' for Sedimentary rocks	$GSI = 8.39\ln(Q\text{-slope}') + 42.19$	0.60
GSI_{chart} & Q-slope' for Igneous rocks	$GSI = 6.23\ln(Q\text{-slope}') + 44.94$	0.58
GSI_{chart} & Q-slope' for all rocks	$GSI = 8.13\ln(Q\text{-slope}') + 44.12$	0.62
GSI_{chart} & GSI_{2013} for Metamorphic rocks	$GSI_{2013} = 0.98(GSI_{chart}) - 0.51$	0.88
GSI_{chart} & GSI_{2013} for Sedimentary rocks	$GSI_{2013} = 1.03(GSI_{chart}) + 1.19$	0.74
GSI_{chart} & GSI_{2013} for Igneous rocks	$GSI_{2013} = 0.95(GSI_{chart}) + 8.41$	0.71
GSI_{chart} & GSI_{2013} for all rocks	$GSI_{2013} = 0.97(GSI_{chart}) + 4.29$	0.75

6. Quantitative Relationships and Errors

Four common statistical metrics, including determination coefficient (R^2), root mean square error (RMSE), mean absolute error (MAE), and mean absolute percent error (MAPE), were used to assess the statistical efficiency of the GSI_{chart} prediction in the training and testing sets [45]. Equations (9)–(11) specify the following performance measures:

$$RMSE = \sqrt{\frac{\sum_{i=1}^{n}(pi - qi)^2}{n}} \tag{9}$$

$$MAPE = \frac{1}{n}\sum_{i=1}^{n}\left|\frac{qi - pi}{qi}\right| \times 100 \tag{10}$$

$$MAE = \frac{\sum_{i=1}^{n}|pi - qi|}{n} \tag{11}$$

Here, p_i and q_i stand in for the i_{th} predicted and expected results, respectively, and n denotes the total number of experiments. Since significant errors are dealt with far more successfully than smaller ones, RMSE is a widely used metric. When the RMSE approaches 0, it means that the prediction error was minimal. However, it does not always guarantee top performance. Additionally, MAE was calculated and is incredibly helpful when data are smooth and continuous [46].

Plotting the residuals vs. the fitted values of the dependent variable allows one to assess the error of the proposed model. The residuals are the discrepancies between the experimental outcomes and the values that the suggested model predicted. The values that have been fitted are those that the proposed model anticipated. Figure 10 illustrates the residuals' balanced and symmetrical dispersion about the horizontal axis.

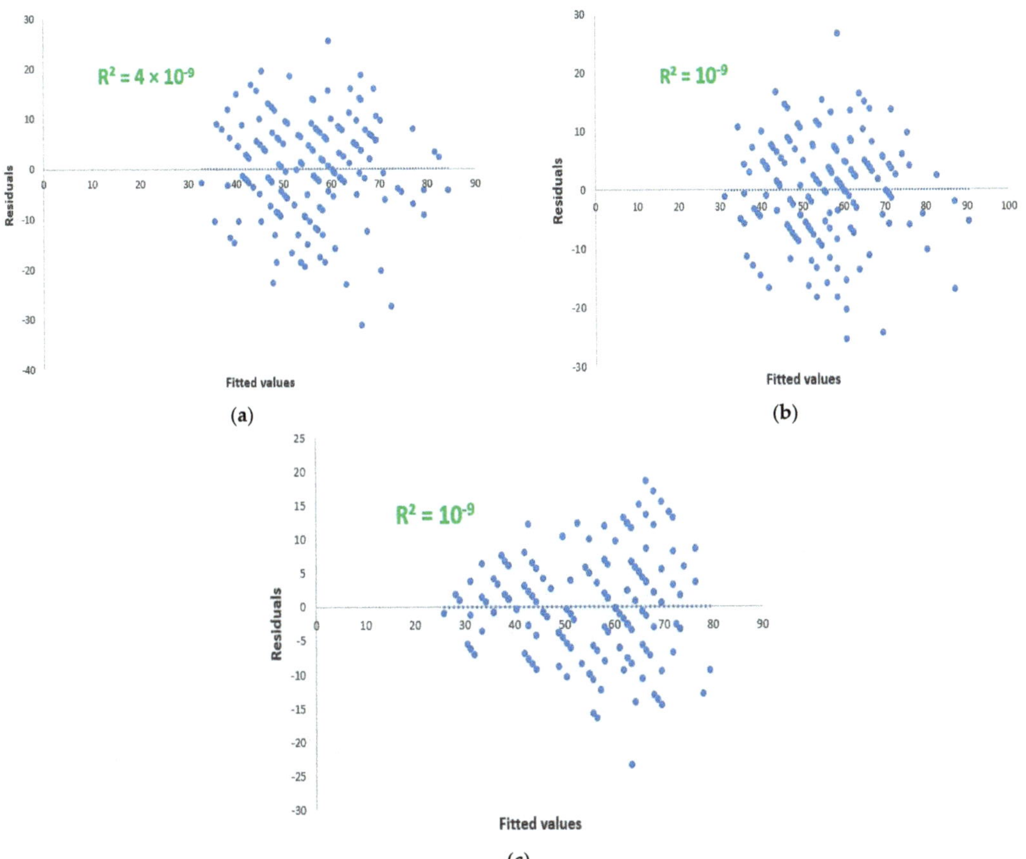

Figure 10. Distribution of the residuals against fitted values: (**a**) between GSI_{chart} and Q-slope', (**b**) GSI_{chart} and Q', (**c**) GSI_{chart} and GSI_{calc}.

Additionally, the residuals' outermost points resemble a circular in shape. The residuals' independent nature and random distribution around the centerline are confirmed by this distribution [47,48]. To make sure of this, the determination coefficient, R-squared, is computed. It turns out to be very nearly equivalent to zero. In other words, the residuals are not dependent on one another. The proposed model's goodness of fit is therefore excellent.

The calculations: These new statistical indicators that provide quantitative connections between GSI_{chart} and Q' and Q-slope', as well as GSI_{2013}, are given in Table 3.

Table 3. Statistical indicators for Q-slope' vs. GSI_{chart}, Q' vs. GSI_{chart}, and GSI_{calc} vs. GSI_{chart} models.

Statistical Metrics	Q-Slope' vs. GSI_{chart}		Q' vs. GSI_{chart}		GSI_{calc} vs. GSI_{chart}	
	Training	Testing	Training	Testing	Training	Testing
RMSE	9.93	10.8	8.5	9.56	7.47	7.85
MAPE	0.16	0.18	0.13	0.16	0.11	0.11
MAE	7.87	8.5	6.69	7.8	5.95	5.9

7. Conclusions

The relationships between the geological strength index (GSI) and the Q-system and Q-slope are explored in this paper. Quantitative correlation analyses have been carried out systematically using study data from various igneous, metamorphic, and sedimentary rocks.

Validated results show that the proposed simplified quantitative correlations accurately reflect the observed relationship between the mentioned parameters.

The statistical measures employed to assess the effectiveness of the model were mean absolute error (MAE), mean absolute percentage error (MAPE), and root means square error (RMSE). These indicators have various relationships: the logarithmic equations provide good forecasting results between GSI_{chart} and Q' and Q-slope', while the linear equations between GSI_{chart} and GSI_{2013} fall in the "stronger prediction" category.

These correlations can be used to assess the surrounding rock mass of various rock types in various places. While these and other connections are likely relevant elsewhere in similar ground conditions, readers are strongly recommended to validate them at their local site before using them. Thus, it is strongly encouraged that engineering geologists and geotechnical engineers have to develop a site-specific correlations chart at various locations.

Author Contributions: S.N.: structure and writing, S.M.D.: calculation, N.B.: in situ measurements and methodology, Á.T.: processing and reviewing, and B.V.: methodology and conceptualization. All authors have read and agreed to the published version of the manuscript.

Funding: The support provided by the Ministry of Culture and Innovation of Hungary from the National Research, Development and Innovation Fund, financed under the TKP2021-NVA funding scheme (project no. TKP-6-6/PALY-2021) is acknowledged.

Institutional Review Board Statement: Not applicable.

Informed Consent Statement: Informed consent was obtained from all subjects involved in the study.

Data Availability Statement: Data will be made available on request.

Conflicts of Interest: The authors declare no conflict of interest.

References

1. Bieniawski, Z.T. *Engineering Rock Mass Classifications: A Complete Manual for Engineers and Geologists in Mining, Civil, and Petroleum Engineering;* Wiley: New York, NY, USA, 1989.
2. Duncan, J.M. State of the Art: Limit Equilibrium and Finite-Element Analysis of Slopes. *J. Geotech. Eng.* **1996**, *122*, 577–596. [CrossRef]
3. Lauffer, H. Gebirgsklassifizierung für den Stollenbau. *Geol. Bauwes.* **1958**, *24*, 46–51.
4. Eberhardt, E. Geological Engineering Practice I—Rock Engineering, Lecture 5 [PowerPoint Slides]. 2017. Available online: https://www.eoas.ubc.ca/courses/eosc433/lecture-material/L5-EmpiricalDesign.pdf (accessed on 1 June 2017).
5. Hoek, E. Reliability of Hoek-Brow estimates of Rock Mass properties and their impact design. *Int. J. Rock Mech. Min. Sci. Geomech. Abst.* **1998**, *35*, 63–68. [CrossRef]
6. Ván, P.; Vásárhelyi, B. Sensitivity analysis of GSI based mechanical parameters of the rock mass. *Period. Polytech. Civ. Eng.* **2014**, *58*, 379–386. [CrossRef]
7. Romana, M.; Serón, J.B.; Montalar, E. SMR Geomechanics Classification: Application, Experience and Validation. In Proceedings of the 10th ISRM Congress, Sandton, South Africa, 8–12 September 2003.
8. Romana, M. New adjustment ratings for application of Bieniawski classification to slopes. In Proceedings of the International Symposium on Role of Rock Mechanics, Zacatecas, Mexico, 2–4 September 1985; pp. 49–53.
9. Tomás, R.; Delgado, J.; Serón, J. Modification of slope mass rating (SMR) by continuous functions. *Int. J. Rock Mech. Min. Sci.* **2007**, *44*, 1062–1069. [CrossRef]
10. Tomas, R.; Cuenca, A.; Cano, M.; García-Barba, J. A graphical approach for slope mass rating (SMR). *Eng. Geol.* **2012**, *124*, 67–76. [CrossRef]
11. Taheri, A.; Tani, K. A Modified rock mass classification system for preliminary design of rock slopes. In Proceedings of the 4th Asian Rock Mechanics Symposium, Singapore, 8–10 November 2006.
12. Taheri, A. A rating system for preliminary design of rock slopes. In Proceedings of the 41st Japan Geotechnical Society Conference (JGS), Kagoshima, Japan, 12–15 July 2006.
13. Santos, A.E.M.; Lana, M.S.; Pereira, T.M. Evaluation of machine learning methods for rock mass classification. *Neural Comput. Appl.* **2022**, *34*, 4633–4642. [CrossRef]

14. Barton, N.R.; Lien, R.; Lunde, J. Engineering classification of rock masses for the design of tunnel support. *Rock Mech.* **1974**, *6*, 189–239. [CrossRef]
15. Hoek, E. Strength of rock and rock masses. *ISRM News J.* **1994**, *2*, 4–16.
16. Bar, N.; Barton, N. The Q-slope Method for Rock Slope Engineering. *Rock Mech. Rock Eng.* **2017**, *50*, 3307–3322. [CrossRef]
17. Somodi, G.; Bar, N.; Kovács, L.; Arrieta, M.; Török, Á.; Vásárhelyi, B. Study of Rock Mass Rating (RMR) and Geological Strength Index (GSI) Correlations in Granite, Siltstone, Sandstone and Quartzite Rock Masses. *Appl. Sci.* **2021**, *11*, 3351. [CrossRef]
18. Hoek, E.; Carter, T.G.; Diederichs, M.S. Quantification of the geological strength index chart. In Proceedings of the 47th US Rock Mechanics/Geomechanics Symposium—ARMA 2013 (ARMA 13–672), San Francisco, CA, USA, 23–26 June 2013.
19. Hoek, E.; Brown, E.T. Empirical Strength Criterion for Rock Masses. *J. Geotech. Eng.* **1980**, *106*, 1013–1035. [CrossRef]
20. Hoek, E.; Brown, E.T. The Hoek-Brown Failure Criterion—A 1988 Update. In Proceedings of the 15th Canadian Rock Mechanics Symposium; University of Toronto: Toronto, ON, Canada, 1988; pp. 31–38.
21. Jianping, Z.; Jiayi, S. *The Hoek-Brown Failure Criterion—From Theory to Application*; Springer: Berlin/Heidelberg, Germany, 2018.
22. Sonmez, H.; Ulusay, R. Modifications to the Geological Strength Index (GSI) and their applicability to the stability of slopes. *Int. Rock Mech. Min. Sci.* **1999**, *36*, 743–760. [CrossRef]
23. Cai, M.; Kaiser, P.K.; Uno, H.; Tasaka, Y.; Minami, M. Estimation of Rock Mass Deformation Modulus and Strength of Jointed Hard Rock Masses using the GSI system. *Int. J. Rock Mech. Min. Sci.* **2004**, *41*, 3–19. [CrossRef]
24. Hudson, J.; Harrison, J. *Engineering Rock Mechanics*; Pergamon: Bergama, Turkey, 1997.
25. Mostyn, G.; Douglas, K. Strength of intact rock and rock masses. In Proceedings of the ISRM International Symposium, Melbourne, Australia, 19–24 November 2000.
26. Renani, H.R.; Cai, M. Forty-Year Review of the Hoek-Brown Failure Criterion for Jointed Rock Masses. *Rock Mech. Rock Eng.* **2022**, *55*, 439–461. [CrossRef]
27. Hoek, E.; Kaiser, P.K.; Bawden, W.F. *Support of Underground Excavations in Hard Rock*; Balkema: Rotterdam, The Netherlands, 1995.
28. Vásárhelyi, B.; Somodi, G.; Krupa, Á.; Kovács, L. Determining the Geological Strength Index (GSI) using different methods. In Proceedings of the ISRM International Symposium—EUROCK 2016, Nevsehir, Turkey, 29–31 August 2016; pp. 1049–1054.
29. Somodi, G.; Krupa, Á.; Kovács, L.; Vásárhelyi, B. Comparison of different calculation methods of Geological Strength Index (GSI) in a specific underground construction site. *Eng. Geol.* **2018**, *243*, 50–58. [CrossRef]
30. Deák, F.; Kovács, L.; Vásárhelyi, B. Geotechnical rock mass documentation in the Bátaapáti radioactive waste repository. *Cent. Eur. Geol.* **2014**, *57*, 197–211. [CrossRef]
31. Somodi, G.; Bar, N.; Vásárhelyi, B. Correlation between the rock mass quality (Q-system) method and Geological Strength Index (GSI). In Proceedings of the Fifth Symposium of the Macedonian Association for Geotechnics, ISRM Specialized Conference, Ohrid, North Macedonia, 23–25 June 2022; pp. 461–468.
32. Bertuzzi, R.; Douglas, K.; Mostyn, G. Comparison of quantified and chart GSI for four rock masses. *Eng. Geol.* **2016**, *202*, 24–35. [CrossRef]
33. Santa, C.; Gonçalves, L.; Chaminé, H.I. A comparative study of GSI chart versions in a heterogeneous rock mass media (Marão tunnel, North Portugal): A reliable index in geotechnical surveys and rock engineering design. *Bull. Eng. Geol. Environ.* **2019**, *78*, 5889–5903. [CrossRef]
34. Winn, K.; Ngai, L.; Wong, Y. Quantitative GSI determination of Singapore's sedimentary rock mass by applying four different approaches. *Geotech. Geol. Eng.* **2019**, *37*, 2103–2119. [CrossRef]
35. Hoek, E.; Marinos, P. Predicting tunnel squeezing. *Tunn. Tunn. Int.* **2000**, *32*, 45–51.
36. Yang, B.; Elmo, D. Why Engineers Should Not Attempt to Quantify GSI. *Geosciences* **2022**, *12*, 417. [CrossRef]
37. Barton, N.; Bar, N. Introducing the Q-Slope Method and Its Intended Use within Civil and Mining Engineering Projects. In *Future Development of Rock Mechanics, Proceedings of the ISRM Regional Symposium Eurock 2015 and 64th Geomechanics Colloquium, Salzburg, Austria, 7–10 October 2015*; Schubert, W., Kluckner, A., Eds.; GEOAUSTRIA: Salzburg, Austria, 2015; pp. 157–162. ISBN 978-3-9503898-1-4.
38. Barton, N.R.; Grimstad, E. *An Illustrated Guide to the Q-System Following 40 Years Use in Tunnelling*; In-House Publishing: Oslo, Norway, 2014.
39. Bar, N.; Barton, N. Q-Slope: An Empirical Rock Slope Engineering Approach in Australia. *Austr. Geomech.* **2018**, *53*, 73–86.
40. Moustafa, E.B.; Hammad, A.H.; Elsheikh, A.H. A new optimized artificial neural network model to predict thermal efficiency and water yield of tubular solar still. *Case Stud. Therm. Eng.* **2022**, *30*, 101750. [CrossRef]
41. Abdolrasol, M.G.; Hussain, S.S.; Ustun, T.S.; Sarker, M.R.; Hannan, M.A.; Mohamed, R.; Ali, J.A.; Mekhilef, S.; Milad, A. Artificial neural networks based optimization techniques: A review. *Electronics* **2021**, *10*, 2689. [CrossRef]
42. Cantisani, E.; Garzonio, C.A.; Ricci, M.; Vettori, S. Relationships between the petrographical, physical and mechanical properties of some Italian sandstones. *Int. J. Rock Mech. Min. Sci.* **2013**, *60*, 321–332. [CrossRef]
43. Cowie, S.; Walton, G. The effect of mineralogical parameters on the mechanical properties of granitic rocks. *Eng. Geol.* **2018**, *240*, 204–225. [CrossRef]
44. IBM Corp. *Released 2015. IBM SPSS Statistics for Windows, Version 23.0*; IBM Corp: Armonk, NY, USA, 2015.
45. Agbulut, Ü.; Gürel, A.E.; Biçen, Y. Prediction of daily global solar radiation using different machine learning algorithms: Evaluation and comparison. *Renew. Sustain. Energy Rev.* **2021**, *135*, 110114. [CrossRef]

46. Shahin, M.A. Use of evolutionary computing for modelling some complex problems in geotechnical engineering. *Geomech. Geoengin.* **2015**, *10*, 109–125. [CrossRef]
47. Khalaf, A.A.; Kopecskó, K. Modelling of Modulus of Elasticity of Low-Calcium-Based Geopolymer Concrete Using Regression Analysis. *Adv. Mater. Sci. Eng.* **2022**, *2022*, 4528264. [CrossRef]
48. Motulsky, H.J.; Ransnas, L.A. Fitting curves to data using nonlinear regression: A practical and nonmathematical review. *FASEB J.* **1987**, *1*, 365–374. [CrossRef] [PubMed]

sustainability

Article

Spatial Effect Analysis of a Long Strip Pit Partition Wall and Its Influence on Adjacent Pile Foundations

Nan Zhou and Jianhui Yang *

School of Civil Engineering and Architecture, Zhejiang University of Science and Technology, Hangzhou 310023, China
* Correspondence: yangjianhui@zust.edu.cn

Abstract: The spatial effect at the end of the foundation pit partition wall is significant, and the displacement of the retaining wall and soil caused by its demolition leads to an additional displacement and bending moment of the adjacent pile foundations, which in turn deteriorates the work behavior of the pile foundation. Taking the project of an open tunnel under a viaduct located in Hangzhou as an example, site monitoring was performed to determine the effect of the demolition of the partition wall on the displacement of the surrounding retaining wall and the soil in the adjacent area and the monitoring data were compared to the finite element analysis results to check the rationality of the finite element model. This model was used to study the influence of the distance from the pile foundation to the partition wall as well as the stiffness of the retaining wall on the displacement and bending moment of the pile foundation during excavation. These results indicate that because of the support effect of the foundation pit partition wall on the retaining wall, the spatial effect at the end of the partition wall is large, and the displacement and bending moments of the pile foundation in the vicinity of the partition wall are lower than those in the far distance. Demolition of the partition wall will increase the displacement and bending moment of adjacent pile foundations, and this effect decreases with increasing distance. The range of influence of the spatial effect at the end of the partition wall is approximately 1.1 times the depth of the foundation pit. When the pile foundation is in the immediate vicinity of the partition wall, the response of the front-row and rear-row piles to the demolition of the partition wall is significantly different. The front row piles are more affected, while the rear row piles are less affected. As the distance increases, the difference in response gradually decreases and tends to be consistent. As the stiffness of the retaining wall increases, the effect of the demolition of the partition wall on the pile foundation decreases. It is recommended that the stiffness of the supporting system near the partition wall be reduced appropriately, and the partition wall should be set at the foundation pit section near the pile foundation, but the response of the foundation pit and the adjacent pile foundation should be paid close attention to when the partition wall is demolished.

Keywords: foundation pit partition wall; spatial effects; adjacent pile foundation; support effect

Citation: Zhou, N.; Yang, J. Spatial Effect Analysis of a Long Strip Pit Partition Wall and Its Influence on Adjacent Pile Foundations. *Sustainability* **2023**, *15*, 10409. https://doi.org/10.3390/su151310409

Academic Editor: Yunfeng Ge

Received: 7 June 2023
Revised: 23 June 2023
Accepted: 29 June 2023
Published: 1 July 2023

1. Introduction

The unloading of foundation pit excavation can cause displacement of the surrounding soil in the direction of the foundation pit, and controlling its environmental impact is a concern for the sustainable development of foundation pit engineering. Long strip foundation pits are often fitted with partition walls, dividing the original pit into several smaller lengths for easy organization of the construction. Heish [1] analyzed the mechanical partition wall mechanism based on the continuous-beam elastic support method, indicating that the partition wall may have a strain-suppressing effect similar to the pit corner on the foundation pit. This indicates that there is a spatial effect at the end of the partition wall that is similar to the corner of a conventional foundation pit. This spatial effect is manifested by the smaller deformation of the foundation pit near the partition wall and the greater deformation away from the partition wall. In the study of the suppression of foundation pit deformation by partition walls, Ou et al. [2] compared engineering cases with and without partition walls and found that partition walls can significantly reduce the deformation of the retaining wall as well as the settlement of the surface at the rear of the wall. Wu [3] counted the deformation of retainment and surface settlement statistically for eleven groups of foundation pits with and without partition walls and found that the deformation of foundation pits with partition walls was significantly smaller than that of foundation pits without partition walls. Hsieh et al. [4,5] used the three-dimensional finite element method to study the influence of parameters such as length, spacing, and thickness of partition walls on the deformation of retaining walls and set out a computational formula for predicting the deformation of retaining walls during excavation of foundation pits with partition walls. Han et al. [6] studied the effect of partition walls on the deformation of foundation pits and adjacent buildings and found that partition walls can greatly reduce the deformation of retaining walls and the settlement of adjacent soil, thereby protecting adjacent buildings. Li et al. [7] studied the effect of partition walls on the settlement of adjacent buildings and found that the higher the stiffness and depth of embedment of the partition walls, the greater the inhibitory effect on the settlement of adjacent buildings. Wu et al. [8] conducted a finite element simulation of the excavation process of a long strip foundation pit that was partitioned using partition walls and found that the smaller the gap distance between the partition walls, the less the horizontal displacement of the retaining wall and the settlement of the soil outside the pit. In the study of spatial effects in foundation pits, Cheng et al. [9,10] compared the results of two-dimensional and three-dimensional analyses of the excavation of foundation pits in areas of soft soil and found that the displacement of the retaining wall at the corner of the pit was significantly greater in the results of the two-dimensional analysis. Finno [11,12] summarized the characteristics of the three-dimensional ground displacement in the surrounding area caused by the excavation through a statistical analysis of a large number of results from the numerical calculation of the engineering of the foundation pit. Fuentes et al. [13] proposed a formula for estimating the effect of corner spatial effects on adjacent buildings in foundation pits and verified it via practical engineering cases. Szepesházi et al. [14] studied the variation of the internal force of the retaining structure in the region of the concave corner during the excavation of the foundation pits. Abbas et al. [15] performed a three-dimensional finite element analysis of retaining wall deformation during the excavation of foundation pits with irregular shapes and found that the short side of the foundation pit is significantly affected by the corner spatial effect, even though the majority of the sectional strain on the long side is close to the results of the plane strain analysis. Russell et al. [16] implemented a three-dimensional analysis function to describe the surface settlement pit around the excavation at the deep foundation pit, and the maximum surface settlement at the rear of the resulting wall is located at the midpoint of the long side of the excavation. Wang et al. [17] conducted finite element analysis and indoor model tests on the spatial effect of slender foundation pits and obtained the spatial distribution pattern of additional soil pressure with retaining wall deformation. Li et al. [18,19] conducted a study of the deformation characteristics of irregular foundation pits using a combination of on-site monitoring and numerical simulations, and found that the pit coner has a restraining effect on the deformation of the diaphragm wall. Liu et al. [20] analyzed the displacement of the retaining wall, surface settlement, and surrounding building settlement data of large deep foundation pits, evaluated the spatial effect of the corners of the foundation pit using the analysis results, and optimized the support structure of the corners of the foundation pit, thereby reducing construction costs. Wu et al. [21] analyzed the deformation law of the retaining wall under the effect of spatial effects based on on-site monitoring data during the excavation process of foundation pits in soft soil areas. They found that the spatial distribution of surface settlement of the foundation pit and displacement of the retaining wall was larger on the long side and smaller on the corners. Liu et al. [22] established a numerical model for the deep foundation pit engineering of buried tunnels, believing that the pit angle of the foundation pit significantly limits the horizontal displacement and support axial force of the retaining wall. Jia et al. [23] used model tests to find that the spatial effect of smaller foundation pits is more pronounced than that of larger ones.

Based on the studies cited above, research on the spatial effects of partition walls has been quite extensive at present, mainly focusing on the distribution of displacement of foundation pit retaining walls and soil under the influence of spatial effects. There is relatively little research on the behavior and response of adjacent pile foundations under the influence of spatial effects. Before the demolition of the partition wall, the displacement of the retaining wall is limited by the support force of the partition wall, reducing the effect of excavation on the behavior of adjacent pile foundations. After the demolition of the partition wall, the supporting force of the partition wall on the retaining wall disappears, and the retaining wall further moves in the direction of the foundation pit. The movement of the soil behind the wall caused by this may lead to additional displacement and bending moments in the pile foundation in the adjacent area. In this paper, on-site monitoring data are used to obtain the surface settlement and deformation of the retaining walls before and after the partition wall of the long strip foundation pit is demolished, from which the rationality of the numerical finite-element model is verified. The model is applied to study the effect of key building variables such as the distance of the pile foundation from the partition wall and the stiffness of the retaining wall on the foundation of the adjacent pile. The spatial effect of the pit partition wall can be used to reduce the impact of excavation on the environment, and the proposal of pit support near the pit partition wall is put forward to achieve the sustainable development goal of the low environmental impact of foundation pit engineering.

2. Monitoring and Models

2.1. Project Overview

The open-cut tunnel is located in Hangzhou City, and it crosses the present viaduct in an east-west direction. The bridge adopts a clustered pile foundation with a pile length of about 55 m, and six bored piles (D = 1.8 m) are arranged under the bearing platform. The closest distance between the north side of the foundation pit and the pile foundation is about 5.1 m, and the closest distance to the south side is about 9.7 m. The planar relationship between the bearing platform and the foundation pit is shown in Figure 1. The excavation depth of the foundation pit is 15.4 m. The retaining structure is bored piers at φ900@1050 mm, and the waterproof structure is high-pressure rotary jet grouting piles at φ800@600 mm. The support inside the foundation pit is made of concrete, steel, and concrete from the top to the bottom of the pit. When excavated to 1.8 m, the first support is set with an 800 mm × 900 mm concrete support, the second support is a φ609 steel support, and the third support is made of concrete. The vertical spacing of the three supports is 5.12 m and 3.45 m from the top to the bottom. A partition wall is placed to divide the foundation pit into east and west areas for construction, which consists of a bored pile and a high-pressure rotary jet grouting pile. The depth of the partition wall is 25 m. The length of the foundation pit in the eastern area is 128 m, while the length of the foundation pit in the western area is 287 m, with a width of 32 m. After the western foundation pit has been excavated down to the bottom of the pit, it will be necessary to excavate the eastern foundation pit. After the east foundation pit is excavated to the bottom of the pit, the partition wall is demolished. The excavation of the eastern foundation pit is decomposed into five working conditions along the depth direction for the construction, as listed in Table 1.

2.2. Monitoring Scheme

The layout of the monitoring points is shown in Figure 1. Arrange eight horizontal displacement monitoring points for the retaining walls on both sides of the foundation pit. Five surface settlement monitoring points are buried behind the walls on both the north and south sides, with distances of 2 m, 5 m, 10 m, 15 m, and 25 m from the foundation pit. There are a total of eight surface settlement survey lines. The tilt measurement point closest to the partition wall (10 m east of the partition wall) is used for analysis. The tilt measurement point on the north side of the pit is marked Q_1, and the tilt measurement point on the south side of the pit is marked Q_2. Surface settlement monitoring points on the north side of the pit are marked as the D_1 survey line, and surface settlement monitoring points on the south side of the pit are marked as the D_2 survey line.

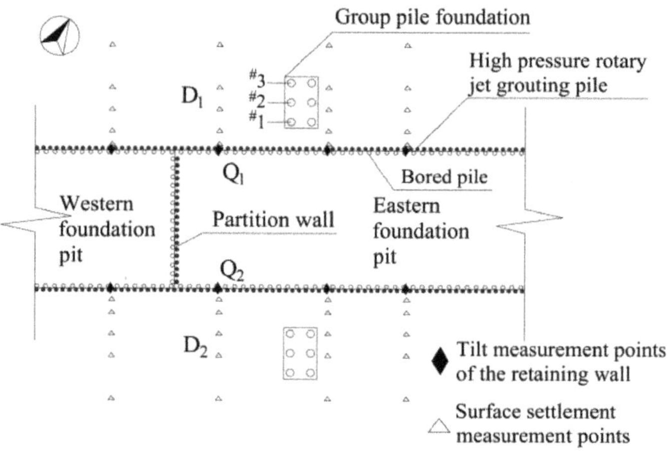

Figure 1. Plane relationship of foundation pit.

Table 1. Working condition division.

Working Condition No.	Construction Content
WC 1	Excavate the first layer of soil up to the crown beam and concrete support bottom.
WC 2	Excavate the second layer of soil to the bottom of the second steel support.
WC 3	Excavate the third layer of soil to the bottom of the third concrete support layer.
WC 4	Excavate soil to the bottom of the foundation pit.
WC 5	Demolish the partition wall.

2.3. Analysis of Monitoring Results

The horizontal displacement of the retaining wall at typical section Q_2 was analyzed, and the measured horizontal displacement values of the retaining wall along the depth direction under different working conditions are shown in Figure 2, with the wall displacement towards the pit being positive. From Figure 2, it can be seen that the horizontal displacement curves of the retaining wall under different working conditions have a similar shape, and the overall shape can be seen to be in the form of an "arc". As the excavation depth increases, the horizontal displacement continues to increase, and the maximum horizontal displacement occurs continuously downward and approaches the bottom of the foundation pit, finally located at a depth of about 14 m. When it was excavated to the bottom of the pit, the maximum horizontal displacement of the retaining wall (condition 4) was found to be 26.43 mm, and following the removal of the partition wall (condition 5), the maximum horizontal displacement was 29.23 mm, an increase of about 11%. This is due to the vanishing of the supporting force on the retaining wall as a result of the demolition of the partition wall in Condition 5, which leads to a further displacement of the retaining wall into the pit.

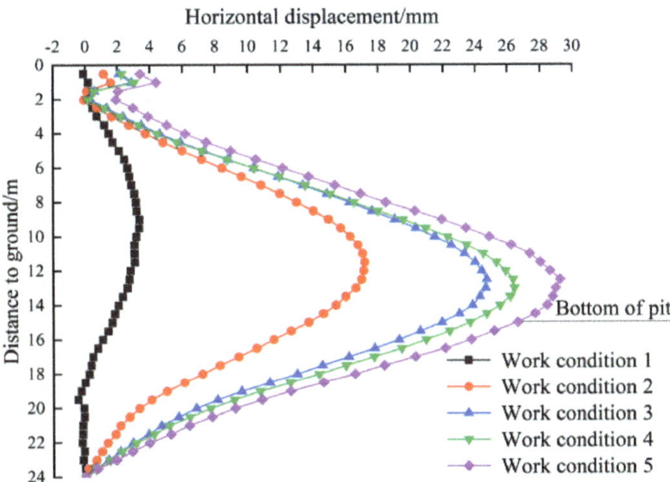

Figure 2. Horizontal displacement of Q_2.

By analyzing the monitoring data from the D_1 and D_2 survey lines to the north and south for surface settlement of the foundation pit, the distribution of surface settlement around the pit is plotted as a function of distance under different working conditions, as shown in Figure 3, with soil settlement as positive. As can be seen in Figure 3, the shape of the curve of the surface settlement distribution as a function of distance under different working conditions is similar, and the overall shape is an "inverted triangle". The amount of settlement increases continuously as the depth of excavation increases, and maximal settlement occurs at a distance of about 15 m from the foundation pit, which is in accordance with the general law of settlement of the surface outside the pit in areas of soft soil. The surface settlement on the D_2 survey line is about 24% greater than that on the D_1 survey line. This is due to the presence of an external haulage road for the earthwork on the south side of the foundation pit, and the heavy traffic load has led to further surface settlement on the south side. When the foundation pit was excavated to the bottom (condition 4), the maximum surface settlement on survey line D_1 was 16.91 mm. After the partition wall was demolished (condition 5), the maximum settlement was 19.09 mm, an increase of about 13%. When the foundation pit was excavated to the bottom (condition 4), the maximum surface settlement on survey line D_2 was 21.95 mm. After the partition wall was demolished (condition 5), the maximum settlement reached 23.59 mm, an increase of about 7%. This is because, after the demolition of the partition wall, the supporting force of the partition wall on the retaining wall disappears, leading to further displacement of the retaining wall towards the pit and exacerbating the soil loss around the pit. As per the provisions of China's GB50497-2009 "Technical code for monitoring of building excavation engineering" on the boundary values of the horizontal displacement at the depth of the retaining wall and the vertical displacement of the surrounding surface of the foundation pit, and combined with hydrogeological conditions in the region, the monitoring scheme for this project determines that the alarm values for the horizontal displacement at depth and for the surface settlement of the retaining wall are both 30 mm and the warning values are both 24 mm. From the analysis above, it can be seen that the maximum horizontal displacement of the retaining wall after the demolition of the partition wall at the Q_2 measurement point was as high as 29.23 mm, which exceeds the warning value and is very near the alarm value. On the D_2 survey line, the maximum surface settlement was 23.59 mm, which is close to the warning value.

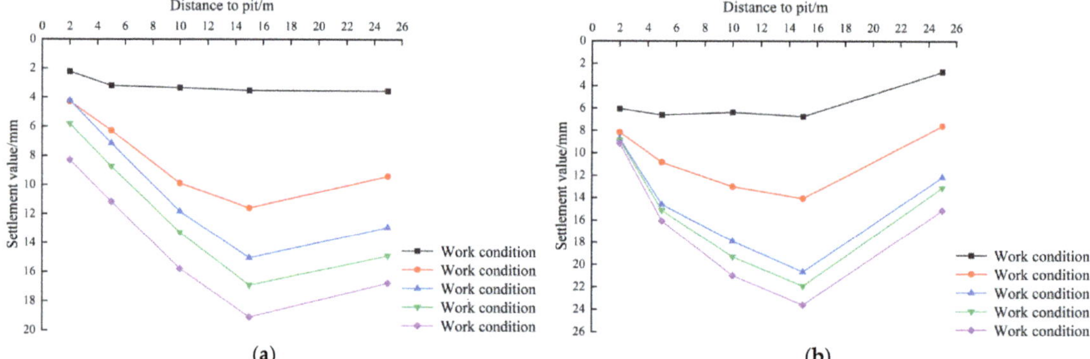

Figure 3. Distribution of surface settlement. (a) D_1 survey line; (b) D_2 survey line.

According to the monitoring data of working condition 5 in Figure 3, Figure 4 shows a curve fit of the relationship between the amount of settlement and the ratio of the distance to the depth of the foundation pit. In agreement with Figure 4, the relation functions between the surface settlement of the D_1 and D_2 survey lines after the partition wall is demolished and the distance from the pit edge are obtained in the form of Equations (1) and (2), respectively:

$$y_1 = -9.52x_1{}^2 + 22.99x_1 + 5.05 \tag{1}$$

$$y_2 = -18.50x_1{}^2 + 36.89x_1 + 5.09 \tag{2}$$

where y_1 and y_2 represent surface settlement on survey lines D_1 and D_2, respectively, and x_1 is the ratio of the distance from the edge of the foundation pit to the measurement point to the depth of the pit.

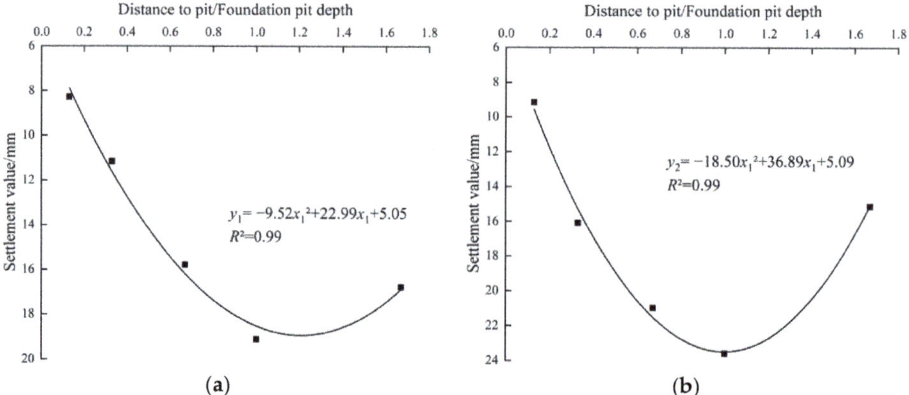

Figure 4. Surface settlement fitting curve. (a) D_1 survey line fitting curve; (b) D_2 survey line fitting curve.

As shown in Figure 4, the goodness of fit coefficient R^2 from Equations (1) and (2) reached 0.99, indicating strong agreement between the fitted relationship function and monitoring data. It is possible that Equations (1) and (2) better reflect the surface settlement pattern on survey lines D_1 and D_2 after the partition wall has been demolished. Based on the results of the calculations in Equations (1) and (2), the point of maximum surface settlement following the demolition of the partition wall is located at a distance of approximately 1 to 1.2 times the depth of the foundation pit relative to the pit edge.

2.4. 3D Finite Element Modeling

Using the finite element software Midas GTS NX, a numerical model of the tunnel foundation pit is established, as shown in Figure 5. Given the boundary effect of the model and the range of spatial effect [24], the size of the model is chosen to be 152 m long (perpendicular to the partition wall direction), 130 m wide (parallel to the partition wall direction), and 85 m deep. The length of the foundation pit in the eastern area is determined to be 80 m, while the length of the foundation pit in the western area is determined to be 72 m, with a width of 32 m and an excavation depth of 15.4 m. Horizontal constraints are imposed on the lateral sides of the model, vertical constraints on the bottom, and a free boundary on the top surface. The computational steps of the simulation are shown in Table 2.

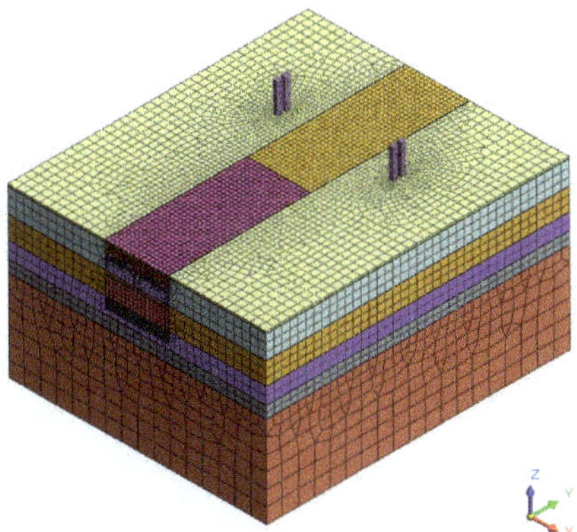

Figure 5. 3D finite element model.

Table 2. Computational procedure.

Simulation Steps	Simulation Content
Step 1	Initial crustal stress equilibrium.
Step 2	Construction of the diaphragm and partition wall.
Step 3	Excavate the first layer of soil in the western area and apply the first support.
Step 4	Excavate the second layer of soil in the western area and apply for the second support.
Step 5	Excavate the third layer of soil in the western area and apply the third support.
Step 6	Excavation to the bottom of the pit in the western area.
Step 7	Excavate the first layer of soil in the eastern area and apply the first support.
Step 8	Excavate the second layer of soil in the eastern area and apply the second support.
Step 9	Excavate the third layer of earthwork in the eastern area and apply the third support.
Step 10	Excavation to the bottom of the pit in the eastern area.
Step 11	Demolish the partition wall.

The classification of soil layers and determination of physical and mechanical parameters based on the geologic survey report are shown in Table 3. Soil simulations were performed using solid elements and a modified constitutive Mohr-Coulomb model. The cushion cap is simulated using solid elements and buried 1.5 m below the ground according to the actual engineering situation. With an elastic modulus of 2.5×10^4 MPa, Poisson's ratio is 0.2. Beam elements were used for the support, waist beam, and link beam. For computational simplicity, the retaining wall was equivalent to a diaphragm wall on the principle of equivalent bending rigidity [25], and plate elements were used in the simulations. Table 4 shows the parameters of the structural calculation.

Table 3. Soil layer calculation parameters.

Soil Layer	Layer Thickness H/m	Bulk Density $\gamma/(Kn \cdot m^{-3})$	Cohesive Force c/kPa	Internal Friction Angle $\varphi/(°)$	Elastic Modulus E/MPa	Poisson's Ratio μ
Plain fill soil	1.2	18.2	11.9	10.1	4.5	0.30
Muddy clay	12.1	16.5	10.2	11.9	4.2	0.30
Silty clay	11.3	17.4	31.9	20.0	4.1	0.30
Clay	8.2	21.8	35.1	27.9	21.3	0.30
Round gravel	6.1	25.2	5.0	42.0	68.3	0.28
Moderately weathered sandstone	46.1	26.1	7.5	42.6	432.0	0.27

Table 4. Structural calculation parameters.

Structure Name	Element Selection	Elastic Modulus E/MPa	Bulk Density $\gamma/(kN \cdot m^{-3})$	Poisson's Ratio μ
Diaphragm wall	Plate element	30.0×10^3	25	0.25
Concrete support	Beam element	25.0×10^3	23	0.20
Concrete waste beam	Beam element	25.0×10^3	23	0.20
Concrete link beam	Beam element	25.0×10^3	23	0.20
Steel support	Beam element	210.0×10^3	77	0.30
Steel waist beam	Beam element	210.0×10^3	77	0.30
Steel link beam	Beam element	210.0×10^3	77	0.30

2.5. Model Rationality Verification

Figure 6 shows a cloud map of horizontal displacements of the retaining walls before and after the demolition of the partition wall. It can be seen in Figure 6 that after the partition wall is demolished, the horizontal displacement of the retaining wall near the end of the partition wall increases significantly. We compared the simulated and monitored values of the measurement point of the horizontal displacement Q_2 of the retaining wall and the surface settlement survey line D_2 before and after the demolition of the partition wall, and the results are shown in Figures 7 and 8. From Figure 7, it can be seen that the simulated horizontal displacement of the retaining walls before and after the partition wall has been demolished is in fundamental agreement with the values obtained from the monitoring data, and maximum displacement occurs at a depth of approximately 14 m, with a difference of 0.017 mm and 0.024 mm between the simulated displacement value and the monitoring value of the maximum displacement. From Figure 8, it can be seen that the surface settlement curves obtained from both the simulation and the monitoring have the shape of an inverted triangle, with maximum surface settlement occurring at a distance of about 15 m from the edge of the foundation pit. The differences between the maximum simulated and monitored maximum surface settlement correspond to 0.586 mm and 0.534 mm, respectively. The agreement between the simulated and monitored values in the two working conditions before and after the partition wall demolition is therefore good, indicating that the numerical model is a reasonable model.

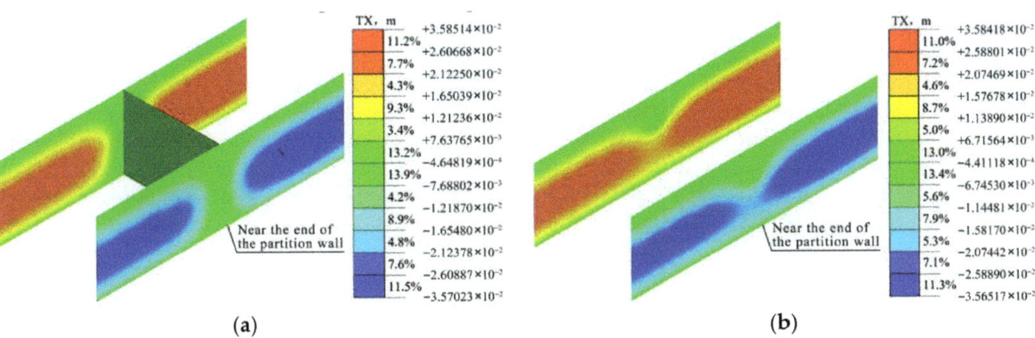

Figure 6. Horizontal displacement cloud chart of the diaphragm wall. (**a**) Before demolition of partition wall; (**b**) After demolition of partition wall.

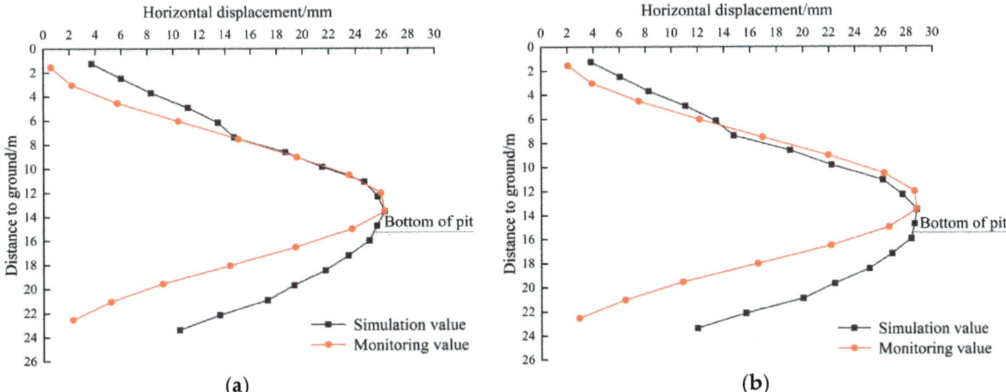

Figure 7. Comparison of simulated and measured values of Q_2. (**a**) Before demolition of partition wall; (**b**) After demolition of partition wall.

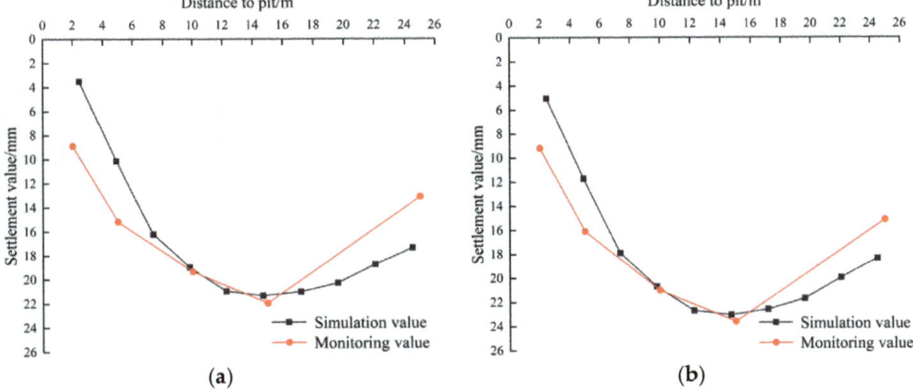

Figure 8. Comparison of simulated and measured values of D_2. (**a**) Before demolition of partition wall; (**b**) After demolition of partition wall.

3. Results

The effects of the distance between the pile foundation and the partition wall as well as the thickness of the retaining wall on the displacement and bending moments of adjacent pile foundations were investigated using the validated finite element model. The northern pile foundation was taken for research purposes due to the proximity of the northern bearing platform to the foundation pit. For the pile group, the foundation of the western pile was marked as #1, #2, and #3 relative to the distance of the foundation pit from near to far. The location relationship is shown in Figure 1.

3.1. Analysis of the Effect of the Distance of the Pile Foundation from the Partition Wall

We compare and analyze seven cases in which the distance L from the centerline of the short side of the bearing platform to the partition wall is 0 m, 6 m, 9 m, 12 m, 15 m, 18 m, and 21 m, and the rest of the parameters are the same as the numerical model discussed above. The horizontal displacement and bending moment distributions of pile #1 are shown in Figure 9. The displacement of the pile foundation is positive when it moves towards the pit, and the bending moment is positive when the north surface of the pile foundation is pulled. The dashed line represents the working condition before the demolition of the partition wall, and the solid line represents the working condition after the partition wall is demolished. Figure 9 shows the curve fit of the relationship between the maximum displacement value of the pile foundation before and after the demolition of the partition wall and the distance, as can be seen in Figure 10. This figure takes the ratio of the distance to the depth of the foundation pit as the horizontal axis and the value of the maximum displacement of the pile foundation as the vertical axis.

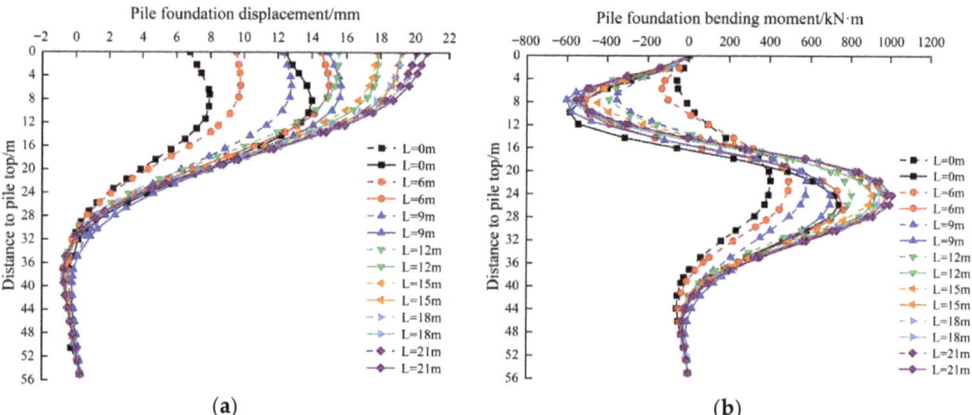

Figure 9. Horizontal displacement and a bending moment of plie1. (**a**) Horizontal displacement distribution; (**b**) Bending moment distribution.

From Figures 9 and 10, it can be seen that the displacement of the pile foundation increases as the distance from the pile foundation to the partition wall increases in both the pre- and post-demolition working conditions of the partition wall. Before the demolition of the partition wall, the reason for this phenomenon was that the partition wall has a supporting effect on the retaining wall, resulting in a small displacement of the retaining wall near the partition wall, while the supporting effect on the retaining wall far away from the partition wall is weak and the displacement is large, indicating a significant spatial effect at the end of the partition wall. After the demolition of the partition wall, this phenomenon occurs because the soil behind the retaining wall has nonlinear deformation characteristics. When the support force of the partition wall disappears, the soil will experience a certain amount of rebound displacement, which is still smaller than the displacement far away from the partition wall when combined with the displacement before demolition. This indicates that in engineering, partition walls can be set in the foundation pit section near the pile foundation to reduce the effect of excavation on the pile foundation.

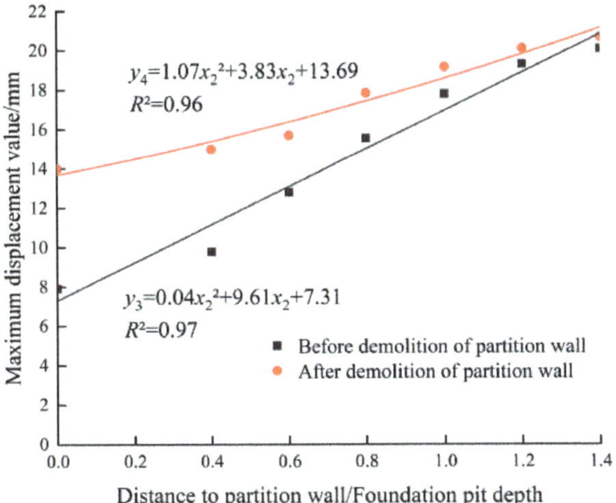

Figure 10. The relationship between the maximum displacement value and distance to the partition wall of pile 1.

In agreement with Figure 10, Equations (3) and (4) show that the relationship functions between the maximum displacement value and the distance of piles #1 before and after the partition wall is demolished.

$$y_3 = 0.04x_2{}^2 + 9.61x_2 + 7.31 \tag{3}$$

$$y_4 = 1.07x_2{}^2 + 3.83x_2 + 13.69 \tag{4}$$

where y_3 and y_4 represent the maximum displacement values of the pile foundation, respectively, before and after the partition wall has been demolished, and x_2 represents the ratio of the distance from the pile foundation pit to the partition wall to the depth of the pit.

In agreement with Figure 10, Equations (3) and (4), it can be seen that the two curves gradually approach one another as the distance increases, indicating that as the distance from the pile foundation to the partition wall increases, there is a gradual decrease in the effect of partition wall demolition on pile foundation displacement. Define the maximum displacement value growth rate as the ratio of the increment of the maximum displacement value of the pile foundation after the demolition of the partition wall to the maximum displacement value of the pile foundation before the demolition. As shown in Figure 10, it is possible to obtain the growth rate of the maximum displacement value of the pile foundation before and after the partition wall is demolished, and the plot of the relationship between displacement growth rate and x can be fitted as shown in Figure 11. From Figure 11, it can be seen that as the distance between the pile foundation and the partition wall increases, the displacement growth rate of the pile foundation gradually decreases before and after the demolition of the partition wall, with first a fast and then a slow trend.

The relationship function between the maximum displacement growth rate of the pile foundation and the ratio of the distance to the depth of the foundation pit obtained from the fit is as follows:

$$R_d = 36.96x_2{}^2 - 107.44x_2 + 79.53 \tag{5}$$

where R_d is the growth rate of the maximum pile foundation displacement value and x_2 represents the ratio of the distance from the pile foundation pit to the partition wall to the depth of the pit.

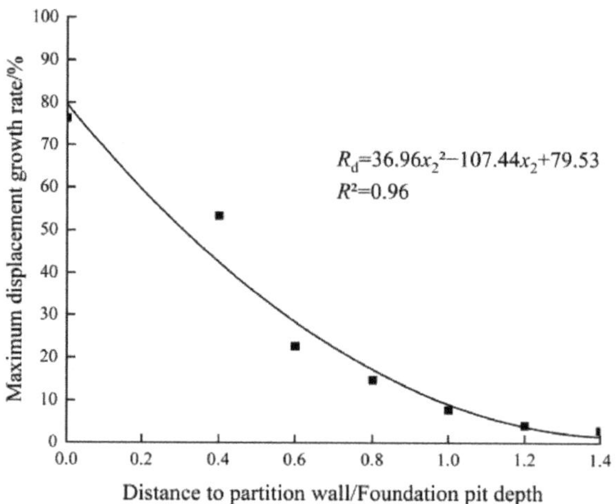

Figure 11. The relationship between the maximum displacement growth rate and distance to the partition wall of pile 1.

Equation (5) allows us to calculate that when the displacement growth rate is 5%, the distance from the pile foundation to the partition wall is approximately 1.14 times the depth of the foundation pit. The results of this study indicate that when the distance between the pile foundation and the partition wall reaches about 1.1 times the depth of the foundation pit, the demolition of the partition wall has little effect on the behavior of adjacent pile foundations, and the range of effect of the spatial effect at the end of the partition wall is about 1.1 times the depth of the foundation pit.

3.2. Differences in Response to the Demolition of the Partition Walls between the Front and Rear Rows of Piles

The curve of the maximum displacement and bending moment increment of the front and rear row piles caused by the demolition of the partition wall with distance is shown in Figure 12. As shown in Figure 12, when the distance from the pile foundation to the partition wall is 0~0.8 depth of the foundation pit, the front row piles have a significantly greater displacement and bending moment increment than the rear row piles. This result indicates that when the distance is closer, the response of the front row and rear row piles to the demolition of the partition wall has a large difference, with the front row piles being more affected and the rear row piles being less affected. When the distance is between 0.8 and 1 times the depth of the foundation pit, it should be noted that the incremental difference in displacement and bending moment values between the front and rear rows of piles begins to decrease significantly. When the distance is between 1 and 1.2 times the depth of the foundation pit, the incremental difference in displacement and bending moment between the front and rear rows of piles further decreases and tends to be consistent. When the distance is between 1.2 and 1.4 times the depth of the foundation pit, the displacement and bending moment increments of the front and rear row piles are basically consistent, indicating that the difference in the response of the front row and rear row piles to the demolition of the partition wall gradually decreases and becomes closer as distance increases. As the spatial effect at the end of the partition wall is significant and the distance increases, the influence of the spatial effect gradually decreases.

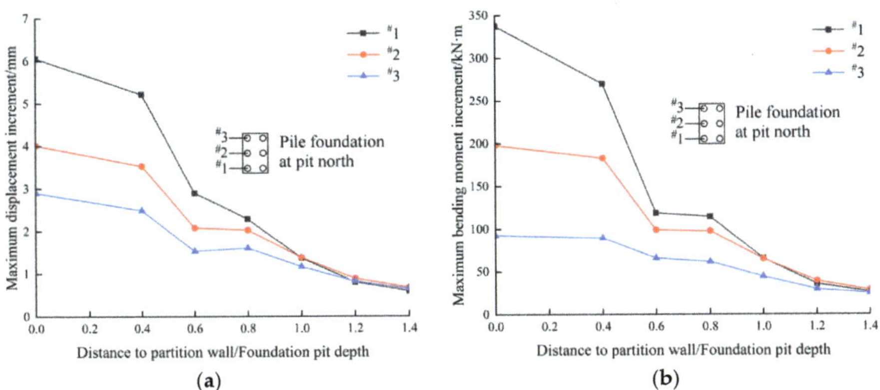

Figure 12. Curve of maximum displacement and bending moment values of the front and rear row pile foundations. (**a**) Maximum displacement increment; (**b**) Maximum bending moment increment.

3.3. Analysis of the Effect of the Retaining Wall Rigidity

Since the stiffness of the foundation pit retaining wall is the main factor affecting the deformation of the foundation pit [26], it inevitably affects the displacement and bending moment of the pile foundation. Due to the positive correlation between the bending stiffness of the diaphragm wall and its thickness, we chose to use the four common cases of diaphragm wall thickness d of 0.6 m, 0.8 m, 1.0 m, and 1.2 m, respectively. The remaining variables were the same as in the above numerical model. Since the demolition of partition walls has the most obvious effect on the behavior of pile foundations when the distance between the pile foundation and the partition wall is relatively close, pile foundation #1 is selected for analysis when L = 0 m in the above simulated working condition, and the distribution of displacement and bending moment is shown in Figure 13. In this figure, the displacement of the pile foundation is positive when it moves towards the pit, and the bending moment is positive when the north surface of the pile foundation is pulled.

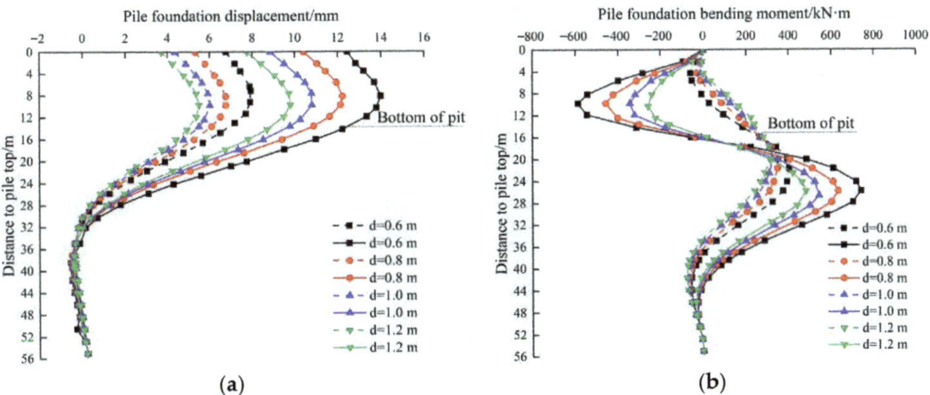

Figure 13. Horizontal displacement and bending moment of pile 1 with different wall thicknesses. (**a**) Displacement distribution; (**b**) Bending moment distribution.

From Figure 13, it can be seen that in the two working conditions before and after the demolition of the partition wall, the displacement and bending moments of the pile foundation decrease with increasing diaphragm wall thickness. Due to the increased stiffness of the diaphragm wall, the deformation of the wall is reduced, and correspondingly, the displacement and bending moment of the pile foundation are decreased. The demolition of the partition wall resulted in a significant increase in the horizontal displacement and bending moment of the pile foundation. The maximum displacement and upper bending moment values both occur at a distance of 10 m from the top of the pile, and the upper and lower bending moments of the pile body have significantly increased. This is because after the partition wall is demolished, the displacement of the retaining wall towards the pit causes the soil to flow towards the pit, resulting in an additional bending moment on the upper pile body. However, the lower pile end is affected by the displacement of the upper pile body towards the pit, resulting in a trend of displacement away from the pit, which forms a thrust on the soil. Due to the penetration of the elevated pile foundation into the rock layer, a significant reverse bending moment occurred in the lower pile body. The increase and growth rate of the maximum displacement and bending moment values are shown in Table 5. From Table 5, it can be seen that as the thickness of the diaphragm wall is gradually increased, there is a gradual decrease in the increment of the pile displacement and bending moment values after the partition wall is demolished, indicating that increasing the stiffness of the diaphragm wall can effectively reduce the effect of partition wall demolition on adjacent pile foundations.

Table 5. Response of pile1 displacement and bending moment caused by the demolition of partition walls with different wall thicknesses.

Thickness/m	Displacement				Bending Moment			
	Before Demolition of the Partition Wall/mm	After Demolition of the Partition Wall/mm	Increment Value /mm	Growth Rate/%	Before Demolition of the Partition Wall /kN·m	After Demolition of the Partition Wall /kN·m	Increment Value /kN·m	Growth Rate/%
0.6	7.93	13.99	6.06	76.39	404.07	741.46	337.39	83.5
0.8	6.75	12.21	5.46	80.74	351.57	632.19	280.62	79.82
1.0	6.01	10.79	4.78	79.52	326.68	543.93	217.25	66.5
1.2	5.5	9.8	4.3	78.27	314.3	483.73	169.43	53.91

4. Discussion

Due to the spatial effect at the end of the partition wall, which is similar to the corner of a conventional foundation pit, the pile foundation adjacent to the partition wall is significantly affected by the spatial effect. Before the partition wall is demolished, the displacement and bending moments of piles near the partition wall are smaller than those far away. After the partition wall is demolished, the displacement and bending moment of the pile foundation increase obviously, but the value is still small after the increase, and the increment decreases with the increase in stiffness of the retaining wall.

With the increasing development of underground space, the depth of foundation pit excavation is also increasing. In order to ensure the safety of foundation pits, the application of large-diameter bored piles, mixing piles, and concrete support is increasing, which leads to high energy consumption and high carbon emissions from foundation pit engineering. This is contrary to the goal of sustainable development. For the foundation pit with a partition wall, the spacing effect of the partition wall can be used to achieve the sustainable development goal of energy savings and reducing the effect of excavation on the nearby environment. It is recommended to take measures such as reducing the number of concrete supports, increasing the horizontal spacing of concrete supports, and reducing the thickness of the retaining wall near the partition wall to reduce the stiffness of the support system. When the foundation pit is close to the pile foundation, it is recommended to set the partition wall near the pile foundation to reduce the response of the pile foundation during excavation. If the deformation requirement of the pile foundation is high, the stiffness of the retaining wall at the end of the partition wall can be appropriately increased to further maintain the stability of the pile foundation's behavior.

5. Conclusions

The focus of this paper was on the effect of the distance from the pile foundation to the partition wall and the stiffness of the retaining wall on the horizontal displacement and bending moment of the adjacent pile foundations during the demolition of the partition wall, and it summarized the law of the difference in response of the front and rear row piles under the partition wall demolition. Based on the excavation project of a long strip foundation pit adjacent to pile foundations in areas of soft soil, monitoring data from the site was used to analyze the effect of partition wall demolition on the retaining wall and the adjacent soil. Finite element software was used to study the effect of the partition walls on the displacement and bending moments of the adjacent pile foundations. The following conclusions can be drawn:

(1) Based on the results of the on-site monitoring, the horizontal displacement and surface settlement of the southern retaining wall of the foundation pit increased by 11% and 12%, respectively, following the demolition of the partition wall, and the values were close to the alarm values. Therefore, the effect of the demolition of the partition walls in the project needs to be considered, especially in situations where there has been substantial deformation before the demolition, and close monitoring of adjacent areas should be carried out during demolition.

(2) The foundation pit partition wall has a strong supporting effect on the retaining wall, rendering the displacement of the retaining wall near the partition wall smaller than that far away. Furthermore, the displacement and bending moments of the pile foundation in the vicinity of the partition wall are correspondingly smaller than those in the far distance, indicating a significant spatial effect at the end of the partition wall. Therefore, the stiffness of the supporting system of the foundation pit near the partition wall can be reduced in the project to realize energy savings and reduce emissions. The demolition of the partition wall will cause the displacement and bending moment of the adjacent pile foundation to increase, but the value after the increase is still small, so it is recommended to set the partition wall in the foundation pit section close to the pile foundation to reduce the adjacent effect of excavation. The effect of demolishing the partition wall on the pile foundation decreases as distance increases, with an effective range of approximately 1.1 times the depth of the foundation pit.

(3) When the distance between the pile foundation and the partition wall is relatively small, there is a significant difference in the behavior response of front row and rear row piles affected by partition wall demolition, manifested as a larger increase in the displacement and bending moment of the piles in the front row of the pile and a smaller increase in rear row pile displacement and bending moment. As the distance increases, the difference in response to partition wall demolition between the front and rear rows of piles gradually diminishes and eventually tends to be consistent.

(4) The effect of partition wall demolition on adjacent pile foundations can be effectively mitigated by increasing the rigidity of the diaphragm wall. As a result of this, When the deformation requirement of the adjacent pile foundation is high, it can be thought of as an appropriate increase in the stiffness of the retaining wall at the end of the partition wall.

Some areas need to be further explored in this study. There are two common methods of demolishing partition walls: demolition after excavation to the bottom of the pit and demolition with the excavation of the pit. For the purposes of this paper, we focus only on the demolition of partition walls after excavation to the bottom of the pit, and future research is needed on the situation where partition walls are demolished with excavation. The project relied on in this article is located in soft soil areas, and the stress state of the rock mass in the project is more complex [27,28]. Further study is needed on the spatial effect of foundation pit partition walls in rock areas and how to adjust the retaining structure to reduce the effect of partition wall demolition [29,30].

Author Contributions: Conceptualization, J.Y.; methodology, J.Y.; validation, N.Z.; investigation, J.Y., and N.Z.; resources, J.Y.; data curation, N.Z.; writing—original draft preparation, N.Z.; writing—review and editing, N.Z.; visualization, N.Z.; supervision, J.Y.; project administration, J.Y.; funding acquisition, J.Y. All authors have read and agreed to the published version of the manuscript.

Funding: This research was funded by the Key Research and Development Program of Zhejiang Province, grant number 2021C01131.

Institutional Review Board Statement: Not applicable.

Informed Consent Statement: Not applicable.

Data Availability Statement: All data included in this study are available upon request by contacting the corresponding author.

Acknowledgments: Thanks are due to Feng Li (Zhejiang University of Science and Technology, Hangzhou) for assistance with the numerical simulation and Zhehao Zhao (Zhejiang University of Science and Technology, Hangzhou) for valuable advice.

Conflicts of Interest: The authors declare that there is no conflict of interest.

References

1. Hsieh, P.G.; Ou, C.Y.; Shih, C. A simplified plane strain analysis of lateral wall deflection for excavations with cross walls. *Can. Geotech. J.* **2012**, *49*, 1134–1146. [CrossRef]
2. Ou, C.Y.; Hsieh, P.G.; Lin, Y.L. Performance of excavations with cross walls. *J. Geotech. Geoenviron. Eng.* **2011**, *137*, 94–104. [CrossRef]
3. Wu, S.H.; Ching, J.; Ou, C.Y. Predicting wall displacements for excavations with cross walls in soft clay. *J. Geotech. Geoenviron. Eng.* **2013**, *139*, 914–927. [CrossRef]
4. Hsieh, P.G.; Ou, C.Y. Mechanism of buttress walls in restraining the wall deflection caused by deep excavation. *Tunn. Undergr. Space Technol.* **2018**, *82*, 542–553. [CrossRef]
5. Hsieh, P.G.; Ou, C.Y. Simplified approach to estimate the maximum wall deflection for deep excavations with cross walls in clay under the undrained condition. *Acta Geotech.* **2016**, *11*, 177–189. [CrossRef]
6. Han, M.; Li, Z.; Mei, G.X.; Bao, X.H.; Jia, J.Q.; Liu, L.L.; Li, Y.Y. Characteristics of subway excavation in soft soil and protective effects of partition wall on the historical building and pile foundation building. *Bull. Eng. Geol. Environ.* **2022**, *81*, 307. [CrossRef]
7. Li, Z.; Han, M.; Liu, L.L.; Li, Y.Y.; Yan, S.H. Corner and partition wall effects on the settlement of a historical building near a supported subway excavation in soft soil. *Comput. Geotech.* **2020**, *128*, 103805. [CrossRef]
8. Wu, T. Influence analysis of partition width of long strip pit on deformation. *J. Undergr. Space Eng.* **2021**, *17*, 807–813.
9. Cheng, C.; Likitlersuang, S. Underground excavation behaviour in Bangkok using three-dimensional finite element method. *Comput. Geotech.* **2018**, *95*, 68–81. [CrossRef]
10. Ahmadi, A.; Ahmadi, M.M. Three-dimensional numerical analysis of corner effect of an excavation supported by ground anchors. *Int. J. Geotech. Eng.* **2022**, *16*, 903–915. [CrossRef]
11. Finno, R.J.; Roboski, J.F. Three-dimensional responses of a tied-back excavation through clay. *J. Geotech. Geoenviron. Eng.* **2005**, *131*, 273–282. [CrossRef]
12. Roboski, J.; Finno, R.J. Distributions of ground movements parallel to deep excavations in clay. *Can. Geotech. J.* **2006**, *43*, 43–58. [CrossRef]
13. Fuentes, R. Influence of corners in excavations on damage assessment. *Geotech. Res.* **2019**, *6*, 91–102. [CrossRef]
14. Szepesházi, A.; Mahler, A.; Móczár, B. Three dimensional finite element analysis of deep excavations' concave corners. *Period Polytech. Civil Eng.* **2016**, *60*, 371–378. [CrossRef]
15. Abbas, Q.; Yoon, J.; Lee, J. Characterization of wall deflection and ground settlement for irregular-shaped excavations with changes in corner configuration. *Int. J. Geomech.* **2023**, *23*, 4022258. [CrossRef]
16. Russo, G.; Nicotera, M.V. A closed form shape function describing 3D settlement field around a deep excavation in sand. *Sci. Rep.* **2022**, *12*, 18528. [CrossRef] [PubMed]
17. Wang, C.; Wang, X. A quantitative analysis of the spatial effects of retaining structure for slender foundation pits. *IOP Conf. Ser. Earth Environ. Sci.* **2018**, *189*, 022036. [CrossRef]
18. Li, Y.; Wang, C.X.; Sun, Y.; Wang, R.C.; Shao, G.J.; Yu, J. Analysis of corner effect of diaphragm wall of special-shaped foundation pit in complex stratum. *Front. Earth Sci.* **2022**, *10*, 794756. [CrossRef]
19. Tan, J.; Zheng, X.Y.; Sun, Y.; Shao, G.J.; Chen, Y.B. Analysis of pit corner effect of special-shaped foundation pit of subway station. *IOP Conf. Ser. Earth Environ. Sci.* **2020**, *558*, 032032. [CrossRef]
20. Liu, L.; Wu, R.; Congress, S.S.C.; Du, Q.; Cai, G.; Li, Z. Design optimization of the soil nail wall-retaining pile-anchor cable supporting system in a large-scale deep foundation pit. *Acta Geotech.* **2021**, *16*, 2251–2274. [CrossRef]
21. Wu, B.; Peng, Y.Y.; Meng, G.W.; Pu, S.Q. Field measurement and analysis of time and space effects of excavation construction of connected deep foundation pits in Ningbo soft soil area. *J. Railw. Sci. Eng.* **2020**, *17*, 82–94.

22. Liu, X.R.; Wang, L.F.; Chen, F.; Yuan, X.F.; Lin, G.Y. Research on deformation characteristics and parameter optimization of inner braced ground wall. *J. Undergr. Space Eng.* **2021**, *17*, 727–738.

23. Jia, M.; Yang, X.H.; Ye, J.Z. Corner effect of active earth pressure for small-sized excavation. *J. Harbin Inst. Technol.* **2016**, *48*, 95–102.

24. Lou, C.H.; Xia, T.D.; Liu, N.W. Spatial effect analysis of the impact of foundation pits on the surrounding environment in soft soil areas. *Chin. J. Geotech. Eng.* **2019**, *41*, 249–252.

25. Li, L.X.; Zhang, Y.J.; Hu, X.B. Analysis of excavation impact of a pit-in-pit based on PLAXIS 3D finite element software. *J. Undergr. Space Eng.* **2016**, *12*, 254–261, 266.

26. Wang, S.R.; Li, Y.; Li, Z.F. Experimental on squeezed pan piles in single and double rows under loading for protecting foundation excavation. *J. China Coal Soc.* **2009**, *34*, 537–541.

27. Wang, S.; Li, D.; Li, C.; Zhang, C.; Zhang, Y. Thermal radiation characteristics of stress evolution of a circular tunnel excavation under different confining pressures. *Tunn. Undergr. Space Technol.* **2018**, *78*, 76–83. [CrossRef]

28. Sun, X.M.; He, M.C.; Liu, C.Y.; Gu, J.C.; Wang, S.R.; Ming, Z.Q.; Jing, H.H. Development of nonlinear triaxial mechanical experiment system for soft rock specimen. *Chin. J. Rock Mech. Eng.* **2005**, *24*, 2870–2874.

29. Xu, C.; Li, Z.; Wang, S.; Wang, S.; Fu, L.; Tang, C. Pullout performances of grouted rockbolt systems with bond defects. *Rock Mech. Rock Eng.* **2018**, *51*, 861–871. [CrossRef]

30. Wang, S.; Xiao, H.; Hagan, P.; Zou, Z. Mechanical behavior of fully-grouted bolt in jointed rocks subjected to double shear tests. *Dynamic* **2017**, *92*, 314–320.

Article

Influence of Gas Pressure on the Failure Mechanism of Coal-like Burst-Prone Briquette and the Subsequent Geological Dynamic Disasters

Ying Chen [1,*], Zhiwen Wang [1], Qianjia Hui [1], Zhaoju Zhang [1], Zikai Zhang [1], Bingjie Huo [1], Yang Chen [2] and Jinliang Liu [3]

[1] Mining Engineering School, Liaoning Technical University, Fuxin 123000, China
[2] Research Center for Rock Burst Control, Shandong Energy Group Co., Ltd., Jinan 250014, China
[3] Huafeng Coal Mine, Xinwen Mining Group Co., Ltd., Taian 271413, China
* Correspondence: chenying@lntu.edu.cn

Abstract: Rock bursts and coal and gas outbursts are geodynamic disasters in underground coal mines. Laboratory testing of raw coal samples is the dominant research method for disaster prediction. However, the reliability of the experimental data is low due to the inconsistency of the mechanical properties of raw coal materials. The utilization of structural coal resources and the development of new coal-like materials are of significance for geodynamic disaster prediction and prevention. This paper studies the failure characteristics and dynamic disaster propensities of coal-like burst-prone briquettes under different gas pressures. A self-made multi-function rock–gas coupling experimental device was developed and burst-prone briquettes were synthesized, which greatly improved the efficiency and precision of the experimental data. The results showed that the burst proneness of the briquette was thoroughly reduced at a critical gas pressure of 0.4 MPa. When the gas pressure was close to 0.8 MPa, both the bearing capacity and the stored burst energy reduced significantly and the dynamic failure duration extended considerably, indicating the typical plastic-flow failure characteristics of coal and gas outbursts. The acoustic emission monitoring results showed that with the increase in gas pressure, the post-peak ringing and the AE energy ratio of coal samples increased, suggesting that the macroscopic damage pattern changed from bursting-ejecting of large pieces to stripping–shedding of small fragments adhered to mylonitic coal. In addition, the transformation and coexistence of coal failure modes were discussed from the perspectives of coal geology and gas migration. This study provides a new method for the scientific research of compound dynamic disaster prevention in burst coal mines with high gas contents.

Keywords: compound dynamic disasters; rock burst; coal and gas outburst; gas pressure; failure mode; briquette

Citation: Chen, Y.; Wang, Z.; Hui, Q.; Zhang, Z.; Zhang, Z.; Huo, B.; Chen, Y.; Liu, J. Influence of Gas Pressure on the Failure Mechanism of Coal-like Burst-Prone Briquette and the Subsequent Geological Dynamic Disasters. *Sustainability* **2023**, *15*, 7856. https://doi.org/10.3390/su15107856

Academic Editor: Antonio Miguel Martínez-Graña

Received: 12 March 2023
Revised: 8 May 2023
Accepted: 9 May 2023
Published: 11 May 2023

1. Introduction

Geodynamic disasters in underground coal mines mainly include rock bursts and coal and gas outbursts. Both pose a serious threat to mining safety and present research areas in the underground coal mining industry [1,2]. The geoconditions that lead to their occurrence and appearance patterns are noticeably different, but they both solely occur in shallow underground mines. After years of research, prediction, monitoring, and prevention technology systems have been gradually established.

In recent years, with the increase in the mining depth, not only are geodynamic disasters more serious and frequent, but they have also become more complicated. For example, a large amount of abnormal gas emissions has been found to occur in coal bursts [3–5] and coal and gas outburst disasters have occurred in high burst-risk coal seams [6,7]. In addition, several coal seams have been identified as highly burst-prone, containing high levels of gas [4–8]. These new geodynamic disaster phenomena are attributed to the deterioration of geological conditions, such as high ground stresses, high temperatures, and softening of the surrounding rock, and have brought great challenges to scientific research and to the current engineering practice of geodynamic disaster prediction and prevention in coal mines.

As two geodynamic disasters can undergo subsidiary transformations or even co-exist, a new research field has been established, namely compound dynamic disasters (CDDs), which include coal bumps along with abnormal gas emissions and rock bursts induced by coal and gas outbursts [8,9]. The study of CDDs focuses on the mutual influence, transformation, and interaction of the two kinds of dynamic disasters, as well as targeted prediction and prevention measures.

Two research methodologies are mainly employed to study compound dynamic hazards. One is case-based field studies, focusing on prediction and prevention technology corresponding to a site's specific geoconditions and mining arrangements and the reverse analysis of the mechanism of disaster occurrence [10,11]. For example, Zhao et al. [12] established a regional CDDs prediction and control technical system based on the characteristics of coal bursts and gas outbursts that occurred in the Pingdingshan coalfield. However, the research results relied highly on disaster data collection which cannot be carried out on a large scale. Another research method uses the theory of rock mechanics to investigate the occurrence process and key factors of disasters based on theoretical analyses or laboratory experiments [13–15]. These two research methods close the gap between theory and practice from the perspective of science and engineering.

For coal burst-dominated disasters, the change in the likelihood of coal bursts is mainly studied from the perspective of dynamic and static loading superposition or burst energy accumulation and dissipation. Dynamic and static load superposition is a traditional theory in rock burst research and has also been introduced into the CCDs research field in recent years [16–18]. One scientific work laid the foundation for prediction technology and determined the formula to calculate the burst tendency index of coal in compound dynamic disasters [19]. From the perspective of energy accumulation and dissipation, Song et al. [20] analyzed the energy of coal samples with a high impact tendency under different gas pressures under uniaxial and cyclic loads; however, the regularity of the results was weak due to data divergence. Xu et al. [21] proposed a modified method to determine coal burst proneness from the perspective of burst energy in cases of compound dynamic disasters dominated by coal bursts; however, only samples weakly prone to burst were involved. In short, the research to determine burst energy is still lacking in data.

Regarding the phenomenon of abnormal gas emissions after a coal burst, current research results show that this is mainly related to the micro-fracture structure of coal and rock, the amount of gas adsorption and analysis conditions, and the environmental temperature [3–5,21–24]. High ground stresses and high temperatures in deep coal seams induce compound dynamic disasters, and the effect of shock waves caused by a rock burst on the gas adsorption ability is the direct result of abnormal gas emissions.

In the research field of the fundamental theory of CDDs, some mechanical models have been established based on the gas–solid coupling constitutive method, and the critical state and influencing factors of the combination and mutual transformation of rock bursts and gas outbursts have been obtained. Noticeable research achievements include a mechanical model for CDDs in circular roadways [25]. Subsequent research on this topic [8] optimized the boundary conditions of the model. However, these theoretical analysis results can only be used as a reference, as they make various assumptions, and it is difficult to measure the relevant parameters.

In recent years, experimental research on CDDs has mainly focused on the impact of gas on burst propensity. However, due to the use of raw coal samples, conclusions drawn from these experimental results are quite different [20,26,27]. Acoustic emission (AE) technology has been widely used in the study of single disasters of rock bursts and coal and gas outbursts [28,29]; however, the effect of gas pressure on the microstructure damage and fracture development of the coal body has not yet been determined.

In this study, the transformation and combination of dynamic disasters under different gas pressures are analyzed using experimental methods. A multi-function solid–gas coupling device was developed and burst-prone briquette specimens were used to improve the reliability and repeatability of the experimental results. The effect and influence of gas pressure on the burst characteristics of coal samples are explored. The energy evolution, catastrophic tendency, and mechanical mechanism of coal deformation under different gas pressures are further explored based on the AE characteristics of coal samples under loading conditions. This study provides new experimental methods and analytical techniques for the study of CDDs and provides the basic theory for the engineering practice of disaster reduction and prevention in deep coal mines with high gas contents.

2. Materials and Methods

2.1. Mold Design

In situ coal samples normally have internal micro defects and joints, leading to a huge inconsistency between samples. In addition, coal with a low burst proneness tends to be soft and heavily fractured, resulting in significant difficulties regarding sample testing. As such, in order to reduce the inconsistency between the testing samples, as well as the scatter in testing data, homogeneous coal briquettes with controlled strengths were prepared for testing.

A 50 mm × 100 mm standard size cylindrical mold was built for producing specimens. The mold consists of a compressive arm, two semi-circular steel tubes, a bottom plate, and three fixing devices to bolt the two semi-circular steel tubes at three different positions (Figure 1). The bottom plate is placed at the bottom of the cylindrical tube. Once the coal powder is poured into the cylindrical mold, the compressive arm is pulled down to compact the powder until the coal specimen is successfully formed. All the components of the mold are made of stainless steel. A schematic diagram of the mold is shown in Figure 1.

Figure 1. Specimen-producing mold. ① Semi-circular steel tube, ② compressive arm, ③ bottom plate, ④ fixing device.

2.2. Specimen Preparation

In order to prepare a coal specimen with appropriate strength and burst proneness, the coal powder and a binder were mixed under appropriate conditions. The coal powder was crushed and ground such that the particle diameter was below 0.3 mm. Then, it was mixed with PS32.5 cement, Hypromellose, and sodium humate powder in different ratios.

As a result of extensive laboratory tests to determine the optimal mixing ratio, the coal specimen was prepared with coal powder, cement, hypromellose, sodium humate powder, and water in the following ratio: 0.9:1.7:0.3:0.3:0.4. The compressive strength of a coal specimen with such a ratio is 7–8 MPa.

The specimen preparation process is shown in Figure 2. The specific process was as follows:

Step 1: The coal powder was ground until all particles had a diameter of less than 0.3 mm and then the coal powder was mixed with cement, hypromellose, sodium humate powder, and water at the optimal ratio mentioned above.

Step 2: The mixture was poured into the mold up to a height of 2 cm and then the material was compressed under 50 kN of compressive stress for 10 min. This process was repeated until the height of the specimen was 10 cm.

Step 3: Once the specimen was successfully formed, it was compressed using the compressive arm and then the sample was removed from the mold.

Step 4: The specimen was placed in a 50 °C oven for over 48 h to dry and harden.

The mixing ratio, mold, pressing stress (50 kN), drying, and hardening environment used in the sample preparation are all the same, so the physical and mechanical properties of the specimen are highly consistent.

Figure 2. Procedure for coal specimen preparation.

2.3. New Multi-Phase (Solid–Gas Coupling) Testing Facility

The new testing facility consisted of the following components. (1) A servo control loading system with a solid–gas coupling device. The maximum vertical loading force of the servo control loading system is 600 kN. The solid–gas coupling device is a sealed chamber that has a maximum gas pressure of 20 MPa. The function of this component is to apply vertical stress to the specimen under a given gas pressure. (2) A gas supply system with a vacuum pump and a gas cylinder. The gas cylinder can provide a maximum gas pressure of 15 MPa. The maximum pressure of the vacuum pump is less than 0.06 Pa. The function of this component is to place the solid–gas coupling device under vacuum and inject gas into the chamber to achieve adsorption saturation of the specimen under a given gas pressure. (3) An acoustic emission (AE) monitoring system. During the experiment, an acoustic emission signal sensor was installed on the surface of the specimen. The acoustic emission signal generated during the loading process was collected and stored through the AE monitoring system. (4) An external monitoring system. This component uses high-speed cameras to record the process of specimen failure. The multi-phase (solid–gas coupling) testing facility has been granted a patent by China National Intellectual Property Administration. The laboratory testing system is illustrated in Figure 3.

2.4. Testing Scheme

The solid–gas coupling testing chamber containing the coal specimen was placed under vacuum before the test in order to inject pure gas. The gas pressure was monitored during gas injection until the pressure reached the required level. In this study, identical experiments were conducted under gas pressures of 0.2, 0.4, 0.6, 0.8, and 1 MPa in order to conduct a comparative analysis. Then, the chamber was left for 24 h to allow the specimen to be fully saturated with gas. Once the specimen was ready, it was tested with the compression machine. Note that the gas pressure was maintained at a constant level for the duration of the test. Burst proneness indicators included the uniaxial compressive strength, the duration of dynamic fracture, and the bursting energy index. In the meantime, an AE monitor was employed to capture the continuous change in AE counts and energy. The detailed procedure for testing was as follows:

Step 1: The coal specimen was placed into the solid–gas coupling testing chamber. An AE sensor was attached to the coal specimen surface within the solid–gas coupling testing chamber. Gas was then injected into the chamber until the gas pressure reached 0.2 MPa.

Step 2: After the coal specimen was immersed in the gas at the required gas pressure for 24 h, burst proneness tests were conducted according to the GB/T25217.2-2010 [30] standard to measure the uniaxial compressive strength, the duration of dynamic fracture and the bursting energy index.

Step 3: The previous two steps were repeated to measure the burst proneness indicators under different gas pressures (0.4, 0.6, 0.8, and 1 MPa).

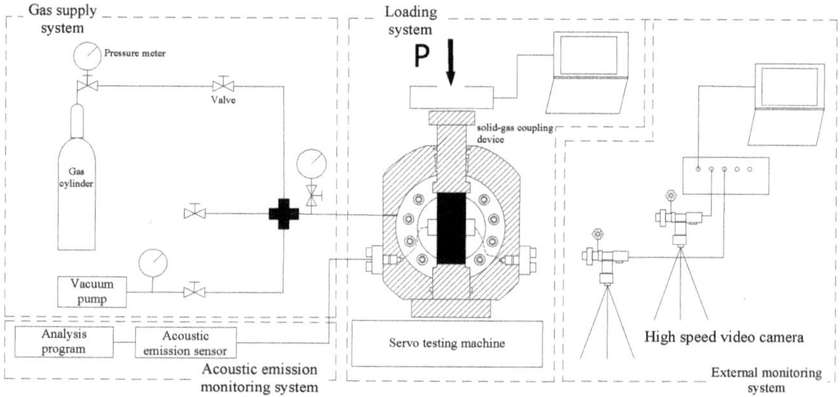

Figure 3. Schematic diagram of the new multi-phase testing facility.

3. Results

3.1. The Effect of Gas on Coal Failure Modes

The macroscopic damage patterns of coal specimens at various gas pressures ranging from 0 to 1 MPa were observed (Figure 4). With no gas present, single-slope shear failure occurred in the specimen, and it was broken into larger blocks that were ejected at the moment of failure. In contrast, when gas was present, more cracks were generated as a result of loading and the size of the cracked bulk tended to decrease. When the gas pressure was 0.2 MPa, some mylonitized coal was observed. When the gas pressure was increased up to 0.4 MPa, the number of cracks increased while the size of the cracks decreased. In the meantime, a large amount of mylonitized coal was formed. The failure of the coal specimen was mainly characterized by the stripping and shedding of small fragments. The test results show that with the increase in gas pressure, the macroscopic damage pattern of coal gradually changes from bursting-ejecting of large pieces to stripping–shedding of small fragments attached to mylonitic coal.

According to the test results, it is hypothesized that the influence of gas on the mechanical properties, failure mode, and even catastrophic characteristics of coal is the combined result of the mechanical response of the inherent properties of coal to an external load and the adsorption softening of gas on the surface of coal matrix particles. There is a critical gas pressure for this effect. When the gas pressure is lower than the critical value, the macroscopic mechanical response of coal is affected by its inherent properties. When the gas pressure exceeds the critical value, the mechanical properties of coal are gradually affected by the gas. Thus, it is very important to determine the critical gas pressure to be able to control the change in the mechanical properties of coal.

Figure 4. Macroscopic damage pattern of coal specimens under different gas pressures.

3.2. The Effect of Gas on Burst Proneness

According to the national standards of the People's Republic of China (GB/T 25217.2-2010) [30], "methods for testing, monitoring and prevention of rock bursts, part 2: classification and laboratory test method on bursting liability of coal", the burst proneness of a coal specimen can be comprehensively measured according to four indexes, including the uniaxial compressive strength, the duration of dynamic fracture, the elastic strain energy index and the bursting energy index. The burst proneness of coal is classified according to its index value (Table 1).

Table 1. Criterion of burst proneness of coal.

Category		I	II	III
Burst Proneness		None	Weak	Strong
Index	Duration of dynamic fracture DT/ms	$DT > 500$	$50 < DT \leq 500$	$DT \leq 50$
	Elastic strain energy index W_{ET}	$W_{ET} < 2$	$2 \leq W_{ET} < 5$	$W_{ET} \geq 5$
	Bursting energy index K_E	$K_E < 1.5$	$1.5 \leq K_E < 5$	$K_E \geq 5$
	Uniaxial compressive strength R_c/MPa	$R_c < 7$	$7 \leq R_c < 14$	$R_c \geq 14$

3.2.1. Uniaxial Compressive Strength

The uniaxial compressive strength refers to the ratio of the failure load of the coal specimen under uniaxial compression to the area of its loading surface. The compressive strength variation curve of coal specimens under different gas pressures is illustrated in Figure 5. The maximum compressive strength of coal specimens without any gas present was 8.38 MPa, which is higher than that with the gas present. This indicated that the burst proneness of coal specimens without gas is weak according to Table 1. At a gas pressure of 0.2 MPa, the compressive strength was reduced by 8.5% to 7.66 MPa. The coal specimen is still weakly burst-prone under these conditions. On the other hand, when the gas pressure was increased to 0.4 MPa, the coal compressive strength was reduced by 28.8% to 5.96 MPa.

The compressive strength was further reduced by 36% to 5.36 MPa when the gas pressure was increased to 0.6 MPa. At gas pressures of 0.8 and 1 MPa, the compressive strengths were 4.99 and 4.60 MPa, exhibiting a 40% and 45% reduction, respectively. As a result, it is apparent that gas absorption could reduce the compressive strength of coal. Considering only the compressive strength index (based on Table 1), the burst proneness of the coal specimen is lost at a critical gas pressure of 0.4 MPa.

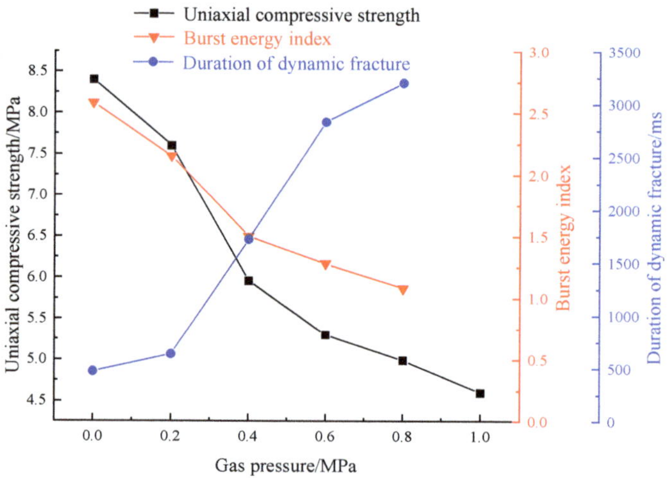

Figure 5. Change in burst proneness at different coal seam gas pressures.

Under gas pressure, expansion stress is generated inside the sample. This weakens the bonding among matrix particles and promotes the development of internal cracks in the sample, resulting in higher damage values at peak strain [31]. This reduces the coal's bearing capacity and increases its softness.

3.2.2. Duration of Dynamic Fracture

The duration of dynamic fracture refers to the time from ultimate strength to complete failure of a coal specimen under uniaxial compression. Figure 5 shows the duration of the dynamic fracture variation curve under different gas pressures. It can be observed that the duration of dynamic fracture increased from 475 ms to 3540 ms as the gas pressure increased from 0 MPa to 1 MPa. When the gas pressure was 0.2 MPa, the duration of the dynamic fracture of the specimen was 673 ms, indicating a loss of burst proneness (based on Table 1). It is therefore concluded that the duration of dynamic fracture increases with the increase in gas pressure. This leads to the gas-saturated coal specimen changing from brittle to tough. The increase in the duration of the dynamic fracture indicates that after the stress peak, the failure process and energy release become slow and moderate, which inhibits bursting. Considering only the duration of dynamic fracture (based on Table 1), the burst proneness of the coal specimen is lost at a critical gas pressure of 0.2 MPa.

3.2.3. Burst Energy Index

The burst energy index refers to the ratio of the accumulated energy before the stress peak to the consumed energy after the stress peak under uniaxial compression. The burst energy index variation curves under different gas pressures are illustrated in Figure 5. It can be observed that the burst proneness tends to decrease as the gas pressure increases, and when there is no gas present, the burst energy index is 2.58, indicating a weak burst proneness (according to Table 1). Following a continuous increase in gas pressure to 0.2 MPa and 0.4 MPa, the burst energy indices were 2.26 and 1.52, respectively, indicating that the coal is burst-prone in both scenarios. Finally, the burst energy indices were 1.28

and 1.08 at gas pressures of 0.6 and 0.8 MPa, respectively, which indicates that the coal is not burst-prone in both of these scenarios. Considering only the burst energy index (based on Table 1), the burst proneness of the coal specimen is lost at a critical gas pressure of 0.6 MPa.

The calculated curve of the burst energy index under different gas pressures is shown in Figure 6 (taking no gas and a gas pressure of 0.6 MPa as an example). Without gas, the accumulated strain energy before the stress peak is equal to the area surrounded by $0C_0Q_0$, and the consumed energy after the stress peak is equal to the area surrounded by $Q_0C_0D_0F_0$. Under a gas pressure of 0.6 MPa, the accumulated strain energy before the stress peak is equal to the area surrounded by $0C_{0.6}Q_{0.6}$, and the consumed energy after the stress peak is equal to the area surrounded by $Q_{0.6}C_{0.6}D_{0.6}F_{0.6}$. It can be observed from Figure 7 that under no gas, the stress increases rapidly after loading. The slope of the curve clearly indicates the hard characteristics of the specimen. The accumulated strain energy is more than that of gas-saturated coal (the area surrounded by $0C_0Q_0$ is larger than that surrounded by $0C_{0.6}Q_{0.6}$). After the stress peak, the stress–strain curve suddenly and sharply decreases, which indicated that the specimen quickly loses bearing capacity, resulting in a strong brittle bursting failure. At the same time, it also shows that the failure after the stress peak consumes less energy, and a large amount of energy is transferred to the kinetic energy of the fragments. Larger fragments are generated after the specimen is destroyed, and ejection occurs at the moment of destruction. With the increase in gas pressure, the strain of the coal specimen increases quickly, and the load capacity decreases significantly after loading, which demonstrates obvious softening compared to the coal specimen that does not contain gas. The accumulated strain energy is less than that of the coal specimen that does not contain gas. After the stress peak, the stress–strain curve decreases gently, indicating a plastic flow failure, which further indicates that the energy consumed after the stress peak increases (the area surrounded by $Q_{0.6}C_{0.6}D_{0.6}F_{0.6}$ is larger than that surrounded by $Q_0C_0D_0F_0$) and the surplus energy used to eject the fragments decreases. As the gas pressure increases, the fragmentation of the specimen increases. The damage pattern of the specimen changes from ejection to stripping–falling.

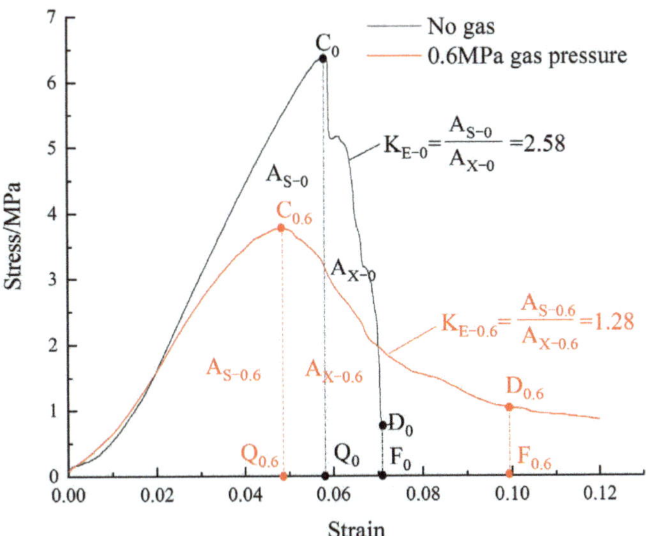

Figure 6. Calculated curve of the burst energy index under different gas pressures.

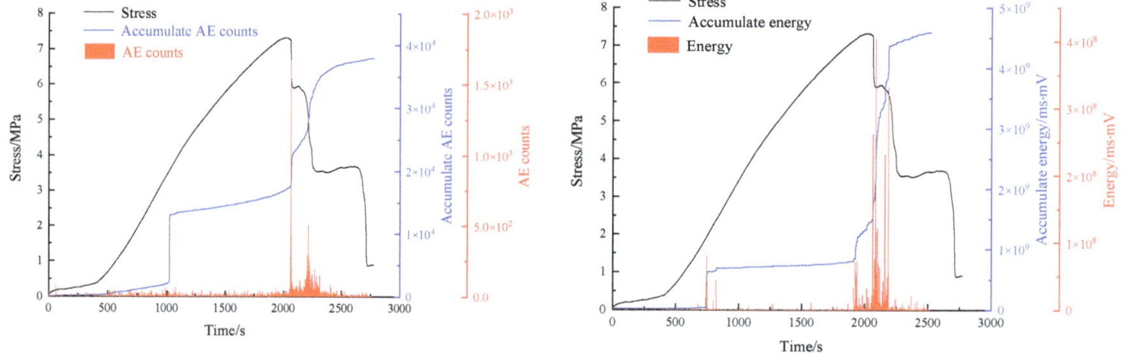

(**a**) AE counts and accumulated counts (**b**) AE energy and accumulated energy

Figure 7. AE counts and energy characteristics under no gas.

3.3. AE Data Analysis for Gas-Saturated Coal Specimen

The AE counts and energy characteristics of the specimen during the loading process are demonstrated in Figures 7 and 8. Significant AE counts were recorded when the axial loading was close to the stress peak, which indicates that the formation of internal microcracks was initiated at the end of the elastic deformation stage. As the stress continued to increase, the cracks propagated until they were fully developed throughout the whole specimen. The maximum AE count and energy appear before and after the stress peak, which illustrates that the failure and released energy are most significant near the stress peak.

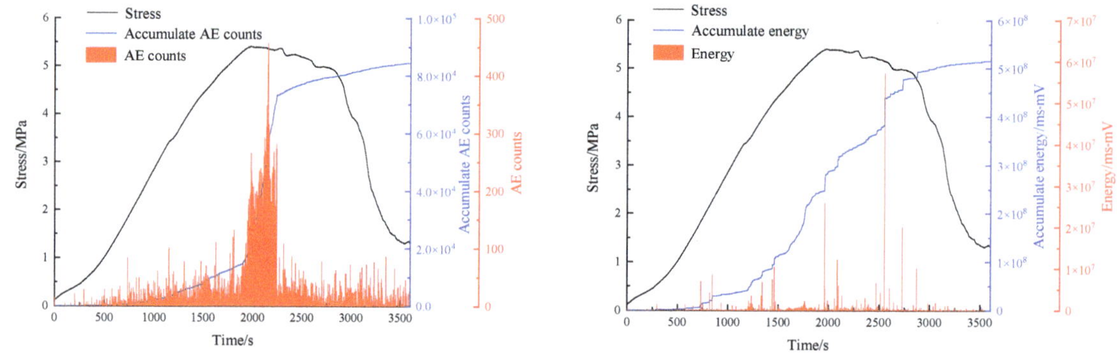

(**a**) AE counts and accumulated counts (**b**) AE energy and accumulated energy

Figure 8. AE counts and energy characteristics under a 1 MPa gas pressure.

The distributions of AE counts and energy under different gas pressures were statistically analyzed, and the results are shown in Figures 9 and 10. When the gas pressure increased from 0 to 1 MPa, the maximum AE count dropped from 1673 to 455 and the corresponding maximum energy dropped from 4.04×10^8 ms·mV to 5.71×10^7 ms·mV. It is also noteworthy that under the influence of the gas, an increased AE count and energy were observed during the residual strength stage of the coal specimen compared to those without gas. The proportion of AE counts after the stress peak increased from 4.4% to 24.2%. The proportion of AE energy after the stress peak increased from 1.3% to 27.6%.

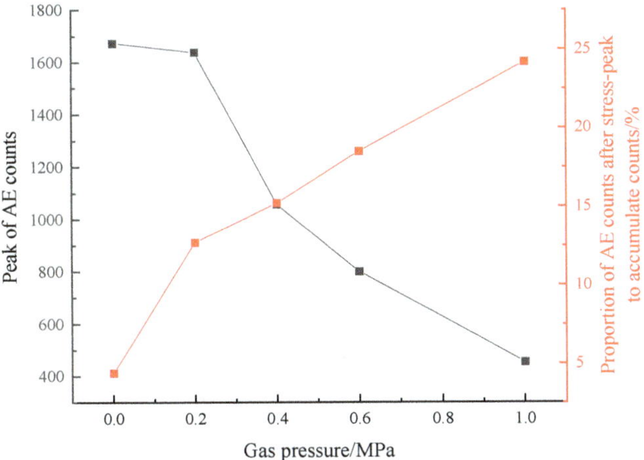

Figure 9. AE count characteristics under different gas pressures.

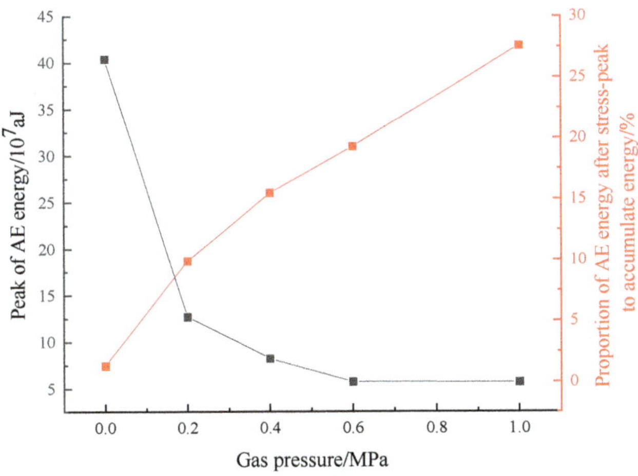

Figure 10. AE energy characteristics under different gas pressures.

Under gas pressure, the amount of gas attached to the surface of the coal matrix particles increases and the cementation continuity between the matrix particles weakens. The irregular micro-strain of the matrix produced by the adsorption of gas leads to relative dislocation between particles [22] and causes the development of micro-cracks, which increases the initial damage to the bearing structure. After a load is placed on the specimen, the damage is further aggravated, which inhibits the accumulation of strain energy. This is the reason why the AE counts and energy are lower at the stress peak of the specimen. With the increase in gas pressure, the proportion of AE counts and energy increases after the stress peak, which indicates that sustained plastic failure occurs in the specimen. This consumes most of the energy, and thus the energy used to produce bursting–ejecting failure decreases. This is the reason why the burst energy index of the specimen decreases with the increase in gas pressure.

4. Discussion

Burst proneness is a natural property of coal. The uniaxial compressive strength, the duration of dynamic fracture, and the burst energy index are important indices to characterize this property. Accurately obtaining their values is of great significance to the prediction and prevention of rock bursts. With the gradual deepening of coal mining, the high-pressure gas contained in coal seams becomes a significant factor that cannot be ignored in rock burst prevention. In this paper, the variation in macroscopic failure characteristics, the compressive strength, the duration of dynamic fracture, and the burst energy index of coal with the increase in gas pressure was determined by extensive testing. The results showed that gas alters the inherent burst proneness of coal and also affects the occurrence potential of rock bursts.

The gas pressure and gas content in deep coal seams are generally high. High gas pressures have a significant impact on the burst proneness of coal, by weakening the strength of coal, prolonging the dynamic failure duration, reducing the burst energy index, and leading to increased strain energy consumption after the stress peak of the specimen. The failure mode of coal changes from brittle bursting to plastic flow. The absorbed gas softens the coal, leading to smaller broken pieces and the appearance of mylonitic coal. Gas desorption causes a large amount of free gas to accumulate, which provides prime conditions for the rock burst to induce a gas outburst. Therefore, deep, high-gas-content coal seams have a lower critical index of catastrophe and an indistinct propensity to cause disasters. These coal seams are easily disturbed by mining, which can induce rock bursts, gas emissions, and outbursts. Destructive damage to the production environment is the result of compound disasters.

This paper analyzed the influence of gas pressure on the failure mechanisms and the burst proneness of coal-like briquettes which are only weakly prone to bursting. The relationship between gas pressure and the mechanical properties of coal samples with a high burst proneness needs to be further studied.

5. Conclusions

(1) A new multi-phase (solid–gas coupling) testing facility composed of a servo control loading system with a solid–gas coupling device, a gas supply system, an acoustic emission (AE) monitoring system, and an external monitoring system has been developed. The preparation of burst-prone coal briquettes was performed for the first time. This provides an experimental basis to determine the influence of gas on the failure mode and burst proneness of specimens.

(2) The gas content has a great influence on the failure mode of coal specimens. With an increase in gas pressure, the macroscopic damage pattern of coal specimens gradually changes from bursting-ejecting of large pieces to stripping–shedding of small fragments adhered to mylonitic coal.

(3) The gas pressure affects the burst proneness index of coal specimens and causes the weakening and loss of the burst properties of coal specimens. The compressive strength of the coal specimen decreases with the increase in gas pressure, and the critical gas pressure for the loss of burst proneness is 0.4 MPa. The duration of dynamic fracture of coal specimens increases with the increase in gas pressure, and the critical gas pressure for the loss of burst proneness is 0.2 MPa. The burst energy index decreases with the increase in gas pressure, and the critical gas pressure for the loss of burst proneness is 0.6 MPa. The duration of dynamic fracture is more sensitive to gas pressure.

(4) The ratio of the AE counts and energy after the stress peak to the accumulated AE counts and energy increases with the increase in gas pressure, which shows that the consumed energy after the stress peak increases. This eventually leads to the loss of burst proneness of the coal specimen.

(5) Based on the test results, it is suggested that when identifying the burst proneness of deep, high-gas-content coal seams, it is necessary to account for the corresponding gas pressure in the environment in the relevant burst proneness tests.

Author Contributions: Methodology, Y.C. (Ying Chen); Validation, Z.Z. (Zhaoju Zhang); Formal analysis, Z.W. and Q.H.; Investigation, B.H.; Resources, J.L.; Data curation, Z.W.; Writing—original draft, Y.C. (Ying Chen); Writing—review and editing, Y.C. (Ying Chen); Visualization, Z.Z. (Zikai Zhang); Funding acquisition, Y.C. (Yang Chen). All authors have read and agreed to the published version of the manuscript.

Funding: This research was funded by the National Natural Science Foundation of China (Grant No. 51874164), Major Scientific and Technological Innovation Projects in Shandong Province (Grant No. 2019SDZY02), and the Science and Technology Plan Announced Bidding Project in Shanxi Province (Grant No. 20191101015).

Institutional Review Board Statement: Not Applicable.

Informed Consent Statement: Not Applicable.

Data Availability Statement: The data used to support the findings of this study are available from the corresponding author upon reasonable request (chenying@lntu.edu.cn).

Conflicts of Interest: The authors declare no conflict of interest.

References

1. Balovtsev, S.V.; Skopintseva, O.V.; Kolikov, K.S. Aerological risk management in preparation for mining of coal mines. *Sustain. Dev. Mt. Territ.* **2022**, *14*, 107–116. [CrossRef]
2. Bosikov, I.I.; Martyushev, N.V.; Klyuev, R.V.; Savchenko, I.A.; Kukartsev, V.V.; Kukartsev, V.A.; Tynchenko, Y.A. Modeling and complex analysis of the topology parameters of ventilation networks when ensuring fire safety while developing coal and gas deposits. *Fire* **2023**, *6*, 95. [CrossRef]
3. Wang, Z.; Yin, G.Z.; Hu, Q.T.; Jin, H.W. Inducing and Transforming Conditions from Rockburst to Coa-l Gas Outburst in a High Gassy Coal Seam. *J. Min. Saf. Eng.* **2010**, *27*, 572–575.
4. Wang, T.; Wang, Z.H.; Liu, H.B.; Guan, Y.P.; Zhan, S.J. Discussion about the mechanism of gas disaster induced by coal bump. *J. China Coal Soc.* **2014**, *39*, 371–376.
5. Sun, X.P.; Ma, Z.Y.; Zhang, F. Analysis on Causes of Abnormal Gas Emission from Working Faces with Rock Burst and Research on Its Treatment Technology. *Coal Technol.* **2021**, *40*, 122–125.
6. Yan, W.W. Cause analysis and prevention technology of abnormal gas emission in working face with rock burst danger. *China Energy Environ. Prot.* **2021**, *43*, 24–27.
7. Zhang, F.; Ma, Z.Y. Mechanism of abnormal gas emission from mining face with rock burst danger and its control. *China Energy Environ. Prot.* **2021**, *43*, 16–20.
8. Li, Z.H.; Zhang, Y.; Liang, Y. Mechanism study on compound disaster of rock burst and gas outburst. *Coal Sci. Technol.* **2021**, *49*, 95–103.
9. Chen, B. Stress-induced trend: The clustering feature of coal mine disasters and earthquakes in China. *Int. J. Coal Sci. Technol.* **2020**, *7*, 676–692. [CrossRef]
10. Gao, B.B.; Ren, C.N.; Dong, Q.; Chen, L.W. Study on dynamic behavior law and microseismic monitoring in stoping process of roadway with high gas and wide coal pillar. *Shock Vib.* **2021**, *2021*, 1–14. [CrossRef]
11. Zhang, J.G.; Lan, T.W.; Wang, M.; Gao, M.Z.; Rong, H. Prediction method of deep mining dynamic disasters and its application in Pingdingshan mining area. *J. China Coal Soc.* **2019**, *44*, 1698–1706.
12. Zhai, C.; Xiang, X.W.; Xu, J.Z.; Wu, S.L. The characteristics and main influencing factors affecting coal and gas outbursts in Chinese Pingdingshan mining region. *Nat. Hazards* **2016**, *82*, 507–530. [CrossRef]
13. Kong, X.G.; Yang, S.R.; Wang, E.Y.; Li, S.G.; Lin, H.F.; Ji, P.F. Fracture evolution and gas emission characteristics of raw coal samples subjected to instantaneous disturbance of impact load. *Chin. J. Rock Mech. Eng.* **2022**. [CrossRef]
14. Zhang, L.; Qi, Q.X.; Ren, T.; Li, X.C.; Gao, K.; Li, C.Y.; Li, X.P.; Yuan, H.H. Study on the damage and fracture characteristics of coal rock based on the X-ray micro-CT scanning technology and statistical strength theory. *Coal Sci. Technol.* **2022**. [CrossRef]
15. Ouyang, Z.H.; Zhang, G.H.; Li, Q.W.; Lai, X.P.; Qin, H.Y.; Zhao, X.D.; Zhou, X.X. Study on the rock burst tendentiousness of coal under different gas pressures. *Arab. J. Geosci.* **2019**, *13*, 179–187. [CrossRef]
16. Jiang, Y.D.; Pan, Y.S.; Jiang, F.X.; Dou, L.M.; Ju, Y. State of the art review on mechanism and prevention of coal bumps in China. *J. China Coal Soc.* **2014**, *39*, 205–213.
17. Amin, M.; Ming, C. Analysis of rockburst in tunnels subjected to static and dynamic loads. *J. Rock Mech. Geotech. Eng.* **2017**, *9*, 1031–1040.
18. Yin, G.Z.; Li, X.; Lu, J.; Li, M.H. Disaster-causing mechanism of compound dynamic disaster in deep mining under static and dynamic load conditions. *J. China Coal Soc.* **2017**, *42*, 2316–2326.
19. Dai, L.; Pan, Y.; Zhang, C.; Wang, A.; Canbulat, I.; Shi, T. New criterion of critical mining stress index for risk evaluation of roadway rockburst. *Rock Mech. Rock Eng.* **2022**, *55*, 4783–4799. [CrossRef]

20. Song, Z.L.; Han, P.B.; Li, W.P.; Yin, G.Z.; Li, M.H.; Kang, X.T. Impact of energy dissipation of coal samples with rockburst tendency from gas in its failure process. *J. China Coal Soc.* **2015**, *40*, 843–849.
21. Xu, J.; Jiang, J.D.; Xu, N.; Liu, Q.S.; Gao, Y.F. A new energy index for evaluating the tendency of rockburst and its engineering application. *Eng. Geol.* **2017**, *230*, 46–54. [CrossRef]
22. Espinoza, D.N.; Pereira, J.-M.; Vandamme, M.; Dangla, P.; Vidal-Gilbert, S. Desorption–induced shear failure of coal bed seams during gas depletion. *Int. J. Coal Geol.* **2015**, *137*, 142–151. [CrossRef]
23. Zhang, L.; Chen, S.; Zhang, C.; Fang, X.; Li, S. The Characterization of Bituminous Coal Microstructure and Permeability by Liquid Nitrogen Fracturing Based on μCT Technology. *Fuel* **2020**, *262*, 116635. [CrossRef]
24. Zhang, L.; Li, J.; Xue, J.; Zhang, C.; Fang, X. Experimental studies on the changing characteristics of the gas flow capacity on bituminous coal in CO_2-ECBM and N_2-ECBM. *Fuel* **2021**, *291*, 120115. [CrossRef]
25. Zhu, L.Y.; Pan, Y.S.; Li, Z.H.; Xu, L.M. Mechanisms of rockburst and outburst compound disaster in deep mine. *J. China Coal Soc.* **2018**, *43*, 3042–3050.
26. Zhang, G.H.; Ouyang, Z.H.; Qi, Q.X.; Li, H.Y.; Deng, Z.G.; Jiang, J.J. Experimental research on the influence of gas on coal burst tendency. *J. China Coal Soc.* **2017**, *42*, 3159–3165.
27. Gao, B.B.; Wang, Z.G.; Li, H.M.; Su, C.D. Experimental study on the effect of outburst-proneness of coal by gas pressure. *J. China Coal Soc.* **2018**, *43*, 140–148.
28. Zhang, X.; Tang, J.; Yu, H.; Pan, Y. Gas pressure evolution characteristics of deep true triaxial coal and gas outburst based on acoustic emission monitoring. *Sci. Rep.* **2002**, *12*, 21738. [CrossRef]
29. Hui, L.; Zeng, C.F.; Dong, Z.; Dong, D. Simulation Experiment and Acoustic Emission Study on Coal and Gas Outburst. *Rock Mech. Rock Eng.* **2017**, *50*, 2193–2205.
30. *GB/T25217.2-2010*; Methods for Test, Monitoring and Prevention of Rock Burst-Part 2: Classification and Laboratory Test Method on Bursting Liability of Coal. General Administration of Quality Supervision. Inspection and Quarantine of the People's Republic of China and China Standardization Administration: Beijing, China, 2010.
31. Ding, X.; Xiao, X.C.; Wu, D.; Lv, X.F.; Pan, Y.S.; Bai, R.X. Study on mechanical constitutive relationship and damage evolution of gas-bearing coal based on initial pore-cracks. *Mater. Rep.* **2021**, *35*, 18096–18103.

 *sustainability*

Article

Proposed Method for the Design of Geosynthetic-Reinforced Pile-Supported (GRPS) Embankments

Rashad Alsirawan *, Edina Koch and Ammar Alnmr

Department of Structural and Geotechnical Engineering, Széchenyi István University, 9026 Győr, Hungary
* Correspondence: alsirawan.rashad@sze.hu

Abstract: Soft soils with unfavorable properties can be improved using various ground-improvement methods. Among these methods, geosynthetic-reinforced pile-supported (GRPS) embankments are considered a reliable option for challenging ground conditions and time-bound projects. Nevertheless, the intricate load transfer mechanism of the GRPS embankment presents challenges due to the multiple interactions among its components. To overcome the limitations of current design methods that do not fully account for all interactions, a simplified design method has been developed for GRPS embankments. This method uses numerical analysis to predict pile load efficiency and geosynthetic tension. In this study, a validated model of the GRPS embankment, which incorporates certain simplifications for design purposes, was adopted. Based on this simplified model, a database of load efficiency and geosynthetic tension was collected to derive the design equations. The design method employed six parameters, namely, pile cap width, pile spacing, embankment height, oedometric modulus of the subsoil, geosynthetic stiffness, and embankment fill unit weight. The design process utilized Plaxis 3D and Curve Expert software. The results showed reasonable agreement between the findings of the proposed design method and the field measurements of eight case studies.

Keywords: simplified design method; load efficiency; geosynthetic tension; geosynthetic-reinforced pile-supported embankments

Citation: Alsirawan, R.; Koch, E.; Alnmr, A. Proposed Method for the Design of Geosynthetic-Reinforced Pile-Supported (GRPS) Embankments. *Sustainability* **2023**, *15*, 6196. https://doi.org/10.3390/su15076196

Academic Editors: Shuren Wang, Chen Cao and Hong-Wei Yang

Received: 24 February 2023
Revised: 23 March 2023
Accepted: 31 March 2023
Published: 4 April 2023

1. Introduction

Geosynthetic-reinforced pile-supported (GRPS) embankments are increasingly being used to provide support for railways, roads, and highways built on weak soils [1]. The use of geosynthetic reinforcement has been shown to increase loads transfer towards piles, resulting in benefits such as the reduction of required piles and the expedition of the construction process. Furthermore, incorporating geosynthetic layers underneath conventional pile-supported (CPS) embankments can address significant settlements, a low bearing capacity of the subsoil, and lateral displacement at the embankment toe [2,3].

By using fewer piles and reducing the use of concrete, GRPS embankments can help reduce embodied carbon (EC), which refers to the carbon dioxide released during the production, transportation, construction, and disposal of materials. As a result, GRPS embankments offer a sustainable and cost-effective solution for infrastructure development on weak soils [3].

This integrated system transfers loads resulting from the embankment weight and surcharge to a firm bearing soil layer. The load transmitted in a GRPS embankment is partitioned into three segments: (i) direct transmission to the piles via the soil arch, (ii) transmission to the piles by means of geosynthetic reinforcement, and (iii) transfer to the soft subsoil. The arching phenomenon observed in the first load segment can be attributed to the relatively higher stiffness of the piles compared to the subsoil. The second load segment induces two phenomena: the geosynthetic experiences membrane tension, and the geosynthetic–soil interface experiences frictional interaction. The third load segment results in the subsoil exerting an upward counter pressure known as "subsoil support" [1,4].

Several design methods are proposed to study the behavior of GRPS embankments. These methods belong to different generations. Terzaghi [5] developed the arching theory based on the trapdoor test. He also observed that shear forces are developed along the frictional surfaces due to differences in stiffness. Russell and Pierpoint's design method [6] used Terzaghi's concept to describe the behavior of the supported embankments, depending on a three-dimensional numerical model. In this method, the support from the soft subsoil is disregarded, and the entire embankment load is borne by the piles. Guido et al. [7] conducted several plate loading tests. This design method assumes that under three-dimensional conditions, the load distributes in the embankment body, forming a pyramid with an angle of 45°, and the geosynthetic carries the pyramid weight completely. Hewlett and Randolph [8] performed a set of three-dimensional model tests in which the soil arch was described as a hemi-spherical dome. This design method assumes that the limit states can occur in the arch crown or in the pile caps. Based on the assumptions of this method, the subsoil support is not considered. Zhang et al. [9] extended the soil-arching concept, representing the hemi-spherical domes as covering the triangular configurations of the piles. Hewlett and Randolph's design method was developed by Low et al. [10] based on a set of laboratory tests. This design method asserts that the arch shape is semi-cylindrical over walls of piles, and the soft subsoil carries part of the applied loads. Russel et al. [11] suggested a new design method using a finite difference analysis of a three-dimensional model in which the subsoil support is considered. Three-dimensional model tests were also carried out by Kempfert et al. [12] to develop a design method which assumes that a multi-shell dome represents the soil arch. The subsoil support is taken into account in this design method. The design method by Collin et al. [13] improved the approach of Guido et al. by creating a stiffened load transfer platform LTP (acting as a beam) with numerous geosynthetic layers to redirect a significant part of the load into the piles. The Nordic Geosynthetic Group [14] theorized that the shape of the soil supported by geosynthetic layers acts as a soil wedge in the design method. A theoretical design method using a semi-circular soil arch was proposed by Abusharar et al. [15]. This method includes the effect of the uniform surcharge and subsoil support, but it is considered mathematically complex.

In the BS 8006 design method [16], two formulas were used to compute the load imparted to the geosynthetic. One was the formula of Marston, which was developed based on plane strain tests on flexible culverts for high embankments; the other formula was Hewlett and Randolph's formula, which assumes that the soil arch is a hemi-spherical dome. Due to groundwater fluctuations, the subsoil support does not count in this method. Van Eekelen et al. [17] changed the amount of load borne by the geosynthetic in the case of Marston's formula. The EBGEO design method [18], which employs a triangular load distribution on the geosynthetic, utilizes the principles of Kempfert et al. [14]. An analytical model was proposed by Van Eekelen et al. [19] for designing GRPS embankments. The arch model is more complicated than other methods and can be described as a model of concentric hemispheres. This method is thought to be conceptually and mathematically challenging for use in design. Both EBGEO [18] and Van Eekelen et al. [19] take the support of subsoil into account but with a linear model.

Zhuang et al. [20] introduced a model to analyze the piled embankment, building upon the work of Hewlett and Randolph. This method proposed some refinements of the failure mechanisms in the crown and pile caps. The design method outlined in the Dutch standard CUR 226 [21] incorporates the arch model developed by Van Eekelen et al. [19]. This method assumes an inverted, triangular distribution of load on the geosynthetic layer when the subsoil provides no support, while a uniform distribution of load is assumed when subsoil support exists. Filz et al. [22] proposed a method of load displacement compatibility (LDC) to analyze GRPS embankments. This method proposed some refinements to Terzaghi's method to estimate the soil arch. The subsoil support is considered in this method. Pham's design method [1] combined the arching theory and the membrane theory. This method proposes the effect of the subsoil in two models (a linear model and a non-linear model). In addition, the frictional interactions between the geosynthetic layers and the surrounding soil are considered in the analysis.

Based on a review of the literature, none of the previously mentioned methods are able to precisely describe the behavior of a GRPS embankment. The moot points are noticeable, such as the soil arch model, load distribution over the geosynthetic layer(s), the role of the soft subsoil, and the subsoil behavior, if it is considered. This is due to the proposed simplifications in each design method, which are attributed to plenty of interactions between the elements of this system.

The aim of this paper is to propose a more comprehensive approach to design, using the finite element method (FEM). The characteristic of this method is the ability to consider the real behavior of underlying soft soil, the frictional behavior of the geosynthetic–soil interface, the consolidation of the soft soil, and all the expected interactions for which the other methods have not been able to complete this mission to different degrees. A validated model of a GRPS embankment with some simplifications is used to perform this study, which focuses on deriving two equations for determining the load efficiency of the piles and the tension in the geosynthetic. Additionally, the proposed design method is evaluated by comparing field measurements from a series of GRPS embankments with the results obtained from these equations to assess their validity.

2. Proposed Design Method

The aim of the method presented in this study is to propose a new concept of design that can avoid some of the drawbacks of experimental and theoretical methods. These analytical design methods differ in the determination of the shape of the soil arch, the magnitude of load applied to the geosynthetic reinforcement, the skin friction at the soil–geosynthetic interface, and the role of the soft subsoil and its real behavior. These variations in the analysis result in a significant difference between the outcomes of these analytical methods. Therefore, the FE method is widely regarded as an effective approach to addressing these challenges and achieving a dependable design.

The incorporation of numerical modeling in the design process contributes to ending the controversy over the moot points regarding the reciprocal behavior between the elements of this system. However, similar to former design methods and in order to avoid complicating the issue, this method is not able to include all the parameters related to embankment fill, soft soil, piles, and geosynthetic reinforcement because of their large number. For the purpose of the design, the following key parameters should be determined: (i) the load efficiency of the pile (E), to derive the pile bearing capacity; (ii) the tension in the geosynthetic (T), to derive the strain and vice versa.

Drawing on the aforementioned approach, the proposed design method employs numerical analysis as its fundamental basis. The following steps outline the methodology employed in the presented design approach:

1. A full-scale model of the GRPS embankment, which included a wide range of field measurements, is described in this study as the starting point.
2. To ensure the reliability of the numerical model, a calibration and validation process was conducted with a high level of accuracy.
3. For design purposes, simplifications were made to the numerical model of the GRPS embankment. These simplifications were carefully selected to reduce computational complexity and improve the efficiency of the analysis while still providing comprehensive and dependable results that are suitable for design.
4. Based on the simplified numerical model of the GRPS embankment, a wide database of the parameters (E; T) was collected through an extended parametric study. The database was analyzed using a mathematical technique to predict two equations that can be used to calculate the key parameters of E and T. Figure 1 shows the key steps of the proposed design method.

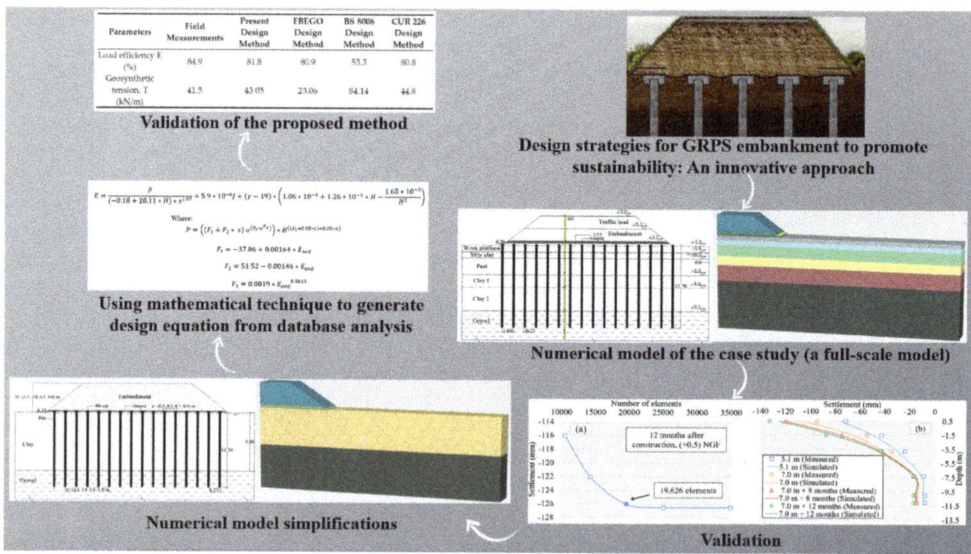

Figure 1. Methodology for a proposed design method.

2.1. Description of GRPS Embankment Model

A full-scale GRPS embankment was built over multiple layers of soft soil (silty clay, peat, and two layers of clay). The soil characterization was completed with six cone penetration tests, three boreholes, and a comprehensive suite of laboratory tests. The groundwater table was located 1.1 m under the surface of working platform, which was built prior to the construction stage to ease the equipment movement. A network of pre-cast concrete piles was driven inside the soft soil to reach a firm layer of gravel located beneath the multiple soft soil layers. The cross-sectional area of the pile was equal to 0.0751 m^2, and the pile spacing was equal to 1.6 m. The LTP was constructed to be 0.7 m thick, and two uniaxial geogrid layers were inserted inside the LTP and located (0.2; 0.4) m over the pile heads. To simulate both the embankment body and the traffic load, the embankment was built in two stages, each with a thickness of 1.9 m. The embankment was located at Virvée swamp as a part of a high-speed railway line that connects Tours and Bordeaux in France [23,24]. Figure 2 depicts the cross-section of the GRPS embankment model. The soil levels were linked to a predetermined zero point (French georeferenced level (*NGF*)) [23].

2.2. Numerical Modeling of GRPS Embankment: Calibration and Validation

The simulation of numerous complicated issues now necessitates the use of numerical modeling. The FE method is a powerful numerical modeling tool that is frequently used in geotechnical engineering [25,26]. Within this study, a new design method for a GRPS embankment is provided based on FE analysis, relying on the merits of "the finite element program".

In this article, the Plaxis 3D program was used for the validation process of the GRPS embankment. To simulate the behavior of the embankment fill, soft soil layers, and gravel layer, the hardening soil (*HS*) model was utilized. Additionally, the piles were represented as embedded beam elements, while the geogrid was modeled using elastoplastic material [24]. To reduce the impact of the boundaries on the accuracy of the simulation, the authors extended the side boundary of the model by 45 m along the x-axis. The model had a depth of 18 m from the ground surface and was fixed in both the vertical and horizontal directions at the bottom, which prevented excessive displacement during testing. Additionally, the vertical boundaries were constrained horizontally to prevent any potential distortion of the results. A mesh sensitivity analysis was applied to choose the mesh size. A medium mesh distribution was used in the simulations, as the settlement behavior was approximately the same in fine and medium distributions. Based on this choice, the number of mesh triangular elements, with 10 nodes and an average size of 0.828 m, was 19,626, as depicted in Figure 3a. This type of mesh element is commonly used in Plaxis 3D simulations to provide a precise representation of the soil and structure geometry. The triangular shape allows for the modeling of complex geometries, and the 10 nodes ensure a high level of precision in the computation of the soil–structure interaction.

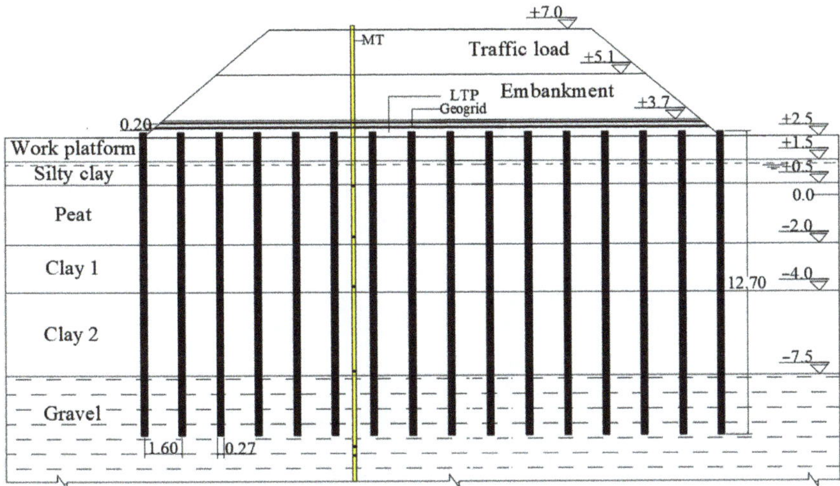

Figure 2. Cross section of GRPS embankment (dimensions in meters)—representative drawing [23].

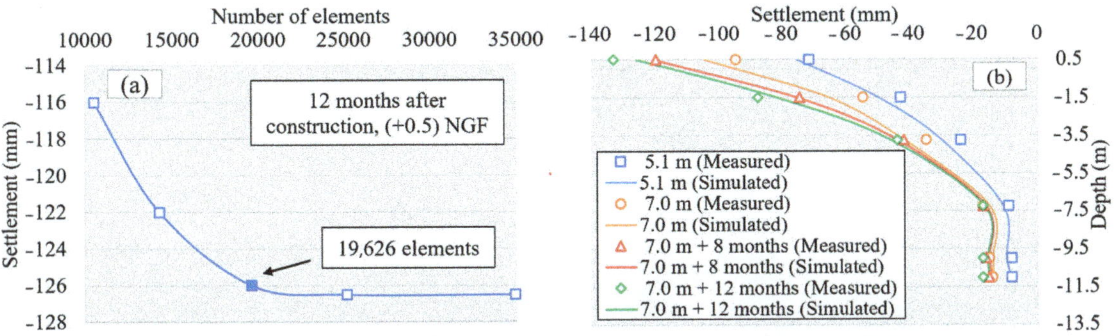

Figure 3. (**a**) The mesh sensitivity analysis. (**b**) Comparison between predicted and measured settlements.

The calibration process involved identifying the optimal values of model parameters to minimize the difference between the numerical predictions and actual measurements. To achieve this, a back analysis approach was employed to optimize the parameters. The initial set of parameters resulted in predicted settlements that were lower than those observed in the field. Through the back analysis, the stiffness parameters of the hardening soil model were refined to produce a better match between the predicted and measured values.

For the validation process, a comparison was made between the settlements predicted from the simulation and those measured in the field with the use of six magnet rings that were fixed on a magnetic probe extensometer (*MT*). The settlements were recorded underneath the embankment body at different depths of 0.5, −1.5, −3.75, −7.25, −10, and −11 m (*NGF*). The field measurements were performed throughout the following stages (Figure 3b): (i) the end of the embankment stage, (ii) the end of the traffic load stage, (iii) at

8 months, and (iv) 12 months after the construction of the full-scale model. Furthermore, another comparison was performed between the vertical stresses measured using the earth pressure cells (EPC_i) and the results from the FE analysis at the pile head (EPC_1) and over the load transfer platform (EPC_2) (see Figure 4). The soil-arching phenomenon resulted in a concentration of stress at the pile head. The instruments in the LTP and the soft soil are illustrated in Figure 2. To obtain additional information regarding the validation process of the full-scale model and the properties of the soft soil layers, embankment fill, geogrid layer, and piles, refer to reference [24].

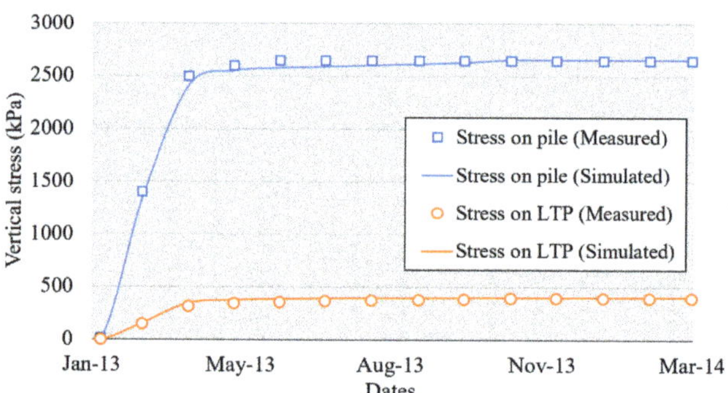

Figure 4. Comparison between predicted and measured vertical stresses.

2.3. A Simplified Model of the Validated GRPS Embankment

Once the validation process for the essential full-scale model was complete, simplifications were made to the validated GRPS embankment model in order to facilitate its generalization. The model was subsequently simplified as follows:

- One normal, consolidated clayey soil layer with a thickness of 9 m was used instead of multiple over-consolidated soft soil layers, as in the original model. The properties of the gravel, clay, and embankment soils are summarized in Table 1.
- The working platform was removed. This is consistent with the majority of the case studies, as noted in the literature.

Table 1. Summary of soil properties.

Basic Parameters	Characters and Units	Embankment Fill	Clay	Gravel
Material model	-	Hardening soil	Hardening soil	Hardening soil
Unsaturated unit weight	γ_{unsat} (kN/m^3)	19	13.5	19
Saturated unit weight	γ_{sat} (kN/m^3)	21	14.5	20
Internal friction angle	φ°	35	29	35
Dilatancy angle	Ψ°	5	0	5
Cohesion	c (kPa)	5	4	10
Reference secant stiffness	E_{50}^{ref} (kN/m^2)	30,000	800	63,000
Reference tangent stiffness	E_{oed}^{ref} (kN/m^2)	30,000	450	63,000
Reference unloading–reloading stiffness	E_{ur}^{ref} (kN/m^2)	90,000	2400	189,000
Exponential power	m	0.5	0.6	0.5
Coefficient of earth pressure at rest	K_0^{nc}	0.426	0.515	0.426
Unloading/reloading Poisson's ratio	v_{ur}	0.2	0.2	0.2
Failure ratio	R_f	0.9	0.9	0.9
Permeability	k (m/day)	0.864	5.55×10^{-4}	1.00
Over consolidation ratio	OCR	-	1	-

Figure 5a illustrates a representative drawing of the GRPS embankment model subsequent to undergoing simplifications, whereas Figure 5b displays the FE mesh of the same model.

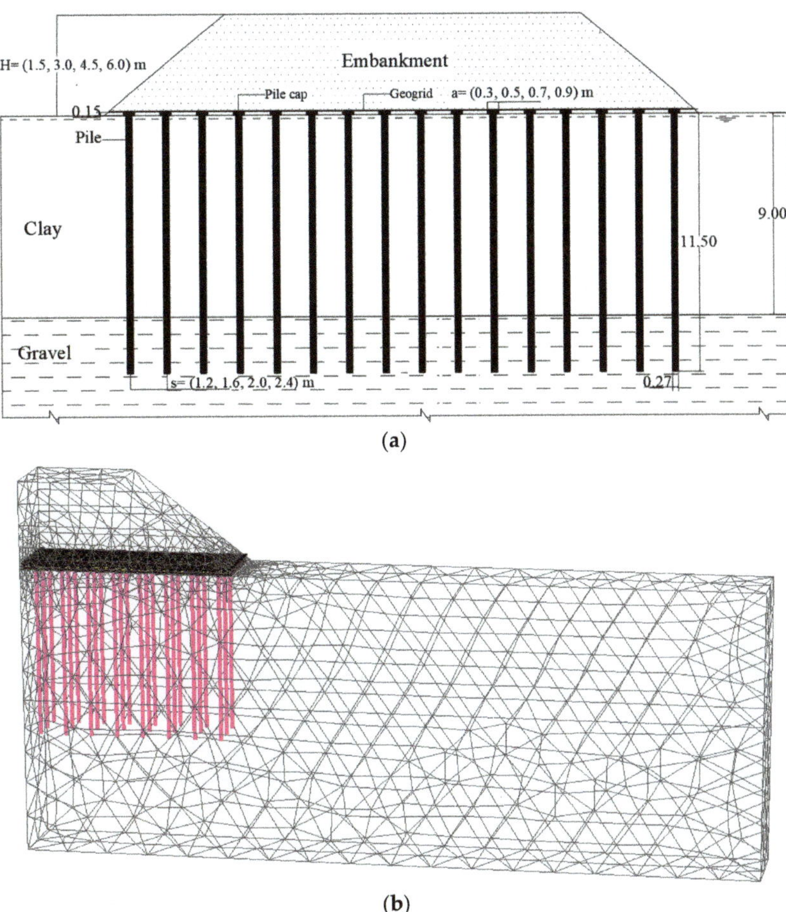

Figure 5. (**a**) Cross section of the simplified GRPS embankment (dimensions in meters); (**b**) FE mesh of the simplified GRPS embankment.

The cover ratio can be governed by adjusting the pile volume or by enlarging the pile cap. The load transfer mechanism is different in two cases. In the first case, the increase in the pile volume simultaneously contributes to an increase in skin friction and a decrease in load efficiency. In the second case, the enlargement of the pile cap leads to an increase in soil arching and consequently an increase in load efficiency, as indicated in Figure 6, while the pile volume is fixed. The tension in the geosynthetic is affected to a small extent due to the change in the load transfer mechanism, as illustrated in Figure 7. The present design method assumes that the pile diameter (a) is fixed at 0.3 m, and the cover ratio can be controlled by the pile cap surface area.

One layer of the biaxial geogrid was utilized in the simplified model. It was observed that inserting one layer of geogrid within the LTP enhances the load efficiency by about 60%, while inserting two layers of geogrid increases the load efficiency by about 63% when compared to an unreinforced piled embankment [27]. In order to achieve a higher load efficiency, the layer of geogrid was positioned directly over the pile heads, which is considered a cost-effective and efficient solution [27]. Table 2 presents the characteristics of the pile material and geogrid.

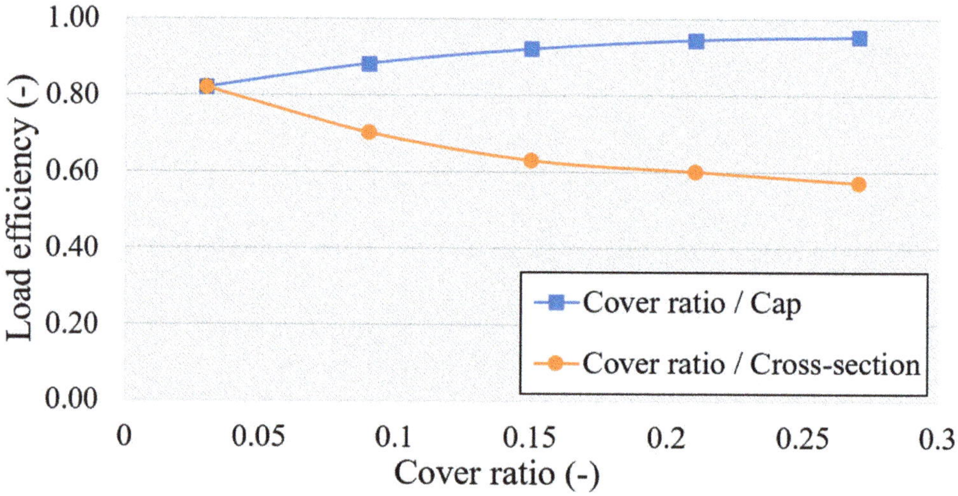

Figure 6. Influence of cover ratio on load efficiency in the two cases.

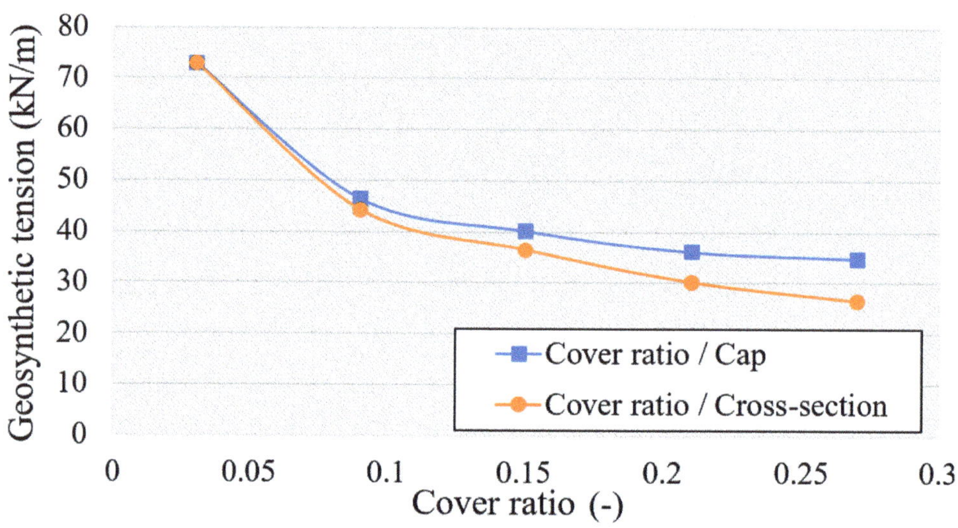

Figure 7. Influence of cover ratio on geosynthetic tension in the two cases.

Table 2. Summary of the pile and geogrid properties.

	Unit Weight γ (kN/m^3)	Young Modulus E (GPa)	Poisson Ratio v (-)	Stiffness J (kN/m)	Length L (m)
Pile	24	20	0.2	-	11.5
Geogrid	-	-	0.2	13,000	-

2.4. Mathematical Derivation of the Design Equations

The present method has been developed based on an extended parametric study obtained from the simplified numerical model. In the beginning of the process, data were collected, including the load efficiency and geosynthetic tension. We ran the Plaxis 3D program over 1000 times with four embankment heights in every calculation process, which was adequate to provide a precise prediction of the values of the dependent variables (*E*;

T). The second step was the analysis of the data in order to derive equations that could be used to predict the values of the dependent variables. These equations were based on a set of the following independent variables: pile cap width (a), pile spacing (s), embankment height (H), oedometric modulus of subsoil (E_{oed}), geosynthetic stiffness (J), and unit weight of the embankment fill (γ). By using the Curve Expert program, a regression model was constructed to describe the interdependence of the independent and dependent variables. Based on the probability concept, the final number of observations for each dependent and independent variable was $4^6 = 4096$. Table 3 contains the presumed values of the independent variables.

Table 3. Presumed values of the independent variables.

Pile cap width (a), m	0.3, 0.5, 0.7, 0.9
Pile spacing (s), m	1.2, 1.6, 2.0, 2.4
Embankment height (H), m	1.5, 3.0, 4.5, 6.0
Subsoil stiffness (E_{oed}) kN/m^2	1000, 4000, 7000, 10,000
Geosynthetic stiffness (J) kN/m	1000, 5000, 9000, 13,000
Unit weight of embankment fill (γ) kN/m^3	17.0, 19.0, 21.0, 23.0

In this methodology, a data structure tree was used in which the root represented the subsequent Equations (9) and (10) of the dependent variables (E; T). There were five levels of parent nodes representing independent variables (H; E_{oed}; J; a; s). The last level of the parent nodes ended in leaf nodes, which represented the last independent variable (γ). The nodes at every level are called siblings [28]. The siblings are integrated in every level to produce a new generation of parents. Herein, in the work mechanism to derive the equation of geosynthetic tension (T) with the following constants (a; E_{oed}; J; s), the unit weight of the embankment fill (γ) affects the geosynthetic tension (T), according to the following equation:

$$T = A + B * \Delta\gamma \tag{1}$$

where A; B represent the coefficients in Equation (1), and $\Delta\gamma = (\gamma - 19)$.

For different values of the embankment height (H), the geosynthetic tension (T) is changed in Equation (1) as shown in Table 4.

Table 4. Influence of embankment height (H) on the geosynthetic tension (T).

$H = 1.5$ m	$T_1 = A_1 + B_1 * (\gamma - 19)$
$H = 3.0$ m	$T_2 = A_2 + B_2 * (\gamma - 19)$
$H = 4.5$ m	$T_3 = A_3 + B_3 * (\gamma - 19)$
$H = 6.0$ m	$T_4 = A_4 + B_4 * (\gamma - 19)$

To include the impact of the height of embankment (H) in the equation of the geosynthetic tension (T), the relationships between A and B (as dependent variables) and H (as independent variable) were found using the Curve Expert program, which are provided as follows:

$$A = F_1 * F_2{}^H \tag{2}$$

$$B = G_1 + G_2 * H + G_3 * H^2 \tag{3}$$

where F_i; G_i represent the coefficients in Equations (2) and (3).

Substituting Equations (2) and (3) into Equation (1), an expression of the geosynthetic tension (T) in terms of the height of embankment (H) and embankment fill unit weight (γ) is provided:

$$T = F_1 * F_2{}^H + \left(G_1 + G_2 * H + G_3 * H^2\right) * (\gamma - 19) \tag{4}$$

To determine how the pile cap width (a) affects the geosynthetic tension (T), the relationships between F_i and G_i (as dependent variables) and a (as independent variable) were found using the Curve Expert program, and the following expression is provided:

$$T = \left(I_1 * exp^{\frac{I_2}{a}}\right) * F_2^{\,H} + (G_1 + G_2 * H + G_3 * H^2) * (\gamma - 19) \tag{5}$$

Similarity, to include the effect of the pile spacing (s) on geosynthetic tension (T), the following equation is used:

$$T = \left((X_1 * exp^{(X_2 * s)}) * exp^{\frac{\frac{X_3}{1+X_4 * exp^{X_5 * s}}}{a}}\right) * (X_6 + X_7 * s)^H + (G_1 + G_2 * H + G_3 * H^2) * (\gamma - 19) \tag{6}$$

Likewise, to find the effect of the geosynthetic stiffness (J) on the geosynthetic tension (T):

$$T = \left(((Y_1 + Y_2 * J) * exp^{((Y_3 + Y_4 * J) * s)}) * exp^{\left(\frac{\frac{Y_5 * J^{Y_6}}{1+(Y_7 * J^{Y_8}) * exp^{(Y_9 * J^{Y_{10}}) * s}}}{a}\right)} \right) * (X_6 + X_7 * s)^H + (G_1 + G_2 * H + G_3 * H^2) \\ * (\gamma - 19) \tag{7}$$

The final step is to determine the effect of the elastic modulus (E_{oed}):

$$T = \left(((Y_1 + Y_2 * J) * exp^{((Y_3 + Y_4 * J) * s)}) * exp^{\left(\frac{\frac{Y_5 * J^{Y_6}}{1+(Y_7 * J^{Y_8}) * exp^{(Y_9 * J^{Y_{10}}) * s}}}{a}\right)} \right) * (X_6 + X_7 * s)^H + \\ (G_1 + G_2 * H + G_3 * H^2) * (\gamma - 19) \tag{8}$$

where:

$$D_1 = Y_1 + Y_2 * J$$
$$D_2 = ((Y_3 + Y_4 * J) * s) = ((Z_1 * E_{oed}^{Z_2} + (Z_3 + Z_4 * E_{oed}) * J) * s)$$
$$D_3 = \frac{\frac{Y_5 * J^{Y_6}}{1+(Y_7 * J^{Y_8}) * exp^{(Y_9 * J^{Y_{10}}) * s}}}{a} = \frac{\frac{(Z_5 + Z_6 * E_{oed}) * J^{Y_6}}{1+((Z_7 + Z_8 * E_{oed}) * J^{Y_8}) * exp^{((Z_9 + Z_{10} * E_{oed}) * J^{Y_{10}}) * s}}}{a}$$
$$D_4 = (X_6 + X_7 * s)^H$$

where I_i; X_i; Y_i; Z_i represent the coefficients in Equations (5)–(8).

Section 5 provides a detailed derivation of the equation used to calculate the load efficiency of the pile (E).

3. Results and Discussion

Initially, this section presents the equations pertaining to load efficiency and geosynthetic tension. Subsequently, a comparison is made between the results obtained from the proposed design method and the field measurements.

3.1. Equations of Load Efficiency and Geosynthetic Tension

The basic equations of this study are (9) and (10), which are worded using the aforementioned methodology, provided in the following equations:

- Load efficiency (E):

$$E = \frac{P}{(-0.18 + 20.11 * H) * s^{1.97}} + 5.9 * 10^{-6} J + (\gamma - 19) * \left(1.06 * 10^{-3} + 1.26 * 10^{-4} * H - \frac{1.65 * 10^{-3}}{H^2} \right) \tag{9}$$

where:

$$P = \left((F_1 + F_2 * s)\, a^{(F_3 * s^{F_4})} \right) * H^{((F_5 + 0.05 * s) - 0.05 * a)}$$

$$F_1 = -37.86 + 0.00164 * E_{oed}$$

$$F_2 = 51.52 - 0.00146 * E_{oed}$$

$$F_3 = 0.0019 * E_{oed}{}^{0.5013}$$

$$F_4 = 12.07 * E_{oed}{}^{-0.2776}$$

$$F_5 = 0.993 + 6.7 * 10^{-6} * E_{oed}$$

- Geosynthetic tension (T):

$$T = D_1 * exp^{D_2} * exp^{D_3} * D_4 + (\gamma - 19) * \left(0.064 * (H + 1.093)^2\right) \tag{10}$$

where:

$$D_1 = 0.078 + 6.25 * 10^{-5} * J$$

$$D_2 = \left(4.95 * E_{oed}{}^{-0.18} - \left(3.95 - 2.54 * 10^{-4} * E_{oed}\right) * 10^{-5} * J\right) * s$$

$$D_3 = \left(\frac{C_1/C_2}{a}\right)$$

$$C_1 = \left(1.132 - 1.63 * 10^{-5} * E_{oed}\right) * J^{-0.14}$$

$$C_2 = 1 + \left(\left(6.167 + 4.09 * 10^{-4} * E_{oed}\right) * J^{0.25}\right) * exp^{-(3.62 - 1.93 * 10^{-5} * E_{oed}) * s * J^{0.0275}}$$

$$D_4 = (1.55 - 0.05 * s)^H$$

To obtain the results, the following input data must be provided with the respective units: pile cap width (a), meters; pile spacing (s), meters; embankment height (H), meters; oedometric modulus of the subsoil (E_{oed}), kN/m^2; geosynthetic stiffness (J) kN/m; and embankment fill unit weight (γ), kN/m^3.

There are many fundamental issues that are considered obvious with this design method:

- This method is regarded as eligible for low- and medium-rise embankments (H = 0.5–6 m).
- The cover ratio can be changed by adjusting the pile cap area, and the ratio (a/s) should be less than 0.75.
- The values of (E; T) are calculated after complete consolidation and a degree of consolidation (U = 1). However, creep behavior is neglected in the design.
- The effect of surcharge load (q) can be compensated for using the formula: $H = q/\gamma$. Based on this, the new height of the embankment (H) must be used in the calculation.
- ASIRI recommendations [29] classify the soil based on the pressiometric modulus for soft soil ($E_M \leq 3$ MPa). In the limitations of the earlier classification, $E_{50} \approx 2\,E_{oed}$ was used for soft soil [30].
- In the present design method, the minimum subsoil oedometric modulus is 0.3 MPa.
- The piles are configured in squared and rectangular patterns. In the case of a rectangular pile arrangement, the equivalent width of the squared area between the piles is proposed to calculate the load efficiency; however, the length of the rectangular area between the piles is utilized to calculate the geosynthetic tension.

Pham introduced the "active depth" concept in which the oedometric modulus of the subsoil can be calculated as follows [31]:

$$E_{oed} = k_s * D_{act} \tag{11}$$

where k_s is the subgrade reaction modulus (kPa/m) and D_{act} is the active depth at which the vertical stress caused by the embankment fill weight and surcharge load is less than 20% of the vertical geostatic stress. It can be calculated using the following simplified Equation [31]:

For clay:

$$D_{act} = 10 * \sqrt[4]{\frac{(s - a)}{6}} \tag{12}$$

In case of various subsoil layers, the oedometric modulus can be calculated by:

$$E_{oed} = \frac{E_{oed.1} * D_1 + E_{oed.2} * D_2 + \cdots + E_{oed.n} * D_n}{D_1 + D_2 + \cdots + D_n} \tag{13}$$

where $E_{oed.1}; E_{oed.2}; E_{oed.n}$ are the oedometric moduli of the several layers, and $D_1; D_2; D_n$ are the thicknesses of the subsoil layers [31].

3.1.1. Sensitivity Analysis

A sensitivity analysis was conducted to assess the impact of the input parameters in Equations (9) and (10) on the load efficiency of the piles (E) and the geosynthetic tension (T). For this purpose, each parameter was increased by 40% while the other parameters remained constant. In the analysis, the following parameters were used ($s = 2$ m; $a = 0.3$ m; $H = 4.0$ m; $E_{oed} = 5.0$ MPa; $J = 6000$ kN/m; $\gamma = 18$ kN/m^3). The results in Table 5 demonstrate the percentage change of (E; T) according to the corresponding increase (40%) of the aforementioned input parameters. It is evident that the most influential parameter on (E; T) is the pile spacing (s).

Table 5. The sensitivity analysis.

Parameter	s (m)	a (m)	H (m)	E_{oed}(MPa)	J (kN/m)	γ (kN/m^3)
Modified value	2.8	0.42	5.6	7.0	8400	25.2
Percentage change of E (%)	−24.8	9.2	4.0	−5.7	2.5	5.6
Percentage change of T (%)	91.1	−26.6	82.0	−9.8	13.1	36.7

3.1.2. Comparison of the Proposed Method Solutions and FE Results

Figure 8a,b compare the outputs of Plaxis 3D (dashed lines) to the results of the design method (solid lines). A remarkable agreement is found, which demonstrates that the proposed design method can effectively predict the values of load efficiency and geosynthetic tension.

3.2. Validation of the Proposed Method

For validation, the field measurements of a series of eight case studies under the limitations of the design method were used. The full-scale models are described briefly in this section. To extend the comparison process, The German standard (EBGEO) [18], Dutch standard (CUR 226) [21], and British standard (BS 8006)—on the basis of Hewlett and Randolph's formula [16]—design methods were inserted in the comparison. A simplified version of these design methods can be found in the literature.

3.2.1. GRPS Motorway Embankment in the Netherlands (Van Eekelen et al., 2020)

As part of the junction rehabilitation between motorway A12 and a local road near Woerden, Netherlands, a new motorway exit was constructed on a reinforced piled embankment, according to a study by Van Eekelen et al. [32]. Shortly after the construction, it was observed that the subsoil support had become negligible. Two geogrid layers with a stiffness (J) of 4611 kN/m were installed within the LTP. A summary of the site conditions and the design parameters is provided herein: embankment height (H) = 1.96 m; unit weight (γ) = 18.3 kN/m^3; surcharge load (q) = 4.2 kPa; pile spacing (s) = 2.25 m; pile cap width (a) = 0.85 m. A 17 m thick layer of soft clay and peat made up the soft subsoil, and the oedometric modulus was E_{oed} = 300 kN/m^2.

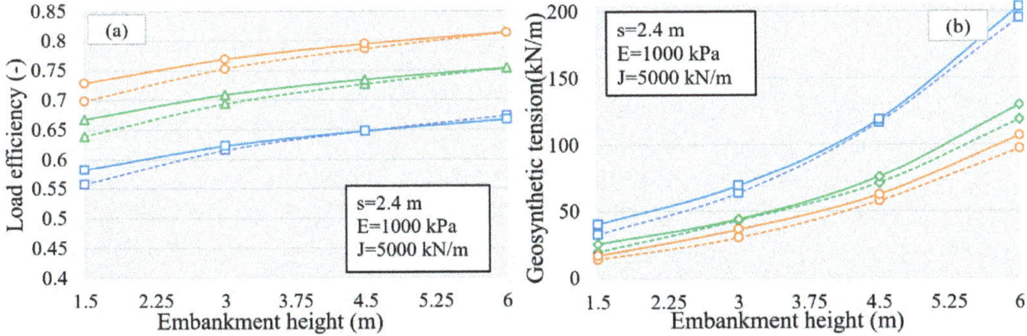

Figure 8. Comparison of Plaxis 3D and equation results: (**a**) load efficiency; (**b**) geosynthetic tension. Dashed line represents the outputs of Plaxis 3D; solid line represents the predicted values of the equation. For blue lines, a = 0.3 m; green lines, a = 0.5 m; orange lines, a = 0.7 m.

Table 6 shows a comparison of the present method with three design methods and the actual field measurements for load efficiency and geosynthetic tension. The results indicate a relatively close agreement with the field measurements. The EBGEO method provides a reasonably accurate prediction of load efficiency; however, it falls short in its estimation of geosynthetic tension compared to the actual measurements. On the contrary, the BS 8006 method overestimates the geosynthetic tension while providing an inaccurate prediction of load efficiency. In comparison, the CUR 226 method's predictions of both load efficiency and geosynthetic tension align well with the field measurements.

Table 6. Comparison of the field measurements with design method results for the Van Eekelen et al. case study.

Parameters	Field Measurements	Present Design Method	EBEGO Design Method	BS 8006 Design Method	CUR 226 Design Method
Load efficiency E (%)	84.9	81.8	80.9	53.3	80.8
Geosynthetic tension, T (kN/m)	41.5	43.05	23.06	84.14	44.8

3.2.2. GRPS Railway Embankment in the Netherlands (Duijnen et al., 2010)

Compressible soils are widespread in the Netherlands. The GRPS embankment was used in this project for a smooth transition between a viaduct with zero settlements and an unsupported embankment. The LTP included three geosynthetic layers with a stiffness (J) = 5237 kN/m. The embankment was constructed using recycled concrete material: embankment height (H) = 2.6 m and unit weight (γ) = 18.3 kN/m^3. The subsoil consisted of two layers of sand and one of soft clay, with an oedometric modulus (E_{oed}) = 3300 kN/m^2 and a thickness of (D) = 8 m. The pile configuration was rectangular in shape. The spacing between the piles was 1.45 × 1.9 m^2, and the width of pile cap (a) = 0.4 m [31,33,34].

The load efficiency and geosynthetic tension are compared in Table 7. The present and CUR 226 methods produce results that are compatible with the measured values. The EBGEO design method relatively underestimates the load efficiency while overestimating the geosynthetic tension relative to the field measurements. The BS 8006 design method underestimates the load efficiency while drastically overestimating geosynthetic tension.

Table 7. Comparison of the field measurements with design method results for the Duijnen et al. case study.

Parameters	Field Measurements	Present Design Method	EBEGO Design Method	BS 8006 Design Method	CUR 226 Design Method
Load efficiency E (%)	76.3	76.5	70.6	63.7	76.3
Geosynthetic tension, T (kN/m)	17.6	15.1	52.37	160.92	16.6

3.2.3. GRPS Highway Embankment in China (Chen et al., 2010)

A GRPS embankment in the southeastern part of Zhejiang Province in China was utilized to elevate the *TJ* (Taizhou Jinyun) highway above numerous layers of soft soil. In this project, pre-stressed tube piles with square caps and an embankment fill consisting of a mixture of clayey soil and gravel were used. A summary of the site condition and the design parameters is provided herein: stiffness of the used geogrid layer (J) = 1500 kN/m; embankment height (H) = 6 m; unit weight (γ) = 21.0 kN/m^3; oedometric modulus of the subsoil (E_{oed}) = 3322 kN/m^2; thickness (D) = 30 m; pile spacing (s) = 2.0 m; width of pile cap (a) = 1.0 m [35].

Table 8 presents a comparison of the measured and calculated load efficiency and geosynthetic tension. The present method, as well as BS 8006 and CUR 226 to a lesser extent, achieve results that are close to the field measurements for load efficiency. However, the tension in the geosynthetic was not recorded during the field measurements. The comparison of the various design methods reveals some variation in the results in this case study.

Table 8. Comparison of the field measurements with design method results for the Chen et al. case study.

Parameters	Field Measurements	Present Design Method	EBEGO Design Method	BS 8006 Design Method	CUR 226 Design Method
Load efficiency E (%)	87.6	88.4	77.6	86.3	82.2
Geosynthetic tension, T (kN/m)	-	27.75	36	47.8	17.6

3.2.4. Full-Scale Model of GRPS Embankment in Korea (Lee et al., 2019)

The performance of a GRPS embankment was evaluated using a full-scale model constructed in a 3 m high, 5 m wide, and 15 m long concrete test box. To simulate the soft subsoil, the voids between the model piles were filled with 0.4 m of polyurethane foam. At a height of 15 cm above the pile caps, one layer of geotextile with a stiffness (J) = 422 kN/m was installed. The oedometric modulus of the polyurethane foam was (E_{oed}) = 1510 kN/m^2. The properties of the poorly- to well-graded embankment fill were as follows: unit weight (γ) = 20.2 kN/m^3, embankment height (H) = 2.55 m, pile spacing (s) = 1.2 m, and finally, width of pile cap (a) = 0.4 m [36].

Table 9 compares the load efficiency and geosynthetic tension. According to the results, the proposed method aligns with the load efficiency measurement results. The EBGEO, CUR 226 and BS 8006 methods underestimate the load efficiency to different degrees. The comparison of the design methods for geosynthetic tension in Table 9 shows some differences in the findings. However, it should be noted that no field measurements were made for geosynthetic tension.

3.2.5. GRPS Highway Embankment in China (Zhao et al., 2019)

The Jin-Bin highway is located in China. A GRPS embankment was proposed to increase the width of the current highway by 5 m on both sides and to extend the four two-way lanes into six or eight two-way lanes. A rectangular arrangement of piles was used in the construction, with a transversal pile spacing (s_1) = 2.5 m and a longitudinal pile spacing (s_2) = 4.5 m. The pile cap width (a) = 1.0 m. On top of the pile caps, a 0.4 m thick crushed stone layer was laid, into which two geogrid layers with a stiffness (J) = 1630 kN/m were inserted. The soil profile consisted of seven layers of soft soil in which the oedometric modulus of these layers (E_{oed}) = 8254 kN/m^2, thickness (D) = 13.5 m, embankment height (H) = 4.8 m, and unit weight (γ) = 19.0 kN/m^3 [37].

Table 9. Comparison of the field measurements with design method results for the Lee et al. case study.

Parameters	Field Measurements	Present Design Method	EBEGO Design Method	BS 8006 Design Method	CUR 226 Design Method
Load efficiency E (%)	76.4	79.3	57.4	67.6	63.6
Geosynthetic tension, T (kN/m)	-	4.79	25.32	46.8	12.74

A comparison between the results of load efficiency and geosynthetic tension obtained from measurements and calculations is provided in Table 10. The current method accurately reflects the measured load efficiency. On the other hand, the EBGEO, CUR 226, and BS 8006 methods show an underestimation of the load efficiency. The geosynthetic tension was not recorded in the measurements. The comparison between the analytical design methods reveals disparities in the results, however, with the BS 8006 method presenting a significantly higher tension in the geosynthetic compared to the other methods.

Table 10. Comparison of the field measurements with design method results for the Zhao et al. case study.

Parameters	Field Measurements	Present Design Method	EBEGO Design Method	BS 8006 Design Method	CUR 226 Design Method
Load efficiency E (%)	66.6	64.1	29.0	37.2	35.4
Geosynthetic tension, T (kN/m)	-	69.9	97.8	1210.5	15.2

3.2.6. GRPS Highway Embankment in China (Liu et al., 2015)

An embankment over a group of grouted gravel piles with two geogrid layers was constructed to support a highway in China. The embankment's fill material was a cohesive soil mixed with approximately 40% fly ash. The soil profile comprised four layers of a silty clay, mud clay, soft clay, and medium silty clay with a full thickness of 27.5 m. Two geogrid layers were installed over the pile caps. A summary of the input data required for the design method is provided herein. The embankment height (H) = 4.6 m, the unit weight (γ) = 19 kN/m^3. The oedometric modulus of the subsoil (E_{oed}) = 2196 kN/m^2. The pile spacing (s) = 2.4 m and the pile cap width (a) = 1.0 m. the geogrid stiffness (J) = 1125 kN/m [31,38].

Table 11 reveals the outcomes of comparing several design methods with the field measurements, demonstrating that the present method provides load efficiency results that are relatively consistent with the field measurements. The EBGEO method underestimates the load efficiency, while the BS 8006 and CUR 226 methods provide results that are relatively close to the measurements. In this study, the tension in the geosynthetic was not included in the measurements. When various design methods are compared, it becomes apparent that the results differ from one another; however, the present approach is in agreement with the CUR 226 method.

3.2.7. GRPS Stockyard Embankment in Brazil (Hosseinpour et al., 2015)

In Rio de Janeiro, Brazil, an embankment supported by a network of geotextile-encased granular columns and one layer of geogrid was constructed to support a stockyard, which is used to store raw materials and coal for steel plate manufacturing. The aim of this project was to decrease the settlement and horizontal displacement and reduce the construction time. Hosseinpour et al. [38] reported the details of the design parameters and site conditions: geogrid stiffness (J) =2000 kN/m, embankment height (H) = 5.35 m), unit weight (γ) = 27.5 kN/m^3, oedometric modulus of the subsoil (E_{oed}) = 5813 kN/m^2, thickness (D) = 11 m, pile spacing (s) = 2.0 m, and pile diameter without cap (d) = 0.7 m [39].

Table 11. Comparison of the field measurements with design method results for the Liu et al. case study.

Parameters	Field Measurements	Present Design Method	EBEGO Design Method	BS 8006 Design Method	CUR 226 Design Method
Load efficiency E (%)	77.7	82.4	62.8	71.4	68.8
Geosynthetic tension, T (kN/m)	-	20.67	48.4	105.8	23.6

Although this case study is out of the limits of the present design method due to the fact that the cover ratio is governed by the pile diameter and not by the cap width, the geosynthetic tension should be very close to the real value measured in the field while the load efficiency is radically different, as previously explained in Figures 6 and 7. Table 12 compares the load efficiency and tension in the geosynthetic. The EBGEO, CUR 226, and BS 8006 methods produce excessively conservative results in terms of load efficiency. The present method slightly overestimates the geosynthetic tension, which is consistent with the results in Figure 7 in which the geosynthetic tension is somewhat higher in the case of a cover ratio controlled by the pile diameter than in the case of a cover ratio controlled by the cap surface area. The EBGEO method is relatively in line with the measured tension in geosynthetic; however, the BS 8006 method significantly overestimates the geosynthetic tension, while the CUR 226 method underestimates it.

Table 12. Comparison of the field measurements with design method results for the Hosseinpour et al. case study.

Parameters	Field Measurements	Present Design Method	EBEGO Design Method	BS 8006 Design Method	CUR 226 Design Method
Load efficiency E (%)	23.4	-	81.8	82.2	84.8
Geosynthetic tension, T (kN/m)	33.6	41.4	26.0	115.3	9.72

3.2.8. GRPS Highway Embankment in China (Liu et al., 2007)

A GRPS embankment was used in this project to support a highway in a northern suburb of Shanghai, China. Pulverized fuel ash was used as an embankment fill. Details of the site conditions and design parameters were reported by Liu et al. [39]: embankment height (H) = 5.6 m, unit weight (γ) = 18.5 kN/m^3, oedometric modulus of the subsoil (E_{oed}) = 6937 kN/m^2, thickness (D) = 16 m, pile spacing (s) = 3.0 m, pile diameter (d) = 1 m, and geosynthetic stiffness (J) = 1180 kN/m [40].

This study is similar to the study conducted by Hosseinpour et al. in that the pile diameter determines the cover ratio. The load efficiency may differ in the current method, but the geosynthetic tension should be close to the actual field measurements. Table 13 presents a comparison of the load efficiency and geosynthetic tension. The EBGEO method underestimates the load efficiency, while the results of the CUR 226 and BS 8006 methods align with the measured value. The current method slightly overestimates the geosynthetic tension, which corresponds to the results in Figure 7. The EBGEO method overpredicts the geosynthetic tension compared to the observed value, while the BS 8006 method overestimates it significantly. On the other hand, the CUR 226 method shows good agreement with the measured geosynthetic tension.

Table 13. Comparison of the field measurements with design method results for the Liu et al. case study.

Parameters	Field Measurements	Present Design Method	EBEGO Design Method	BS 8006 Design Method	CUR 226 Design Method
Load efficiency E (%)	62.6	-	51.1	61.4	57.9
Geosynthetic tension, T (kN/m)	19.97	26.3	53.1	281.3	21.3

4. Conclusions

For the development of a geosynthetic-reinforced pile-supported embankment design, a novel method that addresses all interactions between the components of rigid inclusions has been presented. The method employs the finite element method (FEM) to avoid some of the drawbacks of experimental and theoretical methods. In other words, the proposed method overcomes the moot points related to the reality of the soil arch shape, load magnitude carried by the geosynthetic reinforcement, skin friction along the soil–geosynthetic interface, and the role of the soft subsoil and its real behavior.

A full-scale model of the GRPS embankment was validated. Afterward, certain simplifications were made for design purposes, including the adoption of one soft soil layer and the removal of the working platform. The resulting simplified and validated model of the GRPS embankment was then used to collect data, which were analyzed to derive two equations that can be used to determine the load efficiency of the piles and the tension in the geosynthetic. The design method employed six parameters: pile cap width, pile spacing, embankment height, oedometric modulus of the subsoil, geosynthetic stiffness, and embankment fill unit weight. The cover ratio must be governed by adjusting the pile cap area while keeping the pile diameter constant. For the design process, Plaxis 3D and Curve Expert software were utilized in this study.

The present design method is characterized relatively by the simplicity of the solving of the developed equations, and it covers a wide range of the case studies in the literature.

The proposed design method was compared to various design methods, including EBGEO, CUR 226, and BS 8006, as well as to field measurements of load efficiency and geosynthetic tension across eight case studies. The findings indicate that the current method provides reliable results that are consistent with the measured parameters.

5. Derivation of an Equation for Load Efficiency (E)

The equation below demonstrates how the unit weight of an embankment fill (γ) impacts the load efficiency (E) when considering the constants (a; E_{oed}; J; s).

$$E = \alpha + \alpha_1 * \Delta\gamma \tag{14}$$

In order to account for the influence of the geosynthetic stiffness (J) on the load efficiency equation (E), it is necessary to determine the correlations between α and α_1 (as the dependent variables) and J (as the independent variable). The resulting relationships are presented below:

$$E = (\beta_1 + \beta_2 * J) + \alpha_1 * (\gamma - 19) \tag{15}$$

The given expression below is used to establish the relationship between the embankment height (H) and the load efficiency (E):

$$E = \left(\frac{\lambda_1 * H^{\lambda_2}}{W_T} + \beta_2 * J\right) + \left(\lambda_3 + \lambda_4 * H + \frac{\lambda_5}{H^2}\right) * (\gamma - 19) \tag{16}$$

$$E = (E_1 + E_2) + E_3 \tag{17}$$

The coefficients are denoted by α_i, β_i, and λ_i in Equations (14)–(16). In the present phase of calculation, E_i signifies the components of the load efficiency, while W_T denotes the total weight of the embankment fill and is equal to $(-0.18 + 20.11 * H) * s^{1.97} \approx \gamma_{emb} * H * s^2$.

Owing to some difficulties encountered, a step was carried out by the authors to simplify the derivation process in which the applied load on the pile head $P = \lambda_1 * H^{\lambda_2}$.

In order to determine the influence of the pile diameter (a) on the load efficiency (E), the following equation is provided:

$$E = \left(\frac{\left(\kappa_1 * a^{\kappa_2} * H^{(\kappa_3 + \kappa_4 * a)}\right)}{W_T} + \beta_2 * J\right) + \left(\lambda_3 + \lambda_4 * H + \frac{\lambda_5}{H^2}\right) * (\gamma - 19) \tag{18}$$

To consider the influence of pile spacing (s) on the load efficiency (E), the following equation is employed:

$$E = \left(\frac{(F_1 + F_2 * s) * a^{(F_3 * s^{F_4})} * H^{((F_5 + F_6 * s) + \kappa_4 * a)}}{W_T} + \beta_2 * J\right) + \left(\lambda_3 + \lambda_4 * H + \frac{\lambda_5}{H^2}\right) * (\gamma - 19) \tag{19}$$

Final step involves determining the impact of the elastic modulus (E_{oed}):

$$E = \left(\frac{(F_1 + F_2 * s) * a^{(F_3 * s^{F_4})} * H^{((F_5 + F_6 * s) + \kappa_4 * a)}}{W_T} + \beta_2 * J\right) + \left(\lambda_3 + \lambda_4 * H + \frac{\lambda_5}{H^2}\right) * (\gamma - 19)$$

where:

$$\begin{aligned}
F_1 &= \mu_1 + \mu_2 * E_{oed}\\
F_2 &= \mu_3 + \mu_4 * E_{oed}\\
F_3 &= \mu_5 * E_{oed}^{\mu_6}\\
F_4 &= \mu_7 * E_{oed}^{\mu_8}\\
F_5 &= \mu_9 + \mu_{10} * E_{oed}
\end{aligned} \tag{20}$$

Author Contributions: Conceptualization, R.A.; investigation, A.A.; writing—original draft preparation, R.A.; Numerical modeling, R.A. and A.A.; writing—review and editing, R.A.; supervision, E.K. All authors have read and agreed to the published version of the manuscript.

Funding: This research received no external funding.

Institutional Review Board Statement: Not applicable.

Informed Consent Statement: Not applicable.

Data Availability Statement: Data will be made available upon request.

Conflicts of Interest: The authors declare no conflict of interest.

References

1. Pham, T. Analysis of geosynthetic-reinforced pile-supported embankment with soil-structure interaction models. *Comput. Geotech.* **2020**, *121*, 103438. [CrossRef]
2. Burtin, P.; Racinais, J. Embankment on Soft Soil Reinforced by CMC Semi-Rigid Inclusions for the High-speed Railway SEA. In Proceedings of the 3rd International Conference on Transportation Geotechnics, Guimaraes, Portugal, 4–7 September 2016; Volume 143, pp. 355–362.
3. Mangraviti, V. Displacement-Based Design of Geosynthetic-Reinforced Pile-Supported Embankments to Increase Sustainability. In *Civil and Environmental Engineering for the Sustainable Development Goals*; Antonelli, M., Della Vecchia, G., Eds.; SpringerBriefs in Applied Sciences and Technology; Springer Nature: Berlin/Heidelberg, Germany, 2022; pp. 83–96. [CrossRef]
4. Van Eekelen, S.; Bezuijen, A. Model experiments on geosynthetic reinforced piled embankments, 3D test series. In *2nd European Conference on Physical Modelling in Geotechnics (EUROFUGE-2012)*; Ghent University, Department of Civil engineering: Ghent, Belgium, 2012; pp. 72–83. [CrossRef]
5. Terzaghi, K. *Theoretical Soil Mechanics*; John Wiley and Sons: New York, NY, USA, 1943.
6. Russell, D.; Pierpoint, N. An assessment of design methods for piled embankments. *Ground Eng.* **1997**, *30*, 39–44. Available online: https://trid.trb.org/view/476724 (accessed on 5 March 2022).
7. Guido, V.; Kneuppel, J.; Sweeney, M. Plate loading tests on geogrid reinforced earth slabs. In *Proceedings Geosynthetics 87 Conference*; Industrial Fabrics Association International: New Orleans, LA, USA, 1987; Volume 87, pp. 216–225.
8. Hewlett, W.; Randolph, M. Analysis of piled embankments. *Ground Eng.* **1988**, *22*, 12–18.
9. Zhang, C.; Jiang, G.; Liu, X.; Buzzi, O. Arching in geogrid-reinforced pile-supported embankments over silty clay of medium compressibility: Field data and analytical solution. *Comput. Geotech.* **2016**, *77*, 11–25. [CrossRef]
10. Low, B.; Tang, S.; Choa, V. Arching in piled embankments. *J. Geotech. Eng.* **1994**, *120*, 1917–1938. [CrossRef]

11. Russell, D.; Naughton, P.J.; Kempton, G. A new design procedure for piled embankments. In Proceedings of the 56th Canadian Geotechnical Conference and the NAGS Conference, Winnipeg, MB, Canada, 29 September–1 October 2003; pp. 858–865.
12. Kempfert, H.; Gobel, C.; Alexiew, D.; Heitz, C. German recommendations for the reinforced embankments on pile-similar elements. In *Geosynthetics in Civil and Environmental Engineering*; Springer: Berlin/Heidelberg, Germany, 2004; pp. 697–702. [CrossRef]
13. Collin, J.G.; Han, J.; Huang, J. Geosynthetic-Reinforced Column-Support Embankment Design Guidelines. In Proceedings of the North American geosynthetics Society (NAGS) Conference, Las Vegas, NV, USA, 14–16 December 2005.
14. Nordic Geosynthetic Group. Nordic Guidelines for Reinforced Soils and Fills. 2004. Available online: https://danskgeotekniskforening.dk/sites/default/files/pdf/nordisk_handbok_armerad_jord-engelsk.pdf (accessed on 7 March 2022).
15. Abusharar, S.; Zheng, J.; Chen, B.; Yin, J. A simplified method for analysis of a piled embankment reinforced with geosynthetics. *Geotext. Geomembr. J.* **2009**, *27*, 39–52. [CrossRef]
16. BS 8006. Code of Practice for Strengthened Reinforced Soils and Other Fills. London, UK, 2010. Available online: http://worldcat.org/isbn/0580242161 (accessed on 12 October 2021).
17. Van Eekelen, S.; Bezuijen, A.; Van Tol, A. Analysis and modification of the British Standard BS 8006 for the design of piled embankments. *Geotext. Geomembr.* **2011**, *29*, 345–359. [CrossRef]
18. EBGEO. *Recommendations for Design and Analysis of Earth Structures Using Geosynthetic Reinforcements*; German Geotechnical Society: Munich, Germany, 2011.
19. Van Eekelen, S.; Bezuijen, A.; VanTol, A. An analytical model for arching in piled embankments. *Geotext. Geomembr.* **2013**, *39*, 78–102. [CrossRef]
20. Zhuang, Y.; Wang, K.; Liu, H. A simplified model to analyze the reinforced piled embankments. *Geotext. Geomembr.* **2014**, *42*, 154–165. [CrossRef]
21. *Dutch standard CUR 226*; Design Guideline Basal Reinforced Piled Embankments. CRC Press: Boca Raton, FL, USA, 2016.
22. Filz, G.; Sloan, J.; McGuire, M.; Smith, M.; Collin, J. Settlement and Vertical Load Transfer in Column-Supported Embankments. *Geotech. Geoenviron. Eng.* **2019**, *145*, 04019083. [CrossRef]
23. Briançon, L.; Simon, B. Pile-supported embankment over soft soil for a high-speed line. *Geosynth. Int.* **2017**, *24*, 293–305. [CrossRef]
24. Alsirawan, R.; Koch, E. The finite element modeling of rigid inclusion-supported embankment. *Pollack Period.* **2022**, *17*, 86–91. [CrossRef]
25. Chen, Q. An Experimental Study on Characteristics and Behavior of Reinforced Soil Foundation. Ph.D. Dissertation, Louisiana State University, Baton Rouge, LA, USA, 2007; p. 3361. Available online: https://digitalcommons.lsu.edu/gradschool_dissertations/3361 (accessed on 15 April 2022).
26. Alnmr, A. Material Models to Study the Effect of Fines in Sandy Soils Based on Experimental and Numerical Results. *Acta Tech. Jaurinensis* **2021**, *14*, 651–680. [CrossRef]
27. Alsirawan, R.; Koch, E. Behavior of embankment supported by rigid inclusions under static loads-parametric study. In Proceedings of the 12th IEEE International Conference on Cognitive Infocommunications—CogInfoCom 2021, Online, 23–25 September 2021; pp. 385–389.
28. Karumanchi, N. *Data Structures and Algorithmic Thinking with Python*; CareerMonk: Hyderabad, India, 2016.
29. Simon, S.A. *Recommendations for the Design, Construction and Control of Rigid Inclusion Ground Improvements*; Presses des Ponts: Paris, France, 2012.
30. Brinkgreve, R.; Zampich, L.; Ragi Manoj, N. *Plaxis Connect Edition V20*; Delft University of Technology & Plaxis B: Mekelweg, The Netherlands, 2019.
31. Pham, T.; Dias, D. Comparison and evaluation of analytical models for the design of geosynthetic-reinforced and pile-supported embankments. *Geotext. Geomembr.* **2021**, *49*, 528–549. [CrossRef]
32. Van Eekelen, S.J.; Venmans, A.A.; Bezuijen, A.; Van Tol, A.F. Long term measurements in the Woerden geosynthetic-reinforced pile-supported embankment. *Geosynth. Int.* **2020**, *27*, 142–156. [CrossRef]
33. Van Duijnen, P.; Van Eekelen, S.; Van der Stoel, A. Monitoring of a Railway Piled Embankment. In Proceedings of the 9th International Conference on Geosynthetics, Guarujá, Brazil, 23–27 May 2010.
34. Van der Stoel, A.E.; Brok, C.; de Lange, A.P.; Van Duijnen, P.G. Construction of the first railroad widening in the Netherlands on a Load Transfer Platform (LTP). In Proceedings of the 9th International Conference on Geosynthetics, Guarujá, Brazil, 23–27 May 2010.
35. Chen, R.; Xu, Z.; Chen, Y.; Ling, D.; Zhu, B. Field Tests on Pile-Supported Embankments over Soft Ground. *J. Geotech. Geoenviron. Eng.* **2010**, *136*, 777–785. [CrossRef]
36. Lee, T.; Lee, S.; Lee, I.; Jung, Y. Quantitative performance evaluation of GRPE: A full-scale modeling approach. *Geosynth. Int.* **2019**, *27*, 342–347. [CrossRef]
37. Zhao, M.; Liu, C.; El-Korchi, T.; Song, H.; Tao, M. Performance of Geogrid-Reinforced and PTC Pile-Supported Embankment in a Highway Widening Project over Soft Soils. *J. Geotech. Geoenviron. Eng.* **2019**, *145*, 06019014. [CrossRef]
38. Liu, H.; Kong, G.; Chu, J.; Ding, X. Grouted gravel column-supported highway embankment over soft clay: Case study. *Can. Geotech. J.* **2015**, *52*, 11. [CrossRef]

39. Hosseinpour, I.; Almeida, M.; Riccio, M. Full-scale load test and finite-element analysis of soft ground improved by geotextile-encased granular columns. *Geosynth. Int.* **2015**, *22*, 428–438. [CrossRef]
40. Liu, H.; Ng, C.; Fei, K. Performance of a Geogrid-Reinforced and Pile-Supported Highway Embankment over Soft Clay: Case Study. *J. Geotech. Geoenviron. Eng.* **2007**, *133*, 1483–1493. [CrossRef]

sustainability

MDPI

Article

Dynamic Characteristics of Rock Holes with Gravel Sediment Drilled by Bit Anchor Cable Drilling

Kuidong Gao, Jihai Liu, Hong Chen, Xu Li * and Shuan Huang

College of Mechanical and Electronic Engineering, Shandong University of Science and Technology, Qingdao 266590, China; gaokuidong22@163.com (K.G.); ljhsdust@163.com (J.L.)
* Correspondence: lixu2017@sdust.edu.cn

Abstract: Gravel and sediment frequently build up in holes during the anchor cable installation process, which makes it harder to install the anchor cable and causes reinforcement to fail, which can lead to accidents. In this paper, a bit anchor setup approach is proposed to set up a drill bit at the front of the anchor cable to aid the anchor cable drilling. The feasibility of this approach is established with the aid of DEM-MBD joint simulation and proves the correctness of the simulation model. The major elements affecting drilling effectivity have been studied by the use of 'check + simulation'. The effects exhibit that the axial velocity and working aperture are negatively correlated with the drilling resistance of the anchor cable; the feed price is positively related to the anchor cable drilling resistance; with the increase in anchor cable pitch, the drilling resistance of the anchor cable changes into a hump shape. When the bit is 0.2 m from the backside of the hole, the particle pace vector at the decrease stop of the bit gives a conical distribution. This paper is of fantastic magnitude in relation to the environment-friendly setup of anchor cables and protection against disasters.

Keywords: anchor cable installation; disaster prevention; DEM-MBD joint simulation; bit anchor cable; dynamic characteristics

Citation: Gao, K.; Liu, J.; Chen, H.; Li, X.; Huang, S. Dynamic Characteristics of Rock Holes with Gravel Sediment Drilled by Bit Anchor Cable Drilling. *Sustainability* **2023**, *15*, 5956. https://doi.org/10.3390/su15075956

Academic Editors: Hong-Wei Yang, Shuren Wang and Chen Cao

Received: 16 March 2023
Revised: 28 March 2023
Accepted: 28 March 2023
Published: 29 March 2023

1. Introduction

With the non-stop expansion of mining depth, flooring heave has emerged as one of the foremost traits and varieties of deformation and failure of surrounding rock in gentle rock. And it does great damage to the shape of the roadway. The discount of roadway sections precipitated utilizing ground heave poses a big danger to the security manufacturing of the mine [1–3]. The stress of the surrounding rock of the roadway can be significantly reduced by anchoring the occurrence position of floor heave by the direct bottom control method. Still, the rock mass of a hole wall is easy to collapse, which frequently leads to the accumulation of gravel sediment in the backside of the hole, which brings a huge problem for the installation of the anchor cable [4]. Eventually, this leads to the incidence of disasters.

At present, the most commonly used anchor cable installation method is manual installation, which can be manually adjusted to install the anchor cable to the bottom of the hole when encountering residue accumulated at the bottom of the hole. However, the environment is harsh when the anchor cable is installed, the work intensity of the workers is high and the efficiency is low [5,6]. The method of installing the drill bit in the front section of the anchor cable was proposed. It is of great significance to improve the drilling efficiency and structural optimization of anchor cables.

A number of studies have been conducted on drill bits. Cooley [7] analyzed the heat generation phenomenon in the process of using spiral bits and proposed that the drilling speed, length and diameter of the drill bit are the main factors affecting the drilling temperature. Saha [8] designed a new type of spiral bit, studied the influence of constant feed speed and thrust on the peak temperature of a spiral drill, and analyzed the impact of the spiral blade and tip angle of different types of bit on cutting.

Kim studied the influence of the characteristics of the drill bit and the change in drilling conditions on the drilling force and stress distribution of the drill bit, which furnished a theoretical groundwork for optimizing the geometry of the drill bit [9]. Zhang [10] designed a PRF drill bit. Compared with the traditional twist drill, it was found that the resistance of the improved PRF drill bit was reduced by 23.8% on average under the same cutting conditions. According to the characteristics of bit materials and rock materials, Šporin [11] reveals the wear mechanism of a bit under given rock materials and drilling conditions. Yakym [12] conducted a metallographic analysis of several steel materials commonly used in drill bit production and discussed the methods to prevent premature loss of drill bit workability. Šporin [13] used an optical microscope to analyze the drill bit material's chemical composition and the rock material's mechanical parameters. The research results laid a good foundation for developing new alloy drill bits. Moura [14] developed a new drilling performance prediction model and determined the linear relationship between the 'drillability constant' of the Maurer model and the bit weight on a bit.

However, the drilling bit research mainly focuses on the drilling of solids, and drilling research in granular particles are relatively few.

Scholars have also done a lot of research on the movement of particles. Due to the complexity and randomness of the movement of particulate matter, particulate matter presents some unique properties, different from solids, gases and liquids. Utter [15] conducted the shear test of dense granular materials on the two-dimensional Couette geometric model. It was found that when the thick, fine material is sheared, shear bands and anisotropic force chain networks will be formed. Antony [16] studied the collective behavior of three-dimensional granular media. He found that particles' filling rate and geometric stability are inversely proportional to the friction coefficient and elastic modulus between particles. Daniel Barreto [17] explored the effect of the friction coefficient between particles on the properties of granular materials by the discrete element method. It was found that an increase in friction coefficient can significantly enhance the internal stability of the particle force chain. Senetakis [18] conducted a micro-mechanical sliding test at the contact of broken limestone particles to capture the friction response during steady-state sliding. It was found that there was no significant difference in the friction response between the dry state and the oil-immersed state. Sandeep [19] studied particles' elastic properties and surface morphology's influence on the friction between particles through experiments. It was found that the friction coefficient between particles was negatively correlated with Young's modulus and surface roughness.

Multi-body dynamics (MBD) software has good kinematics analysis ability, but cannot complete the construction of a particle working environment. In 1971, Cundall [20,21] proposed an evaluation approach (DEM) for discrete granular substances, primarily based on molecular dynamics theory. Nowadays, the co-simulation method through DEM-MBD is more mature. Ren [22] used ADAMS-EDEM joint simulation to study the dynamic characteristics of the scraper conveyor under partial load and variable load conditions. The outcomes confirmed that the lateral vibration of the scraper chain used to be positively correlated with the quantity of coal transported, and the longitudinal vibration was once negatively correlated with the quantity of coal transported. Jiang [23] et al. used a DEM-MBD co-simulation to simulate and analyze the dynamic characteristics of the chain drive system under different impact loads and obtained the chainrings and the scraper's vibration under impact load. It provides a reference for the structural optimization and fault diagnosis of scraper conveyors. Shi [24] used DEM-MBD to study ballast damage during tamping. Nowadays, the joint simulation of DEM-MBD has been widely used in engineering machinery [25].

Therefore, combined with the above problems, and drawing on the research results of scholars, this paper proposes a drill anchor cable installation method, and uses DEM-MBD co-simulation and test methods to analyze the main factors affecting the drilling performance of the anchor cable: axial speed, feed rate, anchor cable pitch and working aperture. The dynamic traits of the drill anchor cable setup approach at some points of drilling are studied.

2. Research Method

2.1. Contact Model

In discrete element modeling, the Hertz-Mindlin (no-slip) contact model is most frequently employed. Figure 1 depicts the model's schematic layout.

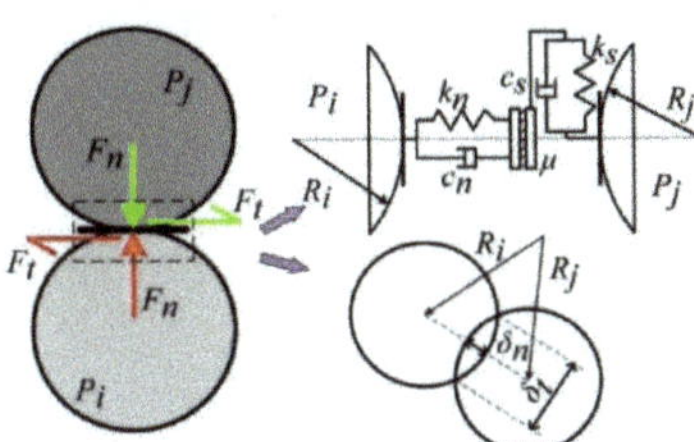

Figure 1. Hertz–Mindlin (no-slip) contact model.

The normal force in the model can be given by Equation (1):

$$F_n = \frac{4}{3}E^*\sqrt{R^*}\delta_n^{\frac{3}{2}} \tag{1}$$

where E^* is the equivalent of Young's modulus, R^* is the equivalent radius and δ_n is the standard overlap.

The identical Young's modulus E^* and equivalent radius R^* are given by Equations (2) and (3), respectively:

$$\frac{1}{E^*} = \frac{\left(1 - v_i^2\right)}{E_i} + \frac{\left(1 - v_j^2\right)}{E_j} \tag{2}$$

$$\frac{1}{R^*} = \frac{1}{R_i} + \frac{1}{R_j} \tag{3}$$

where E_i, v_i, and R_i, E_j, v_j and R_j are Young's modulus, and Poisson's ratio and radius of particles i and j are in contact with each other, respectively.

2.2. DEM Particle Parameters

In the DEM numerical simulation process, DEM particle parameters have an important influence on the motion state of particles, so it is necessary to calibrate the particle-related parameters before simulation. As shown in Figure 2, the relevant parameters of the particle were calibrated through cylinder pulling, slope, rolling and drop experiments. The average diameter of the gravel is 7.4 mm.

The cylinder pulling experiment was used to calibrate the angle of repose of the particle, the slope experiment was used to calibrate the static friction coefficient between gravel and the anchor cable, and the rolling experiment was used to calibrate the rolling friction coefficient between a rock and the anchor cable. The drop experiment was used to calibrate the restitution coefficient of the gravel particle. The DEM parameters after calibration are shown in Table 1. In this table, density, Poisson's ratio and shear modulus are P, σ and G, respectively. The coefficients of recovery, static friction and rolling friction are η, μS and μR, respectively.

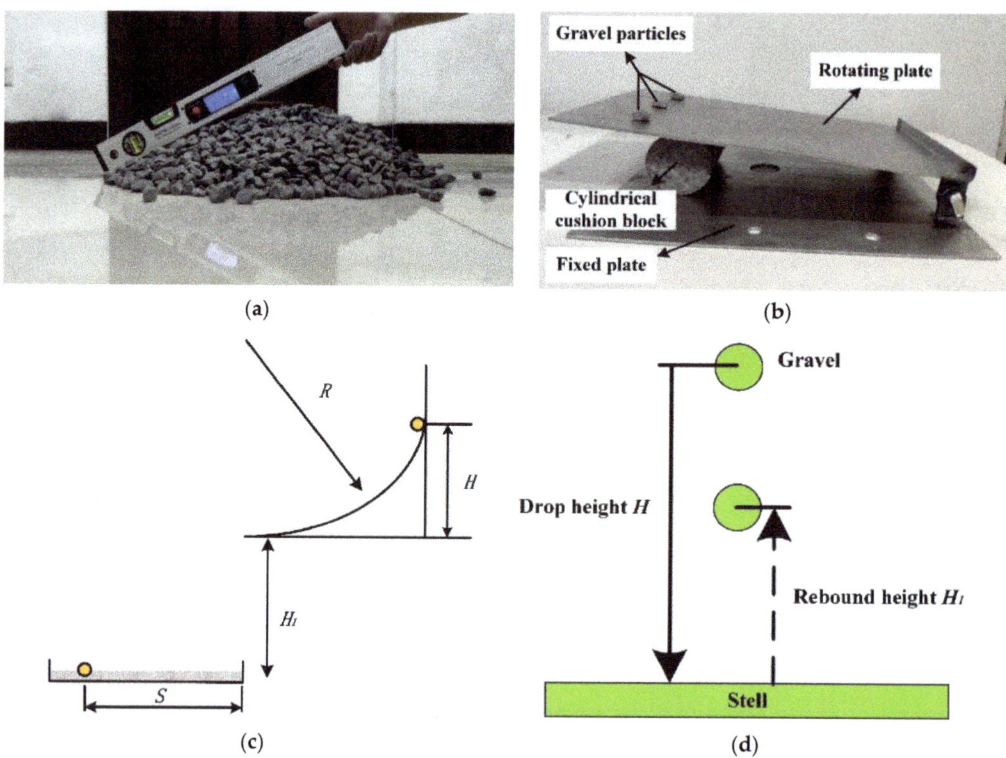

Figure 2. Calibration experiment of DEM particle parameters: (**a**) cylinder pulling experiment, (**b**) slope experiment, (**c**) rolling experiment, (**d**) drop experiment.

Table 1. The parameters of DEM.

	G (GPa)	P (kg/m³)	σ		μR	η	μS
Gravel	11	2700	0.29	Gravel—Gravel	0.2	0.62	0.74
Anchor cable	8.02	7801	0.29	Gravel—Anchor cable	0.25	0.42	0.49

2.3. Construction Model

2.3.1. Build A Test-Bed

The drill bit anchor cable drilling test bench built in this article is shown in Figure 3.

In the experiment, the organizational activities of the mechanical module are regulated by the control module. In the automatic module, the screw motor provides the low feed speed of the anchor cable, and the rotating motor provides the axial velocity of the anchor cable. When the bit anchor cable is drilling, the drilling resistance and torque data are transmitted to the data analysis module through the data acquisition instrument to process and analyze the signal.

2.3.2. Simulation Model

The simulation model is shown in Figure 4. The bottom of the anchor cable in the figure is a conical drill structure. According to the shape and size of gravel sediment in a bottom hole, a suitable gravel particle model was constructed using DEM software. Then, through the DEM software, a cylindrical hole-like workspace was generated. Simulation requirements determined the aperture and height. The DEM generated gravel particles were placed for 5S until they reached a stable state. The particle depth is 0.52 m, the particle diameter is 7.4 mm, and the working aperture is 160 mm. The Hertz–Mindlin (no-slip) contact modle is adopted between particles. The mannequin of the anchor cable is built in MBD software, and DEM-MBD is used for joint simulation.

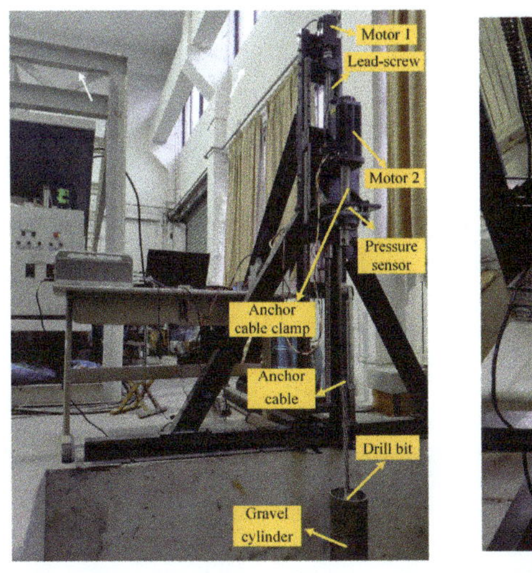

(a)

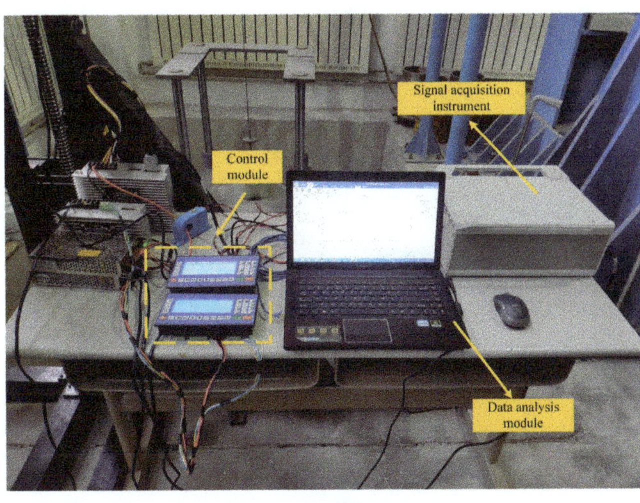

(b)

Figure 3. Test-bed structure diagram; (**a**) Mechanical module; (**b**) Control and data acquisition module.

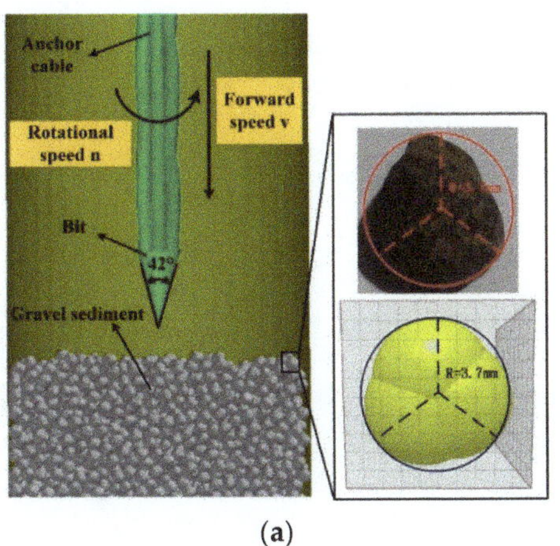

(a)

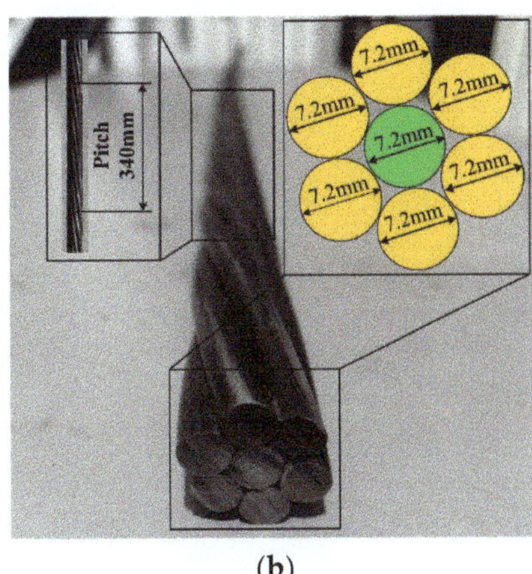

(b)

Figure 4. The bit anchor cable; (**a**) simulation model; (**b**) anchor cable.

2.4. Feasibility Analysis of Bit Anchor Cable Installation Method

In this paper, the feasibility of the proposed anchor cable installation method is verified by discrete element-multibody dynamics co-simulation. The conventional drill-less anchor cable and the drill bit anchor proposed in this paper are compared and analyzed, and the simulation result is shown in Figure 5. The control feed speed is 0.03 m/s and the axial speed is 0.60 rpm. In the process of signal acquisition, the pressure sensor in Figure 3a sensed the pressure signal, and then the signal acquisition instrument in Figure 3b inputted the collected pressure signal to the data analysis module.

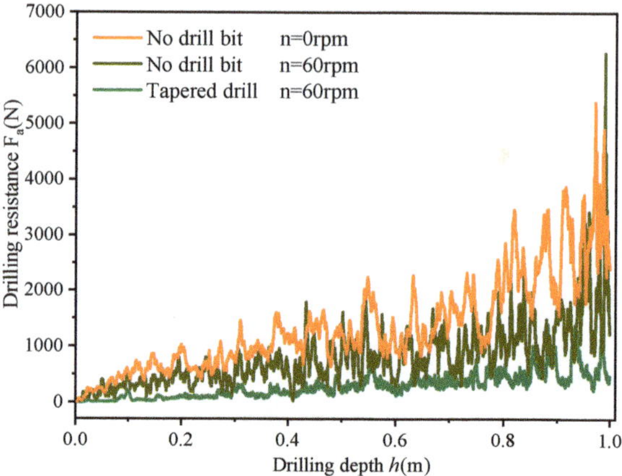

Figure 5. Comparison diagram of drilling resistance.

When the speed is 0 rpm, the drilling resistance is extremely high when drilling without a drill bit anchor cable, and the growth rate of drilling resistance is 2678; when the speed is 60 rpm, the drilling resistance growth rate of the drill-less anchor cable is reduced by 42%, but the drilling resistance of the anchor cable is still large.

When drilling with a drill bit anchor cable, the resistance of the anchor cable is significantly reduced. At 60 rpm, the resistance of the drill bit anchor cable increases by 61%, compared to the drill-less anchor cable during drilling. It can be seen that the drill bit anchor cable proposed in this paper can effectively reduce the drilling resistance during anchor cable installation, and then verify the feasibility of this method.

2.5. Research Program

To further explore the dynamic characteristics of the bit anchor cable under different working conditions, the main factors affecting the drilling performance of the drill anchor cable are analyzed and studied: axial speed, feed rate, anchor cable pitch and cylinder working space aperture. Experimental variables are shown in Table 2.

Table 2. Experimental scheme variable table.

Axial Speed/rpm	Feed Rate/(m/s)	Anchor Cable Pitch/mm	Working Aperture/mm
0, 120, 240, 360	0.03, 0.06, 0.09, 0.12	140, 240, 340, 440, 540	80, 120, 160, 200, 240

3. Results and Discussion

3.1. Axial Speed

The drill bit anchor cable drilling resistance curve is shown in Figure 6. It can be found that as the drilling depth of the drill bit anchor cable increases, the drilling resistance is an upward curve that first increases slowly and then increases rapidly. When the speed is n = 0 rpm, the drilling resistance of the drill bit anchor cable is significantly higher than that of n = 120, 240 and 360 rpm. It can be concluded that the rotational speed has a restraining effect on the resistance of the drill bit anchor cable during drilling. At the same time, it can be seen that the effect on the change of resistance after increasing the speed from 240 rpm to 360 rpm is much less than that of increasing the speed from 120 rpm to 240 rpm.

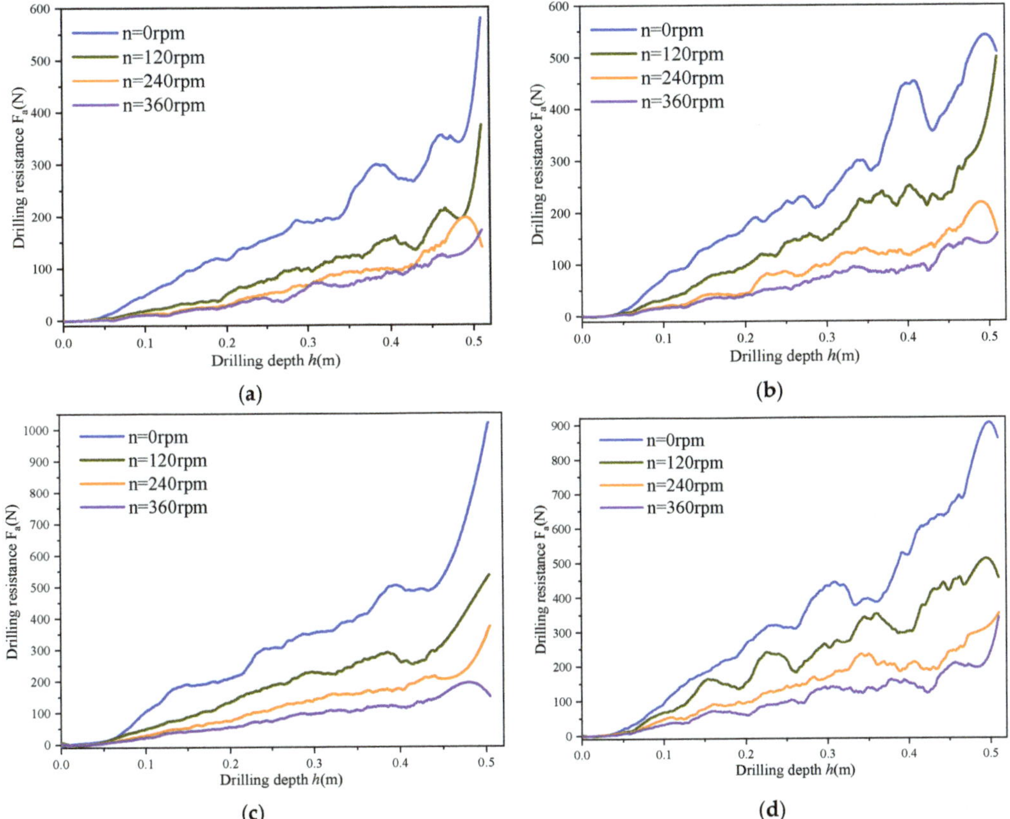

Figure 6. Drilling resistance curves of drill anchor cable under different axial speeds; (**a**) $v = 0.03$ m/s; (**b**) $v = 0.06$ m/s; (**c**) $v = 0.09$ m/s; (**d**) $v = 0.12$ m/s.

To further explore the reasons for the influence of speed on the dynamic characteristics of the drill bit anchor cable during the drilling process, the drill bit anchor cable feed speed was controlled to 0.03 m/s, and the velocity vector of the bottom hole gravel sediment particles under different axial speeds was analyzed. When the drill bit anchor cable is drilled to a depth of 0.4 m, the particle velocity vector plot is shown in Figure 7.

Affected by the screw structure, when the drill bit anchor cable is drilled, the structure of the bolt will constantly pull the surrounding gravel particles upward, so that the particles have an upward-moving trend. The upward movement of the particles will give the downward reaction force to the spiral structure of the drill anchor cable. It can be viewed from Figure 7 that, with the extent of the rotation pace of the drill anchor cable, the variety of upwardly shifting particles around the drill anchor cable and the speed of the particles progressively increase. Therefore, this will give the bit anchor cable a downward thrust. Finally, the drilling resistance of the drill anchor cable in Figure 6 regularly decreases with the amplification in rotation speed.

Figure 8 indicates the simulation and check curves when the feed fee is 0.03 m/s. It can be observed from the parent that, as the rotation pace increases, the drilling resistance of the drill anchor cable progressively decreases. Therefore, the correctness of the simulation is verified.

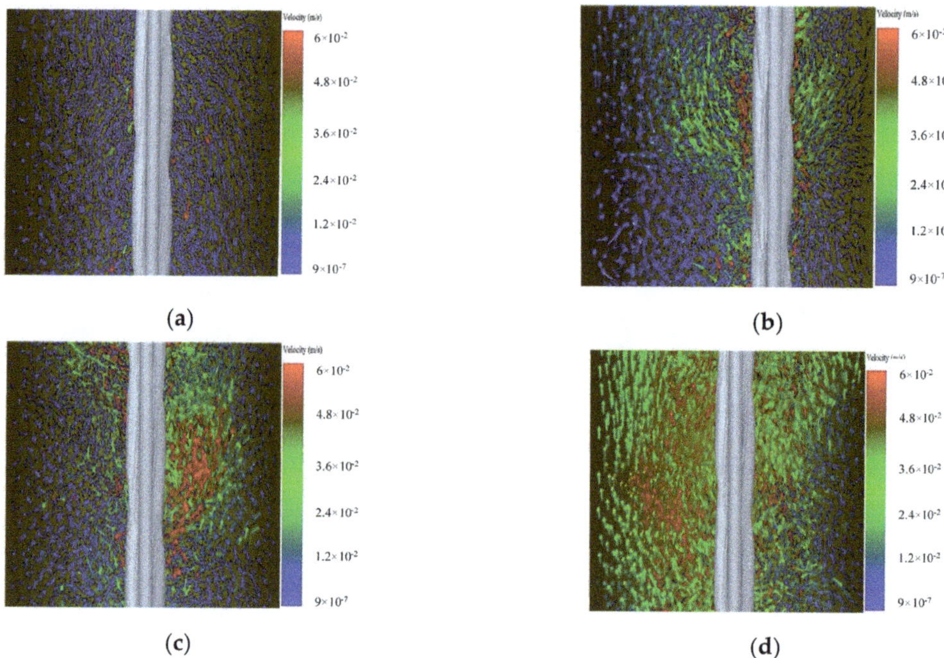

Figure 7. Vector diagram of particle motion around bit anchor cable at different rotation speeds; (**a**) $n = 0$ rpm; (**b**) $n = 120$ rpm; (**c**) $n = 240$ rpm; (**d**) $n = 360$ rpm.

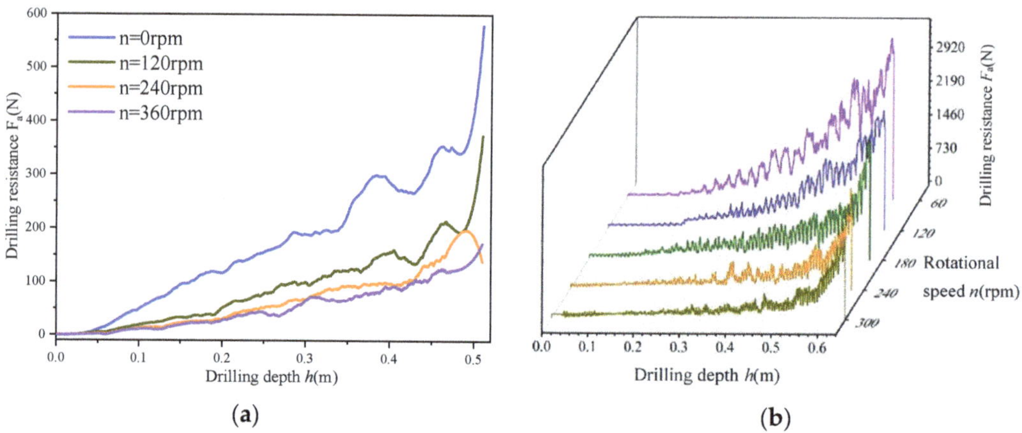

Figure 8. Verify the simulation model: (**a**) Simulation curve; (**b**) Test curve.

3.2. Feed Rate

To explore the influence of feed rate on drill bit anchor cable during drilling, the axial speed of the drill bit anchor cable is set to 0, 120, 240 and 360 rpm, respectively, and different feed speeds are controlled to simulate anchor line drilling. The drilling resistance change curve of the obtained drill bit anchor cable during operation is shown in Figure 9.

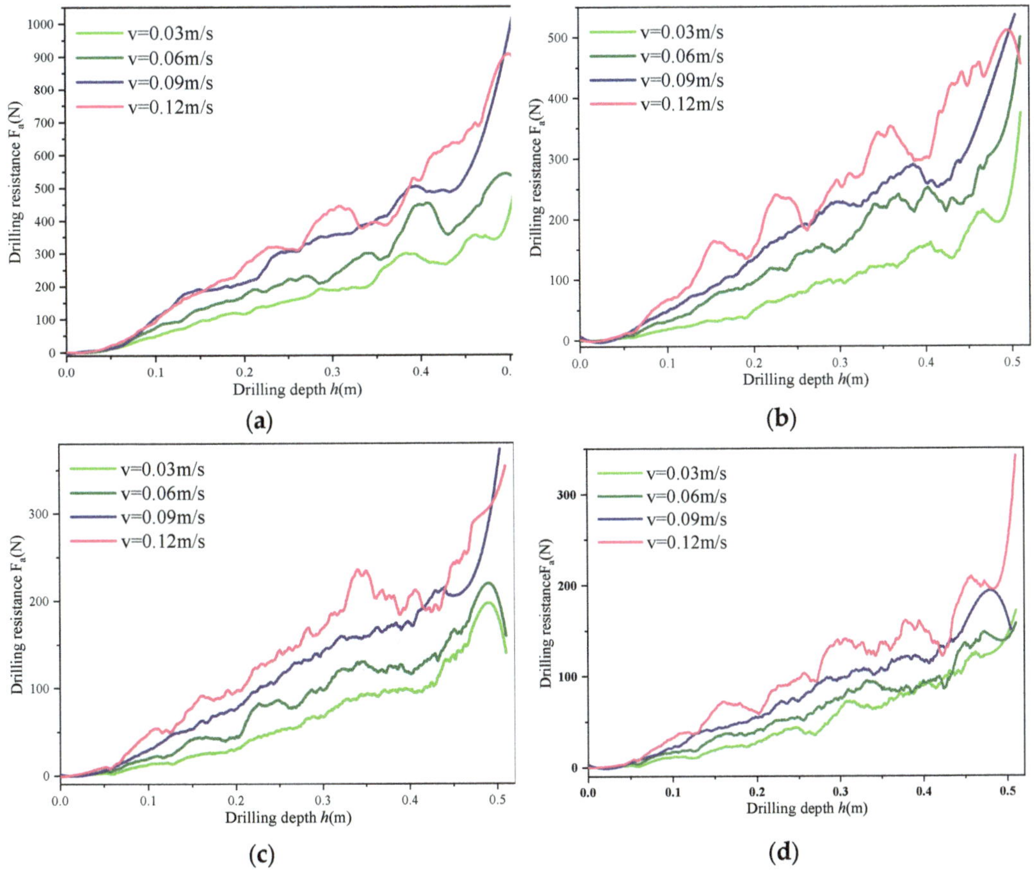

Figure 9. Resistance curve of bit anchor cable drilling below exclusive feeding rates: (**a**) $n = 0$ rpm; (**b**) $n = 120$ rpm; (**c**) $n = 240$ rpm; (**d**) $n = 360$ rpm.

It can be seen from Figure 9 that, under the same axial speed, with the increase of feed speed, the resistance of the drill bit anchor cable during operation gradually increases. To reflect the influence of feed rate on drilling resistance more obviously, the curve of Figure 9 is fitted, the fitted line is $y = a + bx$, and the fitted R^2 is greater than 0.95. The slope is counted, as shown in Table 3, and the data in the Table is plotted as a curve, as shown in Figure 10. It can be considered from the discern that, at the identical speed, the slope of the geared-up straight line of the drilling resistance of the bit anchor cable is positively correlated with the feed charge of the bit anchor cable. The smaller the rotation speed of the drill anchor cable, the stronger the positive correlation between the slope of the fitting line and the feed rate of the bit anchor cable. When the feed charge of the bit anchor cable is 0.12 m/s and the rotation velocity is zero rpm, the slope of the becoming line of the drilling resistance of the bit anchor cable is 715 greater than that at the identical rotation velocity when the feed fee is 0.03 m/s. When the feed charge of the bit anchor cable is 0.12 m/s and the rotation velocity is 360 rpm, the slope of the becoming line of the drilling resistance of the bit anchor cable is solely 183 greater than that at the identical rotation pace when the feed charge is 0.03 m/s.

Table 3. Fitting straight line slope table of drilling resistance.

Rotating Speed (rpm)	Feeding Rate (m/s)	Fitting Straight Line Slope	Rotating Speed (rpm)	Feeding Rate (m/s)	Fitting Straight Line Slope
0	0.03	698	240	0.03	272
0	0.06	933	240	0.06	381
0	0.09	1153	240	0.09	475
0	0.12	1413	240	0.12	555
120	0.03	374	360	0.03	222
120	0.06	595	360	0.06	254
120	0.09	747	360	0.09	328
120	0.12	908	360	0.12	405

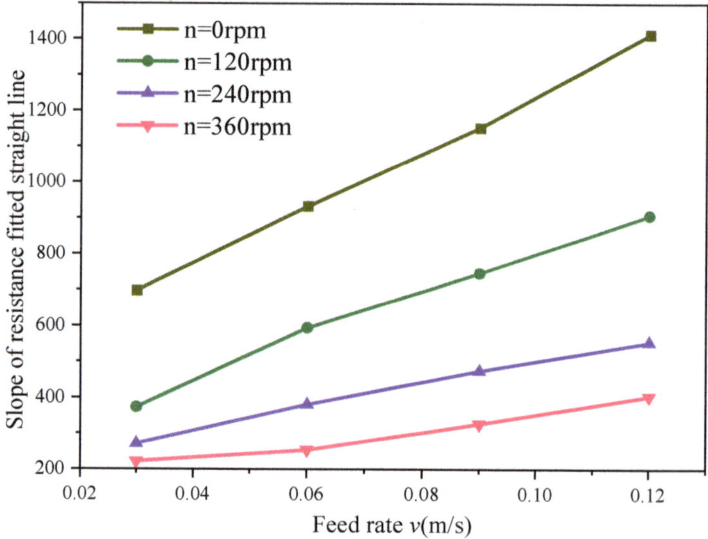

Figure 10. Comparison diagram of fitting straight line slope of bit anchor cable drilling resistance under different feeding rates.

To further explore the influence of feed rate on drill bit anchor cable during drilling, the DEM software program used to be used to export the speed vector layout of particles in the gap when the bit anchor cable used to be simply drilled to the backside of the gap at 360 rpm and with one-of-a-kind feed rates, as proven in Figure 11. When the feed rate is 0.03 m/s, the quicker particles in the gap are spirally disbursed round the drill anchor cable. With the feed price increase, the quicker particles' spiral distribution phenomenon gradually weakens and progressively turns into a vertical motion downwards, alongside the bit anchor cable. The stirring effect of the drill bit anchor cable on the particles in the hole is negatively correlated with the feed rate.

3.3. Anchor Cable Pitch

The stirring effect of the drill bit anchor cable on the particles in the hole is negatively correlated with the feed rate. In addition, the structure also has an unavoidable influence on the resistance of the drill bit anchor cable during drilling, and it is necessary to study the dynamic characteristics at different pitches.

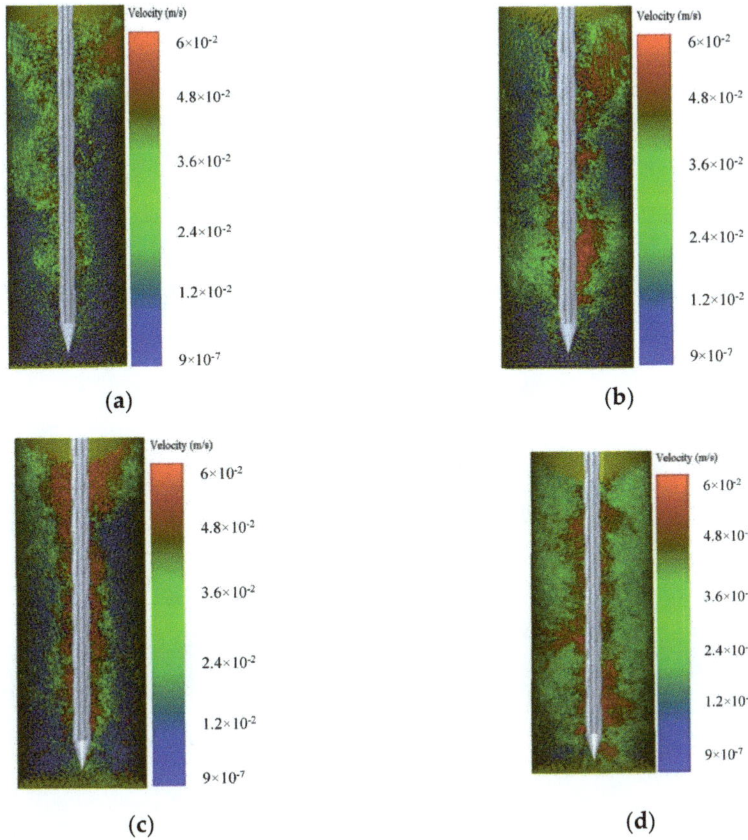

Figure 11. Vector diagram of particle motion state in the hole at different feeding rates: (**a**) $v = 0.03$ m/s; (**b**) $v = 0.06$ m/s; (**c**) $v = 0.09$ m/s; (**d**) $v = 0.12$ m/s.

As shown in Figure 12, five kinds of anchor cables with different pitch were designed, and drilling experiments were conducted on these five kinds of anchor cables, respectively. Then, the drilling simulation was carried out underneath special action parameters. The 340 mm pitch is the common pitch of the anchor cable.

In this paper, the resistance of the drill bit anchor cable under different pitches is linearly fitted during the drilling process, and the fitted R^2 is greater than 0.95. Figure 13 is the slope diagram of the provided straight line of the drilling resistance of the bit anchor cable at the feed rate of 0.06 m/s, and the feed rate of 0.12 m/s under different pitches.

It can be seen from Figure 13 that, when the pitch of the drill bit anchor cable is less than 440 mm, the growth rate of resistance experienced by the drill bit anchor cable during drilling is proportional to the pitch. When the pitch is greater than 440 mm and less than 540 mm, the growth rate of drilling resistance is inversely proportional to the pitch, and when the pitch reaches 440 mm, the growth rate of drilling resistance is the fastest.

Therefore, when the pitch is lower than 450 mm, the resistance value during drilling can be reduced by reducing the screw pitch of the drill bit anchor cable, which is consistent with the research conclusions of the literature [8], which shows that different sizes of the pitch have an important impact on the dynamic characteristics during drilling.

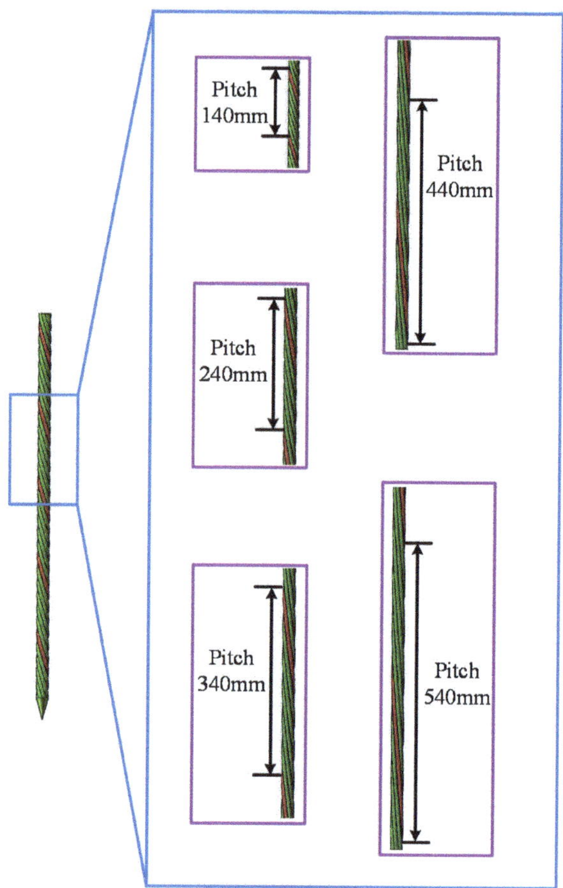

Figure 12. Structural design drawing of bit anchor cable pitch.

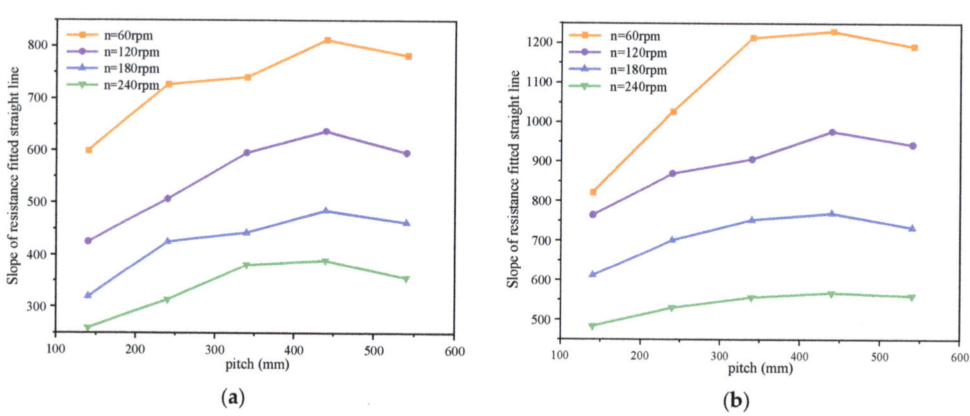

Figure 13. Fitting straight line slope diagram of drilling resistance of bit anchor cable under different pitches: (**a**) $v = 0.06$ m/s; (**b**) $v = 0.12$ m/s.

3.4. Working Aperture

Due to the influence of geological conditions and roadway deformation in different roadways, the aperture required for the installation of bit anchor cables will also be quite different. This section uses an experimental approach to investigate the effect of pore size on anchor cable drilling. The gravel container of the test bench is shown in Figure 14. There are five gravel containers with apertures of 80 mm, 120 mm, 160 mm, 200 mm and 240 mm, respectively. The relevant experimental variables are shown in Table 4. The particle depth in the cylindrical working hole is 800 mm. As shown in Figure 15, the drilling resistance and hindrance torque curves of the anchor cable under this operating condition are shown. At a drilling depth of around 0.5 m, the average of the drilling resistance and hindering torque during drilling is shown in Figure 16.

It can be seen from Figures 15 and 16 that, with the increase in the hole diameter, the drilling resistance and hindrance torque of the anchor cable during the drilling process gradually decrease. When the hole size varies between 80–120 mm and 160–240 mm, the reduction in drilling resistance and hindering torque is relatively small; when the bore size is changed from 120 mm to 160 mm, the drilling resistance and obstacle torque are significantly reduced.

The length of the anchor cable in this experiment is 1.5 m. The stiffness of the anchor cable has a certain relationship with the depth of the anchor cable into the gravel, which will be studied later.

As the drilling depth increases, the tightness between particles gradually increases, which is one reason for the increase in drilling resistance and hindering torque of the bit anchor cable. However, this reason will not lead to a rapid increase in drilling resistance. The reason for the rapid increase in drilling resistance is due to the inability of particles to move longitudinally.

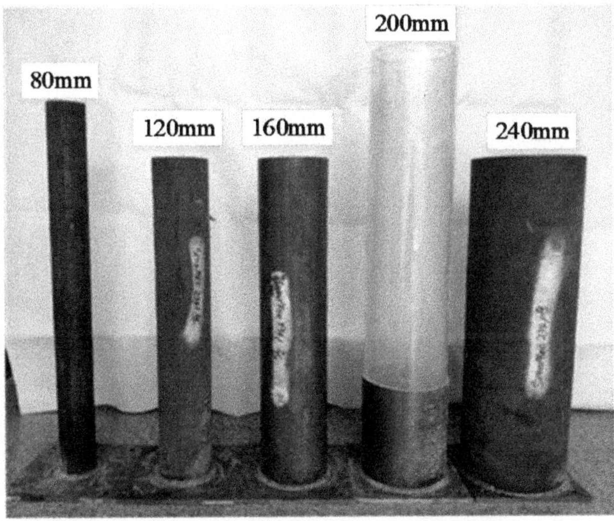

Figure 14. Gravel containers with different working apertures.

Table 4. Aperture simulation variable table.

Feeding Rates (m/s)	Axial Speed (rpm)	Aperture (mm)	Particle Depth in the Hole (mm)
0.01	240, 300	80, 120, 160, 200, 240	800

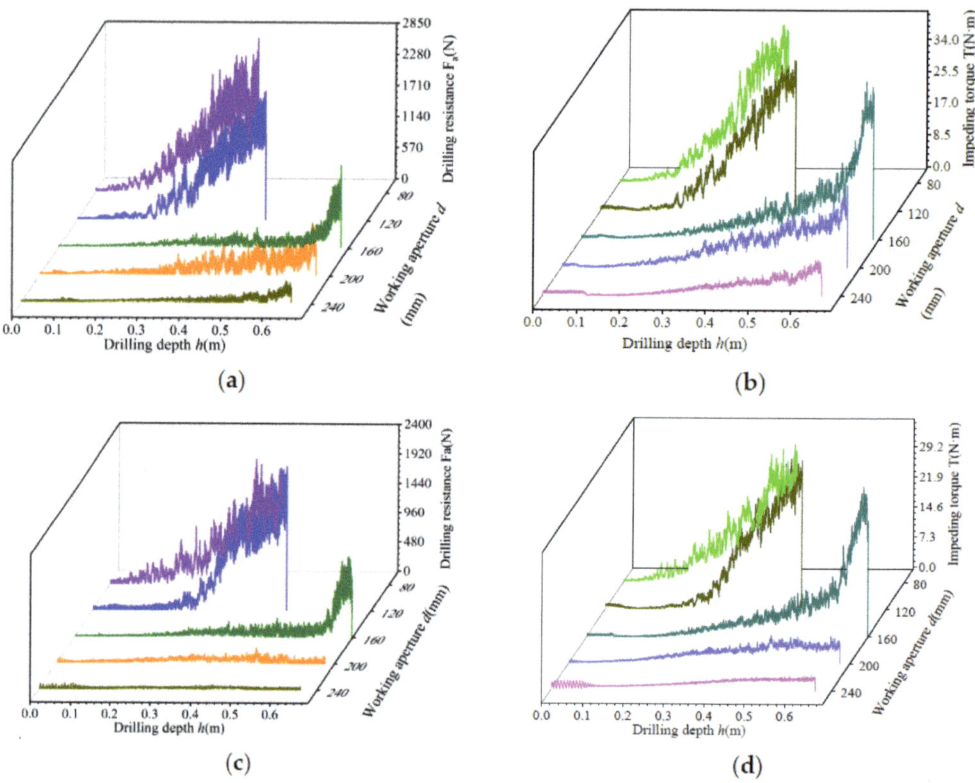

Figure 15. Dynamic features curve of bit anchor cable below amazing apertures: (**a**) *n* = 240 rpm Drilling resistance; (**b**) *n* = 240 rpm Hinder torque; (**c**) *n* = 300 rpm Drilling resistance; (**d**) *n* = 300 rpm Hinder torque.

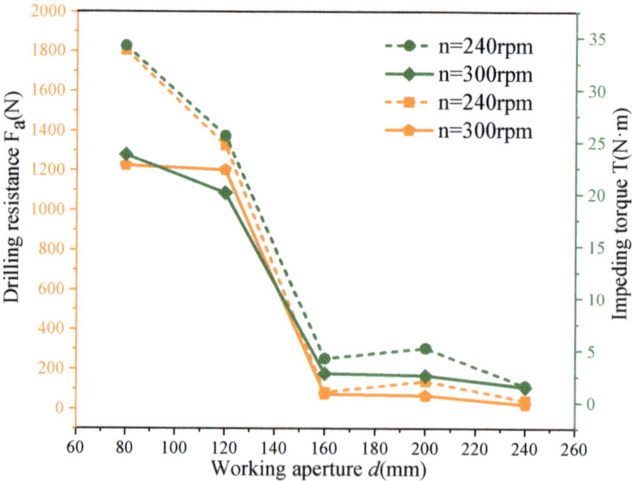

Figure 16. The average value of particle obstruction at different apertures and drilling depths of 0.5–0.55 m.

When the bit anchor cable is drilling, the particles under the bit mainly move in two directions, namely lateral movement and longitudinal movement. However, the space between the bit and the bottom of the hole is limited and, when the bit is near the bottom of the hole, it is difficult for the particles below the bit to move down. The bit anchor cable directly presses against the particles, resulting in a sharp increase in drilling resistance. Figure 17 provides a good illustration of this phenomenon.

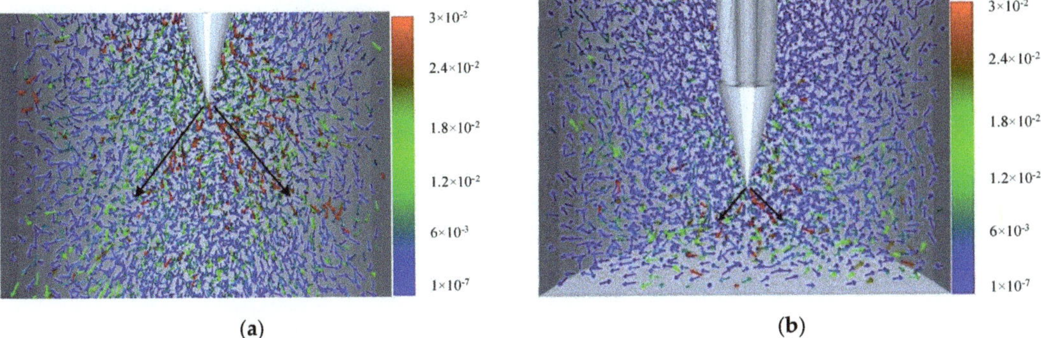

(a) **(b)**

Figure 17. Velocity vector diagram of particles at the lower end of the bit: (**a**) Particle velocity vector diagram when the bit is 0.2 m from the bottom of the hole; (**b**) Particle velocity vector diagram when the bit is 0.05 m from the bottom of the hole.

Figure 17 shows the velocity vector diagram of the particles at the lower end of the drill bit when the drill bit is drilled at different depths. It can be seen from this diagram that, when the drill bit is 0.2 m from the bottom of the hole, the velocity vector of the particles at the lower end of the drill bit show an apparent conical distribution. Under the extrusion of the drill bit, the particles below the drill bit will diffuse horizontally. At this time, the wide variety of particles that diffuse horizontally is large, and the increased fee of the drilling resistance of the drill bit anchor cable is extraordinarily slow. When the bit is 0.05 m away from the backside of the hole, the quantity of particles that endure lateral diffusion decreases sharply, and the drilling stress of the bit anchor cable directly acts on the backside of the gap via the particles, thereby inflicting the drilling resistance of the bit anchor line to push upward sharply.

4. Conclusions

In this paper, a new anchor cable installation method is proposed, and the feasibility of this method is verified. From the perspective of particle motion and anchor cable dynamics, the dynamic characteristics of bit anchor cable was studied. The following conclusions are drawn:

(1)   Affected by using the spiral shape of the drill anchor rod, the axial velocity is negatively correlated with the drilling resistance of the bit anchor cable, and the effects are compared to confirm the correctness of the simulation. The pitch of the bit anchor thread has a hump-shaped relationship with drilling resistance. When the slope is much less than 450 mm, the drilling resistance can be decreased with the aid of appropriately lowering the pitch of the bit anchor cable;

(2)   The feed speed is positively correlated with the resistance encountered during drilling; the degree of influence of the anchor cable on the particles in the hole is inversely correlated with the feed rate during drilling. The slope of the drilling resistance fitting line is positively correlated with the feed velocity, and the smaller the axial speed, the stronger the positive correlation between the slope and the feed speed;

(3)   As the aperture increases, the drilling resistance and the hindering torque of the drill anchor cable gradually decrease. When the bit is 0.2 m away from the bottom of the

hole, the particle velocity vector at the lower end of the bit shows an evident conical distribution, and the particles diffuse horizontally. When the bit is 0.05 m from the bottom of the hole, the number of particles that diffuse laterally decreases sharply.

This paper is of great significance to the efficient installation of anchor cables and the protection against disasters. However, in practice, the anchor cable will have a certain degree of radial bending under the action of drilling resistance. The next step will be to use the finite element method to make the bolt model closer to the actual situation. Since the bit anchor cable needs to drill through gravel, it has certain requirements on the compressive strength of the gravel, which should be less than 30 Mpa.

Author Contributions: Formal analysis, K.G.; funding acquisition, K.G.; methodology, S.H., and X.L.; supervision, H.C., and J.L.; writing—original draft, H.C. and S.H.; writing—review and editing, H.C. and S.H. All authors have read and agreed to the published version of the manuscript.

Funding: This work was supported by the National Natural Science Foundation of China, grant number 52174146; the Project of Shandong Province Higher Educational Young Innovative Talent Introduction and Cultivation Team (Performance enhancement of deep coal mining equipment).

Institutional Review Board Statement: Not applicable.

Informed Consent Statement: Not applicable.

Data Availability Statement: Data were curated by the authors and are available upon request.

Conflicts of Interest: The authors declare no conflict of interest.

References

1. Hao, W.D.; Tao, D. Research on Deformation Control Technology of Broken Soft Rock Pre-mining Roadway. *IOP Conf. Ser. Earth Environ. Sci.* **2019**, *358*, 042047. [CrossRef]
2. Gao, Y.; Wang, C.; Liu, Y.; Wang, Y.; Han, L. Deformation mechanism and surrounding rock control in high-stress soft rock roadway: A case study. *Adv. Civ. Eng.* **2021**, *2021*, 9950391. [CrossRef]
3. Sun, Y.; Li, G.; Zhang, J.; Xu, J. Failure mechanisms of rheological coal roadway. *Sustainability* **2020**, *12*, 1885. [CrossRef]
4. Xu, Y.; Pan, K.; Zhang, H. Investigation of critical techniques on floor roadway support under the impacts of superimposed mining: Theoretical analysis and field study. *Environ. Earth Sci.* **2019**, *78*, 436. [CrossRef]
5. Shi, K.; Wu, X.; Tian, Y.; Xie, X. Analysis of re-tensioning time of anchor cable based on new prestress loss model. *Mathematics* **2021**, *9*, 1094. [CrossRef]
6. Yang, Z.P.; Li, S.Q.; Yu, Y.; Liu, X.; Hu, Y. Study on the variation characteristics of the anchor cable prestress based on field monitoring in a foundation pit. *Arab. J. Geosci.* **2020**, *13*, 1269. [CrossRef]
7. Cooley, R.L.; Barkmeier, W.W. Temperature rise in the pulp chamber caused by twist drills. *J. Prosthet. Dent.* **1980**, *44*, 426–429. [CrossRef]
8. Saha, S.; Pal, S.; Albright, J.A. Surgical drilling: Design and performance of an improved drill. *J. Biomech. Eng. Trans. Asme* **1982**, *104*, 245–252. [CrossRef]
9. Kim, K.W.; Ahn, T.K. Force Prediction and Stress Analysis of a Twist Drill from Tool Geometry and Cutting Conditions. *Int. J. Precis. Eng. Manuf.* **2005**, *6*, 65–72.
10. Zhang, Q.; Wang, J. Geometry, specification, and drilling performance of a plane rake-faced drill point design. *Proc. Inst. Mech. Eng. Part C J. Mech. Eng. Sci.* **2010**, *224*, 369–378. [CrossRef]
11. Šporin, J.; Mrvar, P.; Petrič, M.; Vižintin, G.; Vukelić, Z. The characterization of wear in roller cone drill bit by rock material—Sandstone. *J. Pet. Sci. Eng.* **2018**, *173*, 1355–1367. [CrossRef]
12. Yakym, R.S.; Petryna, D.Y. Analysis of Causes and Preventing Ways of Early Workability Loss of Three-Cone Rock Bit Cutters. *Met. Adv. Technol.* **2020**, *42*, 731–751. [CrossRef]
13. Porin, J.; Balako, T.; Mrvar, P.; Janc, B.; Vukeli, E. Change of the Properties of Steel Material of the Roller Cone Bit Due to the Influence of the Drilling Operational Parameters and Rock Properties. *Energies* **2020**, *13*, 5949.
14. Moura, J.D.; Xiao, Y.; Yang, J.; Butt, S.D. An empirical model for the drilling performance prediction for roller-cone drill bits. *J. Pet. Sci. Eng.* **2021**, *10*, 3287. [CrossRef]
15. Utter, B.; Behringer, R.P. Transients in sheared granular matter. *Eur. Phys. J.* **2004**, *14*, 373–380. [CrossRef]
16. Antony, S.J.; Kruyt, N.P. Role of interparticle friction and particle-scale elasticity in the shear-strength mechanism of three-dimensional granular media. *Phys. Rev. E* **2009**, *79*, 031308. [CrossRef]
17. Barreto, D.; O'Sullivan, C. The influence of inter-particle friction and the intermediate stress ratio on soil response under generalized stress conditions. *Granul. Matter* **2012**, *14*, 505–521. [CrossRef]

18. Senetakis, K.; Sandeep, C.S.; Todisco, M.C. Dynamic inter-particle friction of crushed limestone surfaces. *Tribol. Int.* **2017**, *111*, 1–8. [CrossRef]
19. Sandeep, C.S.; Senetakis, K. Effect of Young's Modulus and Surface Roughness on the Inter-Particle Friction of Granular Materials. *Materials* **2018**, *11*, 217. [CrossRef]
20. Cundall, P.A.; Strack, O.D.L. A discrete numerical model for granular assemblies. *Geotechnique* **2008**, *30*, 331–336. [CrossRef]
21. Cundall, P.A. The measurement and Analysis of accelerations in rock slopes. Ph.D. Thesis, University of London: Imperial College of Science and Technology, London, UK, 1971.
22. Ren, W.J.; Wang, L.; Mao, Q.H.; Jiang, S.B.; Huang, S. Coupling Properties of Chain Drive System under Various and Eccentric Loads. *Int. J. Simul. Model.* **2020**, *19*, 643–654. [CrossRef]
23. Jiang, S.B.; Huang, S.; Zeng, Q.L. Dynamic properties of chain drive system considering multiple impact factors. *Int. J. Simul. Model* **2022**, *21*, 284–295. [CrossRef]
24. Shi, S.; Gao, L.; Xiao, H.; Xu, Y.; Yin, H. Research on ballast breakage under tamping operation based on DEM–MBD coupling approach. *Constr. Build. Mater.* **2021**, *272*, 121810. [CrossRef]
25. Chen, Z.; Xue, D.; Wang, G.; Cui, D.; Fang, Y.; Wang, S. Simulation and optimization of the tracked chassis performance of electric shovel based on DEM-MBD. *Powder Technol.* **2021**, *390*, 428–441. [CrossRef]

Article

Using Electric Field to Improve the Effect of Microbial-Induced Carbonate Precipitation

Jinxiang Deng [1,2], Mengjie Li [1,2], Yakun Tian [1,2], Zhijun Zhang [1,2,*], Lingling Wu [1,2,*] and Lin Hu [1,2]

1   School of Resource & Environment and Safety Engineering, University of South China, Hengyang 421001, China
2   Hunan Province & Hengyang City Engineering Technology Research Center for Disaster Prediction and Control on Mining Geotechnical Engineering, Hengyang 421001, China
*   Correspondence: 130000148665@usc.edu.cn (Z.Z.); 2013000541@usc.edu.cn (L.W.)

Abstract: The precipitation of calcium carbonate induced by *Sporosarcina pasteurii* (*S. pasteurii*) has garnered considerable attention as a novel rock and soil reinforcement technique. The content and structure of calcium carbonate produced through this reaction play a crucial role in determining the rocks' and soil's reinforcement effects in the later stages. Different potential gradients were introduced during the bacterial culture process to enhance the performance of the cementation and mineralization reactions of the bacterial solution to investigate the effects of electrification on the physical and chemical characteristics, such as the growth and reproduction of *S. pasteurii*. The results demonstrate that the concentration, activity, and number of viable bacteria of *S. pasteurii* were substantially enhanced under an electric field, particularly the weak electric field generated by 0.5 V/cm. The increased number of bacteria provides more nucleation sites for calcium carbonate deposition. Moreover, as the urease activity increased, the calcium carbonate content generated under an electric potential gradient of 0.5 V/cm surpassed that of other potential gradient groups. The growth rate increased by 9.78% compared to the calcium carbonate induced without electrification. Significantly, the suitable electric field enhances the crystal morphology of calcium carbonate and augments its quantity, thereby offering a novel approach for utilizing MICP in enhancing soil strength, controlling water pollution, and mitigating seepage. These findings elevate the applicability of microbial mineralization in engineering practices.

Keywords: microbial biomineralization; *Sporosarcina pasteurii*; potential gradient; calcium carbonate

Citation: Deng, J.; Li, M.; Tian, Y.; Zhang, Z.; Wu, L.; Hu, L. Using Electric Field to Improve the Effect of Microbial-Induced Carbonate Precipitation. *Sustainability* **2023**, *15*, 5901. https://doi.org/10.3390/su15075901

Academic Editors: Hong-Wei Yang, Shuren Wang and Chen Cao

Received: 21 February 2023
Revised: 25 March 2023
Accepted: 27 March 2023
Published: 28 March 2023

1. Introduction

The swift advancement of technology and the economy has resulted in the excessive depletion of the Earth's resources. In new construction projects, cement and other cementitious materials are essential [1,2]. However, the production and utilization of cement can readily lead to environmental contamination. In recent years, microbial cement, a novel reinforcement methodology, has captured the attention of countless researchers due to its inherent benefits of environmental conservation, convenience, and pollution-free attributes, aligned with the objective of eco-friendly and green development. Feng et al. [3] investigated a new variety of bacterial-based self-repairing concrete using microbial-induced calcium carbonate precipitation. They fabricated self-healing concrete beam samples incorporating bacteria as a mending agent and polyvinyl alcohol fibers. The outcomes indicated that the flexural strength of beams could be improved by roughly 14%. Sj et al. [4] synthesized the research progress of enhancing the longevity of industrial-by-product-modified concrete and provided insights for the future incorporation of microbial treatment into concrete. Yu et al. [5] utilized microorganisms to hydrolyze organophosphate monoesters to yield phosphate nanomaterials, which exhibited greater stability in structures. The gelation performance of phosphate precipitation induced by microorganisms was exceptional.

Additionally, this technique demonstrated potential applications in removing heavy metals, treating solid waste, and consolidating sandy soil. Zheng and Qian [6] utilized microbial-induced calcification (MICP) technology to prevent cracks in cement-based materials. The outcomes revealed that this technology could increase the area repair rate to 96% and the corresponding waterproof repair resistance to 85%, effectively improving the self-healing efficacy of cracks in later stages of cement-based materials.

Microbial cementation comprises a succession of biochemical reactions that utilize microorganisms to instigate the precipitation of calcium carbonate (MICP) and leverage the precipitated calcite to ameliorate the mechanical characteristics of the rock and soil that necessitate reinforcement. Almajed et al. [7] presented a summary of the applicability, advantages, and disadvantages of MICP in soil treatment and elucidated the mechanism of MICP in both natural and laboratory environments. To realize on-site implementation, they proposed a site-specific method to execute on-site. Feng et al. [8] elaborated on the research findings of MICP technology on recycled aggregate reinforcement and listed the factors influencing the reinforcement effect of recycled aggregate. Post-MICP treatment, the aggregate properties were improved. Gao and Dai [9] suggested optimizing the medium to enhance the cementing effect. Optimizing the medium increased calcium carbonate precipitation at the contact point, enhanced the cementing effect, and increased strength by 1.2–4 times. Lin et al. [10] utilized numerical simulation to determine the dominant calcium carbonate distribution in the pore space and discussed the influence of calcium carbonate distribution on small strain stiffness and permeability. The results indicated that the numerical model could estimate the permeability after MICP treatment. Zhang et al. [11] conducted an experimental comparison of MICP tailings reinforcement technology affected by the seepage field and unaffected by the seepage field. It was discovered that under the influence of the seepage field, the total pores were comparatively reduced, the shear strength was lowered, and the distribution of calcium carbonate was more irregular. This suggests that seepage is one of the primary reasons for the instability of the tailings dam.

MICP is a multi-faceted process with numerous factors that exert influence, including the concentration of reinforcement bacteria, their activity [7,12], the cultural environment in which they grow, the duration and temperature of the culture, and the pH level [13–15]. Presently, the efficacy of MICP is restricted to specific practical applications. To enhance the reinforcement performance and rate, a few researchers have scrutinized calcium carbonate production by varying different culture conditions [16–18]. Among these, Okwadha and Li [19] examined the impact of bacterial species, cementing solution concentration, bacterial concentration, and reaction temperature on calcium carbonate deposition and ascertained that the maximum deposit of calcium carbonate occurs when the concentration of calcium ions is 0.25 mol/L, the concentration of urea is 0.666 mol/L, the bacterial concentration is 2.3×10^8 cells/mL. Several other scholars have evaluated the efficiency of the strengthening bacteria in hydrolyzing urea [20], which facilitates a more uniform strengthening effect. Research has revealed that the alkaline environment yields a higher calcium carbonate content and a better cementing effect than other pH culture conditions. In order to enhance the effectiveness of microbial reinforcement, the primary objective is to elevate the amount of calcium carbonate generated and enhance the crystal composition of calcium carbonate. Numerous studies have demonstrated that the crystal form of calcium carbonate will develop different crystal morphologies in the course of MICP, owing to different reactants [21]; Wei et al. [22] observed five distinct crystal morphologies of calcium carbonate grown in the MICP process via scanning electron microscope (SEM), namely cube, diamond, polyhedral layer, sphere, and irregular shape. By modulating the biological activity of microorganisms, researchers can manipulate calcium carbonate's crystallization rate and alter the resulting crystals' morphology [23]. As a result, promoting the biological activity of reinforcing bacteria can lead to an increase in the quantity of precipitated calcium carbonate precipitation, and improving the effect of microbial reinforcement [24] are effective means of accomplishing the goal.

The promotion of microbial activity can be achieved by implementing an electric field. Since the surface of microbial cells typically carries a charge, applying a potential can prompt microorganisms to undergo biochemical reactions within a specific action dimension created by the generated electric field. Research conducted by some scholars [25,26] has demonstrated that voltage can provide a certain degree of electric oxidation intensity, effectively stimulating microbial growth, enhancing microbial degradation activity, and enabling them to play a more effective role in soil remediation. In soil pollution control, the process of electric field migration can significantly improve the uniform distribution of microorganisms. Mena et al. [27] utilized a direct current in microbial remediation of soil hydrocarbons, applying an electric potential gradient (0–2 V/cm) to the culture of diesel degradation microorganisms, which increased the degradation rate. It was also found that a particular electric field had no significant negative impact on microbial activity. Cheng et al. [28] discovered that the electrokinetic (EK) phenomenon could enhance the mobility and redistribution of mineralized bacteria in heterogeneous rock and soil, thereby significantly promoting bioaugmentation. Weak potential gradients (1 V/cm) enable bacteria to evolve from a loose to a tightly bound state.

Moreover, the electric field promotes bacterial adhesion, fluid contact between anions and cations, organic carbon, and bacteria to stimulate bacterial proliferation. The results also reveal that the electric field is more effective in stimulating bacterial proliferation than transporting bacteria. Few studies have been conducted on applying electric fields in microbial induced mineralization. Therefore, based on the beneficial effect of the electric field on bacteria, the role of an electric field in MICP mineralization reaction was further explored by stimulating the proliferation of mineralized microorganisms and creating more induced precipitation nucleation sites.

This study assesses the viability of utilizing electric fields to enhance microbial-induced mineralization. By examining the effect of different electric fields on the MICP reaction, critical parameters such as the intensity and duration of each potential gradient and the impact of electric field regulation on precipitated calcium carbonate yield were analyzed for their energy efficiency range. Furthermore, to explore the impact of electric fields on microbial-induced mineralization of calcium carbonate, scanning electron microscopy (SEM), Fourier transform infrared spectroscopy (FT-IR), and X-ray diffraction (XRD) were employed.

2. Materials and Methods

2.1. Test Bacteria

The microorganisms used in the experiment (*Sporosarcina pasteurii*, see Figure 1) were purchased from the National Culture Collection and Management Center (No.: ATCC11859). *S. pasteurii* is an anaerobic bacteria [29], rod-shaped, with a bacterial length of 2 to 3 μm and good environmental tolerance. It is a moderately alkaliphilic bacteria growing at 15 to 37 °C. The composition of the liquid medium is 15 g/L of casein (CAS: 91079-40-2), 5 g/L of soya peptone (CAS: 91079-46-8), 5 g/L of sodium chloride, and 20 g/L of urea. The required volume of deionized water was added to the liquid medium, the pH was adjusted to 7.3 [18], the bottle was divided into a 250 mL conical flask, and was finally sterilize in a high-temperature pot at 121 °C for 20 min (urea in the culture medium is filtered and sterilized to remove miscellaneous bacteria) for later use. The solid medium was added with 20 g of agar powder based on the liquid medium. After sterilization in the same step, the solution was poured into a Petri dish to make a flat plate for later use.

2.2. Equipment and Materials

The experiment was divided into two main parts. In the first part, physical and chemical parameters such as bacterial concentration and activity under standard culture conditions were determined by cultivating *S. pasteurii*. The optimal stage for introducing an electric field was identified. In the second part, the effect of the electric field on the mineralization reaction induced by *S. pasteurii* was investigated by introducing an electric field into the cultivated *S. pasteurii*. The flow diagram of the test is shown in Figure 2.

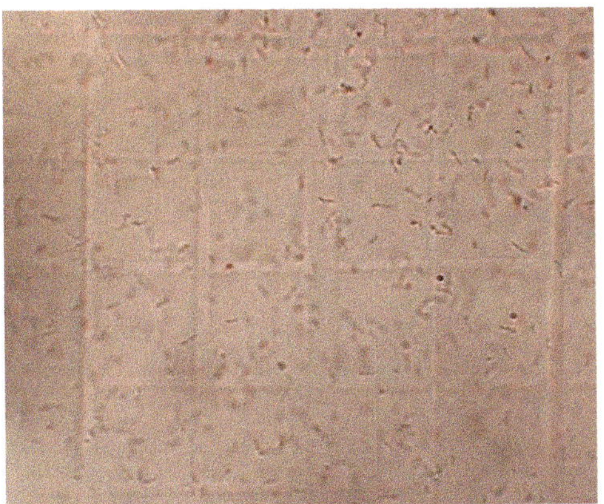

Figure 1. Morphological characteristics of *S. pasteurii*.

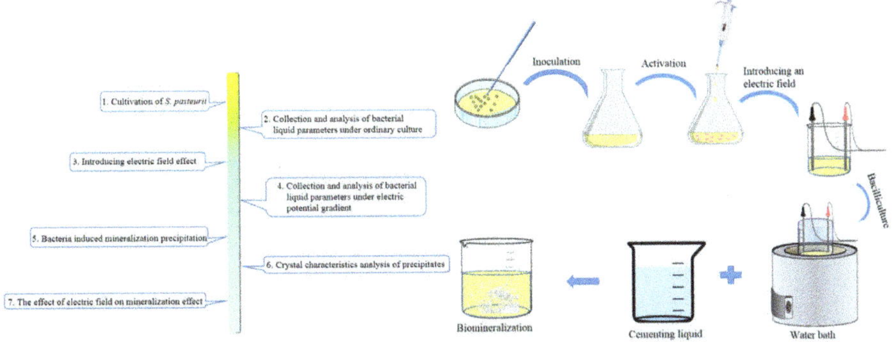

Figure 2. Schematic diagram of test flow.

The main instruments and equipment in the early bacterial culture process include a biochemical incubator (model: LRH-250F), aseptic operation table (model: SW-CJ-1FD), high-temperature sterilization pot (model: GI-54DWS), pH meter (model: pHSJ-3G), conductivity meter (model: TTS-11A), constant temperature oscillation box (model: ZQZY-78AV), protein-nucleic acid analyzer (model: BioPotometer), etc. The leading equipment in the subsequent energization includes a DC power supply (model: MS1002D), graphite electrode, culture tank (self-made microbial constant temperature culture system, See Section 3.2 for details), a peristaltic pump (model: BT100-2J), magnetic stirrer (model: CJJ79-1), and thermometer (Hengyang, China).

2.3. Experiment Method

The steps of this method are as follows. Determine the optical density at 600 nm (OD_{600}) [30] value of the activated and expanded *S. pasteurii* solution with a blank medium using a protein-nucleic acid analyzer and adjust it to 1.0 ± 0.05. Next, inoculate 1.0 mL of the bacterial solution into 100 mL of medium in an Erlenmeyer flask and place it in a constant temperature shaking box at 30°C with 150 r/min. The bacterial solution is sampled every 2 h for 18 times over 36 h (all measurements were performed at least in triplicate, and the data were plotted with Origin and analyzed statistically); three repeatability measurements were taken and recorded when the samples were tested. The corresponding growth curve is determined and plotted. The concentration of bacteria is expressed by the value of OD_{600} measured by the turbidimetric method [30,31]. As the bacteria grow, the medium solution gradually becomes turbid. The bacterial concentration is proportional to the absorbance, which is why the OD value is measured. Researchers commonly use formula (1) to convert specific bacterial counts [30,32].

$$Y = 8.59 \times 10^7 \times Z^{1.3627} \tag{1}$$

Here, Z represents the value of the measured OD_{600}, and Y represents the number of bacteria (cells/mL).

The above equation applies only when the OD_{600} is within 0.2 to 0.8. If the bacterial concentration exceeds the upper limit of the measurement range, the resulting count will include both viable and nonviable cells. Hence, the spreading plate method was employed to accurately determine bacterial growth during cultivation by determining viable cell counts. Additionally, the pH of the bacterial solution was directly measured using a pH meter.

To investigate the impact of electric fields on *S. pasteurii* mineralization, four applied potential gradients of 0.5 V/cm, 0.75 V/cm, 1.0 V/cm, and 1.25 V/cm were established. The bacteria liquid with the most optimal bacterial activity period was selected in the preliminary test and subjected to electric current. After the power was switched on, measurements were taken every 10 min, and plates were coated every 30 min, with 18 samples collected. Each gradient was subjected to electric current for 3 h. After the application of electric fields, the bacterial solution was introduced to the cementing solution to initiate the mineralization process, following which the extent of calcium carbonate precipitation was quantified.

3. Results and Discussion

3.1. Cultivation of S. pasteurii

3.1.1. Bacterial Growth

Figure 3 shows the growth curve of the tested *S. pasteurii* after 36 h of continuous culture. Bacterial growth generally consists of four stages: lag phase, log phase, stable phase, and decay phase. In this experiment, the lag phase is not apparent because the bacteria have been repeatedly cultivated by the research group and thoroughly adapted to the culture environment. As shown in the figure, the bacteria increase faster and become relatively stable after reaching a maximum concentration at around 24 h. During the initial stage of the culture, the pH of the bacterial solution quickly rose from its initial value of 7.3 and attained a value of 9.0 after about 4 h.

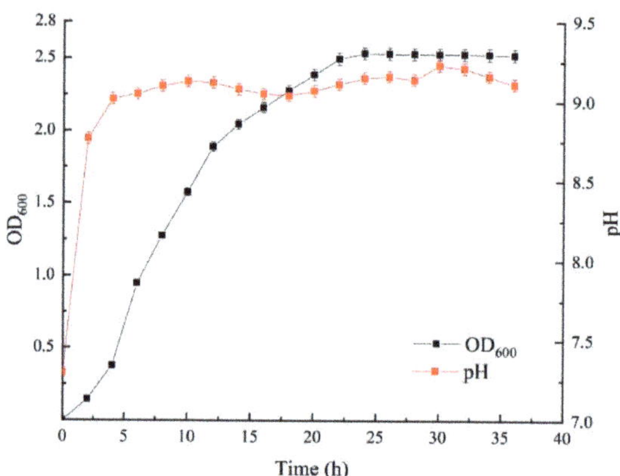

Figure 3. Growth curve of *S. pasteurii* and change trend of pH value.

Due to its alkalophilic nature, the growth rate of *S. pasteurii* is relatively slow in the neutral environment at the beginning of its growth cycle. As the bacteria grow, metabolic activities result in the production of urease. Urea in the culture medium freely diffuses into bacterial cells and is converted into a nitrogen source by urease. Specifically, urease catalyzes the hydrolysis of urea to generate ammonium and carbonate ions following a series of chemical reactions. Firstly, urea is hydrolyzed into Carbamate and ammonia through the action of urease (as depicted in Formula (2)). When exposed to water, Carbamate readily decomposes into carbonic acid and ammonia (as shown in Formula (3)). Subsequently, carbonic acid and ammonia undergo hydrolysis in an equilibrium reaction to produce bicarbonate, carbonate, ammonium, and hydroxide ions (as depicted in Formulas (4) to (6)). As a result of the action of urease, the nitrogen source required for bacterial proliferation is rapidly obtained, allowing the bacteria to enter the logarithmic phase.

Upon the hydrolysis of urea, the subsequent release of ammonium and carbonate ions into the environment engendered a rise in pH levels, creating an alkaline habitat that favored bacterial growth. Upon examination of Figure 3, it can be observed that the proliferation of bacteria exhibited a slow pace during the nascent stages of growth. The pH levels experienced a rapid escalation during this phase, suggesting that the bacteria primarily enhanced the pH levels of their surroundings via the hydrolysis of urea during the initial stages, thus creating an environment for subsequent, expeditious growth. The growth curve of bacteria and the alteration of pH levels of bacterial solution underwent a definite time lag, which could be attributed to the delayed effect of the interplay between the pH levels and bacterial growth, which depends on how *S. pasteurii* obtains energy. Research conducted by Cuzman et al. [33] revealed that the synthesis of ATP and the transportation of *S. pasteurii* are inextricably linked to the production of ammonium ions within the bacteria. The findings indicated that the most appropriate pH level for the normal culture of *S. pasteurii* is roughly 9.25, whereas the intracellular pH value of bacteria hovers around 8.4. Upon the diffusion of urea into the bacteria, the ammonium ions, produced by the hydrolysis of urease, are gradually released into the extracellular space [34]. The bacteria display an accelerated growth rate when the pH levels in vivo and extracellular space differ. However, when the environmental pH exceeds 9.0, the pace of bacterial growth slackens. This is due to the concentration ratio of ammonium ion to ammonia molecule being 70:30 within the bacterial cell, at pH 8.4. The variance in ammonium concentration between the interior and exterior of the cell prompts the ammonium radical to diffuse out of the cell.

On the one hand, the diffusion of ammonium radicals contributes to the alkalinity of the environment, thereby promoting bacterial growth. On the other hand, a portion of the ammonium radicals converts into ammonia molecules and protons. The proton concentration gradient generated by the bacteria is higher than inside the cell, creating an effective proton kinetic potential. Consequently, the cell membrane's proton pump (ATP synthase) is activated to synthesize ATP [33]. This series of energy conversions effectively facilitates proliferation and increases the pH value of the bacterial solution. Due to the time sequence of the biochemical reactions, there is a certain delay between the growth curve of bacteria and the pH value change in the bacterial solution, which underscores the close relationship between bacterial growth and pH value changes.

$$CO(NH_2)_2 + H_2O \xrightarrow{\text{Urease}} NH_2COOH + NH_3 \tag{2}$$

$$NH_2COOH + H_2O \rightarrow H_2CO_3 + NH_3 \tag{3}$$

$$H_2CO_3 \rightarrow HCO_3^{2-} + H^+ \tag{4}$$

$$2NH_3 + 2H_2O \leftrightarrow 2OH^- + 2NH_4^+ \tag{5}$$

$$HCO_3^- + H^+ + 2OH^- \leftrightarrow CO_3^{2-} + 2H_2O \tag{6}$$

3.1.2. Viability and Bacterial Count

S. pasteurii can induce precipitation of calcium carbonate and initiate mineralization reactions, as it secretes urease enzymes with highly efficient hydrolysis capabilities. The level of urease activity and the timing of the addition of the cementing liquid determines the amount of calcium carbonate produced. Whiffin et al. [35] discovered in 2007 that the urease activity of the bacterial solution can be determined by measuring the change in conductivity of the solution. During the hydrolysis of urea by urease, the ion concentration in the bacterial solution rapidly increases, leading to a corresponding increase in conductivity value. The research findings demonstrate a positive correlation between the urease activity of the bacterial liquid and the rate of conductivity change, which can be mathematically represented using conversion Formulas (7) and (8).

$$U = 11.11f \tag{7}$$

$$S = U/OD_{600} \tag{8}$$

In the formula, U is the urease activity of the bacterial liquid (ms/cm/min); f is the rate of change in conductivity (ms/min); S is the unit urease activity (ms/cm/min).

Add 3 mL of the tested bacterial solution to 27 mL of 1.5 mol/L urea and use a conductivity meter to measure the change in conductivity of the mixed solution within 5 min after thorough mixing. The urease activity of the bacterial liquid and its change in unit urease activity can be obtained by conversion, as depicted in Figure 4. This study's lag period is not apparent due to the prior cultivation of *S. pasteurii* in multiple cycles. It can be observed from the figure that the bacterial adaptation to the culture environment is faster in the 0–2 h timeframe, and the bacterial count increases gradually. Between 2 and 16 h, the bacteria begin to grow after an adaptation period. The bacterial cell division leads to a linear increase in the bacterial concentration in the culture medium. This stage corresponds to the logarithmic phase of bacterial culture, during which the growth rate is maximized.

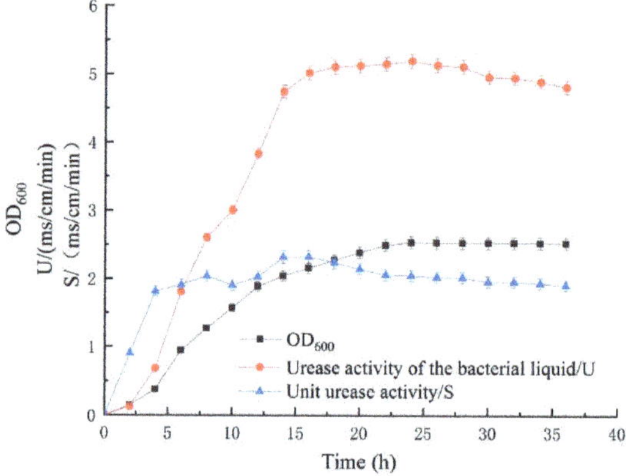

Figure 4. Changes in urease activity during bacterial growth.

After 16 h, stability is achieved as the concentration of bacterial solution reaches its maximum value. The nutrient substances in the culture medium are increasingly consumed, and the toxic products produced by bacterial metabolism lead to a decline in the metabolic capacity of the bacterial cells due to accumulation, causing a decrease in the rate of bacterial proliferation. Over time, the number of newly proliferated and dying bacteria tends to balance gradually. Later, with the influence of unfavourable factors, the decay rate becomes gradually greater than the new ones, causing the bacterial solution concentration to decrease. The trend of the urease activity in the bacterial solution displayed in the figure is strongly correlated with the changes in the concentration of the bacterial solution. As time elapses, the urease activity of the bacterial solution progressively increases along with the bacterial concentration. During this period, the capacity of the bacteria to hydrolyze urea continually intensifies, ultimately resulting in maximal urease activity after approximately 24 h. Additionally, the unit urease activity of the bacterial solution is influenced by its concentration, and its fluctuation is notable, gradually decreasing subsequent to reaching its peak value at 16 h. Notably, the urease activity of the bacterial solution is critical in bacterial mineralization.

In order to investigate alterations in colony-forming units (CFU) of viable bacteria in *S. pasteurii* under standard culture conditions, samples were collected from 10 to 28 h post-culture initiation and cultured using the plate colony counting method. The resultant alteration trend in the viable bacterial count is presented in Figure 5 ($p < 0.05$). Sampling time nodes in the figure comprised logarithmic, late stable periods, and others. It can be observed from the figure that the viable bacterial count increased linearly with the logarithmic phase's growth in the bacterial concentration. The highest viable bacteria count was observed around 14 h, reaching 3.395×10^8 cells/mL. Subsequently, due to unfavourable factors such as bacterial liquid metabolism in the culture environment, the overall bacterial concentration reduced, leading to a decline in viable bacterial proportion and a continuous reduction in the viable bacterial count. The considerable changes in the number of viable bacteria substantially affect the unit urease activity in the bacterial solution.

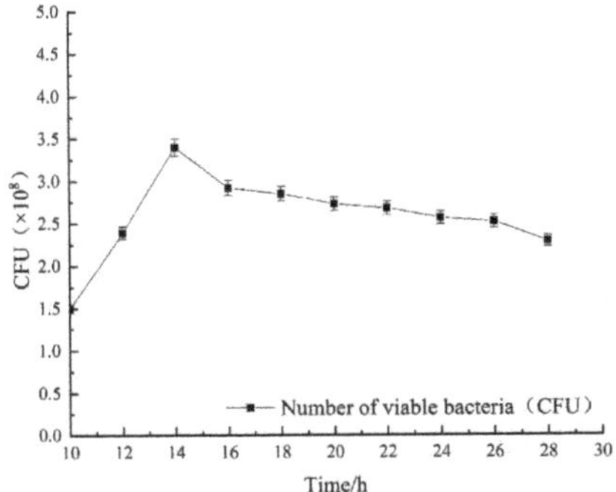

Figure 5. Changes in the number of viable bacteria of *S. pasteurii*.

3.2. The Effect of External Electric Field on Bacteria

To examine the impact of an external electric field on the growth and cementation process of *S. pasteurii*, a custom-made microbial constant temperature culture system was employed for experimentation (Figure 6). Since the optimal cultivation temperature for *S. pasteurii* is approximately 30 °C, a constant temperature water bath with a circulating water pump was employed, with real-time temperature regulation by a thermometer to maintain the culture tank's ambient temperature at 30 °C. A magnetic stirrer was situated beneath the bath heating tank, and the bacterial liquid was stirred by driving previously placed magnetic beads in the culture tank. The potential gradient of the applied electric field was managed at 0.5–1.25 V/cm (voltage ranging from 7.5 V to 18.75 V). During the initial bacterial liquid culture phase, the bacterial liquid (24 h) was analyzed when the concentration reached its maximum, and electrified culture was initiated. The energization duration was set at 3 h, and the sampling frequency was set at 10 min per time. In order to determine changes in the bacterial count during energization, a coating plate test was conducted every 30 min. Two tanks were employed in the experiment, one for energized bacterial culture and the other for bacterial-induced calcium carbonate precipitation under energization.

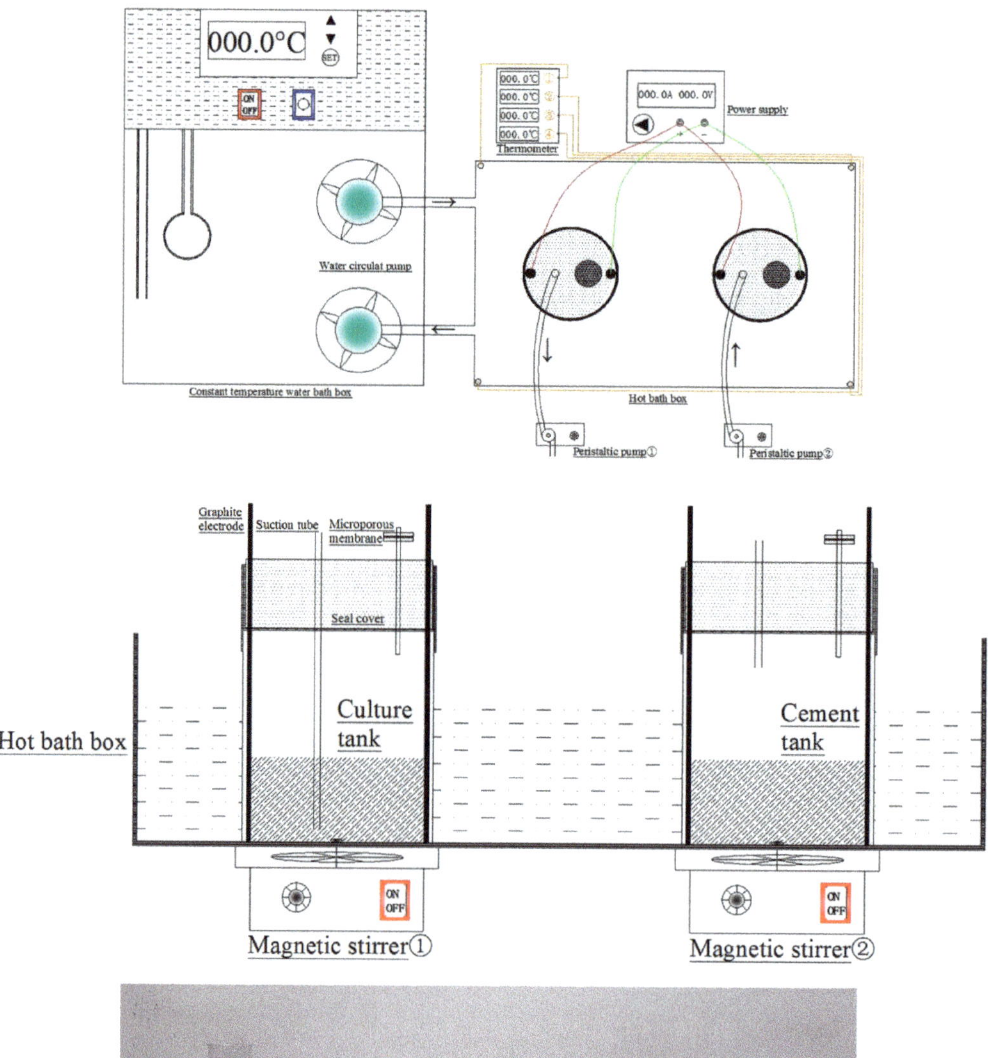

Figure 6. Test devise.

3.2.1. Bacteria Growing in the Electric Field

The growth curve of *S. pasteurii* under different potential gradients is illustrated in Figure 7. The effects of various potential gradients on the bacteria display notable differences. As a result of bacteria's distinct sensitivity to electric fields, the overall growth trend is characterized by an initial increase followed by a decrease. After applying power, electrolysis and bubbles emerge at the contact between the electrode and the bacterial liquid in the culture tank. At the electrode, oxygen evolution occurs at the anode (Equation (9)), and hydrogen evolution takes place at the cathode (Equation (10)).

$$\text{Anode}: 2H_2O - 4e^- \rightarrow 4H^+ + O_2 \uparrow \tag{9}$$

$$\text{Cathode}: 2H_2O + 2e^- \rightarrow 2OH^- + H_2 \uparrow \tag{10}$$

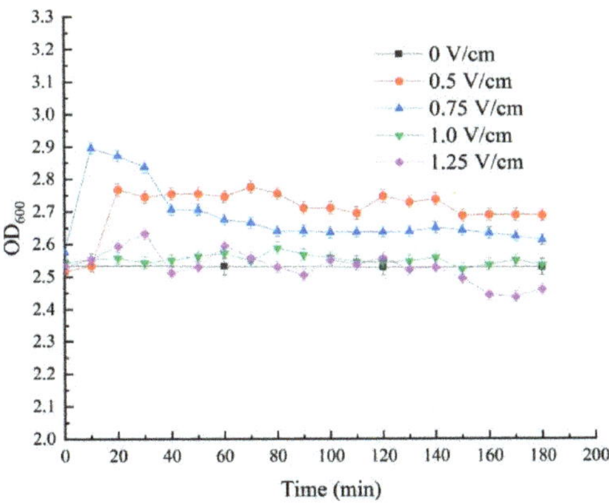

Figure 7. The growth curve of *S. pasteurii* in different potential gradients.

In the initial phase of electrification, the hydrogen and oxygen generated through electrolysis via an external electric field exhibit a specific stimulus on the growth and proliferation of bacteria. Nonetheless, with the amplification of the gradient, the tolerance threshold of the bacteria surpasses. Consequently, the development of bacteria becomes inhibited under the influence of the electric field. The figure displays the usage of the 0 V/cm group as a control group and 1 V/cm gradient as the boundary, the growth curve of bacteria increased obviously under the weak potential gradient, and the concentration increased significantly at the initial stage. However, with the elongation of the energization time, the optical density (OD) value under the 0.75 V/cm gradient begins to decrease. As *S. pasteurii* is an anaerobic bacterium, the oxygen produced at the anode expedites the growth of bacteria.

Nevertheless, with the continued effect of the electric field, the bacterial cells become incapable of tolerating the abnormally high electric potential, leading to changes in the permeability of the cell membrane. The cell membrane is chiefly made up of a double-layered phospholipid molecular arrangement. When an electric field acts on the cell membrane, ionization is accelerated, causing alterations in the osmotic pressure within and outside the membrane, ultimately resulting in bacterial cell rupture and death and thus decreasing the concentration of the bacterial solution. Upon the potential gradient's elevation, the OD value shift range gradually declines. The vehement electrolysis reaction at the electrode and the "inactivation" caused by the direct contact between the bacteria and the electrode are among the reasons for bacterial death. The excessive electric potential exerts an inhibitory impact on bacterial growth. When a gradient of 1.0 V/cm is applied, the growth curve of bacteria is akin to that of the control group. However, it appears unaffected by the electric field from the changing trend. As the gradient rises to 1.25 V/cm, the concentration of bacteria fluctuates significantly, the overall trend being downward, thus impeding bacterial growth. A comparison of the results of the five experimental groups reveals that a potential gradient of 0.5 V/cm is the most suitable gradient for maximizing the growth-promoting effect on *S. pasteurii*. The concentration of the bacterial solution is maximally increased, thus signifying the most suitable potential gradient.

Moreover, in the realm of bacterial response to electric fields, Sun et al. [36] conducted electrophoresis experiments on Escherichia coli and discovered that the bacteria attained the maximal living amount under an electric field of 0.0455 mA/cm. Compared to bacteria cultured under normal conditions, the OD value of the bacterial solution increased by 0.29. Similarly, Zhang et al. [37] explored nitrate removal. They found that low voltage introduced to nitrifying bacteria (500–700 mV) effectively facilitated their proliferation and activity, efficiently removing nitrate from groundwater. Liu et al. [38] proposed the technique of weak microbial current to manage water body eutrophication. By stimulating bacteria with a weak current of 0.2 V, the remediation rate of nitrate nitrogen in the water body was substantially enhanced by boosting bacterial proliferation. These findings demonstrate that introducing an appropriate electric field to bacteria can enhance bacterial accumulation and facilitate subsequent research procedures.

3.2.2. The Influence of Applied Electric Potential on the pH and Viability of Bacterial Liquid

The alteration in pH value impacts the metabolic activity of *S. pasteurii*. The effects of pH on microorganisms primarily manifest in the following aspects: ① Affecting microbial activity, as the most favourable pH for *S. pasteurii* is approximately 9.2. Values that are too high or too low can affect its biological activity; ② modifying the permeability of the cell membrane, which influences the absorption of nutrients by microorganisms in the environment; and ③ enhancing the toxic effect of microbial metabolites on microbes, which can impede the proliferation of microbes and their utilization of nutrients.

Figure 8 illustrates the effect of electrification on the pH value of the bacterial liquid. Fluctuations in pH value are observed in the bacterial liquid under each potential gradient, compared to the control group at 0 V/cm. The best promotion of bacterial regrowth occurs under a gradient of 0.5 V/cm. Due to the high OD_{600} of the bacterial solution used, there is a high consumption of nutrients in the medium, resulting in a limited pH increase under the electric potential gradient. Nonetheless, the overall trend still demonstrates the growth and inhibition of bacteria under each gradient. The change in pH value corresponds to the growth curve, with the bacterial concentration increasing under the gradient of 0.5–0.75 V/cm and the pH value rising accordingly. In the later stages, the bacterial concentration of the bacterial solution stabilizes, and the pH change decreases gradually.

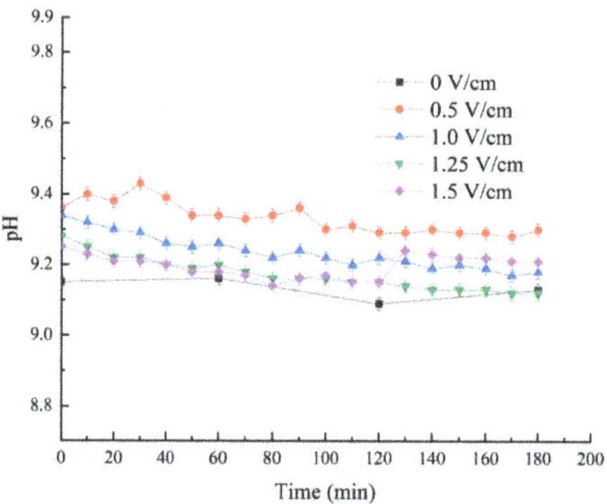

Figure 8. pH of bacterial liquid under different electric potential gradients.

In order to investigate the impact of the applied electric field on the activity of the bacterial liquid, samples were taken every thirty minutes for conductivity measurements. The urease activity was transformed using the conductivity conversion Formulas (7) and (8), and the specific changes are depicted in Figure 9. During the initial stage of electrification, the electric field caused the surface potential of the cell membrane of the bacteria to change, which in turn affected the membrane's permeability. The difference in osmotic pressure inside and outside the bacterial cell enabled the culture environment's nutrients to enter the bacterial body more efficiently. Furthermore, anode electrolysis produced oxygen [39], promoting the proliferation and metabolism of bacteria. This phenomenon was particularly notable at lower potential gradients. At a gradient of 0.5 V/cm, the urease activity increased gradually with the concentration of the bacterial solution. However, in the later stage, the bacteria began to decay progressively due to nutrient depletion, reducing the bacterial liquid's activity. When the potential gradient was 0.75 V/cm, the bacterial concentration increased rapidly due to the promotion effect caused by the initial electric field.

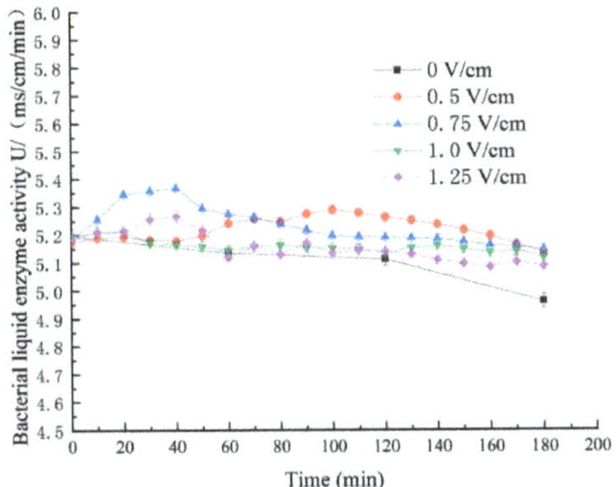

Figure 9. Changes in Enzyme Activity in Bacteria Liquid under Different Electric Potential Gradients.

Nevertheless, the continued electrification led to the rupture of the bacterial cells. The intracellular urease in the bacterial cells was released into the surrounding cultural environment, causing the bacterial cells' breakdown and mixing with the extracellular urease. The total number of urease enzymes increased during the sampling for conductivity measurement, which hastened the measured conductivity value change rate, thus enhancing the activity. Subsequently, urease activity continued to decline due to the inhibition of the electric field.

The results indicate that, according to [40], microbial growth and urease activity are better when the pH value of the microbial solution is within the range of 6.0–10.0. Notably, the microbial growth and urease activity are more favourable at a pH of 7.0–9.0 under normal culture conditions. The pH value can influence the permeability of microbial cell membranes, consequently affecting the absorption and transformation of nutrients by bacteria. Furthermore, the pH value substantially impacts the enzyme production rate and urease activity stability. The study also reveals that microbial growth and urease activity exhibit a strong positive correlation, with better microbial growth resulting in higher urease activity. The optimal pH range for the microbial solution under normal culture conditions is 7.0–9.0, with favourable microbial growth and urease activity. The pH and urease activity of the bacteria solution after the electrified culture have slightly improved compared to existing research results.

Moreover, compared with the bacteria solution under normal culture conditions by other researchers, the pH of the solution after electrification has increased by approximately 0.4. As the bacterial concentration increases, more urease is secreted. Urea is hydrolyzed by urease to form ammonia, which further hydrolyzes to form ammonium and hydroxide ions. According to Ferris et al. [41], hydroxide ions can aid in increasing the pH value surrounding cells, thereby promoting its increase. An elevated pH value can facilitate the transformation of bicarbonate ions into carbonate ions, which, under the influence of exogenous calcium ions, can encourage the formation of calcium carbonate precipitation [42]. Similarly, DeJong et al. [43] have shown that an alkaline environment is more favourable for forming calcium carbonate precipitation.

3.2.3. Effect of Applied Electric Potential on the Survivability of *S. pasteurii*

The *S. pasteurii* cells were found to have a negative charge during the experiment. Previous research suggests that the biological activity of bacteria can be activated by a weak electric field [44]. This is because the hydrogen and oxygen produced by electrolysis in a weak electric field can promote the growth of bacteria, enhance membrane permeability, and accelerate nutrient exchange in the culture environment. In order to investigate the number of viable bacteria at different electric potential gradients as the concentration changes, the bacterial solution was divided into sections every 30 min and plated. The changing trend of viable bacteria is illustrated in Figure 10. The results show that the change in viable bacteria, as measured by CFU, is highly consistent with the changing trend of OD_{600}. During the initial phase of electrification, the bacterial concentration substantially rose under the 0.75 V/cm gradient, leading to a corresponding rapid increase in viable bacteria. However, upon suppression of the electric field, the bacterial concentration and viable bacteria count subsequently decreased. The weak electric field generated by the 0.5 V/cm potential gradient significantly affected the growth activity and cell proliferation of *S. pasteurii*, as evidenced by the relatively stable growth trend of the number of viable bacteria. As the number of bacteria increased, the activity of the bacterial liquid also increased, and the ability of the urease produced by the bacteria to decompose urea was further strengthened, which promoted the proliferation of the pH value of the culture environment. Compared to higher potential gradients such as 1.0 V/cm and 1.25 V/cm, the number of viable bacteria was relatively unchanged due to the suppression of the electric field. A reduced magnitude of potential gradient (e.g., 0.5 V/cm) demonstrated a favorable impact on bacterial proliferation, presenting a potential avenue for further experimental investigations.

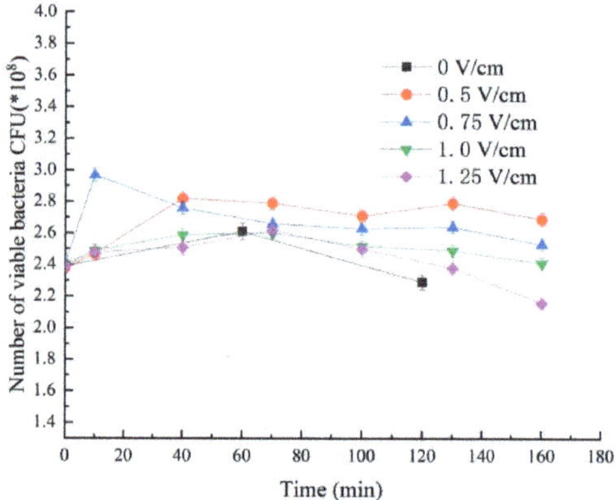

Figure 10. Changes in bacterial survival under different electric potential gradients.

4. Effect of Electric Field on the Process of Bacteria-Induced Calcium Carbonate Mineralization

4.1. The Amount of Calcium Carbonate Produced

S. pasteurii induces mineralization through calcium carbonate precipitation (MICP), which is facilitated by the urease produced by its metabolism. This enzyme hydrolyzes urea, releasing carbonate ions that combine with external calcium ions to form calcium carbonate crystals. The main reactions involved are as follows.

$$Ca^{2+} + CO_3^{2-} \rightarrow CaCO_3 \downarrow \tag{11}$$

The concentration of the cementing liquid, alongside the proportionality of bacterial and cementing liquid (i.e., the amalgamated fluid), will play a pivotal role in precipitation induction. According to present-day research, the optimal amalgamation for forming calcium carbonate crystals and cementing efficacy transpires at a cementing liquid concentration of 1 mol/L. The mixed solution ratio is 1:4. This ratio and concentration yield the highest utilization of effective ingredients in the cementing liquid [45], which is why they are employed in this experiment.

The steps for this method are as follows. Pour 500 mL of cultured bacterial liquid into the aseptic operation table's culture tank and place it into the hot bath. Connect the electrode and circuit and turn on the power. Start the magnetic stirrer and add 2 L of urea and calcium chloride cementing liquid through the peristaltic pump. Observe and document the precipitation reaction process. The bacterial cells play a vital role in inducing mineralization. Besides producing urease and increasing the environmental pH during growth, providing nucleation sites for forming calcium carbonate crystals is crucial. During precipitation, the calcium ions in the cementitious solution first adsorb on the bacterial cell membrane's surface. Subsequently, the area that diffuses into the bacteria generates carbonate ions under the intracellular urease's influence, which are then released outside the membrane and precipitate with calcium ions. As the residue settles, the bacteria become entirely encapsulated, and their inability to grow causes them to die.

The variation in calcium carbonate content resulting from the reaction of a mixed solution under the influence of different electric potential gradients is illustrated in Figure 11. The graph portrays a downward parabolic trend, with 84.22 g as the reference value for the calcium carbonate content when no electrical input is provided (i.e., 0 V/cm). Upon activation, the 0.5 V/cm gradient significantly enhances the efficacy of the bacterial liquid, leading to a 9.78% surge in the final calcium carbonate amount, which reaches 92.45 g. As the potential gradient becomes stronger, the effect of the electric field on the growth and precipitation of liquid bacterial changes from promoting to inhibiting. At 0.75 V/cm, the overall concentration of the bacterial solution increases, resulting in an approximate 4.37% increase in calcium carbonate precipitation compared to the scenario without electrical stimulation. The bacterial decay rate accelerates, causing a 35.80% reduction in the calcium carbonate content under a potential gradient of 1.25 V/cm. The least amount of precipitation is observed under this condition. These findings provide additional evidence that an appropriate external potential can enhance the biological activity of the bacterial liquid, the calcium carbonate precipitation is further stimulated, and the microorganism mineralization process is expedited.

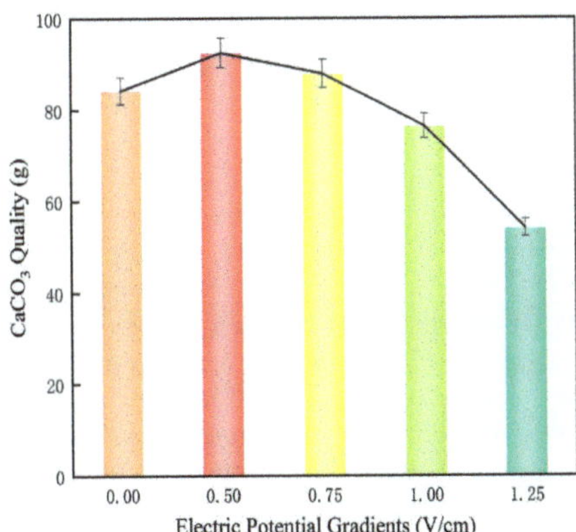

Figure 11. The Effect of Various Electric Potential Gradients on Calcium Carbonate Formation.

4.2. Changes in Crystal Form

4.2.1. Micro-Morphology and Infrared Analysis

The heterogeneous solution was filtrated and desiccated to procure a comprehensive calcium carbonate sediment, and the corresponding calcium carbonate precipitates were isolated. Naked-eye observation revealed that the sediments acquired at different electrical potentials presented different characteristics. Without electric potential, the precipitate exhibited a dense texture, tight cementation between particles, and a flat cross-sectional area of calcium carbonate during sampling. As the electrical voltage escalated, the quantity of calcium carbonate altered, and the adhesion strength between calcium carbonate crystals, the hierarchical ordering of crystal accumulation, crystal structure, and particle size were markedly impacted.

To further investigate the influence of the applied electric field on the microbial-induced calcium carbonate cementation effect, the microscopic crystal structure of calcium carbonate precipitated in specific areas was observed and analyzed using scanning electron microscopy (SEM) and infrared spectroscopy (FT-IR). The surface microstructure of calcium carbonate under different potential gradients is depicted in Figure 12. At a constant temperature of 30 °C, the calcium carbonate deposited without electricity (Figure 12A,B) formed aggregates of various sizes, lacking a distinct morphology in appearance. However, it exhibited a tight connection between particles, dense pore filling, and an impressive cementation effect. Infrared spectroscopy analysis (Figure 12C) indicated that the crystals were a mixture of calcite and vaterite types, as demonstrated by 1421 cm^{-1}, 1082 cm^{-1}, 878 cm^{-1}, 745 cm^{-1}, and 712 cm^{-1} absorption peaks. It is worth mentioning that the peaks observed at 1421 cm^{-1}, 878 cm^{-1}, and 712 cm^{-1} corresponded to the distinctive absorption peaks of calcium carbonate with calcite structure (at V3, V2, and V4, respectively), whereas those at 1082 cm^{-1} and 745 cm^{-1} corresponded to vaterite type calcium carbonate (at V_1 and V_4), suggesting that the crystal structure was primarily composed of calcite. When a gradient of 0.5 V/cm was applied, the calcium carbonate precipitates assumed polygonal shapes with distinct edges and corners (Figure 12D,E), exhibiting several ellipsoid structures. The crystals were cemented together, and the pores at the junctions between the calcium carbonate crystals were expanded, reducing cementation compactness but improving the overall cementation effect. The infrared analysis revealed that the characteristic absorption peaks of calcium carbonate with calcite structure manifested at 1419 cm^{-1} and 712 cm^{-1}, whereas the distinctive peaks of vaterite-type crystals escalated. The absorption peaks at 874 cm^{-1} and 748 cm^{-1}, respectively, at V2 and V4, was mainly due to out-of-plane and in-plane bending vibrations. Finally, when the potential gradient was increased to 0.75 V/cm, the calcium carbonate precipitates began agglomerating in a small volume (Figure 12G,H). Although the crystals were closely connected, large fractured pores were observed between the blocks, and the mutual cementation effect was reduced.

Furthermore, the quantity of ellipsoidal formations within the calcium carbonate crystals exhibits an increase. Many agglomerated ellipsoidal crystals adhere to a limited number of polygonal crystal structures, leading to a division in structural hierarchy. Infrared spectrum analysis (Figure 12I) reveals that most absorption peaks lie at 1404 cm^{-1}, 1082 cm^{-1}, 874 cm^{-1}, and 742 cm^{-1}, all indicative of vaterite calcium carbonate. Compared with a 0.5 V/cm gradient and the absence of electrical influence, the characteristic calcite-type absorption peak at 1420 cm^{-1} shifts, mainly due to the anti-symmetric stretching vibration of the C-O bond under the effect of electrification, resulting in the emergence of a weak calcite-type acromion at 1439 cm^{-1}. Coupled with SEM outcomes, the calcium carbonate formed at a 0.75 V/cm gradient is chiefly of the vaterite type with a small number of calcite crystals. As the potential gradient increases, the spherical structure becomes dominant in the calcium carbonate precipitate crystals formed by the influence of a gradient ranging from 1.0 to 1.25 V/cm (illustrated in Figure 12J,K,M,N). The crystal shapes are regular and spherical, and their distribution is relatively uniform due to the crystal shapes, with no apparent edges or corners. The particles fuse and accumulate, forming a whole with significant pores and poor structural stability. Through infrared spectrum analysis (as demonstrated in Figure 12L,O), the prominent absorption peaks consist of 1401 cm^{-1}, 874 cm^{-1}, 748 cm^{-1}, 745 cm^{-1}, and so on. By consulting the infrared characteristic absorption peak data of various crystal forms, it becomes apparent that only vaterite calcium carbonate exists.

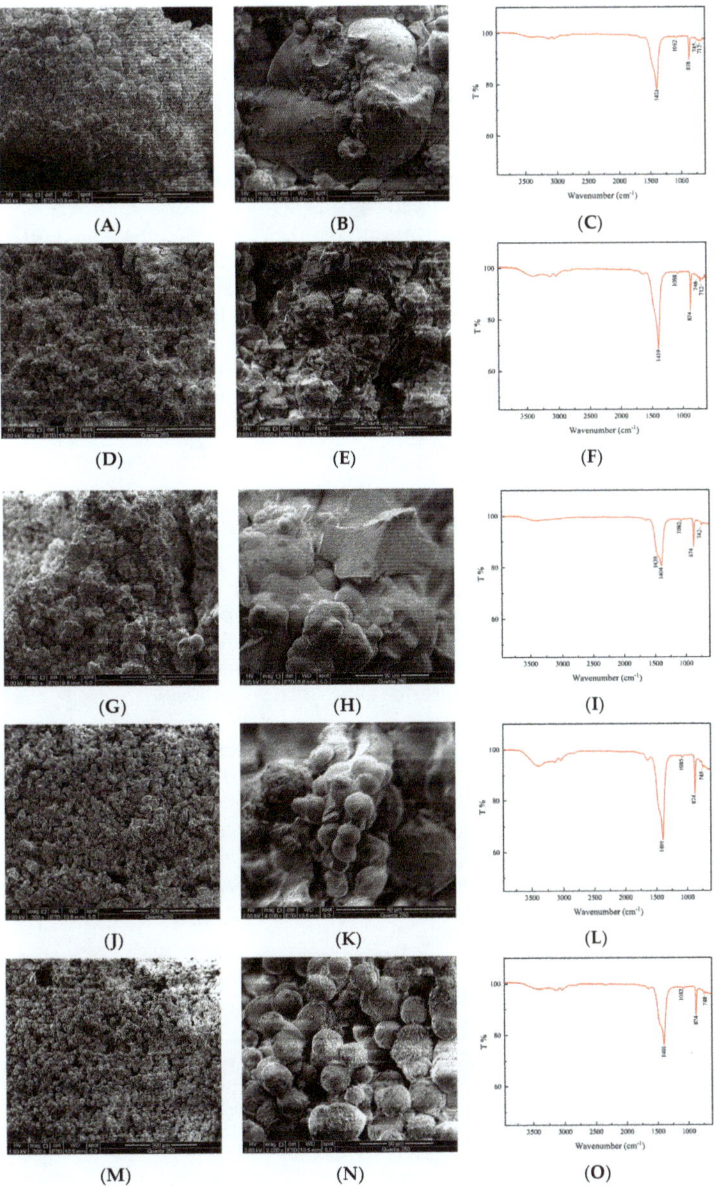

Figure 12. Effects of different potential gradients on the structure of calcium carbonate, (SEM) 0 V/cm: (**A**,**B**); 0.5 V/cm: (**D**,**E**); 0.75 V/cm: (**G**,**H**); 1.0 V/cm: (**J**,**K**); 1.25 V/cm: (**M**,**N**). (FT-IR) 0 V/cm: (**C**); 0.5 V/cm: (**F**); 0.75 V/cm: (**I**); 1.0 V/cm: (**L**); 1.25 V/cm: (**O**).

After comparing and analyzing the infrared spectra obtained from each group, it was discovered that the characteristic peaks of the CO_3^{2-} out-of-plane bending vibration and anti-symmetric stretching vibration underwent a redshift to 878 cm^{-1} and 1421 cm^{-1}, respectively, upon the introduction of a potential gradient and a shift to low wave velocity. The crystal morphology of calcium carbonate underwent a gradual transformation from a mixed crystal form comprising calcite and vaterite to a preponderance of vaterite crystal form. At the same time, the crystal size of calcium carbonate also changed from irregular polycrystalline aggregates to uniform vaterite crystals with a grain size of around 20 µm. These observations suggest that the potential gradient played a significant role in inducing the formation of calcium carbonate.

Calcium carbonate manifests in various structural forms, such as calcite, vaterite, aragonite, and amorphous structures. Of all these forms, calcite boasts the highest stability and consolidation efficacy. Conversely, vaterite proves vulnerable in standard environmental settings, possessing a bad thermodynamic equilibrium. Consequently, as five groups of potential gradients impose their influence, the structural soundness of calcium carbonate produced gradually diminishes. As the potential increases, the crystal structure of calcium carbonate transmutes from stable calcite to unstable vaterite. According to research, the calcium carbonate synthesized via the MICP reaction comprises polyhedral, ellipsoidal, and a few spherical crystals with multiple edges. Such crystals will likely exhibit superior curative properties via microbial cementation [46]. Hence, in practical applications, the curing efficiency of calcium carbonate precipitated by a 0.5 V/cm gradient in this experiment should surpass that of other groups. Moreover, the content of bacteria-induced calcium carbonate is highest under this gradient.

4.2.2. Sediment Composition

In order to investigate the crystal composition of MICP reaction products in an electric field, three different calcium carbonate precipitates (0 V/cm, 0.5 V/cm, and 1.25 V/cm) underwent X-ray diffraction (XRD) analysis, building upon the results of SEM and FT-IR analyses.

The X-ray diffraction pattern depicted in Figure 13 exhibits the calcium carbonate sediment under three distinct potential gradients, with the angle denoting the abscissa and the intensity representing the ordinate. The outcome of the analysis suggests that, in the absence of electricity, the calcium carbonate precipitate crystals consist mainly of calcite, with vaterite accounting for a certain proportion. Most calcium carbonate precipitate crystals comprised calcite when introducing a potential gradient of 0.5 V/cm. As the electric potential gradient increased, the composition of calcium carbonate crystals underwent an observable change, with vaterite, dominant at 1.25 V/cm, providing additional support for the regulation of the curing reaction of microbial-induced calcium carbonate precipitation (MICP) to form calcium carbonate crystals via the introduction of an electric field. Furthermore, the XRD diffraction pattern analysis results complement the trend observed in microscopic observation results of calcium carbonate sediments, indicating a correlation between the two.

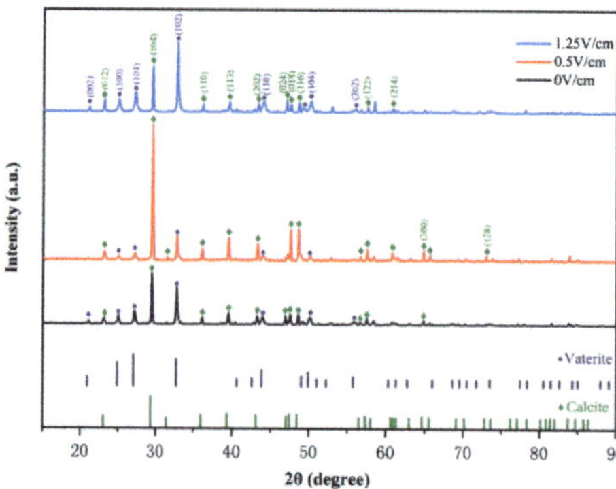

Figure 13. X-ray Diffraction Pattern of Calcium Carbonate Precipitates.

Researchers have posited that the strength growth mechanism of MICP curing is attributed to differences in the crystal structure of calcium carbonate deposits. When calcite serves as the primary material, it yields better reinforcement effects due to its ability to bond more effectively with the surface of particles when present in a higher proportion within the calcium carbonate deposits. According to Gebauer et al. [47], calcite is a more stable polycrystalline form of calcium carbonate and may possess a higher binding strength in clusters than in other types. Zheng et al. [48] found that after using egg white to regulate the crystal form of calcium carbonate in the MICP reaction, the 5% egg white group, which had the highest calcite proportion, exhibited the highest cementing strength. Additionally, Zhang et al. [18] discovered that different concentrations of organic matter are crucial in regulating calcium carbonate's morphology and crystal structure in MICP reinforcement research when another organic matter is added. The calcite crystal form is the primary type of calcium carbonate. When mixed with a certain vaterite crystal form, it can better match and reinforce the pore structure of the sample, enhance the bite force between calcium carbonate and sample particles, and improve the reinforcement effect. Therefore, the calcium carbonate precipitate induced by the electric field has multi-structure characteristics in crystal form. The calcium carbonate induced by the 0.5 V/cm potential gradient has good integrity and a high calcite content, making it highly effective for solidification.

5. Discussion

This study explored the impact of *S. pasteurii* on calcium carbonate production induced by a potential gradient. The findings demonstrate that optimal electrification via a suitable electric potential gradient stimulates bacterial growth, urease activity, nutrient utilization, and urea hydrolysis rate. This, in turn, expedites the induced mineralization of microorganisms, thereby elevating calcium carbonate production. However, elevated potential gradients can stifle bacterial growth and urease activity, causing the induced calcium carbonate content to decline. *S. pasteurii* plays a dual role in microbial-induced calcium carbonate deposition. Firstly, the strain synthesizes urease within the cell, accelerating urea hydrolysis [49,50]. Secondly, *S. pasteurii* can serve as nucleation sites of calcium carbonate, thus promoting the formation of calcium carbonate crystals. The crystal structure of calcium carbonate influenced by microorganisms transforms a potential gradient that acts upon it. The original mixed crystal form of calcite and vaterite gradually shifts to a vaterite-dominated crystal form. The crystal particles transition from an irregular divergent distribution to uniform distribution. Additionally, the microbial mineralization process involves the growth of calcium carbonate crystals. Early stage calcium carbonate crystals facilitate the implantation and formation of the induced calcium carbonate later. Calcium carbonate deposits merge, allowing bacteria and enzymes to effectively aggregate at crystal particles' superposition, offering an additional nucleation site for biological deposition. Gradually, calcium carbonate particles coalesce and chelate into larger particle aggregates.

Although the electric field notably promotes MICP technology, further optimization is necessary to address certain limitations. In terms of energy consumption, the electric field requires a certain amount of energy, including electrodes, power supplies, and other materials, and future considerations should prioritize its economy. Operation and control of the electric field require specific management, including controlling the size and duration of the field, which should be optimized and adjusted throughout the process. Additionally, the test scale may affect the application effect due to laboratory conditions. Thus, the controllable range should be optimized according to the specific situation. Therefore, proper optimization and control of MICP technology by an electric field are essential in enhancing its application effect and economy.

Microbial mineralization is extensively used in the environmental treatment and soil reinforcement due to its ecological compatibility and economic benefits, but some challenges still need to be addressed. For instance, basic research on molecular mechanisms such as high-efficiency urease activity is relatively lacking. Furthermore, it remains unclear if microbial mineralization reactions have a special value in the physiological sense of bacteria themselves. More systematic research is necessary to induce the crystal structure and morphology of calcium carbonate crystals.

This paper explores the impact of electric potential gradient on microbial mineralization based on MICP technology. However, the research duration and experimental ratio gradient level are relatively limited. To better understand the regulation of the electric field on microbial-induced calcium carbonate crystal structure and the change in crystal content, it is necessary to strengthen the metabolic mechanism of microorganisms and promote the application effect of microbial mineralization in various fields.

6. Conclusions

In this study, we employed microbial-induced calcium carbonate precipitation (MICP) technology to enhance the growth and activity alterations of *S. pasteurii* under the most favorable culture conditions with various potential gradients to investigate the impact of the electric field on microbial-induced mineralization. The findings demonstrated the following outcomes: (1) The highest concentration of bacteria was achieved after 24 h, accompanied by the highest level of urease activity. Additionally, the specific number of viable bacteria was determined by coating the plate, and the bacteria solution corresponding to the maximum value of OD_{600} was subjected to an electric charge. (2) An appropriate applied electric field can enhance the activity of *S. pasteurii*, with a notable promotion effect observed under a lower gradient. Moreover, this led to accelerated bacteria growth, improved material utilization rate and urease activity, elevated pH levels within the culture environment, optimized bacteria mineralization, and increased calcium carbonate production. However, with the increase in the potential gradient, the inhibitory effect on the bacterial liquid gradually increased. (3) Calcium carbonate sedimentation was most significant at a gradient of 0.5 V/cm, which was closely linked to the promotion effect of the electric field on the bacterial solution. Furthermore, the crystal shapes of calcium carbonate under different potential gradients were significantly different. The calcium carbonate crystal formed under a low potential gradient was composed of calcite and vaterite, which exhibited substantial cementing effects. In contrast, the calcium carbonate crystal formed under a high potential gradient was primarily vaterite, which exhibited poor thermodynamic stability. Therefore, from an application perspective, the calcium carbonate crystal with a 0.5 V/cm gradient was found to have the most optimal effect, content, and suitable crystal shapes.

7. Significance and Suggestions of Research

The microbial-induced calcium carbonate precipitation (MICP) technology exhibits promising application prospects in various engineering fields. However, there exist several scientific issues that need to be addressed before its practical implementation, such as the by-products generated during the microbial culture process [51], reinforcement uniformity [18], environmental impact [52], process cost, and others [53]. The effect of electric fields on MICP technology is of significant importance. This study introduced an electric field during the precipitation of *S. pasteurii* culture and calcium carbonate. The electric field facilitated the growth and cellular proliferation of *S. pasteurii*. It improved the bacteria's metabolism and enzyme activity, thereby increasing the yield of calcium carbonate and providing more nucleation sites for precipitation. This ultimately reduced the cost of culture materials, making them highly valuable for practical applications. Additionally, the electric field regulated the crystal structure and arrangement of the calcium carbonate induced by *S. pasteurii*, resulting in improved compactness and homogeneity of the cemented precipitation products. Furthermore, it is speculated that the subsequent application of an electric field in MICP reinforcement of porous media could enhance the deposition and mutual cementation between bacteria and calcium carbonate precipitation, thereby improving the cementation performance and shear strength of microbial support of porous media.

Introducing an electric field into MICP technology has widened its potential applications, including surface reinforcement of porous media, such as groundwater and industrial wastewater pollution control, rock and soil improvement, waterproofing, and impermeability. In order to overcome the challenges that impede the development of this technology, more research is necessary to optimize the application of MICP technology, reduce by-product generation, develop new and improved technology, and use computer simulation analysis to enhance the reinforcement effect of MICP. Furthermore, efforts should be directed towards breeding specific urea-dissolving bacteria, stimulating in situ microorganisms to produce a cementation performance, and conducting detailed research to improve and apply MICP technology.

Author Contributions: Conceptualization, J.D. and M.L.; methodology, L.W.; software, L.H.; validation, J.D. and Z.Z.; formal analysis, L.W.; investigation, Y.T.; resources, M.L.; data curation, Y.T.; writing—original draft preparation, J.D.; writing—review and editing, J.D. and M.L.; visualization, L.H.; supervision, Z.Z.; project administration, L.W. and Z.Z.; funding acquisition, L.W. and Z.Z. All authors have read and agreed to the published version of the manuscript.

Funding: This research was funded by the National Natural Science Foundation of China (grant number: 52274167), the Natural Science Foundation of Hunan Province (grant numbers: 2022JJ40374), the Research Foundation of Education Bureau of Hunan Province (grant numbers: 22B0410, 20B496), the Hengyang City Science and Technology Program Project Funding (grant numbers: 202150063769), the Hunan Province's technology research project "Revealing the List and Taking Command" (grant numbers: 2021SK1050).

Data Availability Statement: Not applicable.

Conflicts of Interest: The authors declare no conflict of interest.

References

1. Miller, S.A. The role of cement service-life on the efficient use of resources. *Environ. Res. Lett.* **2020**, *15*, 024004. [CrossRef]
2. Schneider, M.; Romer, M.; Tschudin, M.; Bolio, H. Sustainable Cement Production—Present and Future. *Cem. Concr. Res.* **2011**, *41*, 642–650. [CrossRef]
3. Feng, J.; Chen, B.; Sun, W.; Wang, Y. Microbial induced calcium carbonate precipitation study using Bacillus subtilis with application to self-healing concrete preparation and characterization. *Constr. Build. Mater.* **2021**, *280*, 122460. [CrossRef]
4. Sj, A.; Sg, B.; Msr, A. Influence of biogenic treatment in improving the durability properties of waste amended concrete: A review. *Constr. Build. Mater.* **2020**, *263*, 120170.
5. Yu, X.; Jiang, J.; Liu, J.; Li, W. Review on potential uses, cementing process, mechanism and syntheses of phosphate cementitious materials by the microbial mineralization method. *Constr. Build. Mater.* **2020**, *273*, 121113. [CrossRef]
6. Zheng, T.; Qian, C. Self-Healing of Later-Age Cracks in Cement-Based Materials by Encapsulation-Based Bacteria. *J. Mater. Civ. Eng.* **2020**, *32*, 04020341. [CrossRef]
7. Almajed, A.; Lateef, M.A.; Moghal, A.A.B.; Lemboye, K. State-of-the-art review of the applicability and challenges of microbial-induced calcite precipitation (MICP) and enzyme-induced calcite precipitation (EICP) techniques for geotechnical and geoenvironmental applications. *Crystals* **2021**, *11*, 370. [CrossRef]
8. Feng, C.; Cui, B.; Ge, H.; Huang, Y.; Zhang, W.; Zhu, J. Reinforcement of recycled aggregate by Microbial-Induced mineralization and deposition of calcium carbonate—Influencing factors, mechanism and effect of reinforcement. *Crystals* **2021**, *11*, 887. [CrossRef]
9. Gao, H.; Dai, S. Influence of Culture Medium on Cementation of Coarse Grains Based on Microbially Induced Carbonate Precipitation. *Crystals* **2022**, *12*, 188. [CrossRef]
10. Lin, H.; Suleiman, M.T.; Brown, D.G. Investigation of pore-scale $CaCO_3$ distributions and their effects on stiffness and permeability of sands treated by microbially induced carbonate precipitation (MICP). *Soils Found.* **2020**, *60*, 944–961. [CrossRef]
11. Zhang, Z.-J.; Li, B.; Hu, L.; Zheng, H.-M.; He, G.-C.; Yu, Q.; Wu, L.-L. Experimental study on MICP technology for strengthening tail sand under a seepage field. *Geofluids* **2020**, *2020*, 8819326. [CrossRef]
12. Hoang, T.; Alleman, J.; Cetin, B.; Ikuma, K.; Choi, S.G. Sand and Silty-Sand Soil Stabilization Using Bacterial Enzyme Induced Calcite Precipitation (BEICP). *Can. Geotech. J.* **2019**, *56*, 808–822. [CrossRef]
13. Chek, A.; Crowley, R.; Ellis, T.N.; Durnin, M.; Wingender, B. Evaluation of Factors Affecting Erodibility Improvement for MICP-Treated Beach Sand. *J. Geotech. Geoenviron. Eng.* **2021**, *147*, 4021001. [CrossRef]
14. Chuo, S.C.; Mohamed, S.F.; Setapar, S.; Ahmad, A.; Ibrahim, M. Insights into the Current Trends in the Utilization of Bacteria for Microbially Induced Calcium Carbonate Precipitation. *Materials* **2020**, *13*, 4993. [CrossRef]
15. Lai, Y.; Yu, J.; Liu, S.; Liu, J.; Dong, B. Experimental study to improve the mechanical properties of iron tailings sand by using MICP at low pH. *Constr. Build. Mater.* **2020**, *273*, 121729. [CrossRef]
16. Liu, B.; Zhu, C.; Tang, C.S.; Xie, Y.H.; Shi, B. Bio-remediation of desiccation cracking in clayey soils through microbially induced calcite precipitation (MICP). *Eng. Geol.* **2019**, *264*, 105389. [CrossRef]
17. Xu, X.; Guo, H.; Cheng, X.; Li, M. The promotion of magnesium ions on aragonite precipitation in MICP process. *Constr. Build. Mater.* **2020**, *263*, 120057. [CrossRef]

18. Zhang, Z.J.; Tong, K.W.; Hu, L.; Yu, Q.; Wu, L.L. Experimental study on solidification of tailings by MICP under the regulation of organic matrix. *Constr. Build. Mater.* **2020**, *265*, 120303. [CrossRef]
19. Okwadha, G.D.O.; Li, J. Optimum conditions for microbial carbonate precipitation. *Chemosphere* **2010**, *81*, 1143–1148. [CrossRef]
20. Yu, T.; Souli, H.; Péchaud, Y.; Fleureau, J.M. Optimizing protocols for microbial induced calcite precipitation (MICP) for soil improvement—A review. *Eur. J. Environ. Civ. Eng.* **2022**, *26*, 2218–2233. [CrossRef]
21. Kitamura, M.; Konno, H.; Yasui, A.; Masuoka, H. Controlling factors and mechanism of reactive crystallization of calcium carbonate polymorphs from calcium hydroxide suspensions. *J. Cryst. Growth* **2002**, *236*, 323–332. [CrossRef]
22. Wei, S.P.; Cui, H.P.; Jiang, Z.L.; Liu, H.; He, H.; Fang, N.Q. Biomineralization processes of calcite induced by bacteria isolated from marine sediments. *Braz. J. Microbiol.* **2015**, *46*, 455–464. [CrossRef] [PubMed]
23. Xu, W.; Zheng, J.; Chu, J.; Zhang, R.; Cui, M.; Lai, H.; Zeng, C. New method for using N-(N-butyl)-thiophosphoric triamide to improve the effect of microbial induced carbonate precipitation. *Constr. Build. Mater.* **2021**, *313*, 125490. [CrossRef]
24. Nemati, M.; Greene, E.A.; Voordouw, G. Permeability profile modification using bacterially formed calcium carbonate: Comparison with enzymic option. *Process Biochem.* **2005**, *40*, 925–933. [CrossRef]
25. Gidudu, B.; Chirwa, E.M.N. The combined application of a high voltage, low electrode spacing, and biosurfactants enhances the bio-electrokinetic remediation of petroleum contaminated soil. *J. Clean. Prod.* **2020**, *276*, 122745. [CrossRef]
26. Ossai, I.C.; Ahmed, A.; Hassan, A.; Hamid, F.S. Remediation of soil and water contaminated with petroleum hydrocarbon: A review. *Environ. Technol. Innov.* **2019**, *17*, 100526. [CrossRef]
27. Mena, E.; Villasenor, J.; Canizares, P.; Rodrigo, M.A. Effect of a direct electric current on the activity of a hydrocarbon-degrading microorganism culture used as the flushing liquid in soil remediation processes. *Sep. Purif. Technol.* **2014**, *124*, 217–223. [CrossRef]
28. Cheng, F.L.; Guo, S.H.; Wang, S.; Guo, P.H.; Lu, W.J. Transportation and augmentation of the deposited soil bacteria in the electrokinetic process: Interactions between soil particles and bacteria. *Geoderma* **2021**, *404*, 115260. [CrossRef]
29. Martin, D.; Dodds, K.; Ngwenya, B.T.; Butler, I.B.; Elphick, S.C. Inhibition of *Sporosarcina pasteurii* under Anoxic Conditions: Implications for Subsurface Carbonate Precipitation and Remediation via Ureolysis. *Environ. Sci. Technol.* **2012**, *46*, 8351–8355. [CrossRef]
30. Kang, B.; Zha, F.; Li, H.; Xu, L.; Sun, X.; Lu, Z. Bio-Mediated Method for Immobilizing Copper Tailings Sand Contaminated with Multiple Heavy Metals. *Crystals* **2022**, *12*, 522. [CrossRef]
31. Ramachandran, S.K.; Ramakrishnan, V.; Bang, S.S. Remediation of Concrete Using Microorganisms. *ACI Mater. J.* **2001**, *98*, 3–9.
32. Zhao, Q.; Li, L.; Li, C.; Li, M.; Amini, F.; Zhang, H. Factors affecting improvement of engineering properties of MICP-treated soil catalyzed by bacteria and urease. *J. Mater. Civ. Eng.* **2014**, *26*, 04014094. [CrossRef]
33. Cuzman, O.A.; Richter, K.; Wittig, L.; Tiano, P. Alternative nutrient sources for biotechnological use of *Sporosarcina pasteurii*. *World J. Microbiol. Biotechnol.* **2015**, *31*, 897–906. [CrossRef] [PubMed]
34. Di, P.; Zhi-Ming, L.; Bi-Ru, H.; Wen-Jian, W. Progress on mineralization mechanism and application research of *Sporosarcina pasteurii*. *Prog. Biochem. Biophys.* **2020**, *47*, 467–482.
35. Whiffin, V.S.; Van Paassen, L.; Harkes, M.P. Microbial Carbonate Precipitation as a Soil Improvement Technique. *Geomicrobiol. J.* **2007**, *24*, 417–423. [CrossRef]
36. Sun, X.-T.; Ma, J.; Sun, X.-Y.; Liu, B. Electrolytic Stimulation of *Escherichia coli* by a Direct Current. *Microbiol./Weishengwuxue Tongbao* **2010**, *37*, 1440–1446.
37. Zhang, B.; Liu, Y.; Tong, S.; Zheng, M.; Zhao, Y.; Tian, C.; Liu, H.; Feng, C. Enhancement of bacterial denitrification for nitrate removal in groundwater with electrical stimulation from microbial fuel cells. *J. Power Sources* **2014**, *268*, 423–429. [CrossRef]
38. Liu, H.; Ouyang, F.; Chen, Z.; Chen, Z.; Lichtfouse, E. Weak electricity stimulates biological nitrate removal of wastewater: Hypothesis and first evidences. *Sci. Total Environ.* **2021**, *757*, 143764. [CrossRef]
39. Rabbi, M.F.; Clark, B.; Gale, R.J. In situ TCE bioremediation study using electrokinetic cometabolite injection. *Waste Manag.* **2000**, *20*, 279–286. [CrossRef]
40. Yi, H.; Zheng, T.; Jia, Z.; Su, T.; Wang, C. Study on the influencing factors and mechanism of calcium carbonate precipitation induced by urease bacteria. *J. Cryst. Growth* **2021**, *564*, 126113. [CrossRef]
41. Ferris, F.G.; Phoenix, V.; Fujita, Y.; Smith, R. Kinetics of calcite precipitation induced by ureolytic bacteria at 10 to 20 C in artificial groundwater. *Geochim. Cosmochim. Acta* **2004**, *68*, 1701–1710. [CrossRef]
42. Ng, W.-S.; Lee, M.-L.; Hii, S.-L. An overview of the factors affecting microbial-induced calcite precipitation and its potential application in soil improvement. *Int. J. Civ. Environ. Eng.* **2012**, *6*, 188–194.
43. DeJong, J.T.; Mortensen, B.M.; Martinez, B.C.; Nelson, D.C. Bio-mediated soil improvement. *Ecol. Eng.* **2010**, *36*, 197–210. [CrossRef]
44. Wick, L.Y.; Shi, L.; Harms, H. Electro-bioremediation of hydrophobic organic soil-contaminants: A review of fundamental interactions. *Electrochim. Acta* **2007**, *52*, 3441–3448. [CrossRef]
45. Cheng, L.; Qian, C.; Wang, R.; Wang, J. Study on the mechanism of calcium carbonate formation induced by carbonate-mineralization microbe. *Acta Chim. Sin.* **2007**, *65*, 2133–2138.
46. Rong, H.; Qian, C.-X.; Li, L.-Z. Study on microstructure and properties of sandstone cemented by microbe cement. *Constr. Build. Mater.* **2012**, *36*, 687–694. [CrossRef]
47. Gebauer, D.; Volkel, A.; Colfen, H. Stable prenucleation calcium carbonate clusters. *Science* **2008**, *322*, 1819–1822. [CrossRef]

48. Zheng, H.-M.; Wu, L.-L.; Tong, K.-W.; Ding, D.-X.; Zhang, Z.-J.; Yu, Q.; He, G.-C. Experiment on microbial grouting reinforcement of tailings under the regulation of egg white. *Soils Found.* **2020**, *60*, 962–977. [CrossRef]

49. Muynck, W.D.; Belie, N.D.; Verstraete, W. Microbial carbonate precipitation in construction materials: A review. *Ecol. Eng.* **2010**, *36*, 118–136. [CrossRef]

50. Stocks-Fischer, S.; Galinat, J.K.; Bang, S.S. Microbiological precipitation of $CaCO_3$. *Soil Biol. Biochem.* **1999**, *31*, 1563–1571. [CrossRef]

51. Naveed, M.; Duan, J.; Uddin, S.; Suleman, M.; Hui, Y.; Li, H. Application of microbially induced calcium carbonate precipitation with urea hydrolysis to improve the mechanical properties of soil. *Ecol. Eng.* **2020**, *153*, 105885. [CrossRef]

52. Li, M.; Cheng, X.; Guo, H. Heavy metal removal by biomineralization of urease producing bacteria isolated from soil. *Int. Biodeterior. Biodegrad.* **2013**, *76*, 81–85. [CrossRef]

53. Rahman, M.M.; Hora, R.N.; Ahenkorah, I.; Beecham, S.; Karim, M.R.; Iqbal, A. State-of-the-art review of microbial-induced calcite precipitation and its sustainability in engineering applications. *Sustainability* **2020**, *12*, 6281. [CrossRef]

sustainability

Article

Stability Analysis of Tunnel Surrounding Rock When TBM Passes through Fracture Zones with Different Deterioration Levels and Dip Angles

Mingtao Ji [1,2], Xuchun Wang [1], Minhe Luo [1], Ding Wang [3,4,*], Hongwei Teng [1] and Mingqing Du [1]

1   School of Civil Engineering, Qingdao University of Technology, Qingdao 266520, China
2   Qingdao Metro Group Co., Ltd., Qingdao 266035, China
3   State Key Laboratory for Geomechanics and Deep Underground Engineering,
    China University of Mining and Technology, Beijing 100083, China
4   School of Mechanical Electronic & Information Engineering, China University of Mining and Technology,
    Beijing 100083, China
*   Correspondence: wangdingneil@outlook.com; Tel.: +86-0532-8507-1310

Abstract: In fracture zones, tunneling with a double-shield Tunnel Boring Machine (TBM) presents significant challenges, including deformation overrun of the surrounding rock, TBM jamming, and excavation face collapse. To assure the tunnel construction safety and efficiency, it is necessary and crucial to conduct a stability analysis of the tunnel surrounding rock when a TBM passes through the fracture zones. The tunnels from Jiadingshan Road Station to Anshan Road Station in Qingdao Metro Line 8 are constructed by double-shield TBMs. It inevitably passes through fracture zones with different deterioration levels and dip angles. In this study, based on this construction section, numerical models of fracture zones with different deterioration levels and dip angles were developed to analyze the displacements of tunnel vaults, inverts, and haunches. In addition, the maximum shear stresses of the surrounding rock were analyzed. Finally, the displacement and shear stress variation patterns of the surrounding rock with different deterioration levels and dip angles were obtained. The findings reveal the stability behavior of tunnels under various fracture zones. They can serve as a valuable reference and theoretical foundation for future tunnel construction projects utilizing double-shield TBMs in areas with fracture zones.

Keywords: tunnel; TBM; fracture zone; numerical simulation

Citation: Ji, M.; Wang, X.; Luo, M.; Wang, D.; Teng, H.; Du, M. Stability Analysis of Tunnel Surrounding Rock When TBM Passes through Fracture Zones with Different Deterioration Levels and Dip Angles. *Sustainability* **2023**, *15*, 5243. https://doi.org/10.3390/su15065243

Academic Editors: Hong-Wei Yang, Shuren Wang and Chen Cao

Received: 11 February 2023
Revised: 9 March 2023
Accepted: 14 March 2023
Published: 15 March 2023

1. Introduction

China's escalating urbanization has led to a surge in demand for underground infrastructure, driving the construction of urban subway tunnels. The high level of automation and safety offered by Tunnel Boring Machines (TBMs) has facilitated widespread adoption in this field. In September 2009, Qingdao Metro evaluated the suitability of different TBMs and found that the double-shield TBM method is more suitable for Qingdao's predominantly hard-rock strata. Therefore, the double-shield TBMs have been used to construct tunnels for Qingdao Metro lines 1, 4, 6, and 8. However, as double-shield TBMs are boring through a considerable distance, it is unavoidable to cross adverse geological sections, such as fracture zones. The varying width and dip of the fracture zones and the rock masses' physical and mechanical properties influence tunnel surrounding rock stability. It presents several potential construction challenges, including excavation surface instability, ground settlement overrun, TBM jams, tunnel collapse, and other engineering accidents. Therefore, to assure the tunnel construction safety and efficiency, it is necessary and crucial to conduct stability analysis of the tunnel surrounding rock when a TBM passes through fracture zones.

There have been many studies about the construction risk evaluation of TBM tunneling through fracture zones and technical measures to mitigate risks. Peng et al. [1] and

Xue et al. [2] established the cloud models to assess water inrush risk in tunnels using geological analysis and risk assessment indicators. Both models have been verified and applied under the actual tunnel excavation conditions, which improve the accuracy of the risk assessment of water inrush in the tunnel through water-rich faults. Newman et al. [3] monitored the operating parameters, such as cutter head torque, thrust, and injection of spoil-conditioning additives. The monitoring information can be used to validate the pre-construction geological model and provide valuable assurances for the predicted ground ahead of tunneling operations. Liu et al. [4] and Li et al. [5] suggested a comprehensive prospecting method to ensure safe TBM tunneling to identify adverse geological bodies, including fractured zones and water-bearing structures. In addition, some other practical experiences and techniques, such as consolidation grouting, which have been used to overcome the challenges of TBM tunneling in fracture zones, were investigated [6–8]. However, although there have been considerable advancements in risk assessment and control for TBMs crossing through fracture zones, the existing methods still have some limitations: (1) These methods can only provide qualitative or semi-quantitative information, often lack the quantitative analysis and accurate results. In this case, they can only offer general guidelines for tunnel constructions in such conditions; (2) These methods frequently failed to consider the complex geological conditions and variations in tunnel design, resulting in a lack of adaptability. Therefore, further study is required to understand the influence of fracture zones on tunnel displacement and the surrounding rock's plastic zone.

For the complex geological conditions and changeable surrounding rock behaviors of tunnel construction through fracture zones, the numerical approach is a feasible method to study the influence of fracture zones on tunnel displacement and stress variations. Huang et al. [9] investigated the deformation and failure of the surrounding rock mass triggered by sudden changes in rock quality encountered during tunnel excavation, using the Daliang Tunnel on the Lanzhou-Urumqi railway as a case study. The study combined in situ measurement, theoretical research, and numerical simulation to explore the deformation of the surrounding rock mass in and around fracture zones. The results indicate that the increasing misalignment distance of fracture zones increases the stress level on the tunnel lining. Xie et al. [10] developed a 3D coupled computational fluid dynamics-discrete element method model for fluid–solid interaction systems and successfully simulated the water and mud inrush process in a water-rich fault tunnel. The results revealed the evolution process of water and mud inrush and the formation of an ellipsoidal collapse area inside the fracture zone. Zaheri et al. [11] investigated the impact of strike-slip fault movement on the segmental linings using three-dimensional numerical modeling. The results suggest that the fault dip angle significantly affects the separation of lining segments and denser soils experience more significant tunnel displacement after faulting. The other researchers have also carried out experimental, analytical, numerical, and in situ research for different fracture strata under various conditions, thus explaining the patterns and mechanisms of TBM tunneling in fracture zones [12–22]. However, these studies primarily focused on TBM tunneling through the high in situ stress, deeply buried mountain or hydro tunnels. By contrast, the time of application of double-shield TBMs in urban subway construction is still short. Urban subway tunnels are generally constructed at relatively shallow depths and under complex geological conditions. More specifically, the deformation and stress mechanisms of the tunnel surrounding rock when a TBM tunnels through fracture zones with different characteristics (e.g., different deterioration levels and dip angles) remain immature.

This study addresses this gap by analyzing the fracture zone between Jiading Mountain Station and Anshan Road Station of Qingdao Metro Line 8. Flac3D was employed to establish numerical models of the tunnels, followed by a comprehensive analysis of their stability under different simulation conditions. The study revealed the displacement patterns at critical points of the tunnel, including the vault, invert, and haunch. Additionally, the analysis results illustrate the variation characteristics of shear stress in the surrounding rock under different fracture zone conditions. The findings in this study can provide valuable guidelines for designing and constructing subway tunnels with enhanced safety and efficiency, particularly in areas with intricate geological conditions.

2. Engineering Survey

Qingdao Metro Line 8's tunnel alignment between Jiading Shan Road Station and Anshan Road Station extends over 2459.559 m. It starts at K55 + 102.735 and ends at K57 + 562.294. The tunnel is constructed using parallel boring, with a spacing of approximately 6–10 m between them. The geology of this section is characterized by the presence of plain fill, strongly weathered granite, medium weathered granite, and slightly weathered granite strata. Groundwater is abundant and unevenly distributed, primarily consisting of pore water and fourth-system bedrock fracture water. The Cangkou fault, marked by a strike of 40°, an inclination of 310°, and dips ranging from 50 to 86°, significantly impact this section. The central section of the fault is the widest, reaching a width of 100 m. It is characterized by the development of cataclastic rocks, silty rocks, and mylonite. Multiple fracture zones, influenced by the Cangkou fault, are present in this area, as depicted in Figure 1. The section is excavated using a CSIC φ7032 double-shield TBM, which has a maximum total thrust of 19,396 kN, a maximum speed of 8.2 rpm, a maximum torque of 3947 kN·m, and a stripping torque of 6600 kN·m. The TBM cutter has 25 face cutters of 4830 mm diameter, 8 centrals, and 10 gauge cutters.

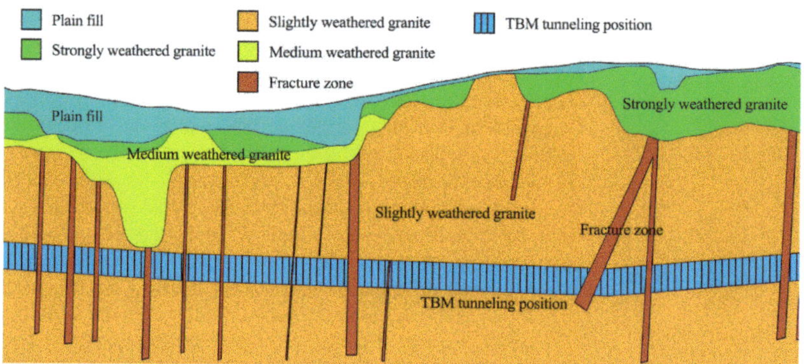

Figure 1. Geological profile of Jiading Mountain Road Station–Anshan Road Station.

3. Numerical Modeling

3.1. Determination of Models

This study investigated the effects of different deterioration levels and fracture zone dip angles on the tunnel's stability. Several assumptions were made, and critical factors were focused on simplifying the numerical modeling. It is assumed that the tunnel structure and surrounding rock are isotropic, homogeneous, and exhibit exemplary elastic-plastic behavior. In the numerical model, the initial ground-stress field was determined by considering only the self-weight of the rock. This approach does not account for the potential influence of groundwater conditions on the ground-stress field.

Based on the geological profile of the Jiadingshan Road–Anshan Road section, numerical models of the section containing the fracture zone were established. The models in this study were composed of layers of plain fill, strongly weathered granite, medium weathered granite, and slightly weathered granite. These models were simplified using a half-tunnel model to take advantage of symmetry. The total width of the model along X direction was set to be three to five times the tunnel diameter according to St. Venant's principle to prevent boundary effects, resulting in a final width of 40 m. Similarly, the model's height along the Z-direction was 60 m, with 30 m each from the tunnel axis to its top and bottom. The length of the model along Y-direction was 120 m. The diameter of the TBM excavated tunnel was 6.3 m. In addition, to make the simulation more representative of the boring process, a TBM shell unit and a liner tube sheet unit were incorporated to simulate the effects of the TBM on the tunnel. The thickness of the shield shell unit was set as 60 mm, and the thickness of the liner unit was set as 300 mm. The model's normal displacement was limited by normal constraints around and at its bottom, with no constraint on its top. The numerical model geometry is illustrated in Figure 2.

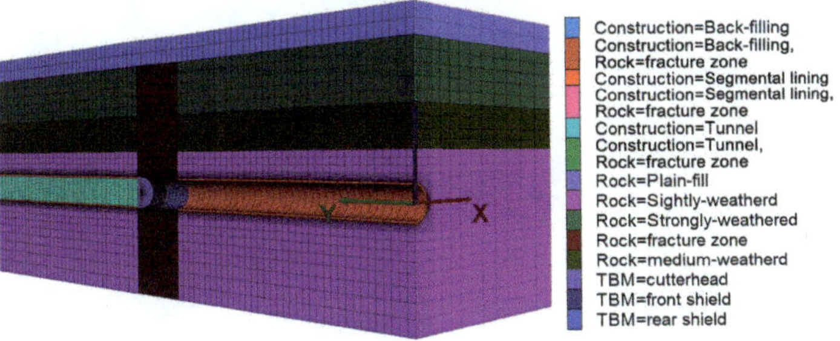

Figure 2. Numerical model geometry.

3.2. Mechanical Parameters of Materials and Boring Steps

The geotechnical deformation in this numerical model follows the Mohr–Coulomb strength criterion. Elastic deformation assumptions were adopted for the liner and grouting, while the TBM cutter and shield body were regarded as rigid materials. Based on the detailed stratigraphic survey report and TBM equipment parameters from Jiading Shan Road Station to Anshan Road Station, Table 1 displays each unit's physical and mechanical parameters. In addition, the parameters after solidification were utilized for the grouting units to simplify the calculation.

Table 1. Physical and mechanical parameters of stratigraphic and TBM equipment units.

Name	Elastic Modulus/GPa	Poisson Ratio	Cohesion/ MPa	Internal Friction/°	Density/ (kg·m^{-3})
Plain fill	8×10^{-3}	0.4	0.3	15	1.65×10^{3}
Strongly weathered granite	1.1	0.35	1.8	30	1.95×10^{3}
Medium weathered granite	3.5	0.25	6.5	49	2.25×10^{3}
Sightly weathered granite	7	0.21	9	58	2.72×10^{3}
liner tube sheet	32.6	0.2	/	/	3.25×10^{3}
TBM cutter head	200	0.3	/	/	5.3×10^{3}
TBM shields	200	0.3	/	/	9.5×10^{4}
Grouting layer	0.35	0.25	/	/	2.5×10^{3}

In order to better simulate the double-shield TBM boring process, a normal surface stress of 5×10^{5} Pa at the excavation face was set to simulate the stress acting on the face by the double-shield TBM. The TBM cutter and shield body moved forward by 1.5 m during each excavation cycle, removing the tunnel, liner, and grouting unit. Subsequently, material parameters were assigned to the grouting unit and liner unit to represent the assembly of the end grouting and tube sheet of the TBM shield. Following these steps, the excavation process was carried out in a cycle until the end of the model excavation.

4. Influence Pattern of the Fracture Zone's Deterioration Levels on the Tunnel Stability

4.1. Scheme Design for Models under Different Deterioration Levels

A fracture zone's width and degree of fragmentation are crucial indicators of its deterioration. In order to examine the impact of deterioration on tunnel stability, twelve sets of models incorporating fracture zones were developed. These models simulated changes in deterioration by using fracture zones with varying widths and degrees of fragmentation (elastic modulus and cohesion). The angle between the fracture zone and the tunnel axis, denoted as θ, was a constant value of 90° in this model. Figure 3 illustrates a schematic representation of the numerical model that considers various levels of deterioration of the fracture zone. Table 2 presents the simulation schemes for different levels of deterioration of the fracture zone.

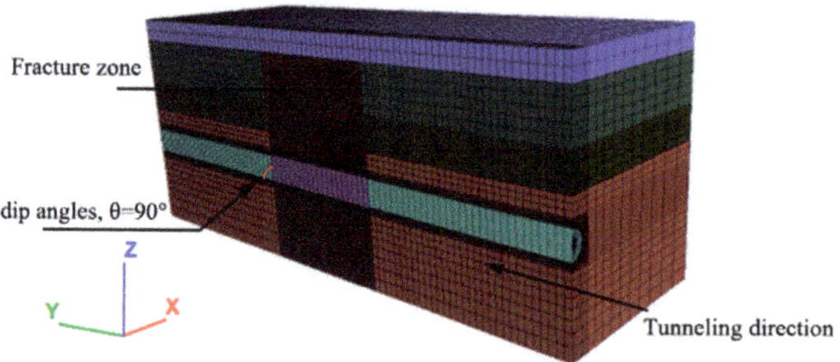

Figure 3. Numerical model considering different deterioration levels of the fracture zone.

Table 2. Simulation scheme sets for different levels of deterioration of the fracture zone.

Simulation Scheme	Dip Angle/°	Width/m	Elastic Modulus/MPa	Poisson Ratio	Cohesion/MPa
1	90	12	600	0.35	0.6
2	90	12	400	0.35	0.55
3	90	12	200	0.35	0.5
4	90	18	600	0.35	0.6
5	90	18	400	0.35	0.55
6	90	18	200	0.35	0.5
7	90	24	600	0.35	0.6
8	90	24	400	0.35	0.55
9	90	24	200	0.35	0.5
10	90	30	600	0.35	0.6
11	90	30	400	0.35	0.55
12	90	30	200	0.35	0.5

4.2. Displacement Patterns in the Tunnel under Different Levels of Fracture Zone Deterioration

After conducting numerical simulations of twelve scheme sets with varying levels of deterioration, the maximum displacements and their corresponding locations at the vault, invert, and haunch were determined accordingly. Table 3 presents the numerically obtained results under various levels of deterioration of the fracture zone.

Table 3 illustrates that the maximum displacement in the tunnel occurs within a range of 60 to 65 m when a double-shielded TBM tunnels through a fracture zone. This displacement occurs regardless of the tunnel vault, invert, or haunch. It implies that the maximum area of tunnel deformation occurs within the fracture zone. This observation is consistent with previous studies, which showed that the stability of tunnels in fractured zones is significantly affected by the degree of fragmentation. In addition, it is reported that the occurrence of the largest displacement is located at the TBM's arrival in the middle of fracture zone [23,24].

The results also indicate that the width of the fracture zone has a limited impact on tunnel displacement. Specifically, it has been demonstrated that when the degree of fragmentation in the fracture zone remains constant, the maximum displacement values at the tunnel vault, invert, and haunch are virtually invariable. Even though the width of the fracture zone may vary, the effects are insignificant. However, with an increase in fragmentation, which is represented by a decrease in the elastic modulus and cohesive stress of the fracture zone rock, the location of the maximum tunnel displacement changes significantly. It is evidenced by the fact that as the degree of fragmentation increases, the maximum displacement values at the tunnel vault, invert, and haunch increase significantly. Furthermore, for a large degree of fragmentation, the maximum displacement of the tunnel surrounding rock increases with the increase in the tunnel width but with a lesser increase in extent.

Table 3. Maximum displacement of tunnel under different levels of deterioration of the fracture zone.

Simulation Scheme	Tunnel Crown/mm	Location/m	Tunnel Invert/mm	Location/m	Tunnel Haunch/mm	Location/m
1	−4.225	Y = 60	4.481	Y = 60	4.359	Y = 60
2	−6.036	Y = 60	6.601	Y = 60	6.322	Y = 60
3	−11.386	Y = 60	12.771	Y = 60	12.057	Y = 60
4	−4.291	Y = 60	4.485	Y = 61	4.432	Y = 60
5	−6.281	Y = 60	6.733	Y = 60	6.568	Y = 60
6	−12.032	Y = 61	13.326	Y = 61	12.843	Y = 60
7	−4.383	Y = 61	4.397	Y = 63	4.500	Y = 60
8	−6.595	Y = 62	6.609	Y = 63	6.632	Y = 60
9	−12.753	Y = 61	13.256	Y = 63	12.956	Y = 61
10	−4.466	Y = 63	4.325	Y = 63	4.492	Y = 60
11	−6.911	Y = 64	6.480	Y = 65	6.610	Y = 61
12	−13.489	Y = 63	13.043	Y = 64	12.949	Y = 62

To further investigate the influence of a fracture zone's width on the stability of tunnels, numerical models with an elastic modulus of 400 MPa and cohesion of 0.55 MPa were screened. The models were utilized to statistic the maximum displacements at the tunnel vault, invert, and haunch of double-shield TBMs traversing fracture zones of varying widths. The results are shown in Figure 4.

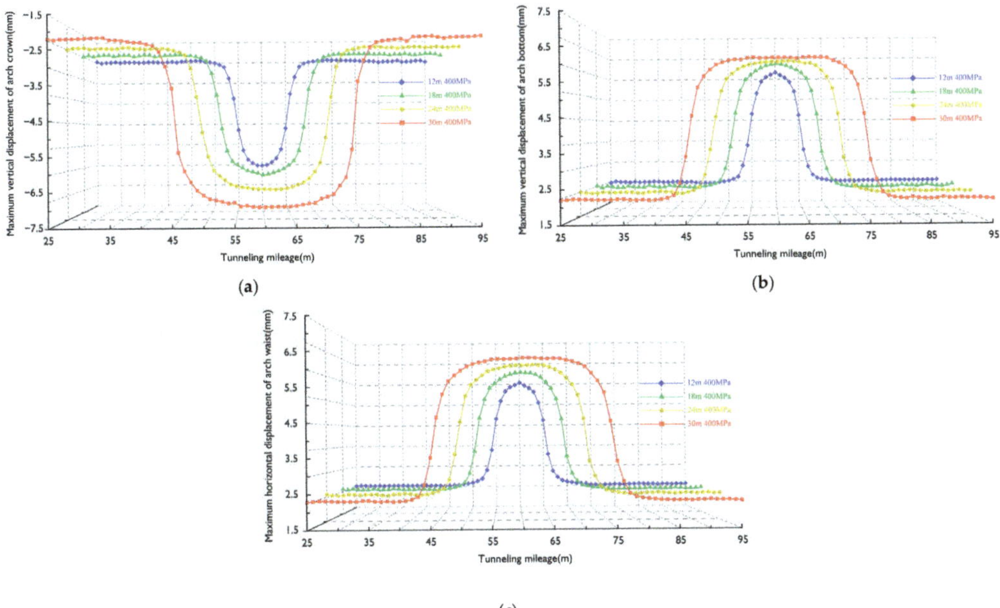

Figure 4. Results of tunnel surrounding rock displacement analysis under different fracture zone widths. (**a**) Maximum vertical displacement of tunnel vault. (**b**) Maximum vertical displacement of tunnel invert. (**c**) Maximum horizontal displacement of tunnel haunch.

Figure 4 illustrates that the vertical displacement value at the tunnel vault is relatively small, approximately 2.20 mm, before the double-shield TBM crosses through the fracture zone. The tunnel invert's heave and the tunnel haunch's horizontal displacement are also relatively minimal. Upon the double-shield TBM's entry into the fracture zone, the displacement at the tunnel vault experienced a significant increase. At the same time, the heave at the tunnel invert and the horizontal displacement at the tunnel haunch also increase rapidly. Once the double-shield TBM has passed through the fracture zone, the displacement at the tunnel vault, invert, and haunch returns to the level before crossing the fracture zone.

According to Figure 4a, the maximum vertical displacement at the tunnel vault increases with an increase in the width of the fracture zone, but the increase is moderate. Meanwhile, the deformation trough of the vault settlement curve also becomes wider, thereby impacting settlement in the non-fracture zone on both sides of the fracture zone. Similar tendencies are also observed from the variation profiles of the heave displacement in Figure 4b and the horizontal displacement in Figure 4c.

Three sets of models were set up with a fracture zone width of 18 m to investigate the influence of the degree of fragmentation on the stability of the tunnel under constant fracture zone width. Similarly, the maximum displacements at the tunnel vault, invert, and haunch were analyzed under different fragmentation, as depicted in Figure 5.

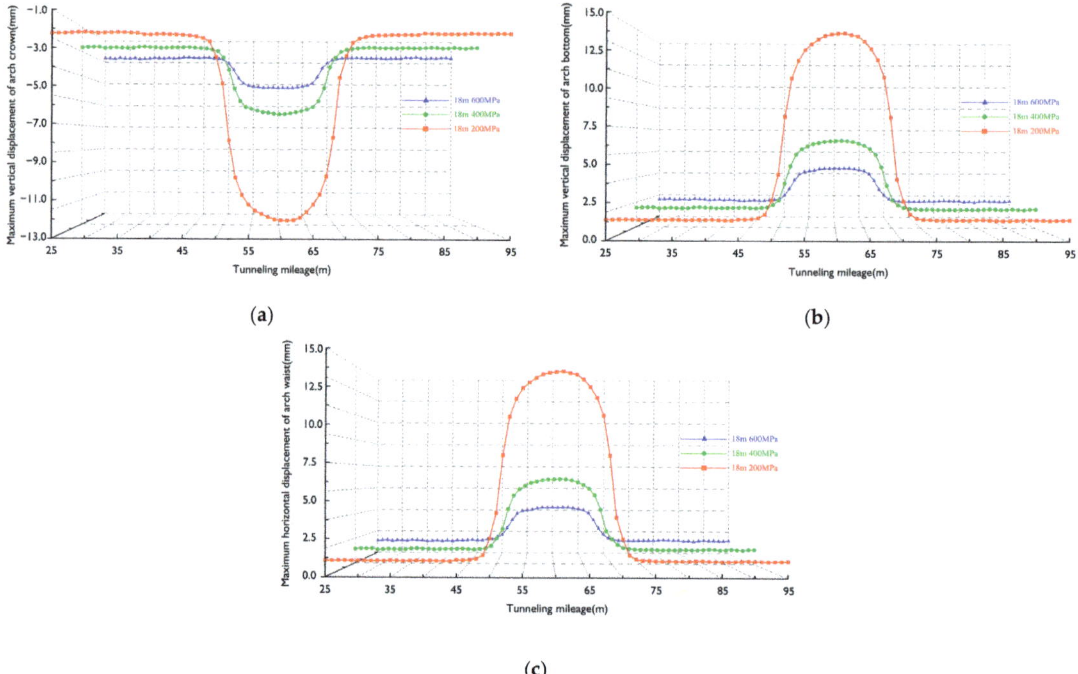

Figure 5. Results of tunnel displacement analysis under different degrees of fracture zone fragmentation. (**a**) Maximum vertical displacement of tunnel vault. (**b**) Maximum vertical displacement of tunnel invert. (**c**) Maximum horizontal displacement of tunnel haunch.

Figure 5 illustrates that the maximum displacements at the tunnel vault, invert, and haunch in the fracture zone area are significantly larger than in the non-fracture zone area. This is consistent with what was previously demonstrated. When the width of the fracture zone remains constant, an increase in the degree of fragmentation, represented by a decrease in the elastic modulus and cohesion of the fracture zone rock, results in a significant increase in the maximum displacement at the tunnel vault, invert, and haunch. As illustrated, the vertical displacement at the tunnel vault increases from 4.29 to 12.03 mm, the vertical heave of the tunnel invert increases from 4.49 to 12.33 mm, and the horizontal displacement of the tunnel haunch increases from 4.43 to 12.84 mm. Instead of impacting the width of the deformation trough, the increase in fragmentation primarily affects tunnel displacement.

In summary, the deterioration of the fracture zone, in terms of width and degree of fragmentation, significantly impacts the tunnel's stability. An increase in the fracture zone's width expands the tunnel deformation range. The degree of fragmentation increases as its relevant mechanical parameters, i.e., the elastic modulus and cohesion, decrease in the fracture zone, leading to a significant increase in the tunnel's deformation displacement value. The maximum displacement value of the tunnel also increases with the increase in the fracture zone width, although the increase is relatively small.

4.3. Shear Stress Patterns in the Surrounding Rock under Different Levels of Fracture Zone Deterioration

Under three-dimensional stress states, shear failure is the most prevalent failure mode for geotechnical structures. Consequently, further analysis of maximum shear stress in the surrounding rock of a double-shield TBM tunnel was conducted. It aims to reveal the effects of various levels of fracture zone deterioration on the shear stress pattern in surrounding rock. The results of this analysis are presented in Figures 6 and 7. Figure 6 illustrates the variations of the maximum shear stress patterns in the surrounding rock with different fracture zone's widths under constant fragmentation degree. By contrast, Figure 7 illustrates the variations of the maximum shear stress patterns in the surrounding rock with different degrees of fracture zone fragmentation under constant fracture zone width.

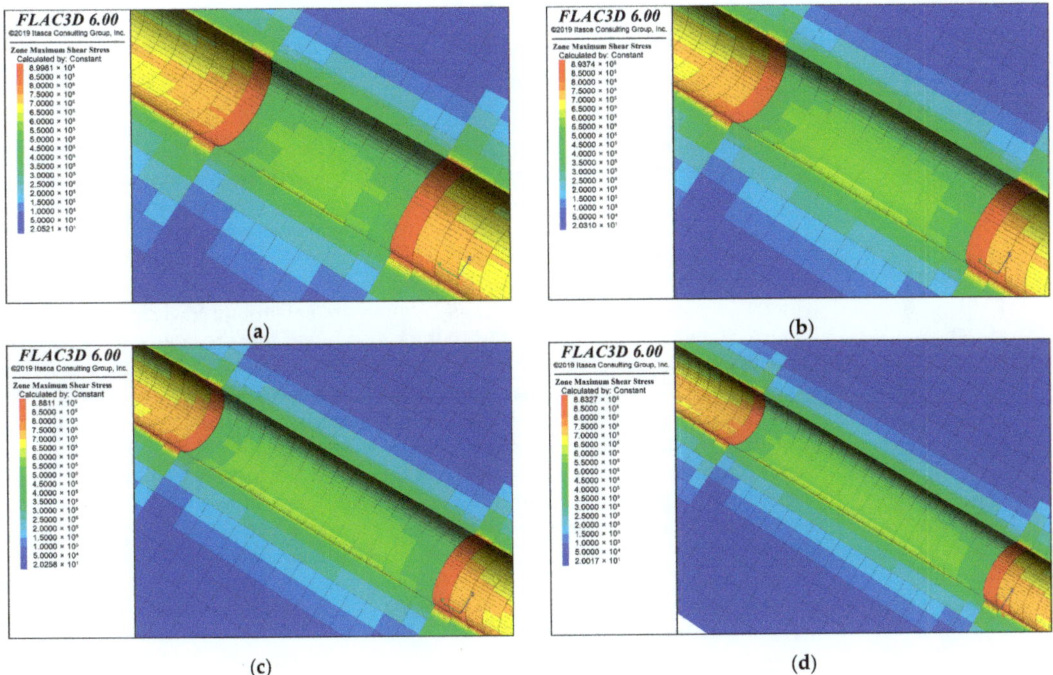

Figure 6. Maximum shear stress patterns in the surrounding rock with different fracture zone widths under constant fragmentation degree (elastic modulus of 600 MPa). (**a**) Width 12 m. (**b**) Width 18 m. (**c**) Width 24 m. (**d**) Width 30 m.

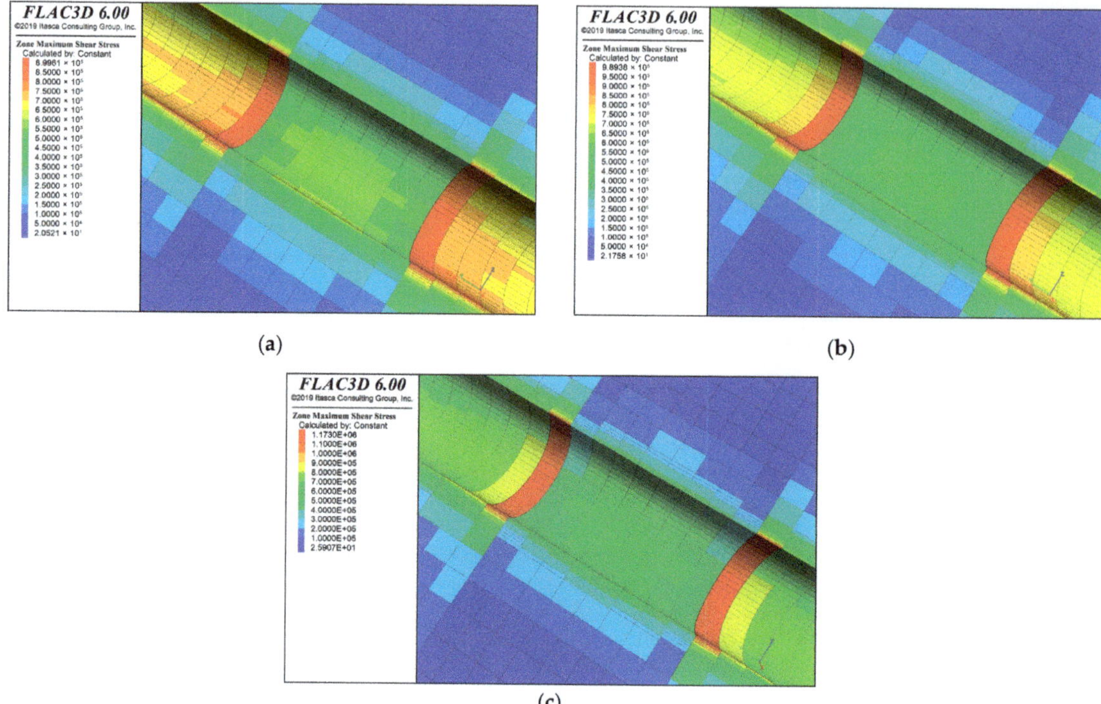

Figure 7. Maximum shear stress patterns in the surrounding rock with different degrees of fracture zone fragmentation under constant fracture zone width (width of 18 m). (**a**) Elastic modulus 600 MPa. (**b**) Elastic modulus 400 MPa. (**c**) Elastic modulus 200 MPa.

As shown in Figure 6, the magnitude of the maximum shear stress in the surrounding rock does not exhibit a significant change with the increase in the fracture zone width under a constant fragmentation degree, reaching a value of approximately 9.89×10^5 Pa. It implies that the fracture zone width minimally influences the maximum shear stress in the surrounding rock. By contrast, as shown in Figure 7, the magnitude of the maximum shear stress in the surrounding rock increases significantly from 9.00×10^5 to 1.17×10^6 Pa with the increase in the degree of fragmentation under constant fracture zone width. This is consistent with the previous study, which also found that the degree of fragmentation of the fracture zone significantly impacts the mechanical behavior of tunnels crossing the fracture zone [25]. The numerical results indicate that various factors, including the degree of fragmentation and fracture width, affect the mechanical characteristics of the tunnel structure. However, the effect of the degree of fragmentation on the shear stress in the surrounding rock is more significant than that of the fracture zone width. The degree of fracture zone deterioration mainly depends on the degree of fracture zone fragmentation, represented by the mechanical properties of the geotechnical body of the fracture zone. Figures 6 and 7 show that the surrounding rock's maximal shear stress occurs near the tunnel invert at both ends. The results reveal that shear failure is most likely to occur at the tunnel invert at both ends of the zone regardless of the fracture width or degree of fragmentation.

5. Influence Pattern of the Fracture Zone's Dip Angles on the Tunnel Stability

5.1. Scheme Design for Models under Different Dip Angles

The stability of a tunnel is also affected by the dip angles of the fracture zone. Numerical models were established considering different dip angles of the fracture zone to investigate the effect of the dip angles on the stability of the surrounding rock. To achieve good control effect, the fracture zone width was kept at 18 m, the deterioration levels of the fracture zone were maintained at 400 MPa in elastic modulus, the Poisson ratio was 0.35, and the cohesion was 0.5 MPa. The schematic representation of the numerical model considering varying dip angles is illustrated in Figure 8. The angle θ between the fracture zone and the tunnel axis ranges from 45° to 135°. Table 4 lists the simulation schemes for seven different fracture zone dip angles.

Figure 8. Numerical model considering different dip angles of fracture zone.

Table 4. Simulation schemes for different dip angles of the fracture zone.

Simulation Scheme	1	2	3	4	5	6	7
Dip angle $\theta/°$	45	60	75	90	105	120	135

5.2. Displacement Pattern in Tunnels under Different Dip Angles of Fracture Zone

The maximum displacement and corresponding locations at the tunnel vault, invert, and haunch were obtained by conducting numerical simulations under seven different dip angles of the fracture zones (as listed in Table 4). The numerical results are summarized in Table 5. Same as above (see Section 3.2), to evaluate the surrounding rock deformation within the fracture zone, the variation profiles of the maximum displacement of the tunnel within the fracture zones and adjacent regions are shown in Figure 9.

Table 5. Maximum displacement of tunnel under different dip angles of fracture zone.

Simulation Scheme	Tunnel Crown/mm	Location/m	Tunnel Invert/mm	Location/m	Tunnel Haunch/mm	Location/m
1	−6.500	Y = 64	7.401	Y = 52	6.854	Y = 59
2	−6.359	Y = 62	7.318	Y = 53	6.752	Y = 59
3	−6.273	Y = 61	6.886	Y = 56	6.623	Y = 59
4	−6.252	Y = 61	6.732	Y = 62	6.536	Y = 59
5	−6.271	Y = 59	6.979	Y = 65	6.569	Y = 59
6	−6.332	Y = 58	7.443	Y = 67	6.651	Y = 59
7	−6.402	Y = 57	7.509	Y = 68	6.779	Y = 59

In comparison with the effect of the deterioration levels of the fracture zone (Table 3, Figures 4 and 5), the dip angles of the fracture zone do not significantly affect the surrounding rock displacement at the tunnel vault, invert, and haunch (Table 5 and Figure 9). However, the variation tendencies are still obvious. Table 5 shows that as the dip angle gradually increase from 45° to 135°, a foregoing decrease and subsequent increase in deformation displacements are observed at the tunnel vault, invert, and haunch. The deformation displacements are minimal when the dip angle of the fracture zone is 90°, the minimum of which are 6.25 mm, 6.73 mm, and 6.54 mm for the tunnel vault, invert, and haunch, respectively. Unlike the variation profiles observed in varying deterioration levels of the fracture zone (Figures 4 and 5), the locations of the maximum vertical displacements at the tunnel vault and invert change with dip angles of fracture zone. However, the maximum horizontal displacement of the tunnel haunch is consistently located at Y = 59 m.

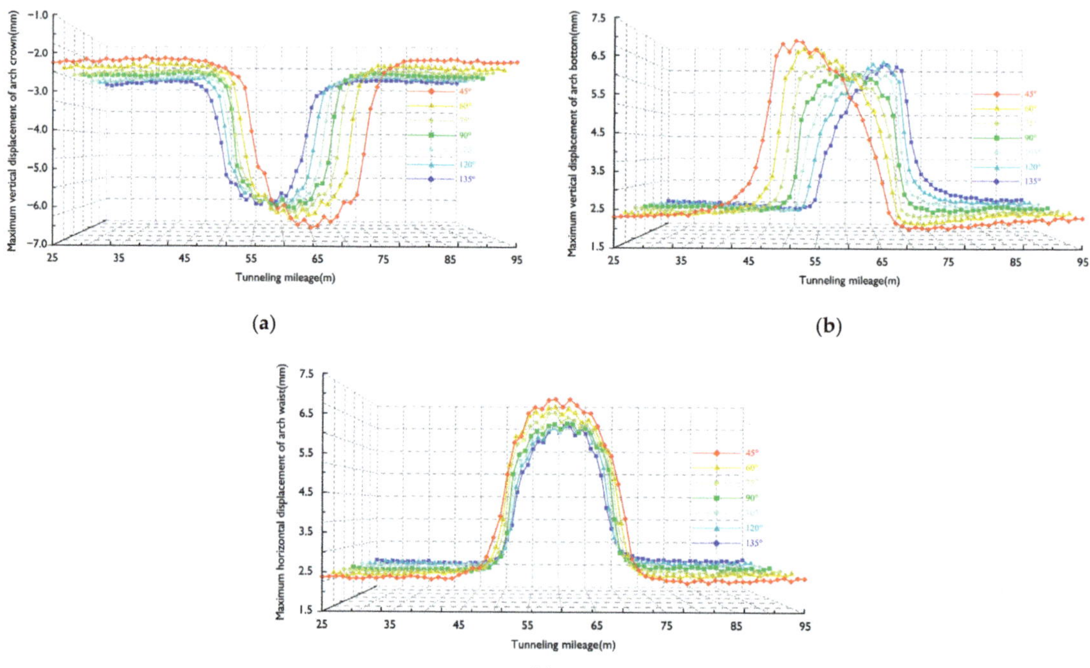

Figure 9. Results of tunnel surrounding rock displacement analysis under different dip angles of fracture zone. (**a**) Maximum vertical displacement of tunnel vault. (**b**) Maximum vertical displacement of tunnel invert. (**c**) Maximum horizontal displacement of tunnel haunch.

As shown in Figure 9, the magnitudes of displacements at the tunnel vault, invert, and haunch in fracture zones are notably higher than those in non-fracture zones. The tunnel vault's settlement and the tunnel invert's heave are closely related to the dip angle of the fracture zone. Surrounding rock deformation occurs preferentially when the excavation surface initially enters the fracture zone (vault or invert). The maximum displacement first decreases and then increases with the increase in the dip angle of the fracture zone. The effect of the dip angle of the fracture zone is more pronounced when the dip angle is acute than obtuse. In addition, it should be noted that the dip angle of the fracture zone does not affect the deformation of tunnel haunch.

5.3. Variation Patterns of the Maximum Shear Stress in the Tunnel Surrounding Rock with Different Dip Angles

In order to analyze the effect of dip angles on the tunnel surrounding rock of a double-shield TBM, seven simulation schemes with different dip angles were conducted (listed in Table 5). The numerical results are shown in Figure 10, which displays the variation patterns of the maximum shear stress in the surrounding rock under different dip angles of fracture zone.

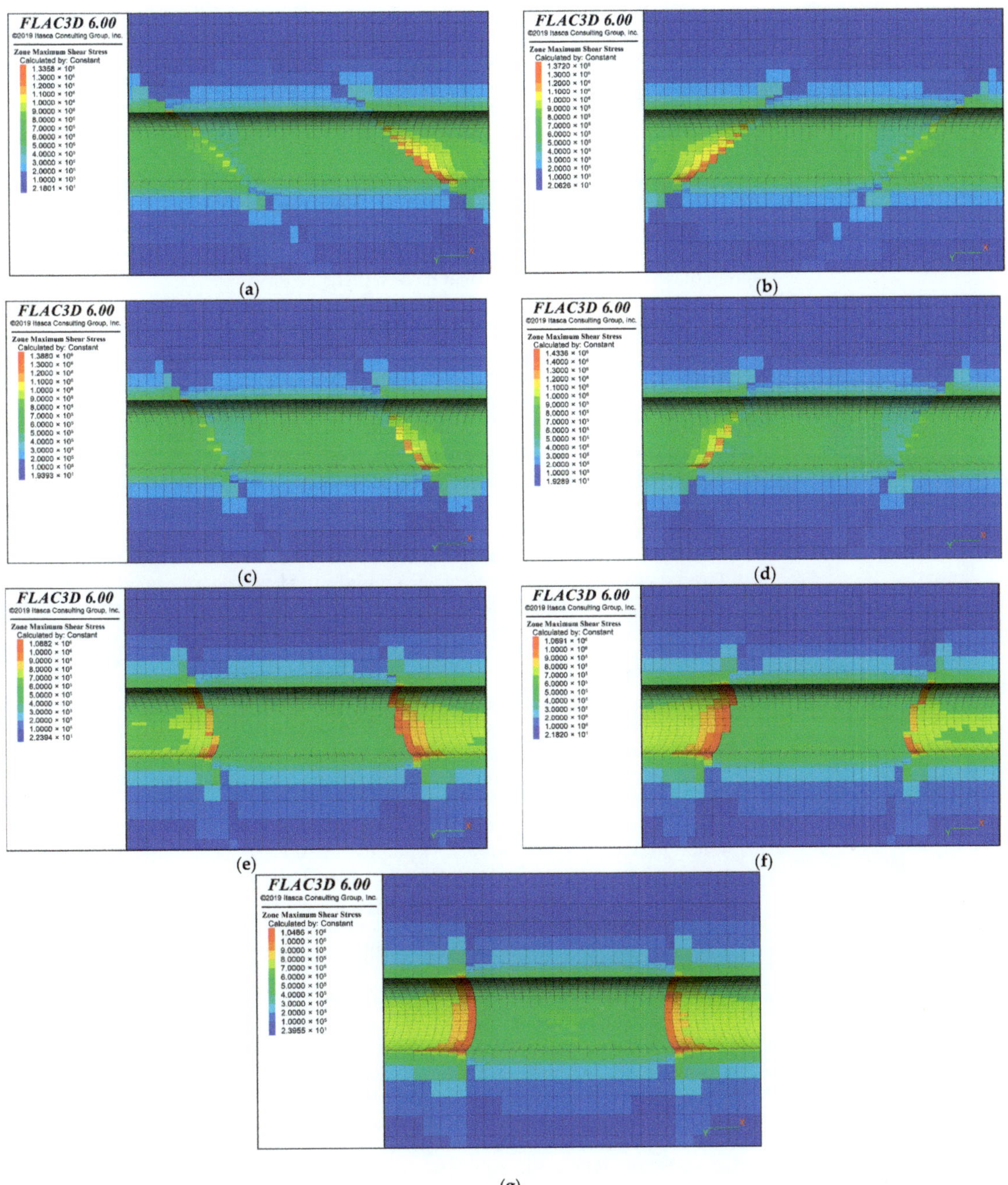

Figure 10. Maximum shear stress patterns in the surrounding rock with different dip angles of fracture zone. (**a**) dip angle 45°. (**b**) dip angle 135°. (**c**) dip angle 60°. (**d**) dip angle 120°. (**e**) dip angle 75°. (**f**) dip angle 105°. (**g**) dip angle 90°.

As shown in Figure 10, the locations of the maximum shear stress in the surrounding rock are all located at the tunnel invert on the edge of the fracture zone, regardless of the dip angle of the fracture zone. This suggests that the edge of the fracture zone is the most critical area for potential damage to the tunnel lining. In addition, it is observed that the magnitudes of shear stresses are higher under obtuse dip angles than that under acute angles. The corresponding magnitudes of the maximum shear stresses were 1.34 MPa and 1.36 MPa, 1.09 MPa, 1.05 MPa, 1.10 MPa, 1.43 MPa, and 1.37 MPa, respectively. This reveals that the maximum shear stress first decreases and then increases with the increase in the dip angle of the fracture zone. The magnitude of the maximum shear stress of the tunnel surrounding rock is the minimum when the dip angle of the fracture zone is 90°. This implies that tunnels crossing through fracture zones with obtuse dip angles are more vulnerable to shear stress failure than those with acute dip angles [26].

6. Discussion on Simplifying Assumptions for the Numerical Simulations

Based on actual engineering project, this study employs numerical simulations to investigate the deformation and stress variation patterns of tunnel surrounding rock and offers theoretical guidance for designing and constructing tunnels when double-shield TBM tunneling through fracture zone. It should be noted that the modeling process involves certain basic assumptions and simplifications, which may influence the research results:

(1) The initial stress field is assumed to be only affected by the weight of the surrounding rock itself. However, in actual situation, the stress field of underground rocks is affected by various factors, such as tectonic movement, geological history, and seismic activity. This assumption did not consider the influence of other factors and ignored factors such as rock displacement and groundwater level changes. In future research, a more realistic initial stress field model will be established based on actual observation data to improve the accuracy of the simulation results.

(2) The parameters of the grouting unit are based on the assumption that the grout has solidified. This assumption overlooks the grout's rheological properties and duration, which may not fully reflect the actual behavior of the material. In the future, more data on the grouting parameters will be obtained through laboratory experiments or field tests to better describe the grouting process.

(3) In this study, the Mohr–Coulomb strength criterion was used to describe rock and soil deformation. However, although this model has been widely used in many practical situations and proven effective in previous studies, its assumption cannot sufficiently reflect the complex behavior of rock and its variations in the deformation parameters, which is an aspect we need to improve further. Therefore, our future study will rely on obtaining more realistic material strength data through laboratory experiments and field tests. Additionally, new mechanical models will be developed to describe different types of rocks and strain conditions in more detail and accuracy.

7. Conclusions

In this paper, based on the urban subway construction section between Jiadingshan Road Station and Anshan Road Station of the Qingdao Metro Line 8, we employed numerical modeling to study tunnel surrounding rock's deformation and stress variations when TBMs tunnel through fracture zones. A comprehensive stability analysis was conducted to investigate the variation patterns of the deformation displacements and the maximum shear stresses in the tunnel surrounding rock of a double-shield TBM under different deterioration levels and dip angles of the fracture zone. Overall, in terms of the deterioration level in the fracture zone, the degree of fragmentation significantly impacts the stability of a double-shield TBM tunnel. The other indicator, the width, has less of an impact on it. The fracture zone's dip significantly affects specific sections of the tunnel. The detailed conclusions are as follows:

(1) The deterioration level of fracture zone significantly influences the stability of a double-shield TBM tunnel. With the increase in the degree of the fracture zone fragmentation,

the magnitudes of the maximum displacements at the tunnel vault, invert, and haunch increase significantly. Under constant deterioration level, as the fracture zone width increases, the deformation displacements of the surrounding rock at the tunnel vault, invert, and haunch increase slightly, and the scope of the influence of the fracture zone deformation increases.

(2) It is observed that the locations of the maximum shear stress in the surrounding rock are all located at the tunnel invert when the double-shield TBM tunnels through the fracture zone. Furthermore, it is found that the fracture zone width has a minimal impact on the maximum shear stress in the tunnel surrounding rock. As the degree of fragmentation increases, the maximum shear stress of the surrounding rock increases.

(3) The numerical results reveal that the maximum vertical displacements at the tunnel vault and invert exhibit a foregoing decrease and subsequent increase as the dip angle of the fracture zone increases. However, the dip angle of the fracture zone has little influence on the horizontal displacement at the tunnel haunch.

(4) The maximum shear stress in the surrounding rock exhibits a foregoing decrease and subsequent increase as the dip angle of the fracture zone increases. The shear stress is minimal when the dip angle of the fracture zone is 90°. In addition, it is observed that the magnitudes of shear stresses under obtuse dip angles are higher than that under acute angles.

Based on actual engineering project, this study employs numerical simulations to investigate the deformation and stress variation patterns of tunnel surrounding rock and offers theoretical guidance for designing and constructing tunnels when double-shield TBMs tunnel through fracture zone. It should be noted that the modeling process involves certain basic assumptions and simplifications, which may influence the research results. Furthermore, the analysis of stress and deformation in the surrounding rock did not consider the influence of groundwater on the fracture zone. Therefore, a coupled flow-solid analysis of the fracture zone will be conducted in future research works.

Author Contributions: Conceptualization, M.J., M.L. and X.W.; methodology, M.J. and M.L.; field test, M.J. and H.T.; data curation, M.J. and H.T; data analysis, M.J. and H.T.; writing—original draft preparation, M.J. and M.L.; writing—review and editing, X.W., M.D. and D.W.; supervision, X.W. and D.W.; project administration, M.J., X.W., D.W. and M.D.; funding acquisition, M.J., X.W., D.W. and M.D. All authors have read and agreed to the published version of the manuscript.

Funding: This research was financially supported by the National Natural Science Foundation of China (grant no.: 52108371 and grant no.: 52204115), and the Shield/TBM Construction Risk Consultation Project in Qingdao Metro Line 8.

Institutional Review Board Statement: Not applicable.

Informed Consent Statement: Not applicable.

Data Availability Statement: All data included in this study are available upon request by contact with the corresponding author.

Acknowledgments: Thanks are due to Guangzhao Zhang (Qingdao University of Technology, Qingdao, China) for assistance with the numerical simulation, and Zelin Lu (Qingdao University of Technology, Qingdao, China) for valuable advice.

Conflicts of Interest: The authors declare that there is no conflict of interest.

References

1. Peng, Y.; Wu, L.; Zuo, Q.; Chen, C.; Hao, Y. Risk Assessment of Water Inrush in Tunnel through Water-Rich Fault Based on AHP-Cloud Model. *Geomat. Nat. Hazards. Risk* **2020**, *11*, 301–317. [CrossRef]
2. Xue, Y.; Li, Z.; Li, S.; Qiu, D.; Su, M.; Xu, Z.; Zhou, B.; Tao, Y. Water Inrush Risk Assessment for an Undersea Tunnel Crossing a Fault: An Analytical Model. *Mar. Georesour. Geotechnol.* **2019**, *37*, 816–827. [CrossRef]
3. Newman, T.; Hueso, O.; Martinez Goirigolzarri, M. Effects of Changing Geology on the Performance of a Thames Tideway Tunnel Boring Machine. *Proc. Inst. Civ. Eng.-Geotech. Eng.* **2022**, 1–14. [CrossRef]

4. Liu, B.; Guo, Q.; Liu, Z.; Wang, C.; Nie, L.; Xu, X.; Chen, L. Comprehensive Ahead Prospecting for Hard Rock TBM Tunneling in Complex Limestone Geology: A Case Study in Jilin, China. *Tunn. Undergr. Space Technol.* **2019**, *93*, 103045. [CrossRef]

5. Li, S.; Nie, L.; Liu, B. The Practice of Forward Prospecting of Adverse Geology Applied to Hard Rock TBM Tunnel Construction: The Case of the Songhua River Water Conveyance Project in the Middle of Jilin Province. *Engineering* **2018**, *4*, 131–137. [CrossRef]

6. Bayati, M.; Hamidi, J.K. A Case Study on TBM Tunnelling in Fault Zones and Lessons Learned from Ground Improvement. *Tunn. Undergr. Space Technol.* **2017**, *63*, 162–170. [CrossRef]

7. Lin, P.; Yu, T.; Xu, Z.; Shao, R.; Wang, W. Geochemical, Mineralogical, and Microstructural Characteristics of Fault Rocks and Their Impact on TBM Jamming: A Case Study. *Bull. Eng. Geol. Environ.* **2022**, *81*, 64. [CrossRef]

8. Park, J.; Ryu, J.; Choi, H.; Lee, I.-M. Risky Ground Prediction Ahead of Mechanized Tunnel Face Using Electrical Methods: Laboratory Tests. *Ksce, J. Civ. Eng.* **2018**, *22*, 3663–3675. [CrossRef]

9. Huang, J.; Wei, X.; Luo, Y.; Gong, H.; Liu, T.; Li, X. Analysis of the Deformation Characteristics of the Surrounding Rock Mass of a Deep Tunnel During Excavation Through a Fracture Zone. *Rock Mech. Rock Eng.* **2022**, *55*, 7817–7835. [CrossRef]

10. Xie, Q.; Cao, Z.; Sun, W.; Fumagalli, A.; Fu, X.; Wu, Z.; Wu, K. Numerical Simulation of the Fluid-Solid Coupling Mechanism of Water and Mud Inrush in a Water-Rich Fault Tunnel. *Tunn. Undergr. Space Technol.* **2023**, *131*, 104796. [CrossRef]

11. Zaheri, M.; Ranjbarnia, M.; Dias, D.; Oreste, P. Performance of Segmental and Shotcrete Linings in Shallow Tunnels Crossing a Transverse Strike-Slip Faulting. *Transp. Geotech.* **2020**, *23*, 100333. [CrossRef]

12. Do, N.-A.; Dias, D.; Oreste, P.; Djeran-Maigre, I. Behaviour of Segmental Tunnel Linings under Seismic Loads Studied with the Hyperstatic Reaction Method. *Soil Dyn. Earthq. Eng.* **2015**, *79*, 108–117. [CrossRef]

13. Zhou, Z.; Chen, Z.; Wang, B.; Jiang, C.; Li, T.; Meng, W. Study on the Applicability of Various In-Situ Stress Inversion Methods and Their Application on Sinistral Strike-Slip Faults. *Rock Mech. Rock Eng.* **2023**. [CrossRef]

14. Han, K.; Wang, L.; Su, D.; Hong, C.; Chen, X.; Lin, X.-T. An Analytical Model for Face Stability of Tunnels Traversing the Fault Fracture Zone with High Hydraulic Pressure. *Comput. Geotech.* **2021**, *140*, 104467. [CrossRef]

15. Li, L.; Han, Y.; Wang, J.; Jin, Q.; Xiong, Y.; Chong, J.; Ba, X.; Fang, Z.; Wang, K. Study on Critical Safety Distance Between the Shield Tunnel and Front Fault Fracture Zone in Urban Metro. *Geotech. Geol. Eng.* **2022**, *40*, 5667–5683. [CrossRef]

16. Wang, S.; Li, C.; Li, D.; Zhang, Y.; Hagan, P. Skewed pressure characteristics induced by step-by-step excavation of a double-arch tunnel based on infrared thermography. *Tech. Vjesn.* **2016**, *23*, 827–833. [CrossRef]

17. Wang, S.; Li, D.; Li, C.; Zhang, C.; Zhang, Y. Thermal Radiation Characteristics of Stress Evolution of a Circular Tunnel Excavation under Different Confining Pressures. *Tunn. Undergr. Space Technol.* **2018**, *78*, 76–83. [CrossRef]

18. Fan, H.; Li, L.; Chen, G.; Liu, H.; Gao, J.; Li, C.; Peng, X.; Zhou, S. Analysis Method of the Water Inrush and Collapse in Jointed Rock Mass Tunnels: A Case Study. *Eng. Anal. Bound. Elem.* **2023**, *146*, 838–850. [CrossRef]

19. Bao, X.; Xia, Z.; Ye, G.; Fu, Y.; Su, D. Numerical Analysis on the Seismic Behavior of a Large Metro Subway Tunnel in Liquefiable Ground. *Tunn. Undergr. Space Technol.* **2017**, *66*, 91–106. [CrossRef]

20. Anato, N.J.; Assogba, O.C.; Tang, A.; Youssouf, D. Numerical Investigation of Seismic Isolation Layer Performance for Tunnel Lining in Shanghai Soft Ground. *Arab. J. Sci. Eng.* **2021**, *46*, 11355–11372. [CrossRef]

21. Chen, J.; Liu, L.; Zeng, B.; Tao, K.; Zhang, C.; Zhao, H.; Li, D.; Zhang, J. A Constitutive Model to Reveal the Anchorage Mechanism of Fully Bonded Bolts. *Rock Mech. Rock Eng.* **2023**, *56*, 1739–1757. [CrossRef]

22. Luo, T.; Wang, S.; Zhang, C.; Liu, X. Parameters Deterioration Rules of Surrounding Rock for Deep Tunnel Excavation Based on Unloading Effect. *Dyna* **2017**, *92*, 648–654. [CrossRef]

23. Wang, L.; Hong, K.; Quan, Y.; Zhao, H.; Zhao, X.; Yang, Z.; Zhou, Z. Model Test of TBM Tunnel Crossing Large Dip Angle Fault Zone under High In Situ Stress. *Geofluids* **2022**, *2022*, 9735532. [CrossRef]

24. Abdollahi, M.S.; Najafi, M.; Bafghi, A.Y.; Marji, M.F. A 3D Numerical Model to Determine Suitable Reinforcement Strategies for Passing TBM through a Fault Zone, a Case Study: Safaroud Water Transmission Tunnel, Iran. *Tunn. Undergr. Space Technol.* **2019**, *88*, 186–199. [CrossRef]

25. Li, H.; Li, X.; Yang, Y.; Liu, Y.; Ma, M. Structural Stress Characteristics and Joint Deformation of Shield Tunnels Crossing Active Faults. *Appl. Sci.* **2022**, *12*, 3229. [CrossRef]

26. Zheng, Y.; Wu, K.; Jiang, Y.; Chen, R.; Duan, J. Optimization and Design of Pre-Reinforcement for a Subsea Tunnel Crossing a Fault Fracture Zone. *Mar. Georesour. Geotechnol.* **2023**, *41*, 36–53. [CrossRef]

Article

Nonlinear Finite Element Analysis of a Composite Joint with a Blind Bolt and T-stub

Jincheng Hua [1,*], Xinwu Wang [2], Huanhuan Liu [3,*] and Haisu Sun [2]

1   School of Civil Engineering, Henan University of Science and Technology, Luoyang 471032, China
2   Henan International Joint Laboratory of New Civil Engineering Structures, Luoyang University of Science and Technology, Luoyang 471023, China
3   School of Civil Engineering and Architecture, Wuhan University of Technology, Wuhan 430070, China
*   Correspondence: hjc990516@163.com (J.H.); lhhzsc@163.com (H.L.)

Abstract: A detailed nonlinear finite element model was established based on completed experiments to investigate the behavior of a blind-bolted T-stub composite joint that connects a composite beam to a concrete-filled square tube column. This was accomplished by comparing the experimental results and the finite element simulation results using the hysteresis curve, failure mode, plastic deformation and strain development of the T-stub to ensure the reliability and accuracy of the finite element model. A parametric study was carried out on the base model to expand the library of test data. It was observed from the comparison that the proposed nonlinear FE model predicted the behavior of the composite joint. The wall thickness of the column and reinforcement ratio had a significant influence on the ultimate bending moment of the composite joint and the performance of the composite joint was mainly controlled by the reinforcement ratio when the concrete slab was under a positive bending moment. The flange of the T-stub, the web of the T-stub and the axial compression ratio had little effect on the performance of the composite joint.

Keywords: blind bolt; concrete-filled steel tubular column; T-stub; composite joint; finite element analysis

Citation: Hua, J.; Wang, X.; Liu, H.; Sun, H. Nonlinear Finite Element Analysis of a Composite Joint with a Blind Bolt and T-stub. *Sustainability* **2023**, *15*, 4790. https://doi.org/10.3390/su15064790

Academic Editors: Hong-Wei Yang, Shuren Wang and Chen Cao

Received: 8 February 2023
Revised: 23 February 2023
Accepted: 7 March 2023
Published: 8 March 2023

1. Introduction

The damage investigation of the Beiling earthquake in the United States in 1994 and the Kobe earthquake in Japan in 1995 showed that the welded rigid joints of the steel-frame beams and columns were seriously damaged, while semi-rigid joints formed by bolted connections were relatively rare. This led researchers to re-recognize beam–column connection joints and to conduct a large amount of research on semi-rigid joints composed of connectors (such as end plates, T-shaped steel, angle steel, etc.) and bolts. For instance, J. Lee [1] carried out monotonic loading tests on square steel tubular columns connected by one-sided bolted T-shaped steel, established a three-dimensional finite element model, compared the test results with the finite element simulation results and evaluated the joints in accordance with European specifications. This shows that this type of node is a semi-rigid connection. Furthermore, Wang Jingfeng [2] carried out four full-scale model tests of concrete-filled steel tubular columns with single-sided bolted end-plate connections, established a reliable finite element model and studied the influence of multiple factors on the moment-bearing capacity and rotational stiffness of composite joints by changing parameters. H. T. Thai [3] studied the performance of single-sided bolted end-plate connections between concrete-filled steel tubular columns and steel–concrete composite beams through finite element software and studied the performance of joints with shear studs and the reinforcement ratio by changing the parameters.

At present, there are three problems in the research on most semi-rigid joints. First, steel columns are mostly H-shaped and the research on steel columns with closed sections, such as square steel-tube columns, is less developed. When faced with steel columns with closed sections, such as square steel-tube columns [4], it is necessary to weld or install hand holes on the column wall, which greatly reduces the construction efficiency. Furthermore, the welding quality is difficult to ensure, as it is affected by many factors and the seismic performance of the joints is uneven [5]. In recent years, the defects caused by welding have been overcome with the development of unilateral high-strength bolts [1]. Blind high-strength bolts only need to be installed from one side, which is convenient for construction and solves the problem that ordinary high-strength bolts are difficult to apply to closed sections, such as square steel-pipe columns. A second problem is that, in the selection of connectors, domestic and foreign scholars seem to prefer to use end-plate connections [6] and the research on the use of T-shaped steel as the connector is relatively undeveloped. Finally, most of the tests did not consider the combined effect of concrete floor slabs. As the main load-bearing member, the reinforcement in the slab has a great impact on the seismic performance of the joint [3].

In order to expand the test database on semi-rigid joints with the use of T-shaped steel as the connector and launch the corresponding design standards, this paper establishes a refined three-dimensional finite element model of single-sided bolted T-shaped steel connection square steel tubular columns and steel–concrete composite beam joints based on completed tests and verifies the reliability and accuracy of the finite element simulation results from multiple perspectives with the test results. Based on the parameters of the basic model, the influence of the T-steel web, T-steel flange, square steel tubular column wall, axial compression ratio and reinforcement ratio on the seismic performance of this type of composite joint are analyzed.

2. Experimental Program

2.1. Test Specimens

The joint specimens in the test were selected from the edge column joints of the typical eight-layer steel frame structure. The main design parameters are shown in Figure 1a. The T-stub was used in the connector; the specific size of T-stub is shown in Figure 1b. In the process of its assembly, the T-stub steel was connected with the square steel-tube column through the ordinary 10.9 friction high-strength bolts, after which the beam was connected with the T-stub steel through the nested single-sided bolts, the diameters of which were 16 mm. The steel tube column featured internal partition (all the stiffeners and internal partitions in the specimen were 10 mm) and the diameter of the holes in the plate was 100 mm for the pouring of core concrete. The width of the concrete floor was 1000 mm, the thickness of the concrete protective layer was 15 mm and the double-layer bidirectional reinforcement is adopted. The HRB335, rebar measured 8 mm in diameter and 150 mm in spacing. According to the partial shear design, the round-headed shear bolts, which had diameters of 16 mm, heights of 85 mm and spacing of 150 mm, were arranged along the center line of the beam. The material properties of steel are shown in Table 1. The measured average compressive strength of concrete cube was 34.7 Mpa and the elastic modulus was 28,514 N/mm^2.

Table 1. Material test results.

Component	T mm	f_y MPa	f_u MPa	E GPa	A (%)
Column	10	345	491	209	30
web of beam	6	280	442	196	35
flange of beam	9	252	440	199	32
web of T-stub	9	271	447	227	35
flange of T-stub	14	268	447	197	33

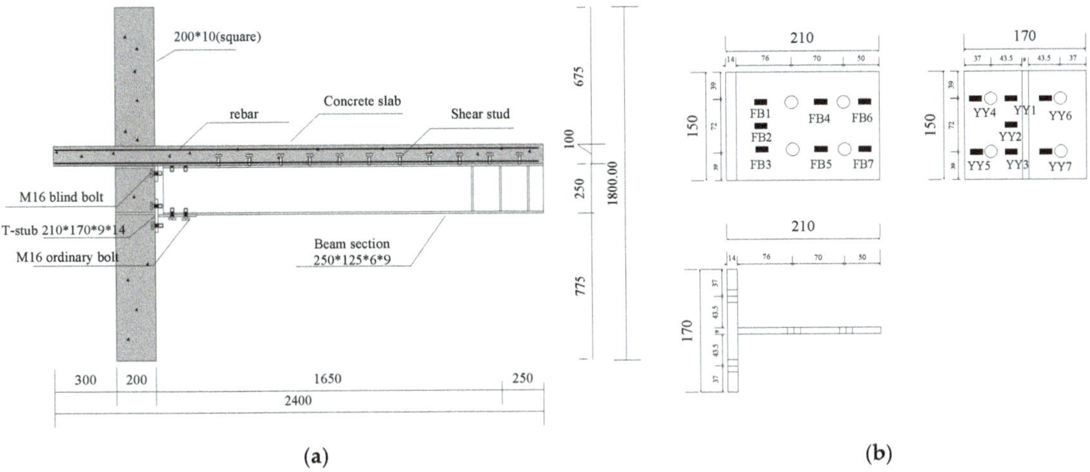

(**a**)   (**b**)

Figure 1. Parameters of composite joint. (**a**) design parameters. (**b**) specific size of T-stub.

2.2. Loading Device and Loading System

The schematic diagram of the test model and the loading site are shown in Figures 2 and 3, respectively. The bottom of the square steel-tube column of the specimen was connected with unidirectional hinge support; in turn, the hinge support was connected to the rigid ground through anchor bolts. The top of the column and the end of the beam connected with the 2000-kilonewton hydraulic servo actuator and the 1000-kilonewton hydraulic servo actuator, respectively. During the loading test, the hydraulic servo actuator, with a capacity of 2000 kN, was arranged to apply vertical load with an axial compression ratio of 0.25, which remained constant throughout the test. The hydraulic servo actuator, with a capacity of 1000 kN, was used to apply the load to each beam end in the vertical direction.

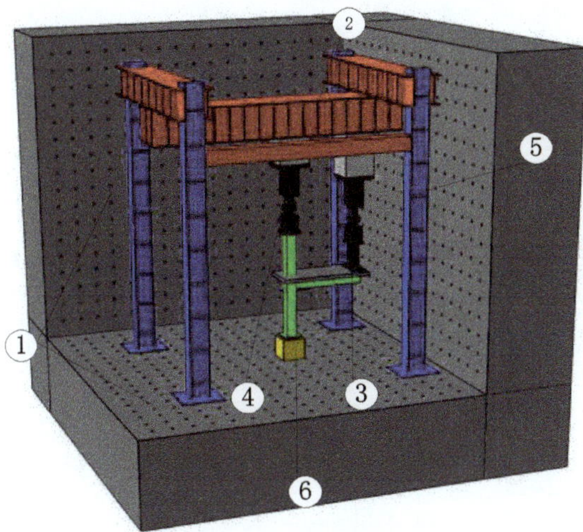

Figure 2. Test model. Note. 1. Rigid foundation and reaction wall; 2. beam of reaction frame; 3. hydraulic servo actuator of 1000 kN capacity; 4. hydraulic servo actuator of 2000 kN capacity; 5. column of reaction frame; 6. unidirectional hinge support.

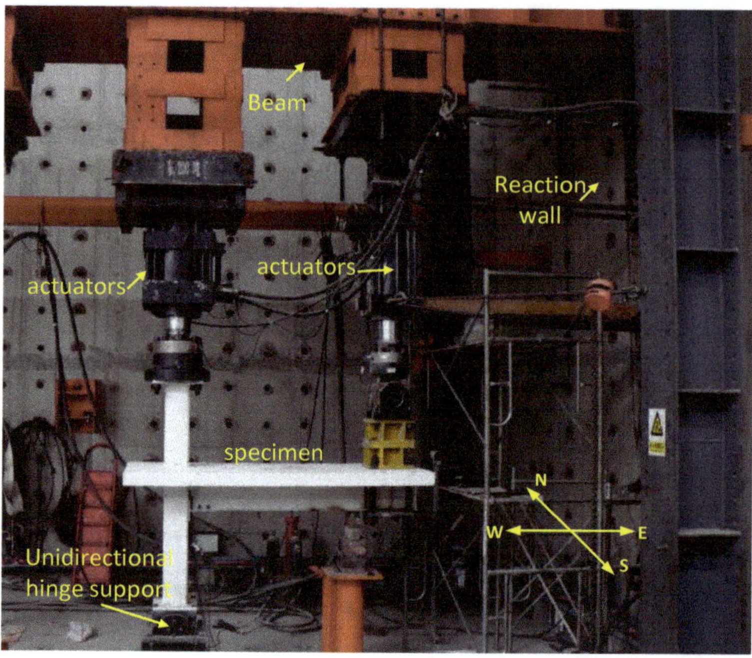

Figure 3. Loading site.

The loading system featured displacement control. Before the specimen yielded, one-thousandth of the beam length, namely 1.65 mm, was taken for each stage and each stage was cycled once. When the component reached the yield strain, which was measured in the material test, the specimen was considered to yield and the displacement at this time was defined as yield displacement Δ_y. In order to prevent the rapid development of concrete cracks, the specimens were loaded at times of yielding displacement, i.e., $1\Delta_y$, $1.5\Delta_y$, $2\Delta_y\cdots$. Each level was cycled three times. The hydraulic servo actuator protrusion was positive and the indentation was negative. When the load dropped to 85% of the ultimate load or the component failed, the test ended.

3. Nonlinear Finite Element Model

3.1. Material Modeling

The double-line strengthening elastic–plastic model and two flow plastic models are mainly used in the material modeling of steel in finite element programming. Relevant studies [7,8] demonstrate that the two material models can reflect the actual situation of steel. For the steel used in this paper, we adopted the double-line strengthening elastic–plastic model. In the test's preparation stage, a section was cut from the same batch of steel for processing the test piece; the research institute was entrusted to carry out the material performance test and the material parameters of each component of the finite element were defined according to the material properties measured by the material performance test. The elastic modulus of steel used for each component was 2.06×105 MPa and Poisson's ratio was 0.3. For all of the steels, we adopted von Mises' yield criterion and an isotropic hardening rule [9]. Figure 4 shows the stress–strain curves of each steel material.

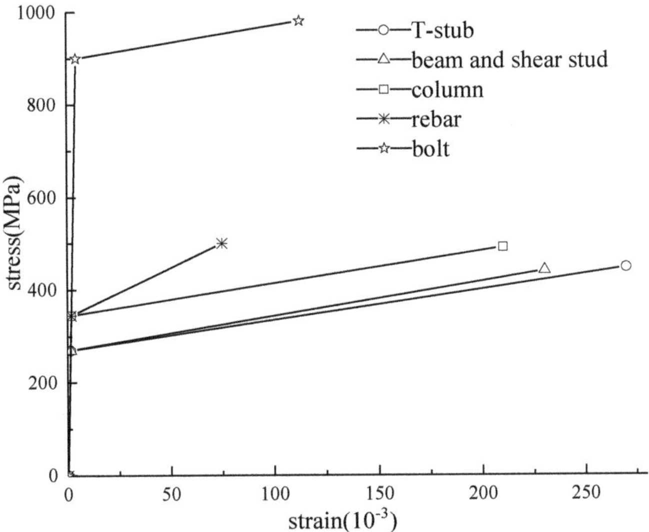

Figure 4. Stress-strain curves for steel materials.

The smeared cracking model, brittle cracking model and damaged-plasticity model are given in ABAQUS for concrete simulation. In this paper, the damaged-plasticity model was used to simulate the concrete performance, which made it possible to input a multilinear uniaxial compression stress–strain curve and provided a universal capability for the analysis of concrete under monotonic or cyclic loading [9,10], based on a damage-plasticity algorithm. The stiffness–recovery coefficients $\omega_t = 0$ and $\omega_c = 0.25$ were introduced, where ω_t and ω_c control the stiffness recovery from pressing to pulling and from pulling to pressing, respectively.

The concrete in the model featured the stress–strain relation curve shown in Figure 5, according to the following formulae:

$$\sigma = (1 - d_c)E_c\varepsilon \tag{1}$$

$$d_c = 1 - \frac{\rho_c n}{n - 1 + x^n}x \leq 1 \tag{2}$$

$$d_c = 1 - \frac{\rho_c}{\alpha_c(x - 1)^2 + x}x > 1 \tag{3}$$

$$\rho_c = \frac{f_{c,r}}{E_c\varepsilon_{c,r}} \tag{4}$$

$$n = \frac{f_{c,r}}{E_c\varepsilon_{c,r} - f_{c,r}} \tag{5}$$

$$x = \frac{\varepsilon}{\varepsilon_{c,r}} \tag{6}$$

where α_c is the parameter value of the descending section of the stress–strain curve of concrete under uniaxial compression, which was taken according to Table C.2.4 in Appendix C of the *Code for Design of Concrete*; $f_{c,r}$ is representative value of uniaxial compressive strength of concrete, which can be taken as f according to actual structural analysis needs f_c, f_{ck}, f_{cm}; $\varepsilon_{c,r}$—The corresponding peak compressive strain of concrete relevant to uniaxial compressive strength, $f_{c,r}$, is taken according to Table C.2.4 in Appendix C of *Code for Design of Concrete*; d_c is damage-evolution parameter of concrete under uniaxial compression. This was calculated from Appendix C of the *Concrete Design Code* [11]. It should be noted that since the main objective of this paper is to study the seismic performance of semi-rigid composite joints with the T-stub as connector, the core concrete in the steel tube does not introduce the concrete stress–strain curve under triaxial compression. In addition, since the age of concrete on the test day did not reach 28 days, the value given in Figure 5 is lower than the measured strength.

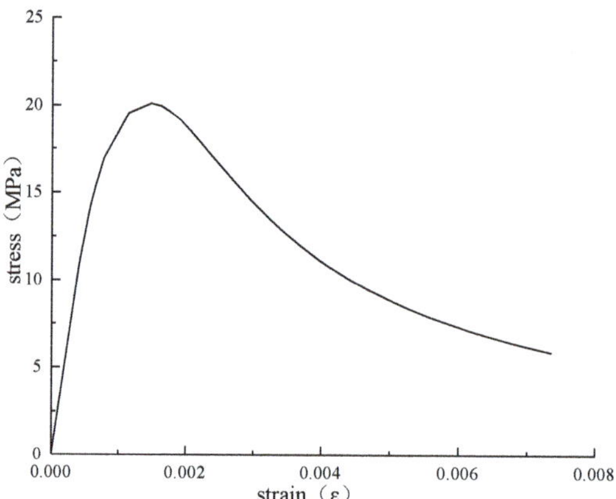

Figure 5. Stress–strain curves for concrete.

3.2. Finite Element Type and Mesh

In order to obtain reliable results, the finite element model requires the selection of appropriate element types. Except for the reinforcing bars, all components were modeled by eight-node solid elements (C3D8R), which prevented shear locking [12] or modeling the steel reinforcement; a two-node linear truss element (T3D2) was used.

The mesh design used the symmetry of the geometrical and mechanical properties of the T-stub-and-bolts model. Many attempts have been made in mesh-convergence studies to obtain a reasonable mesh that can obtain reliable results while reducing the computation time. Based on the mesh-convergence studies, the bolt-mesh seed measured 5 mm, the T-stub and shear-stud-mesh seed measured 7 mm and the rest of the components measured 35 mm. Appropriate mesh encryption was carried out for the contact parts. It should be noted that in the subsequent parameter analysis, due to the change in the component size, the mesh size was also fine-tuned to achieve better convergence. Figure 6 shows the overall mesh division of the components and Figure 7 shows the mesh of each component. The blind bolts and ordinary bolts were simplified and the nut, rod and head of the bolt were modelled as only one part [12,13].

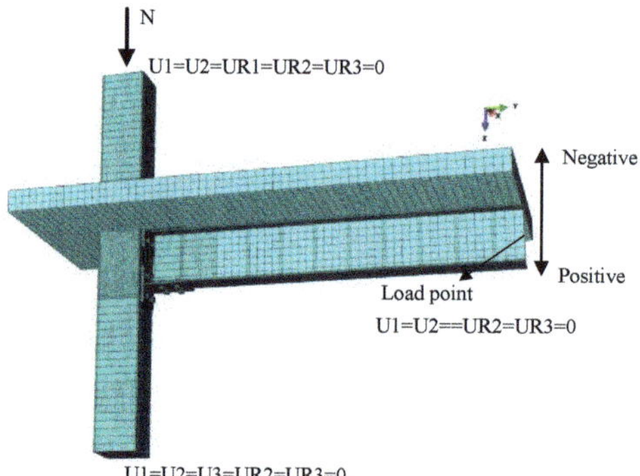

Figure 6. Three-dimensional FE model of composite joints.

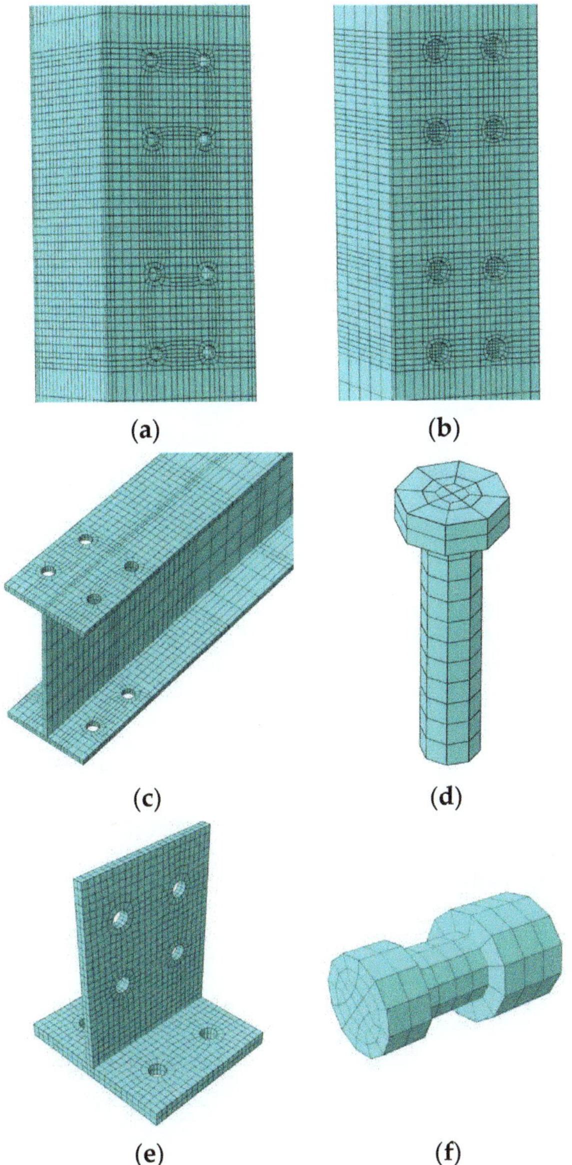

Figure 7. Three-dimensional idealization of the connection components. (**a**) Column; (**b**) core concrete; (**c**) beam; (**d**) shear stud; (**e**) T-stub; (**f**) bolt.

3.3. Contact Interaction

There were complex contact and interaction relationships in the composite joints and the results of the finite element analysis depended on the accurate modeling of the contact between these components. There were two main types of contact in this composite joint. The first was the contact between steel and concrete; the second was the contact between steel and steel. Surface contacts were established on two matched surfaces to simulate the contact between steel and concrete, using hard contact in the normal direction and the friction formula "penalty" in the tangential direction, with a friction coefficient of 0.5. The contact between steel and steel was achieved by creating a general contact in the initial analysis step. The friction coefficient is 0.3.

The column-top center, column-bottom center and beam-end section center were taken as the control points, while the column-top plane, column-bottom plane and beam-end I-shaped section were taken as the control surfaces. The interaction was established through coupling, so as achieve the imposition of boundary conditions and loading.

For the components embedded in the concrete slab during the test, such as parts of the T-stub, parts of the ordinary bolts and blind bolts, rebars and shear studs, these components were taken as the embedded region, the concrete slab was taken as the host region and the embedded region command was used to achieve the interaction between the above components and the concrete slab. This technique eliminated the translational degrees of freedom of the embedded nodes and made them correspond with those of the host element, so as to achieve the perfect combination between the components and the surrounding concrete [1,12,14–18].

3.4. Boundary Conditions and Load

In order to obtain better simulation results, the finite element model should adopt the same boundary conditions and loading regime as the test. Corresponding boundary conditions were applied to the top of the column, the bottom of the column and the end of the beam, as shown in Figure 6.

The application of the load was divided into six analysis steps: (1) a small bolt load (100 N) was applied to the bolt using the Bolt Load option in ABAQUS, so that each contact relationship could be established smoothly; (2) the bolt load was applied to the target value of 105 kN; (3) the preload of the bolt was changed to the fixed bolt length; (4) the axial compression was applied to the target value of 514 kN; (5) according to the yield displacement $\Delta_y = 18.6$ mm obtained on the test, the beam end was controlled by displacement and the vertical load was applied on the each beam end.

4. Comparison

4.1. Comparison of Curves

The comparison of the moment–rotation skeleton curve under cyclic load between the simulation results and the test results is shown in Figure 8. The moment–rotation hysteresis curve is shown in Figure 9.

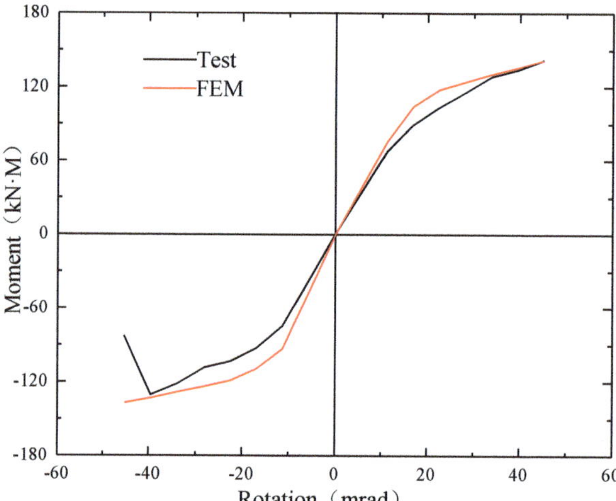

Figure 8. Comparison of moment–rotation skeleton curves.

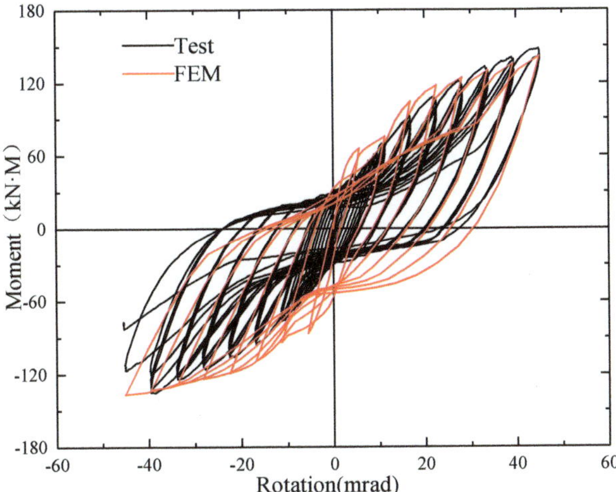

Figure 9. Comparison of moment–rotation hysteresis curves.

As can be seen from the comparison figure, there are some differences between the finite element simulation curve and the test curve, mainly in the following points:

(1) On the test, the curve of the specimen had an obvious descending section, but this phenomenon was not simulated with the finite element method.

(2) In the skeleton curve comparison diagram, the ultimate bending moment values are similar, but there are some differences when the positive and negative rotation are between 15 mrad and 40 mrad. There were some initial imperfections in the concrete slab on the test, but the concrete slab in the finite element was homogeneous; therefore, the test's bending moment value was lower than that of the finite element simulation value under the same rotation.

(3) It can be seen that the initial rotational stiffness was different in the comparison of the skeleton curves. In the finite element analysis, the size of the specimen was accurate, the bolt and the screw hole were aligned strictly according to the central axis and the initial imperfections of the components were ignored. Therefore, the initial rotational stiffness in the finite element analysis was relatively large.

(4) It can be seen from the hysteresis curve comparison diagram that the curve was relatively full in the positive direction, which was due to the fact that the effect of the bolt slip was not well-simulated in the finite element simulation.

4.2. Comparison of Failure Mode

As shown in Figure 10, along with the fracture of T-stub web and flange and the concrete drop from the concrete slab, the hysteresis curve showed an obvious downward section and the test ended. In the finite element analysis, the maximum stress of the T-stub almost reached the ultimate strength obtained in the material property test and the web presented slight local buckling, which was similar to the experimental phenomenon. This shows that the finite element simulation results reflect the failure mode of the specimen better.

4.3. Futher Comparison

Although the moment–rotation curve and failure mode of the test and the finite element had a high degree of similarity, this was not sufficient to fully verify the accuracy of the finite element model. The T-stub had obvious plastic deformation in the test. The evaluation of the deformation of the T-stub and the stress change in the key measurement points can be used to further verify the accuracy of the finite element model.

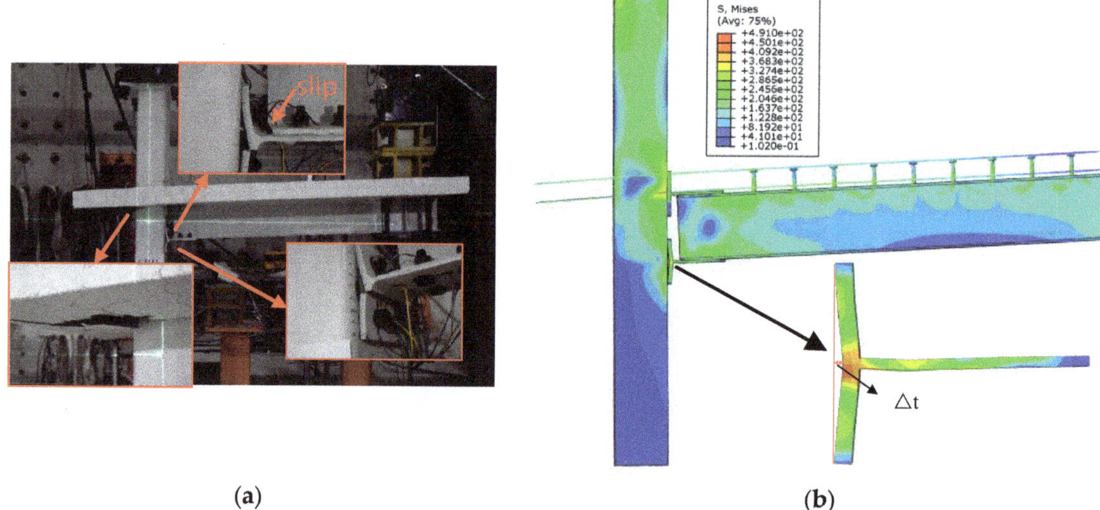

(a)　　　　　**(b)**

Figure 10. Comparison of failure model. (**a**) Failure mode. (**b**) Location of Δ_t.

The plastic deformation of the T-stub was mainly reflected in the pulling up of the flange and the web buckling. The gap width between the flange and the wall of the column, Δ_t, shown in Figure 10, was recorded during the test. In the finite element analysis, this value was the difference between the displacement deformation of the corresponding point of T-stub and the displacement deformation of the wall of the column. The relevant data were extracted and compared with the data recorded during the test. The comparison figure is shown in Figure 11, where the abscissa is the load level and the ordinate is Δ_t [14].

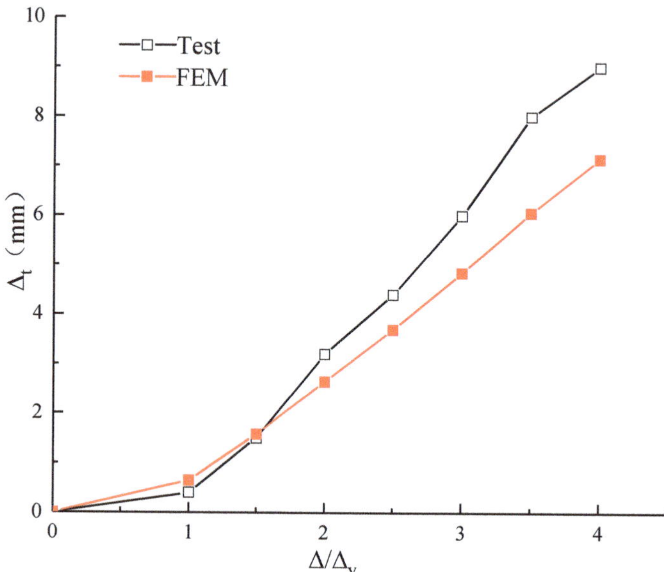

Figure 11. Comparison of Δ_t.

It can be seen from the figure that the gap width between the T-stub and the wall of the column was large on the test. The reason for this is that the actual thickness of the T-stub used in the test was less than 14 mm due to manufacturing errors and grinding during processing, which led to greater plastic deformation.

It was inevitable that some strain gauges would be damaged due to the large deformation of the components in the process of the test. Therefore, the complete data measured at the junction of the flange and the web of the T-stub (the measurement point YY3 shown in Figure 1 were selected to compare with the strain development data of the corresponding measurement points extracted through the finite element simulation results. The comparison figure is shown in Figure 12. Each stage was cycled three times after the specimen yield on the test, while each stage was cycled once in the finite element analysis.

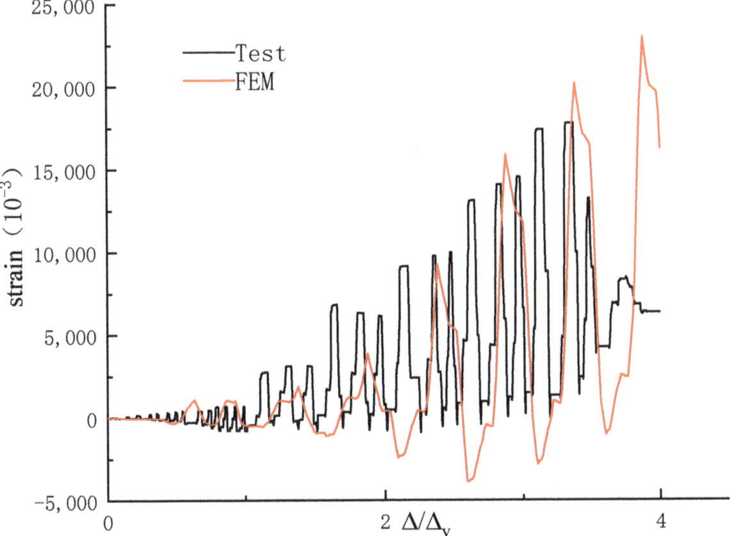

Figure 12. Comparison of strain.

There was almost no compressive strain in the middle and late stages of the test due to the residual deformation of the T-shaped steel on the test. However, this situation was not well-simulated in the finite element, resulting in a great difference in compressive-strain area. In general, the results of the finite element analysis better reflect the development of the strain on the test.

Table 2 shows the relevant indexes, their ratio on the test and the finite element simulation. Apart from the negative initial rotational stiffness, the simulation results of other indexes were similar.

In this section, it is shown that the finite element simulation results can better predict the relevant seismic performance of this type of composite joint after ignoring the initial imperfections and errors of size by evaluating the hysteresis curve, skeleton curve, failure mode, plastic deformation of the T-stub and strain at key measurement points.

Table 2. Comparison of indicator table.

	Test Value V_t	Finite Element Value V_f	Ratio V_t/V_f
yield rotation (mrad)	31.18	33.83	0.92
yield moment (kN·M)	124.38	130.59	0.95
positive ultimate bending moment (kN·M)	144.39	141.81	1.02
positive initial rotational stiffness (kN·M/rad)	6323.88	6747.94	0.94
negative initial rotational stiffness (kN·M/rad)	6788.09	8268.25	0.82
total energy consumption (KJ)	20.08	26.00	0.77
equivalent damping coefficient	0.162	0.157	1.03
ductility coefficient	1.45	1.33	1.09

5. Parameter Study

According to the finite element simulation results in Section III, the seismic performance of the semi-rigid composite joints with the use of the T-stub as the connector was studied more widely and the test database was expanded by changing the relevant key parameters in the base model. The main parameters to be changed are shown in Table 3, according to which the reinforcement ratio could be changed by changing the diameters of the rebar, which were 6 mm, 8 mm, 10 mm and 12 mm, respectively. The italics in the table represent the basic parameters and the main calculation results are shown in Table 4.

Table 3. Parameter range table.

Parameter Types	Parameters
web of T-stub (mm)	7 9 11 13
flange of T-stub (mm)	12 14 16 18
wall thickness of column (mm)	8 10 12 14
axial compression ratio	0.1 0.25 0.4 0.55
reinforcement ratio (%)	0.45 0.80 1.26 1.81

Table 4. Table of simulation results.

	Yield Rotation (mrad)	Yield Moment (kN·M)	Positive Ultimate Bending Moment (kN·M)	Negative Ultimate Bending Moment (kN·M)	Positive Initial Rotational Stiffness (kN·M/rad)	Negative Initial Rotational Stiffness (kN·M/rad)	Total Energy Consumption (KJ)	Equivalent Damping Coefficient	Ductility Coefficient	Δ_t (mm)
Base	33.83	130.59	136.64	141.1	6747.94	8268.25	26	0.157	1.33	7.15
FB7	32.68	118.28	132.51	126.75	6627.95	8159.15	25.24	0.159	1.38	5.54
FB11	34.63	136.51	139.81	147.79	6801.01	8396.85	25.86	0.154	1.30	7.51
FB13	34.6	140.81	144.14	151.06	6786.86	8351.28	26.63	0.153	1.30	7.36
YY12	34.42	129.7	127.56	138.98	6743.28	7973.69	25.62	0.163	1.31	8.49
YY16	32.68	129.76	142.06	141.69	6679.55	8443.75	26.89	0.159	1.38	6.65
YY18	33.04	130.51	143.41	144.22	6701	8597.01	26.44	0.154	1.36	5.48
ZB8	33.17	115.51	121.4	123.56	6516.128	8165.68	23.54	0.163	1.36	8.69
ZB12	34.85	146.44	148.27	156.67	7153.72	8790.48	28.57	0.160	1.29	6.46
ZB14	35.09	156.96	157.91	166.21	7587.34	9117.11	31	0.170	1.28	5.67
GJ6	32.47	118.13	134.76	129.54	6396.24	8061.32	26.3	0.168	1.39	7.16
GJ10	34.45	143.16	138	151.53	7153.08	8357.5	26.42	0.152	1.31	7.57
GJ12	/	/	/	/	7655.9	8521.7	/	/	/	
ZY1	33.54	130.7	136.21	140.53	6628.5	8268.69	25.73	0.159	1.34	7.32
ZY4	33.88	132.61	139.05	143.91	6680.46	8573.8	26.41	0.159	1.33	7.51
ZY55	34	131.56	135.18	141.81	6693.65	8240.14	26.12	0.159	1.33	7.52

Variations in the T-stub's size have a great influence on its plastic development. Therefore, the Δ_t between the flange of the T-stub and the wall of the column mentioned in the previous section were also compared as an index. In addition, the load corresponding to the rotation of 0.045 rad is taken as the ultimate strength, since the FEM curve has no descending section in the analysis [19].

5.1. Web Thickness of T-Stub

As shown in Table 4, the thickness of the webs had little influence on the performance of composite joints in the range where the web thickness of the T-stub is not less than the flange thickness of the beam (9 mm) in this set of models. The yield's bending moment and the positive and negative ultimate strength were significantly reduced when the web thickness was 7 mm. These were 90.6%, 93.4% and 89.8% of those of the base model, respectively.

The most obvious change was the local buckling degree in the web of the T-stub with the change in web thickness. Local buckling can be clearly seen in Figure 13, while there was no obvious local buckling in the web of the T-stub in FB11 and FB13. It is worth noting that the material's failure strength was reached in the part of the middle of the web when the thickness of the web was 7 mm, indicating that the reduction in the web thickness may have caused the fracture to occur there.

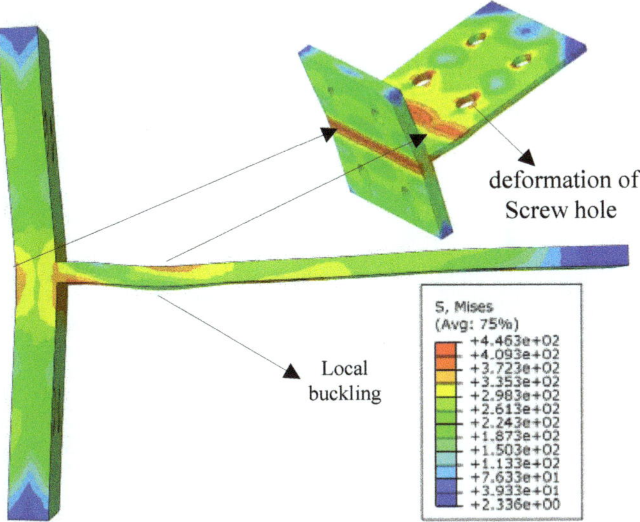

Figure 13. Stress chart of T-stub in FB7 model.

Each index had a different degree of increase with the increase in web thickness. In general, the overall performance of the composite point was the best when the web thickness of the T-stub was 11 mm. Stiffeners can be selected to improve the performance of T-stub steel [20–23].

5.2. Flange Thickness of T-Stub

It can be seen from Table 4 that the flange thickness of the T-stub had little influence on the overall performance of the composite joints. The following differences were noted.

(1) The flange of the T-stub was one of the main components under negative loading. The negative initial rotational stiffness was significantly increased with the increase in flange thickness. These increases were of 3.7%, 5.9% and 7.8%, respectively, compared with the YY12 model.

(2) The plastic deformation of the flange also differed in line with the increase in the flange thickness of the T-stub. The value of Δ_t was reduced by 15.8%, 21.7% and 35.5%, respectively, compared with the YY12 model. Figure 14 shows the plastic deformation of the T-stub in the different models.

(3) The negative ultimate bending moment of the YY12 model, which was 90.4% of that of the base model, decreased greatly. The positive and negative ultimate bending moments did not change greatly with the increase in thickness in the other models.

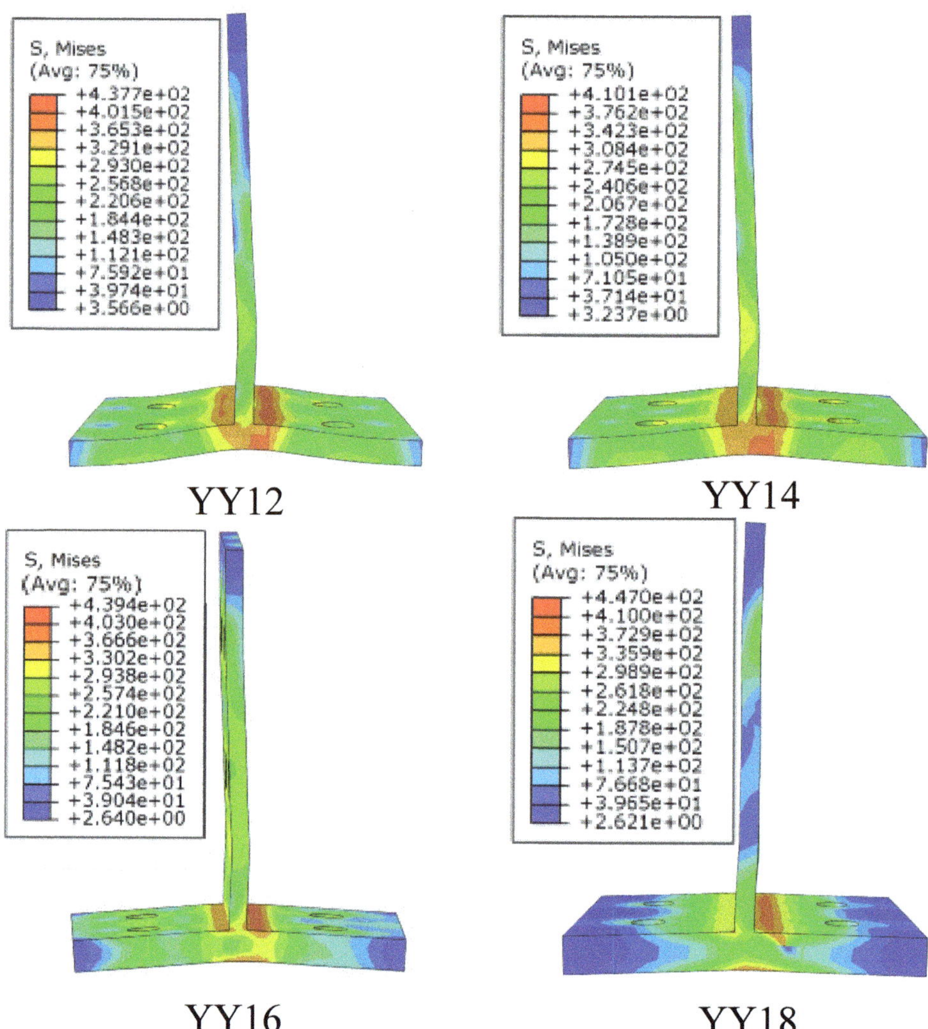

Figure 14. Comparison of levels of plastic deformation.

Excessive flange thickness had a limited effect on the improvement in the overall performance of the combined joints. According to the data of this group of models, the optimal flange thickness of the T-stub was 16 mm.

5.3. Wall Thickness of Column

The web and flange of the square steel-tube column were carefully considered in the component method. For the square steel-tube column with inner baffle, the compression stiffness of the web can be considered as infinite.

Furthermore, it can be seen from the data in Table 4 that all the indexes except the ductility coefficient and Δt increased greatly with the increase in the wall thickness of the column. Figure 15 shows the moment–rotation skeleton curve to more clearly reflect the variation in each index with the wall thickness of the column. It can be seen from Figure 15 that the increase of positive and negative ultimate bending moments gradually decreases with the increase in the wall thickness of column. Figure 16 shows the stress cloud diagram of the column in the models ZB8 and ZB14. It can be clearly seen that the stress near the screw hole decreased greatly with the increase in thickness. The cylindrical surface covered by the lower T-stub did not reach the yield stress. In Figure 16, the plumping up on the cylindrical surface covered by the T-stub in the deformation diagram of the ZB8 model was observed. This was not prominent in the other three models. The results show that the plastic deformation of the column in the composite joints decreased greatly with the increase in the wall thickness of the column, even though there was no plastic deformation in ZB14. Figure 16 also shows that the beam in ZB14 had a relatively obvious plastic deformation around the screw hole, which was not obvious in the other models.

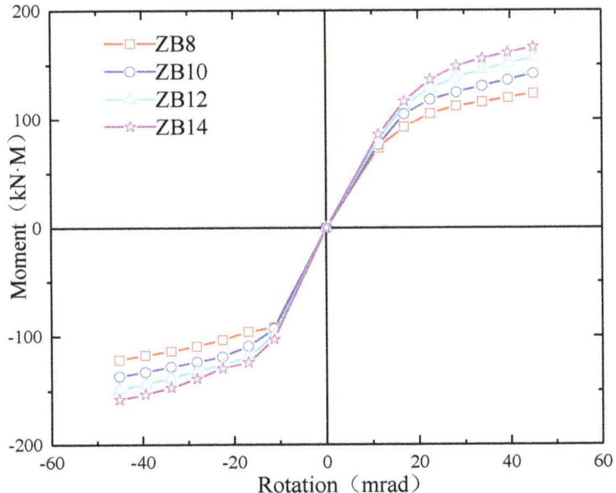

Figure 15. Moment–rotation skeleton curve of ZB group.

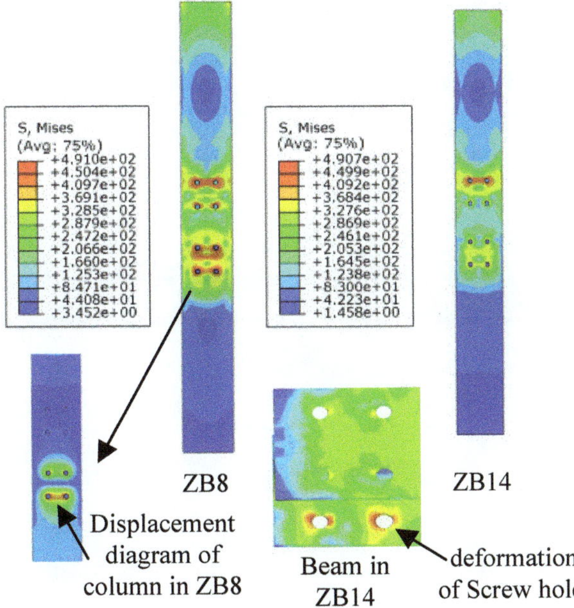

Figure 16. Stress chart of column in ZB8 and ZB14.

The total energy consumption of the composite joint increased in line with the increase in the wall thickness of the column, but the ductility coefficient and Δ_t decreased. In general, the increase in the wall thickness of the column greatly increased the amount of steel, resulting in the waste of materials, while the increase in each index was limited. According to the simulation results, the 10-mm wall thickness of the column used on the test can meet the demand.

5.4. Axial Compression Ratio

Regarding the range of the axial compression ratio in this study, the influence of the axial compression ratio on the overall performance of the composite joints was almost negligible. However, the performance of the composite joint with the axial compression ratio in the range of 0.55~1 could not be observed. Nevertheless, according to the regularity of the simulation results in this group, it can be predicted that each index will slowly decrease with the increase in the axial compression ratio.

5.5. Reinforcement Ratio

In this group of models, the load at the beam end and the mesh deformation at the joint were too large with the increase in the rebar diameter, which made the model convergence more difficult. Therefore, the calculation of the GJ12 model was not completed. Figure 17 shows the moment–rotation skeleton curves of the rebars with diameters of 6 mm, 8 mm and 10 mm. As can be seen from the figure, the rebar was one of the main stressed members under positive loading; the positive initial stiffness and ultimate bending moment capacity were greatly improved with the increase in the steel-bar diameter. However, the change in the rebar diameter had little effect on the negative initial stiffness and ultimate bending moment capacity because the concrete floor was the main compressive component under negative loading and the rebar diameter had a relatively limited effect on the overall performance.

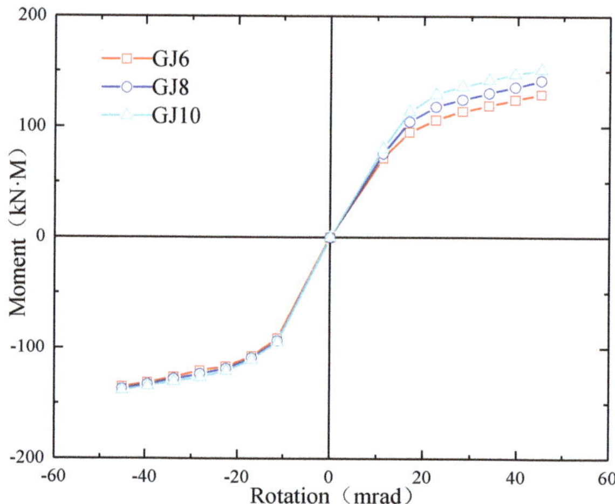

Figure 17. Moment–rotation skeleton curve of GJ group.

6. Conclusions

In order to expand the test database of composite joints with T-stub steel as the connector and to formulate the design criteria, based on the completed tests, this paper established a reliable finite element model of semi-rigid composite joints of square steel tubular columns and steel–concrete composite beams connected by blind bolted T-stub steel through the large finite element software, ABAQUS. Based on the limited research in this paper, the following conclusions can be drawn:

(1) It was shown that the finite element model proposed in this paper can better reflect the real situation of specimens through the comparison of the hysteresis curve, failure mode, strain growth and other aspects and can be used to analyze the performance of these semi-rigid composite joints under cyclic load.

(2) The web thickness of the T-stub and the axial compression ratio have little influence on the overall performance of the composite joints, but if the web thickness of the T-stub is smaller than the flange thickness of the beam, the positive and negative ultimate bending moments of the composite joint are significantly reduced. In our study, it is possible that the failure occurred in the web of the T-stub. Stiffeners can be selected to improve the performance of T-stub steel.

(3) The increase in the flange thickness of the T-stub can limit the development of plastic deformation, which is referred to as Δ_t in this paper. At the same time, the ultimate bending moment and initial rotational stiffness can be significantly improved. However, the improvement in YY18 is very limited compared with that in YY16.

(4) The positive initial rotation stiffness and the positive ultimate bending moment capacity of the composite joints are mainly controlled by the reinforcement ratio. This is because the concrete slab practically does not participate in the work under positive loading and the rebar mainly provides resistance. The increase in the reinforcement ratio leads to a small increase in the corresponding negative indexes.

(5) The increase in the wall thickness of the square steel-tube column has a great influence on the overall performance of composite joints. The positive and negative initial rotational stiffness, the positive and negative ultimate bending moment and energy dissipation are greatly increased. However, the increase in amplitude is gradually reduced and the ductility and Δ_t are reduced.

Author Contributions: Software, J.H.; Formal analysis, J.H.; Supervision, X.W., H.L. and H.S. All authors have read and agreed to the published version of the manuscript.

Funding: This research was funded by NSFC project grant number 51678264, Central Plains technological innovation leading talents grant number 214200510002 and Henan University Science and Technology Innovation Team grant number 21IRSTHN010.

Data Availability Statement: Data is unavailable due to privacy.

Acknowledgments: The research reported in this paper is part of Project 21IRTSTHN010 and Project 212102310969, supported by the Henan Province, China.

Conflicts of Interest: The authors declare no conflict of interest.

References

1. Lee, J.; Goldsworthy, H.M.; Gad, E.F. Blind bolted T-stub connections to unfilled hollow section columns in low rise structures. *J. Constr. Steel Res.* **2010**, *66*, 981–992. [CrossRef]

2. Wang, J.; Spencer, B. Experimental and analytical behavior of blind bolted moment connections. *J. Constr. Steel Res.* **2013**, *82*, 33–47. [CrossRef]

3. Thai, H.-T.; Uy, B. Finite element modelling of blind bolted composite joints. *J. Constr. Steel Res.* **2015**, *112*, 339–353. [CrossRef]

4. Alostaz, Y.; Schneider, S.P. Analytical behavior of connections to concrete-filled steel tubes. *J. Constr. Steel Res.* **1996**, *40*, 95–127. [CrossRef]

5. Yang, Y.; Li, H. Study on the static performance of H-beam to square tubular column connections with internal stiffening. *J. Xi' Univ. Archit. Technol. Soc. Sci.* **2017**, *49*, 14–21+28. (In Chinese)

6. Özkılıç, Y.O. Cyclic and monotonic performance of stiffened extended end-plate connections with large-sized bolts and thin end-plates. *B. Earthq Eng.* **2022**, *20*, 7441–7475. [CrossRef]

7. Li, W.; Ma, D.Y.; Xu, L.F.; Qian, W.W. Performance of concrete-encased CFST column-to-beam 3-D joints under seismic loading: Analysis. *Eng. Struct.* **2022**, *252*, 0141–0296. [CrossRef]

8. Zhang, J.-C.; Huang, Y.-S.; Chen, Y.; Du, G.-F.; Zhou, L.-J. Numerical and experimental study on seismic behavior of concrete-filled T-section steel tubular columns and steel beam planar frames. *J. Central South Univ.* **2018**, *25*, 1774–1785. [CrossRef]

9. Lu, D. *Study on Nonlinear Finite Element Analysis Method of Concrete-Filled Steel Tube Structure under Cyclic Loading*; Chongqing University: Chongqing, China, 2021. (In Chinese)

10. Gil, B.; Bayo, E. An alternative design for internal and external semi-rigid composite joints. Part II: Finite element modelling and analytical study. *Eng. Struct.* **2008**, *30*, 232–246. [CrossRef]
11. *GB50010-2010*; Code for Design of Concrete Structures. China Building Industry Press: Beijing, China, 2010. (In Chinese)
12. Ataei, A.; Bradford, M.A.; Valipour, H.R. Moment-Rotation Model for Blind-Bolted Flush End-Plate Connections in Composite Frame Structures. *J. Struct. Eng.* **2015**, *141*, 04014211. [CrossRef]
13. Gil, B.; Goñi, R. T-stub behaviour under out-of-plane bending. I: Experimental research and finite element modelling. *Eng. Struct.* **2015**, *98*, 230–240. [CrossRef]
14. Takhirov, S.; Popov, E. Bolted large seismic steel beam-to-column connections Part 2: Numerical nonlinear analysis. *Eng. Struct.* **2002**, *24*, 1535–1545. [CrossRef]
15. Özkılıç, Y.O. A Cem Topkaya, Extended end-plate connections for replaceable shear links. *Eng. Struct.* **2021**, *240*, 112385. [CrossRef]
16. Özkılıç, Y.O. Interaction of flange and web slenderness, overstrength factor and proposed stiffener arrangements for long links. *J. Constr. Steel Res.* **2022**, *190*, 107150. [CrossRef]
17. Özkılıç, Y.O.; Bozkurt, M.B.; Topkaya, C. Mid-spliced end-plated replaceable links for eccentrically braced frames. *Eng. Struct.* **2021**, *237*, 112225. [CrossRef]
18. Gil, B.; Goñi, R.; Bayo, E. Experimental and numerical validation of a new design for three-dimensional semi-rigid composite joints. *Eng. Struct.* **2013**, *48*, 55–69. [CrossRef]
19. Li, W.; Han, L.-H. Seismic performance of CFST column to steel beam joints with RC slab: Analysis. *J. Constr. Steel Res.* **2011**, *67*, 127–139. [CrossRef]
20. Özkılıç, Y.O. The effects of stiffener configuration on stiffened T-stubs. *Steel Compos. Struct.* **2022**, *44*, 489–502.
21. Özkılıç, Y.O. The capacities of thin plated stiffened T-stubs. *J. Constr. Steel Res.* **2021**, *186*, 106912. [CrossRef]
22. Almasabha, G. Gene expression model to estimate the overstrength ratio of short links. *Structures* **2022**, *37*, 528–535. [CrossRef]
23. Chen, Z.; Gao, F.; Wang, Z.; Lin, Q.; Huang, S.; Ma, L. Performance of Q690 high-strength steel T-stub under monotonic and cyclic loading. *Eng. Struct.* **2023**, *277*, 115405. [CrossRef]

Review

A State-of-the-Art Review and Numerical Study of Reinforced Expansive Soil with Granular Anchor Piles and Helical Piles

Ammar Alnmr *, Richard Paul Ray and Rashad Alsirawan

Department of Structural and Geotechnical Engineering, Széchenyi István University, 9026 Győr, Hungary
* Correspondence: alnmr.ammar@hallgato.sze.hu

Abstract: Expansive soils exist in many countries worldwide, and their characteristics make them exceedingly difficult to engineer. Due to its significant swelling and shrinkage characteristics, expansive soil defies many of the stabilization solutions available to engineers. Differential heave or settlement occurs when expansive soil swells or shrinks, causing severe damage to foundations, buildings, roadways, and retaining structures. In such soils, it is necessary to construct a foundation that avoids the adverse effects of settlement. As a result, building the structure's foundations on expansive soil necessitates special consideration. Helical piles provide resistance to uplift in light structures. However, they may not fully stabilize foundations in expansive soils. A granular anchor pile is another anchor technique that may provide the necessary resistance to uplift in expansive soils using simpler methods. This review and numerical study investigate the fundamental foundation treatments for expansive soils and the behavior of granular anchors and helical piles. Results indicate that granular anchor piles performed better than helical piles for uplift and settlement performance. For heave performance, the granular anchor and helical piles perform nearly identically. Both achieve heave reductions greater than 90% when L/H > 1.5 and D = 0.6 m.

Keywords: expansive soil; swelling; stabilization; foundations; uplift force; granular anchor pile; helical pile; numerical simulation

Citation: Alnmr, A.; Ray, R.P.; Alsirawan, R. A State-of-the-Art Review and Numerical Study of Reinforced Expansive Soil with Granular Anchor Piles and Helical Piles. *Sustainability* **2023**, *15*, 2802. https://doi.org/10.3390/su15032802

Academic Editors: Hong-Wei Yang, Shuren Wang and Chen Cao

Received: 18 December 2022
Revised: 28 January 2023
Accepted: 31 January 2023
Published: 3 February 2023

1. Introduction

Expansive soil is a problematic soil that damages civil engineering structures worldwide [1]. Earthquakes, expansive soils, landslides, hurricanes, tornadoes, and floods are the six most dangerous natural hazards, according to Baer [2], with expansive soils tied for second place with hurricane wind/storm surge in terms of economic losses to buildings. During periods of excessive moisture, expansive soil swells, causing a structure to heave. Expansive soil shrinks when soil moisture decreases, leading to construction settlement [3]. Expansive soil can also create lateral displacement by applying pressure to the vertical face of a foundation, basement, or retaining wall. According to Bowles [4], McOmber and Thompson [5], Nelson et al. [6], and Walsh et al. [7], the soil expands and shrinks in a zone ranging in depth from one meter to more than 20 m below the ground surface. This zone comprises the depth of seasonal variation in moisture; thus, structural damage is due to volumetric changes here, commonly referred to as the "active" or "unstable" zone [8,9]. If volume changes in the active zone of expansive soils occur near a foundation, they cause structural damage. Expansive soil swells excessively when wet and shrinks excessively when dry. Without warning, it may generate large fissures at the surface; the fissures can be 20 cm wide and 4 m deep [10].

As previously stated, swelling soils cause large-scale damage to civil engineering structures due to volumetric increases accompanied by a loss of strength during wet seasons and shrinking during summer [11]. Buildings crack, roads become rutted, and retaining structures deteriorate [12]. These soils are found in almost every country worldwide and challenge geotechnical engineers everywhere [13].

Some researchers have worked on improving soil behavior with special additives [14–19]. Others studied the benefits of alternative foundations, especially deep foundations, to resist the damaging effects of expansive soil [20–25]. The Department of the Army [26] highlighted the importance of foundations in expansive soils. It recommended choosing cost-effective foundations to minimize structural distress and differential movement between structural elements.

Helical and granular anchor piles are two types of deep foundations used at a construction site to support and stabilize structures. They are frequently employed when conventional foundation techniques, like deep concrete foundations, are neither feasible nor practical [27]. The helical and granular anchor piles have proven their effectiveness in cohesive [28–31] and cohesionless soils [32–37]. They are economical, quick to install [27], and environmentally friendly, as they do not pollute the soil or water [38,39]. Furthermore, as shown in Table 1, helical and granular anchor piles provide pullout resistance in a variety of practical applications that are primarily exposed to tensile loads [34,40–42].

Table 1. Practical applications of helical and granular anchor piles (compiled by the authors based on [34,40–42]).

Application	Helical Piles	Granular Anchor Piles
Retaining Walls	✔	under base only
Slope and Landslide Stabilization	✔	X
Tie-down Structures (concrete dam, offshore wind, uplift slab)	✔	✔

Moreover, helical and granular anchor piles are viable alternatives to traditional anchoring methods. Granular anchor pile installation is low-cost and does not necessitate sophisticated equipment [27,37]. Comparing granular anchor piles to a commonly accepted practice, such as helical piles, can be useful in identifying the position of the granular anchor piles within the broad spectrum of anchoring methods [37].

Numerical and experimental studies have demonstrated the behavior of helical [43–50] and granular anchor piles [25,32,51–54]. However, studies have not examined their anchoring capabilities in expansive soil [55]. Furthermore, the studies did not emphasize comparative studies of the two types of expansive soils.

This study reviews the behavior of both types of piles in expansive soils, followed by a numerical comparison using the PLAXIS 3D software. The primary objectives of this work are to (1) review popular solutions for mitigating the adverse effects of expansive soils on engineering facilities, (2) review research on the behavior of helical piles and granular anchor piles in expansive soils, and (3) compare the granular anchor piles to helical piles based on (2) with the help of numerical modeling using PLAXIS 3D.

2. Research Method

To fully grasp the topic, we applied a mixed review technique and numerical study to assess the performance of traditional and special foundations in expansive soil [56] by:

- surveying the database and selecting keywords,
- assembling and screening relevant papers,
- comparing the different foundations and identifying research gaps,
- applying numerical modeling to compare helical and granular anchor piles.

Section 7.1.2 discusses numerical modeling methodology in detail.

Surveying the Databases and Keywords

All documents were searched, including journal articles, books, conference papers, and proceedings. The most relevant reports were found in three literature databases: Google Scholar, Web of Science (WoS), and Scopus.

Figure 1 depicts the keywords of the existing knowledge domain in the foundations on expansive soil using VOSviewer. The VOSviewer displays the keyword network as a distance-based representation. Each keyword represents a node in this network, and links provide the connections. The distance between two nodes determines the strength or weakness of a relationship. A greater distance indicates a weaker link between keywords, whereas a smaller distance implies a stronger link [57]. The link strength of a node relates to the sum of all the connection strengths attached to it. The different colors represent different research years, and the size of the nodes generated refers to the number of publications where the phrase was first used [57]. Our findings will be highlighted in the following sections, which compare the types and suitability of foundations and primary building solutions for expansive soils and deep foundations. We will focus on the behavior of helical and granular anchor piles and their role in reducing heave.

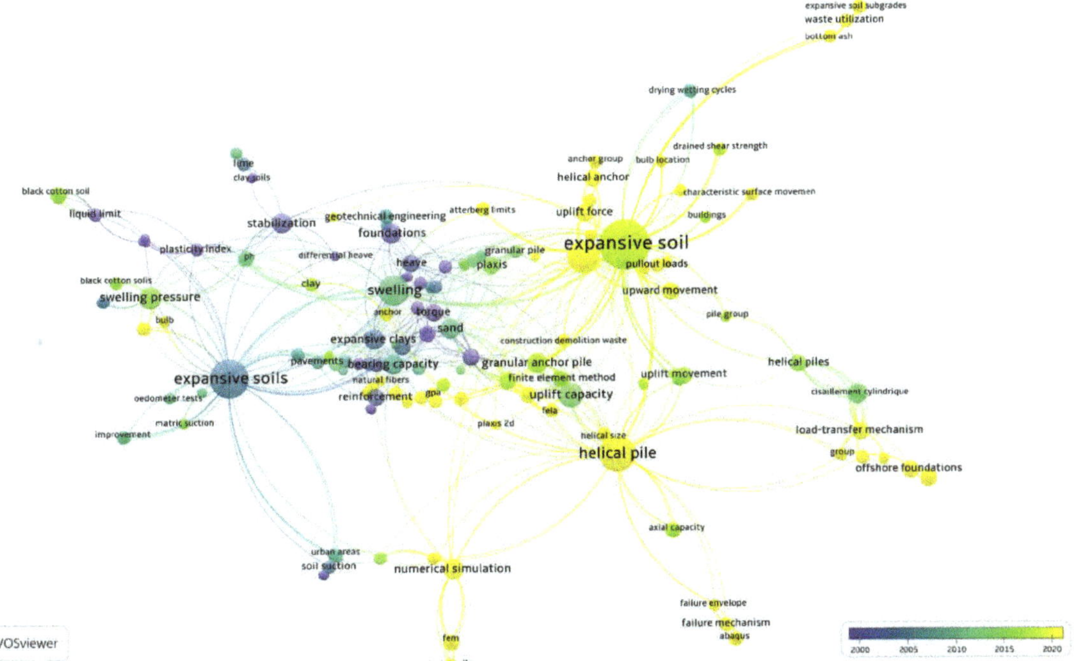

Figure 1. A keyword network.

3. Field Identification of Expansive Soil

Soil type identification in the field usually requires determining the index properties of the soil, such as color, texture, and plasticity, without requiring special equipment. Engineers can modify expansive soils' behavior by mechanical, thermal, chemical, and other means. As a result, it is essential to investigate expansive soil's physical and engineering properties, primarily when it is used as a construction material or for foundation purposes.

Adem and Vanapalli [58] observed that swelling soils exhibit surface fissures. They may absorb considerable amounts of water through the fissures during rainfall or local site alterations (such as water pipe, sewer, or storm drain leakage). The added moisture creates a soft, heavy, and sticky clay. The clays can shrink and stiffen as they dry, resulting in ground shrinkage (volume reduction) and cracking.

Various classification methods in the laboratory evaluate index properties that infer expansive soil behavior. Typical tests include Atterberg limits and clay size percent to classify a soil's expansion potential as low, medium, high, or extremely high. Soils classified as CH or CL in the USCS or A-6 or A-7 in the AASHTO classification systems are considered expansive soils in general [59].

Holtz and Kovacs [60] proposed three essential components to identify swelling damage to structures:

- The soil contains montmorillonite (a highly active mineral with a high swelling potential found in clay).
- The natural water content of the soil is close to its plastic limit.
- A source of water is available for potentially swelling soil.

Expansive soils are classified using a variety of systems. In Table 2, Bowels [4] summarized the findings of Holtz [61] and Dakshanamurthy and Raman [62] to classify the swelling potential of expansive soils. Table 2 shows the potential changes in soil volume as a function of the liquid limit (LL) and the plasticity index (PI).

Table 2. Potential changes in the soil volume as a function of liquid limit (LL) and plasticity index (PI) (compiled based on [4]).

Liquid Limit (LL%)	Plasticity Index (PI%)	Potential for Volume Change
20–35	<18	Low
35–50	15–28	Medium
50–70	25–41	High
>70	>35	Very high

AASHTO [63] specifies a method that determines whether the soil is expansive and predicts the magnitude of swelling. The soil's Atterberg Limits correlate to the natural soil suction during construction, as shown in Table 3.

Table 3. Potential for volume change as a function of liquid limit (LL), plasticity index (PI), and soil suction (compiled based on [63]).

Liquid Limit (LL%)	Plasticity Index (PI%)	Soil Suction (kPa)	Potential for Volume Change
<50	<25	<144	Low
50–60	25–35	144–383	Medium
>60	>35	>383	High

Consequently, a comprehensive foundation design in expansive soils requires a site-specific geotechnical investigation with specialized laboratory testing to identify the index properties (Atterberg limits, moisture content, soil suction), swelling potential, and swelling pressure (e.g., ASTM D4546 [64]). According to Chen [65], the last test is the most important and reliable one for evaluating expansive soils.

4. Types and Suitability of Foundations

There are two types of foundations: shallow foundations (individual [isolated] or combined footing, strip, stiffened mat) and deep foundations (drilled pier, helical piles, granular anchor piles).

According to Jones and Jefferson [66], the principal types of foundations utilized in expansive soils worldwide include pile and beam or pier and beam systems (Figure 2), reinforced rafts, and modified continuous perimeter spread footings, summarized in Table 4.

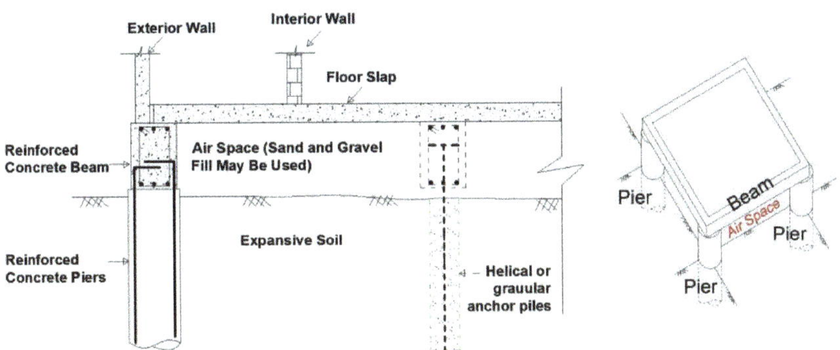

Figure 2. Pier, helical pile or granular anchor pile and beam foundations (compiled by the authors based on [59]).

Table 4. Different types of foundations used in expansive soils (compiled based on [66]).

Type of Foundation	Philosophy of Design	Advantages	Disadvantages
Pier, helical pile, or Granular anchor pile and beam (Figure 2)	Isolate structure from expansive movement by mitigating swell using anchoring to create stable layers	Utilized in a wide range of soils; effective in high-swell potential soils	The design and construction processes are relatively complex. Specialized contractors are required.
Raft or stiffened raft	Protects the structure from differential settlements by providing a rigid foundation.	Reliable on soils with a moderate swell potential; no special building equipment is required.	Only works for building relatively simple layouts; comprehensive construction quality control is required.
Modified continuous perimeter footing or deep trench fill foundations	Same as raft or stiffened raft foundation—includes stiffened perimeter beams.	No specialized equipment is required for this simple construction.	Ineffective in highly expansive soils or tree-influenced zones.

In low-swelling soils (PI < 15), standard shallow foundations are frequently used [26] when the footing angular rotation (deflection/span length) ratios (Δ/L) are 1/600 to 1/1000 or the differential movement <1 cm.

Stiffened mat foundations (Figure 3) will support buildings in expansive soil (PI $\geq$ 15), where the expected differential movement could be as high as 10 cm. The mats' stiffening beams considerably reduce differential distortion [26]. Table 5 displays the beam spacing and depth according to the type of mat.

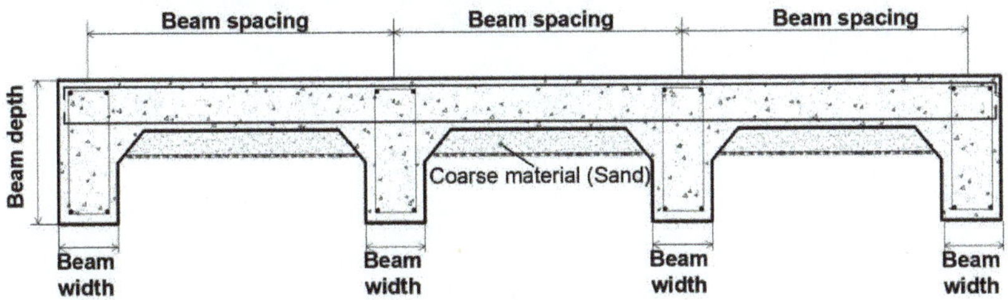

Figure 3. Stiffened mat foundations (compiled by the authors based on [26]).

Table 5. Beam spacing and depth according to the type of mat (compiled based on [26]).

Type of Mat	Beam Depth, cm	Beam Spacing, m
Light	40 to 50	6 to 4.5
Medium	50 to 65	4.5 to 3.6
Heavy	65 to 75	4.5 to 3.6

If appropriately designed and erected, a pile or beam-on-a-drilled-shaft foundation will adapt to a wide range of soil conditions and tend to reduce the effects of heaving soil. Deep foundations can support nearly any superstructure with low differential soil movement. They can achieve shaft deflection/spacing ratios of less than 1/600 [26].

5. Fundamental Building Solutions for Expansive Soils

First, this section briefly presents the basic problem-solving methods for expansive soil foundations. Then it addresses using deep foundations to resolve expansive soil problems that are more complex.

Peck et al. [67]; Bowels [4]; Murphy [68]; and Zumrawi et al. [69] suggest three main techniques to prevent structural damage to newly constructed structures caused by expansive soils.

5.1. Reduce or Prevent Swelling

There are three methods to reduce or prevent swelling in soil:

1. Removing and replacing the expansive soil: If the layer of moderately expansive soil is immediately below the foundation, remove it and replace it with improved soil. When correctly compacted, the replacement layer will distribute loads better and reduce the adverse effects of swelling on the foundation (Lytton et al. [70]; Rao et al. [71]; Murthy and Praveen [72]; Walsh et al. [73]; Ahmed [74]; Srinivas [75]). The effectiveness of the removal and replacement method depends on the thickness and soil type of the replacement layer. A thin, impervious cap may prevent surface water infiltration into the underlying expansive clay. In contrast, a granular replacement layer may encourage deeper wetting of the remaining expansive soil.

2. Changing the soil's characteristics: Gromko [76] provides several ways to reduce, if not eliminate, heave in expansive soils. Economics and workability will dictate the selection of one of the following strategies:

   - Controlling the level of compactness: Gromko [76] concurs that maintaining the degree of compactness is one of the most practical and cost-effective ways to reduce heave in expansive soil. A soil's swelling potential diminishes when compacted on the high side, possibly 3–4% above the optimum moisture content. However, in cases where the overall heave exceeds 35 mm, a slab on grade will not perform well.
   - Stabilization through chemicals: Chemical stabilization of expansive soil with various stabilizers, such as fly ash, lime, or cement, has dramatically minimized heave; however, contractors use lime stabilization more than any other chemical agent to stabilize expansive soil. Mixing 4% to 8% lime with plastic clay reduces the plasticity index of the topsoil layer while increasing its load-bearing capacity (Gromko [76]; O'Neill and Poormoayed [77]; Bowels [4]; Prusinski and Bhattacharja [78]; Moayed et al. [79]; Belabbaci et al. [80]; AL-TAIE et al. [81]; and Mahedi et al. [82]).
   - Pre-moistening of expansive soil: pre-moistening is another method for increasing the soil moisture content by immersing an area in water. Jeyapalan et al. [83] stated that submerging the expansive soil in water before building attains most of the estimated heave. Slow water seepage through highly plastic soil, on the other hand, may make this time-consuming. A 10–15 cm thick layer of sand, coarse gravel, or granular soil on top of the area will, according to Gromko [76], provide

the contractor with an excellent working surface during and after pre-moistening. This layer reduces evaporation, adds a minor surcharge, and creates a level, uniform subgrade.

3. Controlling the soil's water content: One of the most effective ways to minimize the heave of expansive soil is to manage its moisture content. Moisture control technologies applied around the perimeter of structures will reduce wetting or drying under the foundation. Impermeable barriers (such as retaining walls and geotextile membranes), proper drainage systems, and vegetation control will maintain moisture levels [76,77,84]).

5.2. Creating a Flexible Building Style and Designing a Resilient Foundation System

Allowing the soil to swell within cavities built into the foundation's base (the waffle slab, Figure 4) allows a flexible building style with a robust and stiff foundation system. According to Chen [65], this method has been tested in limited cases in the Denver area with mixed results. Helical or granular anchor piles can be used to support this system.

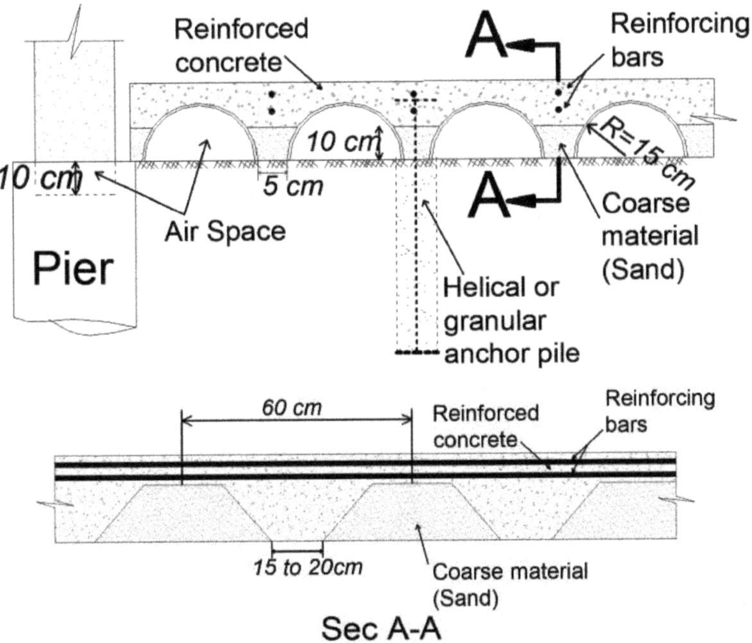

Figure 4. Design of waffle slab with cavities on expansive soil (compiled by the authors based on [65]).

5.3. Isolating the Structure from the Expansive Soil Environment

Short piers or piles lift suspended floor slabs above the active zone of swelling (Figure 5). The harmful movement will not reach the floor slab, keeping the structure intact. Support beams and piers combine to provide an effective foundation system. Deep foundation alternatives appear in the following sections.

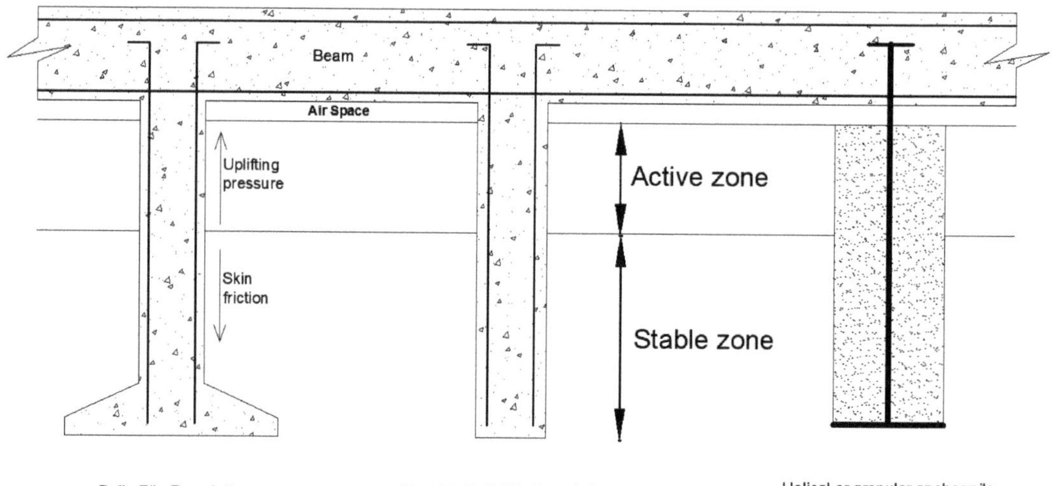

Figure 5. Grade beam and piles system (compiled by the authors based on [65]).

6. Deep Foundations in Expansive Soils

Deep foundations transfer structural loads to competent soils well below the ground surface. They may gradually transfer loads into the soil with depth through skin friction or rely on a hard-bearing stratum at their base. For large buildings, deep foundations resist uplift and overturning during strong winds or earthquakes. Pile foundations also limit settlement and effectively resist lifting forces from expansive soils.

Once below the active moisture zone (Figure 5), expansive soils offer moderate strength and do not contain free water. They become suitable materials for driving or drilling piles. As a result, designers often consider deep foundations as an option. Deep foundations provide an attractive solution when a structure requires minimal settlement or resilient support. This review examines two very effective methods to offer the advantages of deep foundations at a low cost (i.e., helical and granular anchor piles). Large areas such as highways are prohibitively expensive, and cut-off walls or shoulder dressings are a better alternative (Dafalla and Shamrani [85]). In some cases, with extremely high swell potential, expansive soil treatment with cement and lime or partial soil replacement is unavoidable.

Table 6 presents the most popular types of pile foundations. The installation impact provides a simple grouping for piles, i.e., displacement, small displacement, and non-displacement. The three levels of displacement refer to the soil that lies in the path of the installation. Displacement piles push the soil aside as they thrust into the ground. Small displacement piles are hollow or have a thin cross-section, so they deflect a smaller volume of soil during installation. Non-displacement piles remove soil from their path so that the neighboring soil remains undisturbed. Sorochan [86] states that in-situ cast piles work best because driving or vibrating piles may cause further difficulties. Helical and granular anchor piles create little to no soil displacement during installation. They are uniquely suited to expansive soils, as discussed in the following sections.

Table 6. The most popular types of pile foundations (compiled based on [26]).

Classification	Type	Description
Displacement	Timber Precast concrete Steel circular/rectangular Tapered timber/steel	Driven piles have a solid circular or rectangular cross-section or a closed bottom end. Piles that have been hammered or jacked into position
Small displacement	Precast concrete Prestressed concrete Steel H section Steel circular/rectangular Screw (helical pile)	Open-end cylinder, rectangular, H section, or screw configuration pile with a small cross-section
Nondisplacement	Drilled shaft Tubes filled with concrete Precast concrete Injected cement mortar Steel section Granular anchor pile	Piles of concrete are placed in open boreholes drilled by using a rotary auger, baling, grabbing, airlift, or reverse circulation methods.

6.1. Granular Anchor Pile

A granular anchor pile is a relatively new foundation technology for reducing expansive clay heave and enhancing foundation performance [87,88]. It is a variation on the traditional granular pile where an anchor carries a tensile load. The pile concept (Figure 6) uses the lower clay as an anchor point. As the expansive clay tries to lift the foundation, the inner rod feels tension (T). The lower granular column gains traction through friction against the (non-expanding) clay. Additional lateral stresses contribute to the sand's strength throughout [89].

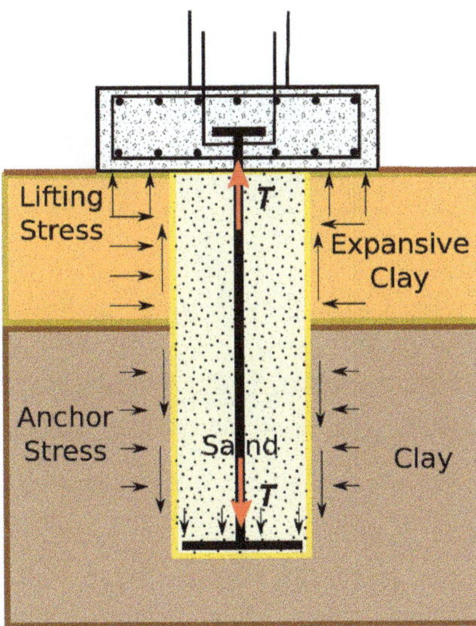

Figure 6. The concept of a granular anchor pile and the numerous forces acting on the foundation (compiled by the authors based on [89]).

6.1.1. Granular Anchor Pile Installation

Borehole preparation consists of drilling and casing to prevent surrounding soil from entering the hole. Next, insert a mild steel anchor rod with one end connected to the anchor plate into the borehole, resting at the bottom. Finally, fill it with granular material and compact it inside the borehole. A well-graded blend of locally available crushed stone aggregate and sand is typically employed. The filling-compacting process proceeds in layers where compaction effort produces uniform density [52].

6.1.2. Previous Research on Granular Anchor Piles

Rao et al. [89] and Phanikumar et al. [90] reported the results of a field-scale test program evaluating the pullout behavior of granular anchor piles buried in expansive clay beds. The compacted clay bed held single piles with 100, 150, and 200 mm diameters and lengths of 500, 750, and 1000 mm. Preparation of the clay included compaction at 15% moisture content (optimum = 27%), then installing the test pile, followed by water inundation and heaving. Researchers also investigated one group configuration to evaluate its effectiveness. They found:

- The granular anchor piles significantly reduced both the magnitude and rate of heave. Compared to an unreinforced expansive clay bed, the full-depth granular anchor pile reduced heave by 70%, 87%, and 92% for 100, 150, and 200 mm diameters, respectively. Relative movement along the pile-soil interface mobilized uplift resistance, reducing heave. A larger interface surface area (diameter) created stronger uplift resistance and less heave, as shown in Figure 7. The results of laboratory-scale model granular anchor piles installed in expansive clay beds corroborate this performance [91]. Granular piles also act as drain paths due to their high hydraulic conductivity, reducing the time for moisture stabilization and heaving equilibrium [92,93].
- Increasing the length of the granular anchor piles increases the resistance to upward movement, which is consistent with Vashishtha and Sawant [94]. For granular anchor piles with lengths of 500, 750, and 1000 mm, the uplift load required to generate a 25-mm heave increased from 9 to 12, then 14 kN. The uplift load of the three different granular anchor piles with a diameter of 200 mm is shown in Figure 8.
- The uplift resistance or failure pullout load increased as the diameter of the granular anchor piles increased, which is consistent with Vashishtha and Sawant [94]. The increase results from a larger area for skin friction resistance along the pile-soil interface. For example, a 25-mm upward movement generated resistances of 11, 11.5, and 14.2 kN for piles with 100, 150, and 200 mm diameters, respectively. Figure 9 shows the uplift load of the three piles, each with a length of 1000 mm.
- The pile group configuration consisted of five "passive" piles surrounding a sixth test pile. The test only loaded the center pile, while the surrounding piles contributed to the clay bed's general stiffness and moisture equilibrium. Compared to the single pile, the center pile produced a higher uplift resistance for a given upward movement. The center pile resisted a pullout load of 18 kN, compared to 12 kN for individual piles, showing a 50% improvement. Figure 10 shows the uplift load of both tests.

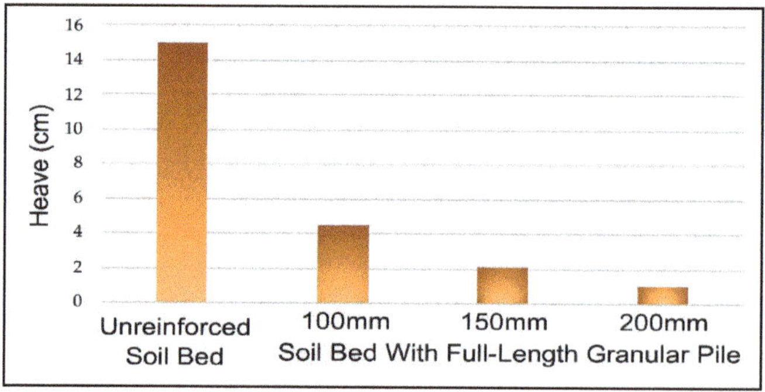

Figure 7. Clay bed heave after 100 days for various diameters of 1000 mm granular anchor pile length (compiled by the authors based on [89]).

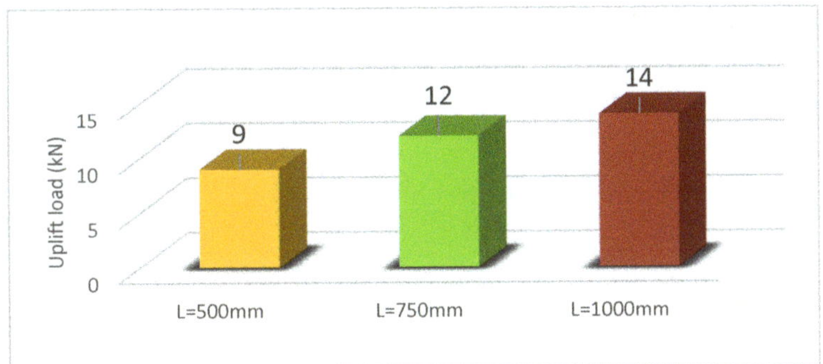

Figure 8. The uplift load of the three lengths of granular anchor piles with 200 mm diameter is calculated at 25 mm of upward movement (compiled by the authors based on [89]).

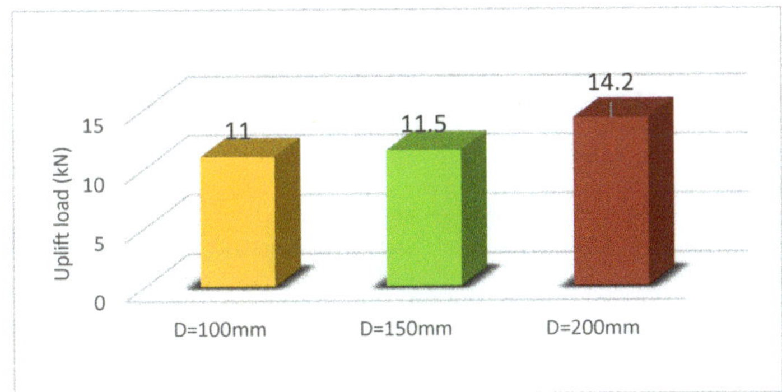

Figure 9. The uplift load of three granular anchor piles, 1000 mm in length, is calculated at 25 mm of upward movement (compiled by the authors based on [89]).

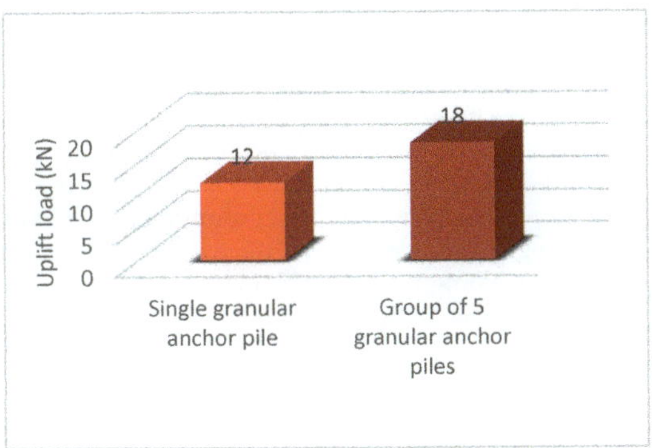

Figure 10. The uplift load of a single granular anchor pile and the center pile within a group with 150 mm diameter and 1000 mm length (compiled by the authors based on [89]).

To evaluate the compressive load response, Srirama Rao et al. [95] conducted field-scale plate load tests on granular anchor piles constructed in expansive clay beds. The dimensions of the granular anchor piles were the same as in the studies by Rao et al. [89] and Phanikumar et al. [90]. The test configurations varied: (1) unreinforced expansive clay beds; (2) expansive clay beds reinforced with a single granular anchor pile; and (3) expansive clay beds reinforced with a three-pile group. The plate could load both the clay bed and the piles at the same time or only the piles. The diameter and length of the pile group studied were kept constant at 150 mm and 1000 mm, respectively. For the single anchor piles, the embedment ratio varied between 2.5 and 7.5, while the diameter varied from 100–200 mm. Before testing, the researchers flooded the testing bin to encourage saturation and heaving. The findings revealed that, for a 25-mm settlement, the single pile required three times the compressive stress compared to the unreinforced bed.

Ismail et al. [96] analyzed granular anchor pile foundations in expansive soil using PLAXIS 2D and PLAXIS 3D software. The study investigated single and multiple pile behavior using a range of pile diameters and lengths. Their results corroborated previous testing and highlighted the benefits of increased pile diameter. The analyses also demonstrate that placing a group of piles under a footing rather than a single pile can decrease the efficiency of the granular anchor pile foundation system. In addition, pile spacing within the 2d to 4d range did not affect pile group efficiency significantly.

Sivakumar et al. [97] investigated the ultimate pullout capacity, load-displacement response, and failure mode of granular anchor piles. These authors also pointed out how to integrate this anchor properly into ordinary civil engineering construction. The experiments mentioned in this study follow in three parts.

The first section compares the ultimate pullout capacity of granular anchors to that of traditional concrete anchors cast in situ. These tests were carried out at Queen's University Belfast (QUB) on old filled deposits with an experimental program evaluating two variables: granular anchor pile lengths (L) of 0.5, 1.0, and 1.5 m and granular anchor pile diameters (D) of 0.07 and 0.15 m.

The second part of the investigation examined the performance of granular anchors on a lodgment fill deposit at Trinity College Dublin's Santry Sports Grounds. They used two lengths (L) (0.45, 1.62 m) and diameters (D) (0.15, 0.20 m).

The third part of the study included laboratory model tests at QUB. Soft, firm, stone-free clay (undrained shear strength cu = 30 kPa) was packed into a wooden box with dimensions of $1.2 \times 0.7 \times 0.7$ m. The following three column configurations were investigated: (1) D = 0.035 m and L = 0.7 m with a single plate at the bottom of the column; (2) D = 0.035 m and L = 0.7 m with a plate at mid-height and a second plate at the bottom of the column; (3) D = 0.035 m and L = 0.35 m with a single plate at the bottom of the column.

For the first section, the pullout capacities of the concrete and granular anchors of $L \times D = 0.5\ m \times 0.07$ m were 5.2 and 5.5 kN, respectively, according to the results. Both anchors failed due to shaft resistance deployed along the column's length. The soil surrounding the concrete anchor did not experience any considerable displacement (either heave or subsidence) until it reached a failure state. Around the granular anchor, the soil gradually heaved as the anchor load increased incrementally to failure. The concrete anchor ($L \times D = 0.5\ m \times 0.15$ m) failed at 8.0 kN capacity due to a loss of shaft resistance and ductile behavior. The corresponding granular anchor failed at 6.7 kN, exhibiting abrupt behavior. The granular anchor with $L \times D = 1.0\ m \times 0.07$ m exhibited ductile failure with localized end bulging, whereas the concrete anchor failed in shaft resistance due to rapid pullout. The concrete and granular columns mobilized pullout capacities of 16.3 and 16.1 kN, respectively.

The second part of the experimental results on short anchors with lengths of $L \times D = 0.5\ m \times 0.196$ m and $0.45\ m \times 0.148$ m failed due to shaft resistance, mobilizing a 12 kN pullout capacity. The pullout capacity increased as anchor length and/or diameter increased. Anchors achieved pullout capacities of 39, 42, and 44 kN with L = 0.96, 1.0, and 1.3 m and D = 0.219 m, respectively. The pullout capabilities for 0.168 m diameter anchors were 33, 40, and 42 kN for L = 0.8, 1.47, and 1.62 m, respectively.

The third section shows that increasing the length and presence of a multiple-plate anchor system increases the granular anchor pile's pullout capability (Figure 11).

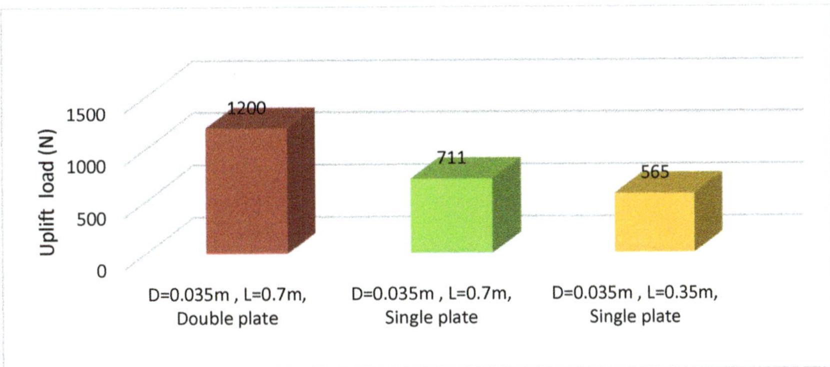

Figure 11. The uplift load of granular anchor piles (double and single plate) is calculated at 25 mm of upward movement (compiled by the authors based on [97]).

Sivakumar et al. [97] came to an important conclusion: when the ratio of L/D > 7, localized bulging causes the granular anchor piles to collapse. It is particularly effective in transferring applied weight to strata at depth. However, short granular anchor piles failed due to shaft resistance and had a pullout capability similar to traditional cast-in-situ concrete anchors. Granular anchors have the advantages of being quick to install, low in cost, and able to withstand applied loads immediately after installation.

Krishna and Murty [98] tested the pullout capacities of granular anchor piles and conventional concrete piles in the lab and in the field and compared the results from both. Tests conducted in unsaturated and saturated conditions aided in the evaluation of capacity reduction. Results of the pile load tests appear in Figures 12–14.

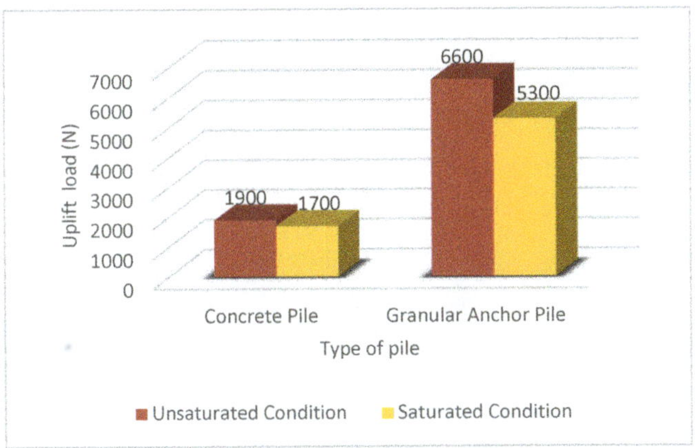

Figure 12. The uplift load of granular anchor piles in the laboratory (the pile diameter is 50 mm and the pile length is 200 mm) (compiled by the authors based on [98]).

According to the results of the pullout tests conducted in the laboratory and the field under unsaturated and saturated conditions, granular anchor piles have a pullout resistance that is about three times that of identical concrete piles. According to the laboratory test results, the pullout resistance of granular anchor piles decreases by roughly 14% compared to the unsaturated state, while that of concrete anchor piles decreases by 26%. According to field experiments, granular anchor piles are reduced by roughly 25%, while concrete anchor piles are reduced by 32%.

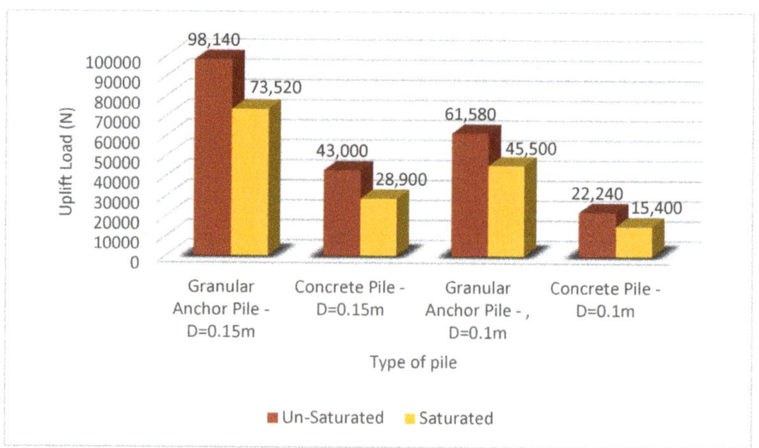

Figure 13. Uplift load of granular anchor piles and concrete piles in the field for two different diameters with 1 m pile length (compiled by the authors based on [98]).

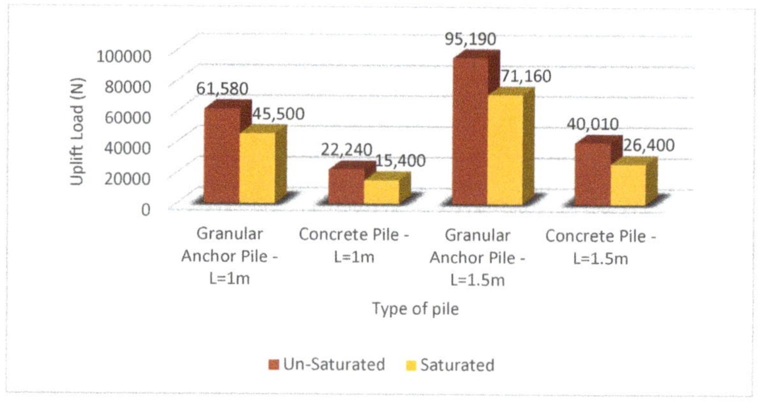

Figure 14. Uplift load of granular anchor piles and concrete piles in the field for two different lengths with a pile diameter of 0.1 m (compiled by the authors based on [98]).

Due to the complete packing and lateral displacement of granular fill caused by ramming, granular anchor piles have a high pullout resistance. The close interaction and interlocking of the sides of the borehole with such lateral displacement of granular fill are questionable in the case of concrete piles.

To study the behavior of the granular anchor pile in expansive soil, Aljorany et al. [87] used a laboratory test on an experimental model, as well as numerical modeling and analysis using PLAXIS 2D software. The effects of various parameters, such as granular anchor pile diameter, granular anchor pile length, footing diameter, non-expansive clay layer thickness (stable zone), and expansive clay layer thickness, were investigated. The results demonstrated the effectiveness of granular anchor piles in decreasing expansive soil heave. Based on the findings, three independent variables affect the heave of granular anchor piles: the L/D ratio, the L/H ratio, and the B/D ratio (where L is the length of a granular anchor pile, D is the diameter of a granular anchor pile, H is active soil thickness, and B is footing diameter/width). With a ratio of L/H = 1.0, the pile reduced heave by 38%. The reduction increased to 90% when the pile penetrated the stable zone to achieve L/H = 2.

Johnson and Sandeep [99] examined the effect of the relative density of the granular fill and pile diameter on pullout capacity. They compared encased and non-encased granular piles using a laboratory model test on black cotton soil (expansive soil). They performed pullout tests on a 30 mm-diameter anchor pile in a clay bed with a water content of 40%. The relative density of the granular fill varied from 50% to 70%. They also varied the pile diameter from 30 to 50 mm. A 30% increase in capacity resulted from the larger pile diameter.

Muthukumar and Shukla [100,101] conducted laboratory model tests to examine the heave reduction of soil due to granular anchor piles with and without encasement and helical anchor piles. The testing conditions consisted of: (1) a clay bed with no piles; (2) a granular pile without an encasement; (3) a granular pile with an encasement; and (4) a helical pile. All pile installations (2,3,4) reduced heave in the clay bed. The granular pile with encasement outperformed the one without, verifying the results of Roy et al. [66]. The encasement resists swelling forces and adds to the uplift resistance. Increasing the stiffness of the encasement produced higher confining stresses and greater resistance. Their granular piles performed better than helical piles, as shown in Figure 15.

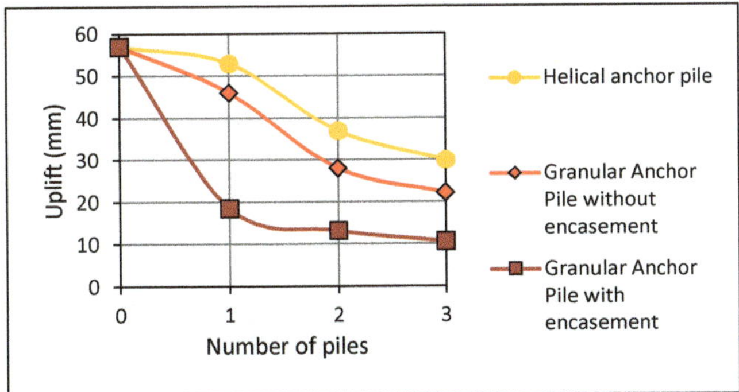

Figure 15. Comparing the behavior of granular anchor piles with and without encasement and helical anchor piles against uplift (compiled by the authors based on [100]).

Sharma [53] used PLAXIS 3D finite element software to conduct a numerical analysis to estimate the uplift performance of granular anchor piles in expansive soil. He applied an upward displacement of 10% of the pile diameter to the top. The expansive soil did not swell during the simulation, so volume change did not contribute to the model. He examined the effects of diameter (D), length (L), number of piles, pile spacing (s), and soil modulus (E_s). The simulations revealed that the pile's uplift capacity increases with increasing length and diameter (Figure 16a,b) and soil modulus (Figure 17). Single piles and pile groups showed similar behavior. Furthermore, the analysis showed that the ideal length-to-diameter ratio (L/D ratio) is between 10 and 13, beyond which uplift resistance increases only marginally.

Abbas [102] built a 30 cm-diameter and 55 cm-high testing tank to examine granular anchor piles in expansive clay. The clay mix consisted of 50% bentonite and 50% natural clay. Laboratory test results revealed the clay mix to have a liquid limit of 98, 15% free swell potential, and 210 kPa confined swell pressure. The expansive clay layer extended to a depth (H) of 25 cm, with a 20-cm-thick sand layer below it. Anchor piles reached the top of the sand or penetrated (L_s) 50–100 mm into it. The surface foundation measured 25 cm in diameter, while the piles varied between 20 and 50 mm. The effects of pile length (L), diameter (D), and depth of penetration into the sand layer influenced the heave resistance of the pile. The L/H, L/D, and Ls/H ratios provided dimensionless comparisons for performance.

Sharma and Sharma [103] studied the reaction of model granular anchor piles installed in poor clayey soil to pullout forces by altering pile length, diameter, and the relative density (RD) of granular fill material in a lab environment. They also studied the performance of crushed construction debris added to stabilize the clay. They modified the granular pile by reinforcing it with a geogrid. Four different combinations of clay and piles resulted. Their test results showed that longer piles and broader diameters increased pile capacity. Adding reinforcement with geogrids also increased capacity.

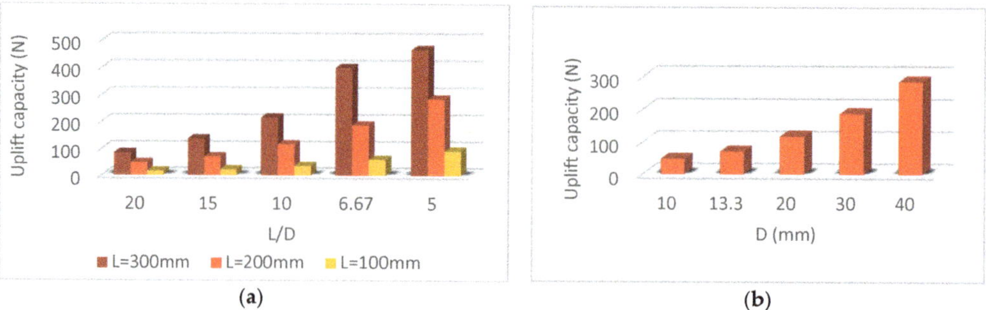

Figure 16. (**a**) Effect of length and L/D ratio; (**b**) Effect of diameter for L = 200 (compiled by the authors based on [53]).

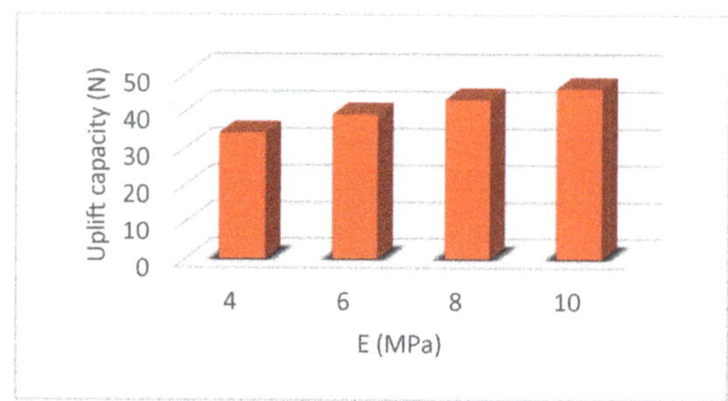

Figure 17. Effect of modulus of elasticity (compiled by the authors based on [53]).

Khan and Gaddam [104] discussed several scale model tests in their work to better understand the heave and pullout behavior of the granular anchor pile foundation system. They investigated the effect of the spacing by using two granular piles at different separations. Compared to a reinforced expansive soil bed without granular material, the swell potential was lowered to around 68.09% for the 50-mm-diameter granular anchor pile (Figure 18a). The swell potential shrank with closer pile spacing (Figure 18b). According to Rao et al. [89], the 2D pile spacing produced the best results. A 50-mm-diameter granular anchor pile improved its pullout capacity by almost 400% (Figure 19).

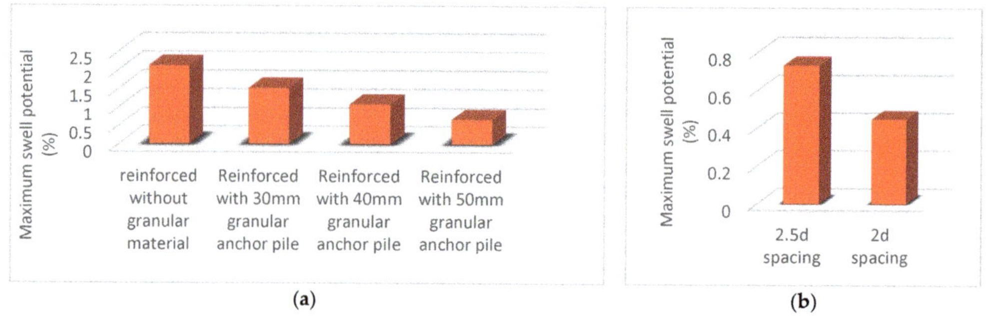

Figure 18. (**a**) Maximum swell potential for different diameters of granular anchor pile (**b**) Variation of swell potential with spacing between two granular anchor piles (compiled by the authors based on [104]).

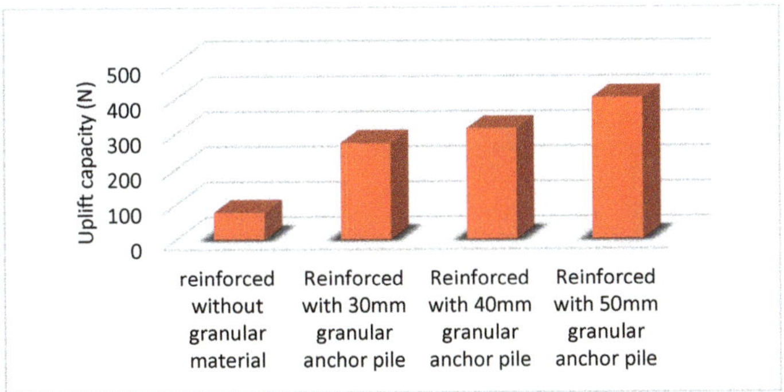

Figure 19. Uplift capacity for different diameters of granular anchor piles (compiled by the authors based on [104]).

6.2. Helical Piles

Helical piles offer deep foundation support in a prefabricated package. Unlike micro and driven piles, they have spiral blades that screw into the soil and provide compressive and tensile capacity.

The lead component of the helical pile is rotated into the soil first. Multiple helical blades attach to the square shaft; their size, shape, and number might vary. The lead is followed by several extensions, depending on the required torque or depth. Several extensions bolted together will achieve the desired depth of penetration. The top of the pile connects to the foundation system [105]. Figure 20 shows the components of a helical pile.

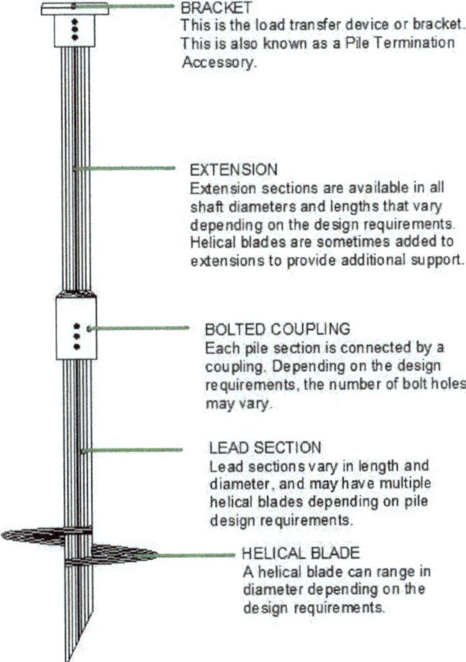

Figure 20. Helical pile components (compiled by the authors based on [105]).

6.2.1. Helical Pile Installation

Helical piles install in a relatively straightforward manner. They require some special equipment, similar in size to a hydraulic excavator. While helical piles are easy to install, they require careful attention when subsurface conditions change or block installation. A truck-mounted auger or hydraulic torque motor attached to a backhoe, forklift, skid-steer loader, or other hydraulic equipment will turn the helical pile shaft into the soil [105]. Figure 21 shows typical installation equipment. The hydraulic motor is the system's main component and applies torque (or rotating force) to the top of the helical pile. Motor speeds are usually slow to reduce disturbance by the blades during installation. Motors for helical pile installation typically deliver a torque of 6 to 100 kN-m or more [105]. The helix (helical bearing plate) installs below the depth of the active zone at a minimum of 5.4 kN-m of torque [106]. The torque motor should be able to rotate clockwise and counterclockwise and have an adjustable rotating speed. The equipment should have enough stability to maintain position and alignment during installation. Installation torque provides an estimate of pile capacity [105]. The relationship between capacity and torque has been thoroughly established empirically [107] and theoretically [108]. A torque indicator records torque levels during installation.

Figure 21. Installation of a helical pile (compiled by the authors based on [www.idealfoundations.com.au] (accessed on 22 June 2022)).

Pack [106] states that over 130,000 square shaft helical piers have provided remedial repair and foundations for new facilities in expansive soils since 1986, including the Front Range's highly expansive bedrock. There have been no reported failures or poor performance of square shaft helical piers that were accurately specified and installed. The guiding principles for using square helical piers are as follows:

(1) Install square-shaft helical piers with a minimum installation torque of 5.4 kN-m to ensure they penetrate below the active soil zone. The point at which the square shaft helical pier will not penetrate or advance deeper into the formation due to the density or hardness of the material is known as the "refusal depth."
(2) Allow only a single helix lead section, so there is no significant bearing in the active zone.
(3) Even when there is no dead weight, the small contact area of the square shaft reduces uplift forces on the pier to levels that eliminate heave.
(4) Uplift forces on the pier remain negligible due to the smooth square shaft surface.
(5) Water must not flow down the shaft's sidewalls to the soil's bearing depth. Migrating water may activate the lower soil.
(6) Using IBC and ISO 9001-listed square shaft helical piers ensures that proper material is supplied and installed for expansive soil conditions.
(7) Trained and experienced contractors can correctly install the square shaft helical piers in expansive soils.

6.2.2. Previous Research on Helical Piles

Mahmoudi and Ghanbari [35] acknowledge that there are few studies on the behavior of helical piles in expansive soils; therefore, more research is needed. In the following sections, a review of research will be conducted to determine the most critical aspects of the behavior of helical piles in expansive soils and highlight important considerations for future research.

Chao et al. [109] calculated the length required to achieve a 25 mm allowable movement for a variety of deep foundations (straight shaft pier, belled pier, helical pier) with a diameter of 25 cm installed in expansive soil for ideal and poor drainage conditions at various expansion potential (EP) values. The helical pier consisted of a single helix and a 3-inch-diameter steel shaft, with a minimum dead load of 50 kN on the piers. Figure 22 shows the required pier lengths for ideal and poor drainage conditions. The required pier length must increase about fivefold to overcome poor drainage, as illustrated in Figure 22. As a result, the required pier design lengths decrease dramatically with ideal surface drainage. However, such ideal surface drainage conditions rarely occur. Surface ponds form around structures, at least in localized areas, during times of irrigation. Helical piers, as opposed to belled piers, are effective under conditions of high expansion potential (EP) and poor drainage. This results from the helical piers' negligible uplift from skin friction.

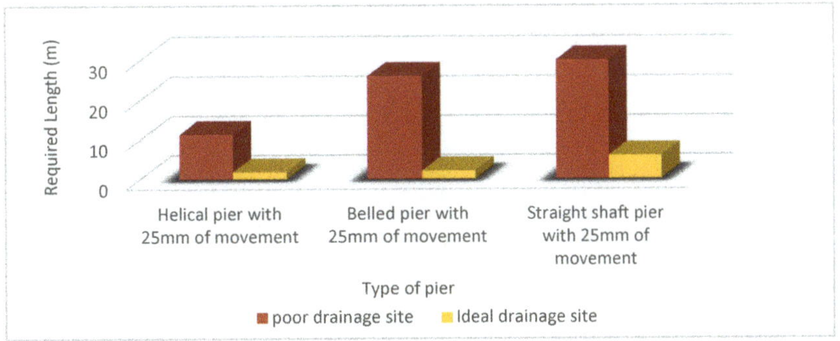

Figure 22. The required pier lengths for both ideal and poor drainage conditions (compiled by the authors based on [109]).

Al-Busoda and Abbas [110] mixed in soil additives during helical pile installation in expansive soil to examine their effects on upward movement and pullout force. The percentages of additives ranged from 0.5% to 6%. The helical piles extended 15 cm into the soil (L/D = 27) with a double helix (dh = 20 mm). Adding 3% silica fume, 3% coal fly ash, and 6% hydrated lime reduced the uplift movement of helical piles by more than 50%. Furthermore, a 3% mixture of (3:1) silica fume/coal fly ash or a 2% mixture of (1:1) hydrated lime and cement reduced uplift movement by more than 50%.

Al-Busoda and Abbas [111] examined helical pile models (dh = 15, 20 mm) drilled in an expansive soil bed over a sandy soil layer. They compacted a 200-mm-thick sand layer to 40% or 80% relative density. Above it, they compacted a 300-mm-thick layer of expansive soil. Helical piles (L = 350, 400, 450 mm; dh = 15, 20 mm) penetrated the sand layer with one or two helixes. Water then infiltrated around the helical pile from the bottom of the sandy soil to the surface of the expansive soil via four drains. They found that the upward movement of helical piles was reduced as the depth of the sandy layer, the helix diameter, and the number of helixes increased. In addition, increasing the relative density of the sand layer improved anchorage and reduced upward movement.

Al-Busoda and Abbas [112] conducted a study similar to [111] but focused on a single and a group of helical piles embedded in expansive soil. They used the same parameters as their previous study [111], except the pile lengths were 150, 200, and 300 mm, and they did not penetrate the sand. Four piles in a square pattern formed the group model. The length (L), diameter of helix (dh), and number of helices significantly impacted upward movement. The deeper helical piles with larger L/D ratios showed increased pullout capacity compared to the shallower piles, where D is the shaft diameter. The maximum upward movement of a group of helical piles was less than that of a single pile.

In the case of a single helical pile, increasing the L/D ratio for ordinary (no helix) and helical piles reduced the pile uplift movement caused by expansive soil, as illustrated in Figure 23a,b and Figure 24a,b. The long piles are anchored in a deeper soil layer, even though it is still within the active zone. When the L/D ratio increased from 27 to 53, upward movement decreased by 67% for plain piles, whereas single helix and double helix plates decreased by 84% and 77%. For short piles (L/D, L/H small), the ordinary pile demonstrated less uplift than the single or double helix piles, confirming Pack's idea [106] that a helix within the active zone defeats the purpose of the helical design. The maximum uplift movement reduces with increasing penetration depth (larger L/H) due to the less expansive soil near the bottom of the layer and the action of the helical pile, as illustrated in Figure 25a,b.

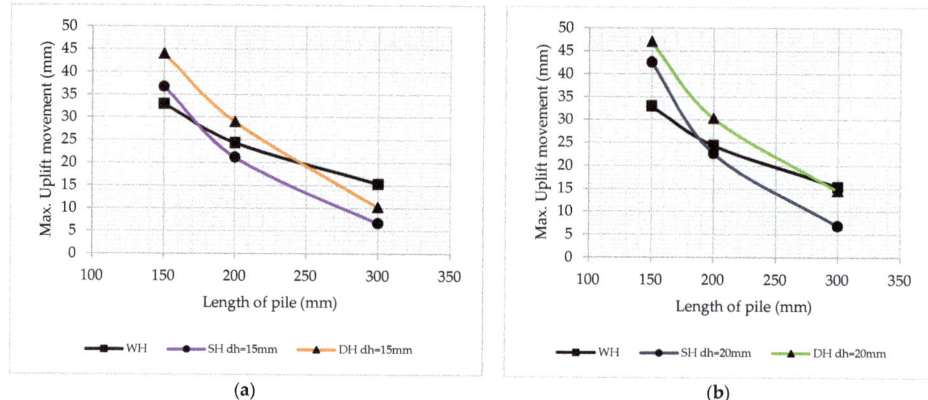

(a)   (b)

Figure 23. Maximum uplift movement of helical piles of various lengths with helix diameters (**a**) dh = 15 mm (**b**) dh = 20 mm (where WH means without helix, SH means single helix, Spmax means maximum uplift movement, and dh means the diameter of the helix) (compiled by the authors based on [112]).

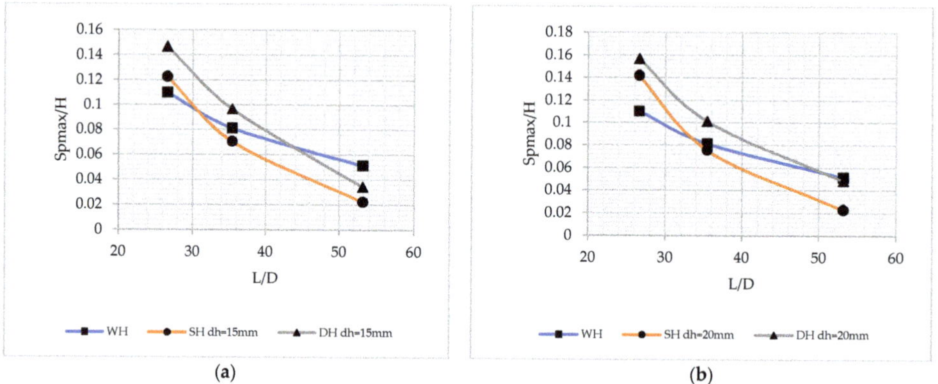

(a)   (b)

Figure 24. Variation of the Spmax/H ratio with the L/D ratio of helical piles of various lengths and helix diameters (**a**) dh = 15 mm (**b**) dh = 20 mm (where WH means without helix, SH means single helix, Spmax means maximum uplift movement, and dh means helix diameter) (compiled by the authors based on [112]).

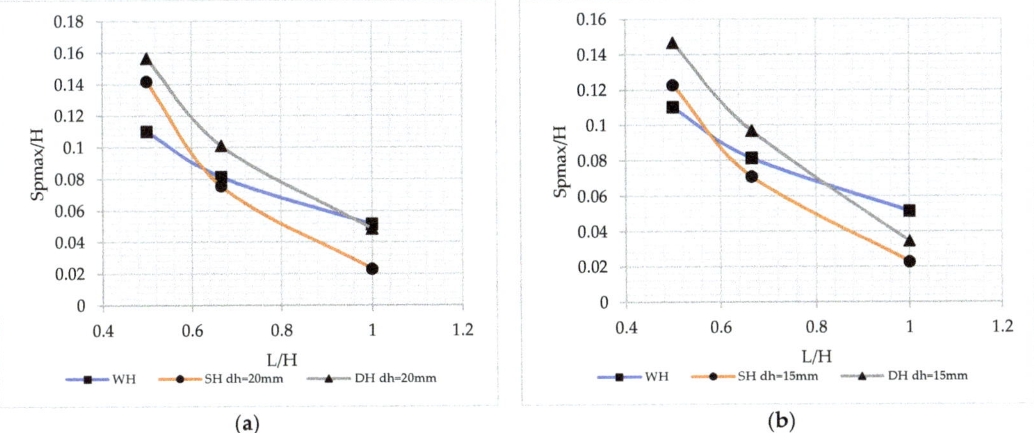

(a)   (b)

Figure 25. Variation of the Spmax/H ratio with the L/H ratio of helical piles of various lengths and helix diameters (**a**) dh = 20 mm (**b**) dh = 15 mm (where WH = without helix, SH = single helix, Spmax = maximum uplift movement, dh = helix diameter) (compiled by the authors based on [112]).

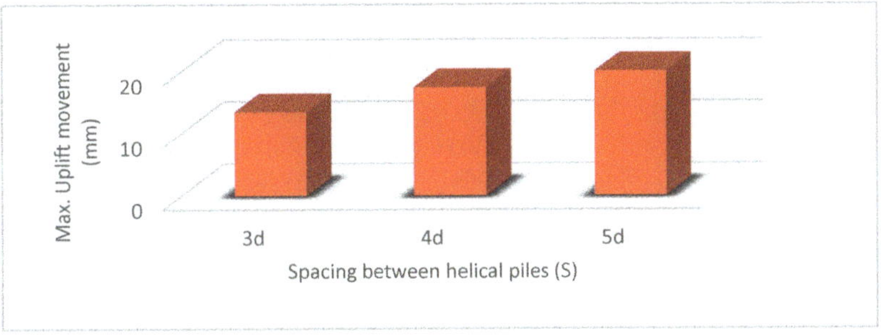

Figure 26. Maximum uplift movement for a pile group with L/D = 35 and a single helix (DH = 20 mm) (compiled by the authors based on [112]).

In the case of a group of helical piles, similar to single helical piles, increasing the L/D ratio for a group of helical piles reduces the pile uplift movement caused by expansive soil. The long piles provide resistance by using the deeper soil layer. Even if it is within the soil's active zone, it has much less moisture and does not expand. When increasing L/D from 27 to 53 for spacing (S = 3 DH), heave is reduced by 87–91% and 70–79% for single- and double-helix plates, respectively. For all L/D ratios, the helical pile group with a single helix demonstrated greater resistance than the group with a double helix. By lifting, the presence of helix plates in the active zone works against the pile's purpose. Upward movement increases when the pile spacing increases, as shown in Figure 26. The benefits of confinement stress reduce as pile spacing increases, generating lower pullout capacity.

Al-Busoda et al. [41] investigated inclined and vertical helical piles under the base of retaining walls. The helical pile was made up of three 0.30 m diameter plates. Two at the pile shaft's base provided vertical support in sandy soil. They had a separation distance five times their diameter (5 x dh). The third helix plate formed a rigid link with the footing of the retaining wall. The cross-section of the pile shaft measured 0.10 × 0.10 m, with two effective lengths (10 and 16 m). The authors found the best solution by placing two vertical helical piles and one inclined helical pile with an inclination of 30° from vertical. They chose a pile length to soil depth ratio of L/H = 3.2. In this case, the vertical movement was reduced by 94% and the lateral movement by 70%. In general, helical piles beneath retaining walls resist and control vertical movement but require inclined piles to resist lateral movement.

Albusoda and Abbase [42] modeled helical piles with PLAXIS 3D-2013 software. They applied the hardening soil model with volumetric strain expansion to mimic heave. A 44 m-high communication tower rested on an expansive soil layer 5 m deep. Wind loads and swelling soil generated uplift and lateral forces that required a pile foundation. The helical piles extended into dense, sandy soil beneath the expansive soil. Their effective lengths (7.50, 9.50, and 11.50 m) provided increasing resistance to uplift and lateral forces. Various patterns of piles created better or worse solutions to reduce the vertical and lateral movement of the tower. This study indicated that helical piles with double helix plates require L/D ratios of 118, 113, and 102 for helical pile groups of 2×2, 3×3, and 4×4 to achieve zero-uplift movement of the tower foundations. The L/H ratio could be achieved (1.52, 1.37, and 1.20) for helical pile groups of 2×2, 3×3, and 4×4, with double helix plates for zero-uplift. The pile groups also provided good resistance to lateral movement in general.

Mulyanda et al. [113] tested uplift loading on helical piles with diameters of 15, 20, and 25 cm. They also placed plates with different diameters on the same shaft. Their principal findings were:

1. As the diameter of the helix increases, so does the uplift capacity in a nearly-linear relationship.
2. Tapered plate configurations have capacities similar to their average diameter if the largest plate is on top.

Their study consisted of only five tests, so any further generalizations would not be appropriate.

7. Comparison between Helical and Granular Anchor Piles

The review study finds that helical and granular anchor piles are current alternative methods that mitigate the adverse effects of expansive soils. Most research has shown that they effectively reduce heave and increase bearing capacity.

Granular anchor piles are simple to install and do not require specialized expertise or equipment. However, they take longer to install because they require excavation and then backfilling with gravel soil, as opposed to the faster helical piles, which install directly both vertically and at an incline but require special equipment for installation and torque measurement.

The measured torque helps to predict the bearing capacity. It is a critical indicator of reaching the stable soil zone below the active zone when torque exceeds about 5.4 kN-m (Pack [106]). The indicator ensures consistent performance of the helical piles, which is less predictable for granular anchor piles.

Although few studies compare the performance of granular anchor piles with that of helical piles, Muthukumar and Shukla's study [100] found that granular anchor piles outperform helical piles. Since granular soil replaces some expansive soil, it provides more substantial resistance to uplift forces. Adding lateral confinement due to swelling contributes to greater strength by increasing friction at the pile-soil interface. Muthukumar and Shukla [100] did not consider the effects of soil disturbance resulting from their installation of helical piles. The disturbance would reduce swelling pressures in the active soil zone and anchor resistance at depth. These two actions would tend to cancel each other out.

Using laboratory model tests, Joseph et al. [37] compared the uplift capacity of granular anchor piles to helical piles. According to the findings, the granular anchor pile outperformed the helical pile. However, their study was conducted on medium-density sand rather than expansive soil.

7.1. Numerical Analysis Comparison Using PLAXIS 3D

This study used the program PLAXIS 3D to evaluate the effectiveness of both types of foundations in reducing heave caused by expansive soil. The analyses determined the pullout/pulldown loads of the helical and granular anchor piles. The numerical performance of piles using PLAXIS agrees with the results of laboratory and field tests [24, 114]. Problem complexity and the challenges of laboratory and field tests make numerical analysis an appealing method for the comparative parametric study of the targeted piles in this work [115].

7.1.1. Problem Description

The investigated problem consists of a shallow square footing (1.0 × 1.0 × 0.5 m) resting on an expansive soil layer and reinforced with a single granular anchor or helical pile. Varying lengths and a diameter of 0.6 m serve as input to the analysis (Table 7). The active zone of the expansive soil layer extends from the surface to a 4-m depth, where stable, saturated, dense sand lies beneath with a thickness of 10 m.

Table 7. Problem dimensions in this study.

Diameter (D) (m)	Granular Anchor Pile Length (L) (m)	Cap Width (B) (m)
0.6	4 7 10	1

7.1.2. Methodology

Figure 27 shows a cross-sectional view of unreinforced and reinforced soil with helical and granular anchor piles. The 3D model calculated heave for three lengths of granular anchor piles and helical piles where saturation occurred from the top (water infiltration) or from the bottom (rising ground water). Pullout load and structural load capacities allowed for realistic design comparisons.

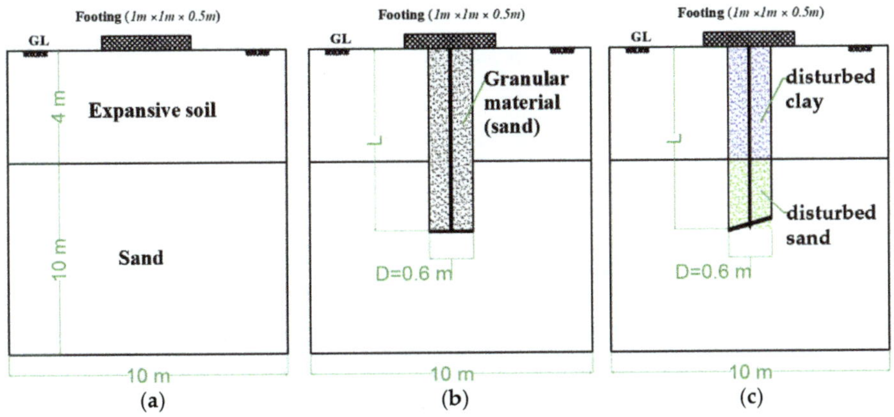

Figure 27. Cross-sectional view of (**a**) unreinforced soil, (**b**) granular anchor pile, and (**c**) helical pile.

Input Data

Model geometry appears in Figure 27 and Table 7. The model domain measured 10 m long, 10 m wide, and 14 m deep. The borehole option created the first layer of expansive soil with a thickness of 4 m and the second layer of sand 10 m thick. The volume for helical and granular anchor piles occupied the middle of the model. Soil properties simulated the construction process, where fill sand replaced expansive soils as the granular pile developed. Both piles contained beam and plate elements as well.

The hardening soil model represented the expansive clay and sand layers and the sand fill used in the granular pile. Anchor plates and the surface footing contained steel and concrete. Both behaved elastically. Tables 8 and 9 list all materials and constitutive models as cited by the Swiss Standard [116]; Xiao et al. [117]; Adem and Vanapalli [118]; and Pack [119]. Table 10 contains the properties of sand and expansive soil following disturbance from the helical pile installation [120].

Table 8. Soil properties in finite element analysis.

Model Parameter	Expansive Soil (Undrained Behavior)	Sand (Drained Behavior)	Granular (Drained Behavior)
γ_{unsat} (kN/m^3)	15.3	16	19
γ_{sat} (kN/m^3)	18.4	19	21
E_{50}^{ref} (kN/m^2)	5000	40,000	50,000
E_{oed}^{ref} (kN/m^2)	5000	40,000	50,000
E_{ur}^{ref} (kN/m^2)	15,000	120,000	150,000
C' (kN/m^2)	17	0.1	0.1
φ' (°)	20	37	38
ψ (°)	0	7	8
ν_{ur} (−)	0.3	0.25	0.2
m (−)	1	0.5	0.5

Table 9. Steel and concrete properties considered in finite element analysis.

Model Parameter	Helix Plate (Linear Elastic)	Shaft (Linear Elastic)	Concrete Foot (Linear Elastic)
Unit weight (kN/m^3)	78	78	24
E (kN/m^2)	200,000,000	200,000,000	30,000,000
v (−)	0.3	0.3	0.15
Thickness (m)	0.01	-	0.5
Dimension (m)	-	0.1 × 0.1	1 × 1

Table 10. Parameters of disturbed sand and clay after the implementation of a helical pile.

Model Parameter	Expansive Soil (Undrained Behavior)	Sand (Drained Behavior)
γ_{unsat} (kN/m^3)	15.3	16
γ_{sat} (kN/m^3)	18.4	19
E_{50}^{ref} (kN/m^2)	3500	20,000
E_{oed}^{ref} (kN/m^2)	3500	20,000
E_{ur}^{ref} (kN/m^2)	10,500	60,000
C' (kN/m^2)	8	0.1
φ' (°)	19	30
ψ (°)	0	1
ν_{ur} (−)	0.3	0.25
m (−)	1	0.5

Boundary Conditions

Horizontal displacements are assumed to be zero at the lateral boundaries. In addition, both horizontal and vertical displacements equal zero at the bottom. This assumption corresponds to natural behavior in which surrounding soil at large horizontal distances functions as horizontal fixities [121]. After assigning all necessary inputs to the model, Plaxis generated the mesh shown in Figure 28. A coarse mesh is used for both types of piles, with some refinement around the piles (the number of soil elements is 14,526 and the number of nodes is 22,207).

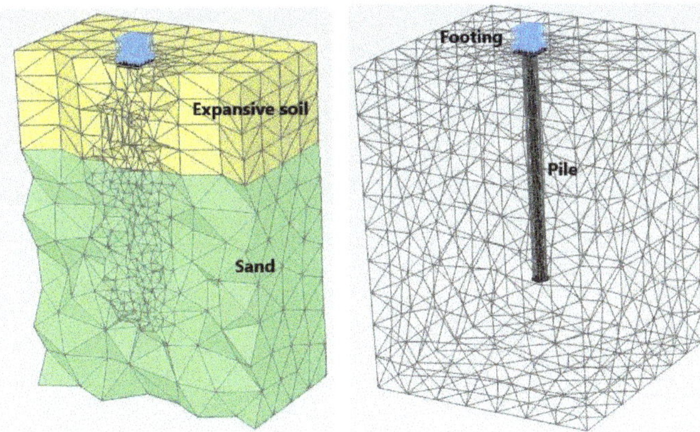

Figure 28. Generated mesh.

Initial Condition and Calculation Phases

The water table was assumed to be at the base of expansive soil when the initial conditions were assigned. Then, using Jacky's formula, initial stresses were calculated by assuming coefficients of earth pressure at rest, K0 = 0.536 and K0 = 0.4 for expansive soil and sand, respectively, where K0 = 1-sin.

In this numerical analysis, there were seven stages of construction. The soil behavior during all phases is plastic because the duration of construction (i.e., consolidation) does not impact behavior in this problem.

- In the first step, soil volume deactivation created the borehole for helical or granular anchor pile installation. Anchor plate, anchor rod, and granular anchor material, or disturbed sand and clay, are activated.
- The second analysis step activated the footing plate.
- The third analysis step applied the load to the footing (40 kN/m^2).
- During the fourth step, a volumetric strain of 8% was applied to each part of the ex-pansive soil volume from top to bottom to simulate heave.
- The next step continued in one of three possible directions:
- (A) The same volumetric strain of 8% was applied from the bottom upward.
- (B) An upwardly prescribed displacement of 25 mm on the surface footing generated the tensile resistance of the pile.
- (C) A downward displacement of 25 mm on the surface footing produced compres-sive resistance in a pile.

The change in volumetric strain mimics the heave of the clay in the analysis (positive volumetric strain). It is related to the degree of saturation in the expansive soil. Complete swelling occurs at a water content of 30%, following an S-shaped curve as determined by Tripathy et al. [122]. For highly plastic clays with porosities ranging from 0.4 to 0.6, the degree of saturation equivalent to 30% moisture content would be around 90% [123]. As a result, the moisture-swell function can provide 100% swelling at a saturation level of approximately 90% (Figure 29). Al-Shamrani and Dhowian [124] demonstrated that data from the triaxial compression test predicted field measurements of surface heave and reported that the results of the traditional oedometer test are about 1/3 as accurate as the actual surface heave. As a result, the maximum volumetric strain is 8%, which is 1/3 of the maximum free swell value obtained by Thakur and Singh [125]. This study uses an 8% positive volumetric strain for comparison.

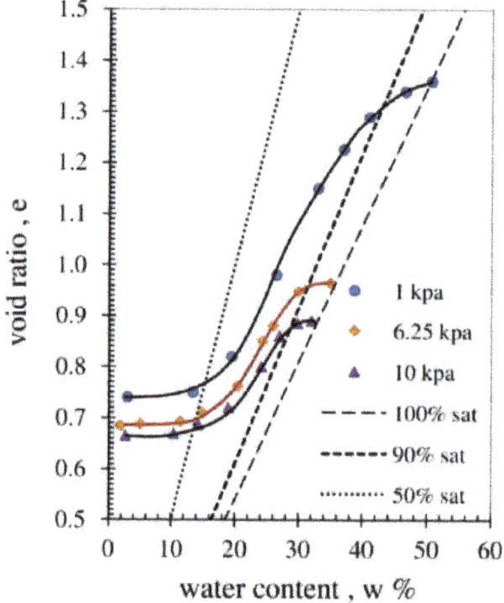

Figure 29. S-shape curves for different surcharges [123].

7.2. Comparison between Helical and Granular Anchor Piles: Results of Numerical Analysis

7.2.1. Pullout Load Comparison

Increasing the granular anchor pile and helical pile length increases the pullout load required to resist an upward movement (25 mm). Table 11 shows the pullout loads for granular anchors and helical piles. Note from Table 11 that the granular anchor pile achieves an 18% to 25% better performance than the helical pile. The granular anchor pile has a larger contact area with the soil, giving it better friction resistance. The helical pile disturbs and weakens the soil during its installation, reducing resistance. The pullout behavior of the three different lengths of granular anchors and helical piles with a diameter of 0.6 m is shown in Figure 30.

Table 11. Pullout load comparison between HP and GAP.

L (m)	GAP Pullout Load (kN)	HP Pullout Load (kN)	(GAP-HP)/HP (%)
4	159	135	17.8
7	271	222	22.1
10	315	253	24.5

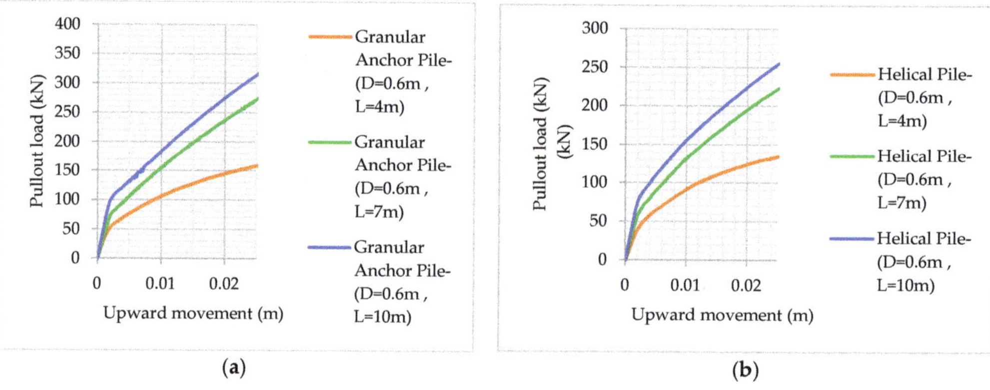

(a) (b)

Figure 30. Pullout behavior of (**a**) granular anchor piles and (**b**) helical piles.

7.2.2. Compressive Load Comparison

Increasing the pile lengths increases the load resistance for a 25 mm displacement. Table 12 shows the loads for both granular anchors and helical piles. Note from Table 12 that the granular anchor pile achieves a 4% to 11% better performance than the helical pile. Figure 31 presents the compressive load behavior of the three lengths of granular anchors and helical piles with a diameter of 0.6 m.

Table 12. Compressive load resistance of GAP and HP.

L (m)	GAP Capacity (kN)	HP Capacity (kN)	(GAP-HP)/HP (%)
4	241	230	4.8
7	290	262	10.7
10	315	284	10.9

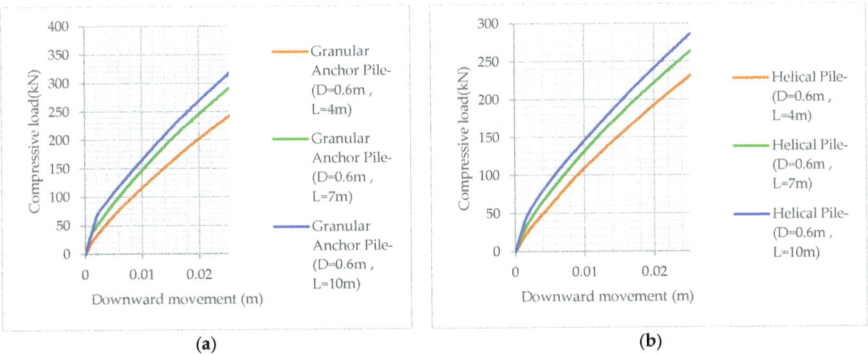

(a) (b)

Figure 31. Compressive behavior of (**a**) granular anchor piles and (**b**) helical piles.

7.2.3. Heave Comparison

This section presents heave reductions from three lengths of granular anchor and helical piles. The analysis examines the effects of saturation going downward from the top (rainfall infiltration) versus upward from the bottom (groundwater rise). The heave comparison looks at movement as each successive meter of the active zone saturates until it is completely saturated. Figure 32 illustrates the possible directions of moisture migration.

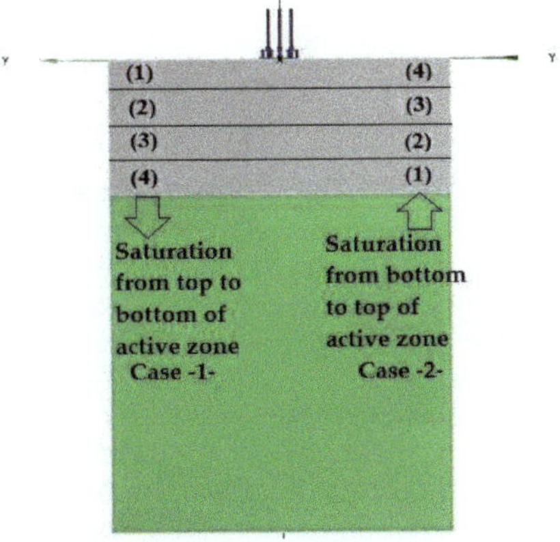

Figure 32. Numbering of active zone parts for the two cases of saturation (where the numbers (1–4) represent the sequence of saturation of the partial layers of the expansive soil).

Tarting Saturation from the Top (Case 1)

One can observe from Figure 33 that the soil expands as each successive meter of the active zone saturates. The piles behave similarly when both remain in the active zone (L = 4 m). If the piles penetrate the stable soil below (L = 7, 10 m), heave reduces drastically (by 1/5), with the two types of piles exhibiting slightly different behavior. At the final stage of saturation, both piles resist heaving nearly equally, with granular anchor piles slightly better. They perform better under more confining stresses. The stresses generate higher interface friction for the granular anchor piles and hold the helix more tightly for the helical piles. The behavior is consistent with Muthukumar and Shukla's study [100]. However, there are some minor differences due to the smaller scale of their tests. Figure 34 further highlights the benefits of penetration into the stable zone.

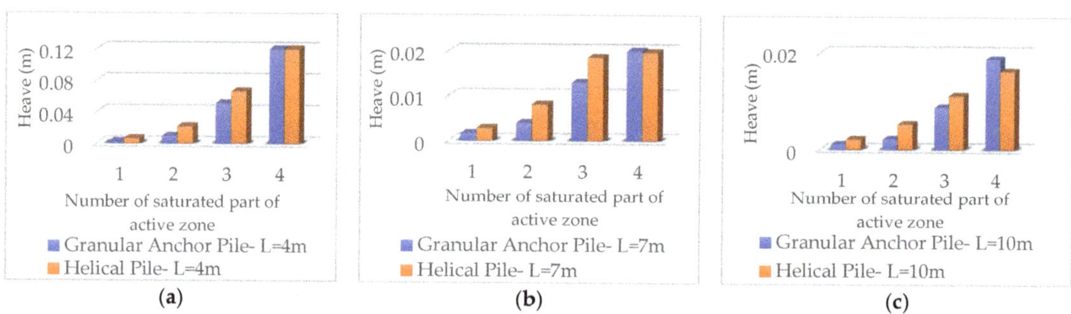

Figure 33. Heave of both helical and granular anchor piles for Case 1. (**a**) L = 4 m; (**b**) L = 7 m; (**c**) L = 10 m.

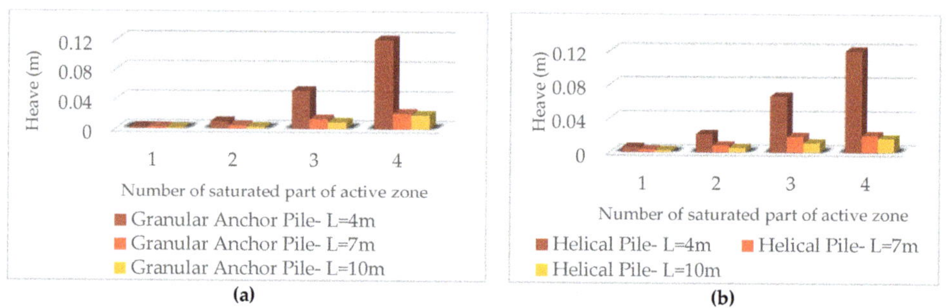

Figure 34. Effect of length on the heave of both helical and granular anchor piles for Case 1. (**a**) granular anchor pile; (**b**) helical pile.

Starting Saturation from the Bottom of Active Soil (Case 2)

Figure 35 demonstrates a fundamental difference between the two pile types. As the initial saturation occurs, the granular pile loses some resistance and the helical pile does not. However, in the following wetting stages, the granular pile does not change, whereas the helical one does. Either way, the final heave for the non-penetrating piles (4 m) is less than the top-down saturation (see Figure 35a vs. Figure 33a). Saturation from the bottom (Figure 35b,c) produces roughly half the heave as seepage from the top (Figure 33b,c) for penetrating piles. For Figure 35b,c, the granular pile yields with the first stage of moisture migration, followed by the helical pile in later stages.

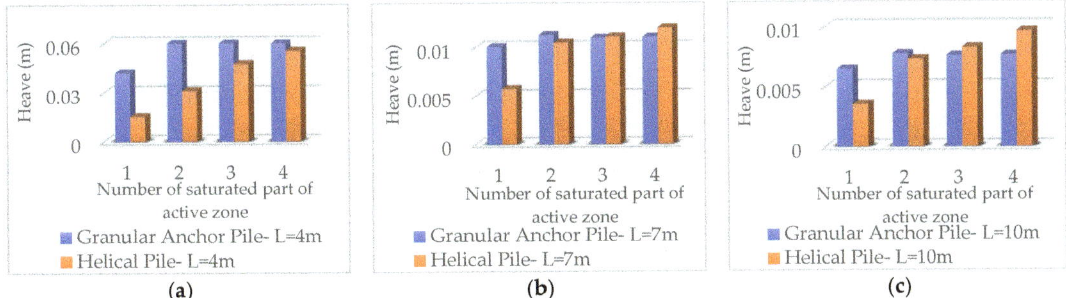

Figure 35. Heave of both helical and granular anchor piles for Case 2. (**a**) L = 4 m; (**b**) L = 7 m; (**c**) L = 10 m.

Figure 35 shows a heave comparison for each stage. In this case, the effectiveness of the helical pile and granular anchor pile becomes close, and the effectiveness of the granular anchor pile becomes better at full saturation. Figure 36 further highlights the benefits of penetration into the stable zone. The percentage of heave reduction when increasing the length from 4 to 7 m and at full saturation of the active zone is 77% and 85% for the helical pile and granular anchor pile, respectively.

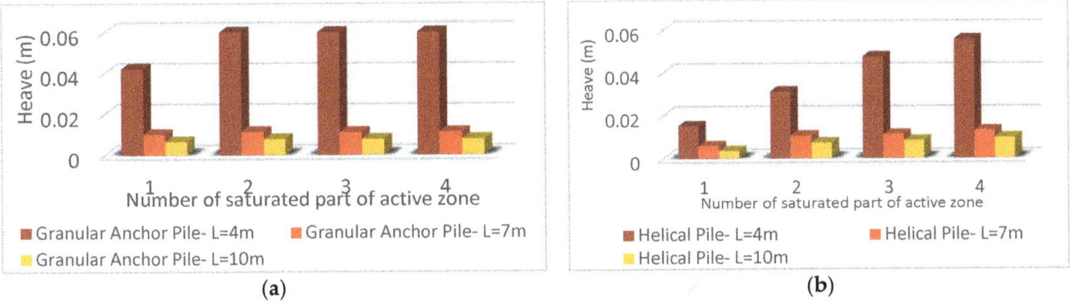

Figure 36. Effect of length on the heave of both helical and granular anchor piles for Case 2. (**a**) granular anchor pile; (**b**) helical pile.

Table 13 compares the heave values of the two types of piles and the percentage of heave reduction for each type compared to the no-pile scenario. At full saturation, the reduction percentages are close for both types of piles; when L/H > 1.5, where H is the thickness of the active zone, and the reduction percentage exceeds 90%.

Table 13. Heave values of the two types of piles as well as the percentage of heave reduction for each type.

Reduction Heave of Helical Pile (%)	Reduction Heave of Granular Anchor Pile (%)	Heave (m)			Length (m)	Case
		Unreinforced Soil	Helical Pile	Granular Anchor Pile		
54.9	51.7	0.268	0.121	0.129	4	
92.6	92.2	0.268	0.020	0.021	7	Case-1-
94.0	93.0	0.268	0.016	0.019	10	
78.6	71.7	0.260	0.055	0.073	4	
95.1	95.7	0.260	0.013	0.011	7	Case-2-
96.3	97.0	0.260	0.010	0.008	10	

Table 14 shows the comparison between helical and granular anchor piles, straight piles, and under-reamed piles, considering the review and numerical study.

Table 14. Comparison between piles considering the review and numerical study.

Property	Granular Anchor Pile	Helical Pile	Straight Shaft Piles	Under-Reamed Piles
Pullout load	Excellent	Good	Fair	Good
Downward load	Fair	Poor	Good	Excellent
Heave	Excellent	Excellent	Good	Very good
Combining concrete elements	Good	Good	Very good	Very good
Cost	Inexpensive	Expensive	Moderately Inexpensive	Moderately Expensive
Damage during installation	No Damage	Exposed to damage	No damage	No damage
lateral loads	Weak	Weaker	Good	Good
Type of load transfer	Tip resistance and skin friction	Tip resistance	Skin friction and tip resistance	Tip resistance and skin friction
Direction of installation	Vertical direction	All directions	Vertical direction	Vertical direction
Installation time	Installed with no cure time, allowing for quick project implementation.	Installed with no cure time, allowing for quick project implementation.	Concrete takes 2–4 weeks to cure.	Concrete takes 2–4 weeks to cure.
Environmental Effects	Steel can be removed and reused, reducing waste.	Steel can be removed and reused, reducing waste.	Non-reusable after installation.	Non-reusable after installation.

8. Advantages of Granular Anchor Piles and Helical Piles

Helical and granular anchor piles are foundations used at a construction site to support and stabilize structures. They are frequently employed when conventional foundation techniques are neither feasible nor practical [27]. Granular and helical anchor piles enjoy several social and environmental advantages. One advantage is that they can be installed quickly with less disruption. In urban areas, construction may disrupt local businesses and residents.

Also, helical and granular anchor piles are frequently more cost-efficient than other foundation types, making them more affordable for projects of all sizes [27,37]. Developers now have a simpler way of building new structures, which lowers economic risk and promotes development and economic growth. Finally, helical and granular anchor piles reduce material demand, making them more environmentally friendly than other foundations. Concrete is a significant contributor to greenhouse gas emissions during production, but it is absent from these foundations. Additionally, the smaller environmental impact of these foundations will mitigate the effects of construction on the surrounding environment. In general, helical and granular anchor piles can support the development of cost-effective, efficient, and sustainable infrastructure, which can benefit society.

9. Conclusions

The conclusions and recommendations based on the literature review and numerical study for the use of granular anchor piles and helical piles are as follows:

- Granular anchors and helical piles offer practical solutions to expansive soil foundation problems by limiting heave and providing structural support.
- According to field and laboratory studies, the granular anchor pile appears to be on par with or better than some currently employed tension-resistant foundation systems, such as concrete straight shaft piles, belled piles, and helical piles.
- The two types of failure reported were shaft failure and localized swelling failure at the granular anchor pile's base. Short granular anchor piles experienced shaft failure, while long granular anchor piles experienced localized bulging failure.
- The ideal length-to-diameter ratio is between 10 and 13, after which there is no considerable increase in the uplift resistance. The diameter of the granular anchor pile influences the uplift resistance more than the length. Uplift resistance increases as the granular fill's elastic modulus and relative density increase. The granular anchor pile's uplift capacity decreases as the moisture content of the surrounding soil increases.

- Granular anchor pile groups installed in expansive soil reduce heave better than single piles, but their ideal spacing and arrangement require additional research.
- Soil modifiers added during helical pile installation in expansive soil (adding 3% silica fumes, 3% coal fly ash, 6% hydrated lime, 3% mixture of silica fumes to coal fly ash [3:1] or a 2% mixture of hydrated lime: cement [1:1]) reduces the uplift movement of helical piles by more than 50%.
- Helical pile length, number of helixes, and their diameter significantly impact the amount and rate of upward movement. The deeper helical piles with larger L/D ratios showed pullout capacity greater than the shallower piles. The maximum upward movement of a group of helical piles is less than a single pile.
- Installing square shaft helical piers with a minimum installation torque of 5.4 kN-m, or to refusal, ensures that the helix (helical bearing plate) embeds below the depth of seasonal moisture change (active zone).
- A single helix will embed below the active zone, while a second helix above it may cause unwanted lift.
- The granular anchor pile outperforms the helical pile in resisting pullout and compressive forces. The degree of performance improvement increases with pile length, reaching up to 24.5% for uplift force and 11% for downward force.
- In numerical study case 1, as the active zone became more saturated, the granular anchor pile's relative effectiveness decreased compared to the helical pile. When both piles increase from 4 to 7 m long and the active zone is fully saturated, the heave is reduced by approximately 84%.
- In numerical study case 2, when the piles are entirely within the active zone, the helical pile resists heaving better than the granular anchor pile. Heaving decreases significantly as pile length penetrates beyond the active zone. During full saturation of the active zone, the piles with 7 and 10 m lengths reduced heave by 77% and 85% for the granular anchor and the helical pile, respectively.
- Most numerical studies of the behavior of helical or granular anchor piles in expansive soils ignored the change in suction in expansive soils. Instead, they imposed a volume change inside the effective area, an approximation that does not represent field conditions. Future studies will need to account for changes in suction and effective stress and their effect on the volume change of the expansive soil. Our program will conduct future research on this topic.

Author Contributions: Conceptualization, A.A. and R.P.R.; investigation, A.A.; writing—original draft preparation, A.A. and R.A.; numerical modeling, A.A., R.P.R. and R.A.; writing—review and editing, R.P.R.; supervision, R.P.R. All authors have read and agreed to the published version of the manuscript.

Funding: This research received no external funding.

Informed Consent Statement: Not applicable.

Conflicts of Interest: The authors declare no conflict of interest.

References

1. Stoll, S.C.; Henning, S.R.; Bagley, A.D.; Wieghaus, K.T. Foundation Damage Assessments and Structural Repairs. In *Forensic Engineering*; American Society of Civil Engineers: Denver, Colorado, 2022; pp. 166–174. ISBN 9780784484548.
2. Baer, D.H. *Building Losses from Natural Hazards: Yesterday, Today and Tomorrow*; JH Wiggins Co.: London, UK, 1978.
3. Karim, M.R.; Rahman, M.M.; Nguyen, H.B.K.; Kazmi, S.; Devkota, B.; Karim, R.; Rahman, M.; Bao, H.; Nguyen, K. Accounting for Expansive Soil Movement in Geotechnical Design—A State-of-the-Art Review. *Sustainability* **2022**, *14*, 15662. [CrossRef]
4. Bowles, J.E. *Foundation Analysis and Design*; McGraw-Hills Inc.: New York, NY, USA, 1988; ISBN 0070067767.
5. McOmber, R.M.; Thompson, R.W. Verification of Depth of Wetting for Potential Heave Calculations. In *Advances in Unsaturated Geotechnics*; 2000; pp. 409–422. Available online: https://ascelibrary.org/doi/10.1061/40510%28287%2928 (accessed on 15 March 2022).
6. Nelson, J.D.; Overton, D.D.; Durkee, D.B. Depth of Wetting and the Active Zone. In Proceedings of the Expansive Clay Soils and Vegetative Influence on Shallow Foundations, London, UK, 8 October 2001; pp. 95–109.

7.   Walsh, K.D.; Colby, C.A.; Houston, W.N.; Houston, S.L. Method for Evaluation of Depth of Wetting in Residential Areas. *J. Geotech. Geoenvironmental Eng.* **2009**, *135*, 169–176. [CrossRef]

8.   Teodosio, B.; Kristombu Baduge, K.S.; Mendis, P. A Review and Comparison of Design Methods for Raft Substructures on Expansive Soils. *J. Build. Eng.* **2021**, *41*, 102737. [CrossRef]

9.   Ijaz, N.; Ye, W.; ur Rehman, Z.; Dai, F.; Ijaz, Z. Numerical Study on Stability of Lignosulphonate-Based Stabilized Surficial Layer of Unsaturated Expansive Soil Slope Considering Hydro-Mechanical Effect. *Transp. Geotech.* **2022**, *32*, 100697. [CrossRef]

10. Fulzele, U.G.; Ghane, V.R.; Parkhe, D.D. Study of Structures in Black Cotton Soil. In Proceedings of the IRF International Conferece on Advances Sciences Engineering & Technology, Pune, India, 16 October 2016; Volume 4, ISBN 978-93-86291-14-1.

11. Briggs, K.M.; Loveridge, F.A.; Glendinning, S. Failures in Transport Infrastructure Embankments. *Eng. Geol.* **2017**, *219*, 107–117. [CrossRef]

12. Simons, K.B. Limitations of Residential Structures on Expansive Soils. *J. Perform. Constr. Facil.* **1991**, *5*, 258–270. [CrossRef]

13. Zamin, B.; Nasir, H.; Mehmood, K.; Iqbal, Q. Field-Obtained Soil-Water Characteristic Curves of KPK Expansive Soil and Their Prediction Correlations. *Adv. Civ. Eng.* **2020**, *2020*, 1–13. [CrossRef]

14. Dang, L.C.; Khabbaz, H.; Ni, B.J. Improving Engineering Characteristics of Expansive Soils Using Industry Waste as a Sustainable Application for Reuse of Bagasse Ash. *Transp. Geotech.* **2021**, *31*, 100637. [CrossRef]

15. Medina-Martinez, C.J.; Sandoval-Herazo, L.C.; Zamora-Castro, S.A.; Vivar-Ocampo, R.; Reyes-Gonzalez, D. Natural Fibers: An Alternative for the Reinforcement of Expansive Soils. *Sustainability* **2022**, *14*, 9275. [CrossRef]

16. Taleb, T.; Unsever, Y.S. Study on Strength and Swell Behavioral Change and Properties of the Clay–Fiber Mixtures. *Sustainability* **2022**, *14*, 6767. [CrossRef]

17. Tiwari, N.; Satyam, N.; Puppala, A.J. Effect of Synthetic Geotextile on Stabilization of Expansive Subgrades: Experimental Study. *J. Mater. Civ. Eng.* **2021**, *33*, 04021273. [CrossRef]

18. Tiwari, N.; Satyam, N. Coupling Effect of Pond Ash and Polypropylene Fiber on Strength and Durability of Expansive Soil Subgrades: An Integrated Experimental and Machine Learning Approach. *J. Rock Mech. Geotech. Eng.* **2021**, *13*, 1101–1112. [CrossRef]

19. Tiwari, N.; Satyam, N.; Puppala, A.J. Strength and Durability Assessment of Expansive Soil Stabilized with Recycled Ash and Natural Fibers. *Transp. Geotech.* **2021**, *29*, 100556. [CrossRef]

20. Li, H.; Wang, Y.; Yin, Z.; Al-Soudani, W.H.S.; Fattah, M.Y.; Ziyara, H.M.; Albusoda, B.S. An Experimental Study of the Load Carrying Capacity of Straight Shaft and Underreamed Piles in Expansive Soil. *IOP Conf. Ser. Mater. Sci. Eng.* **2021**, *1067*, 1–12. [CrossRef]

21. Onur, M.İ.; Bıçakcı, M.; Kardoğan, P.S.Ö.; Erdağ, A.; Aghlmand, M. Laboratory model design for deep soil mixing method. *Adv. Civ. Archit. Eng.* **2022**, *13*, 59–69. [CrossRef]

22. El-Samea, A.; Hassan, W.; Mowafy, Y.M.; El-Naiem, A.; Abdo, M.; Towfeek, A.R. numerical analysis of shallow foundation on expansive soil. *J. Al-Azhar Univ. Eng. Sect.* **2022**, *17*, 480–501. [CrossRef]

23. Mansour, M.A.; El Naggar, M.H. Optimization of Grouting Method and Axial Performance of Pressure-Grouted Helical Piles. *Can. Geotech. J.* **2021**, *59*, 702–714. [CrossRef]

24. Ziyara, H.M.; Albusoda, B.S. Experimental and Numerical Study of the Bulb's Location Effect on the Behavior of under-Reamed Pile in Expansive Soil. *J. Mech. Behav. Mater.* **2022**, *31*, 90–97. [CrossRef]

25. Roy, R.; Rao, C.N.; Mouli, S.S. Evaluation of Heave Behavior by Numerical Modeling of Granular Pile Anchor in Expansive Soil. *Lect. Notes Civ. Eng.* **2023**, *297*, 103–115. [CrossRef]

26. Department of the Army USA Technical Manual TM 5–818–7, Foundations in Expansive Soils 1983. A: STPAGE2.PDF (army.mil).

27. Das, B.M.; Shukla, S.K. *Earth Anchors*, 2nd ed.; J. Ross Publishing: Fort Lauderdale, FL, USA, 2013; ISBN 978-1-60427-077-8.

28. Li, W. *Centrifuge Modeling and Large Deformation Analyses of Axially Loaded Helical Piles in Cohesive Soils*; University of Alberta: Edmonton, AB, Canada, 2022.

29. Venkatesan, V.; Mayakrishnan, M. Behavior of Mono Helical Pile Foundation in Clays under Combined Uplift and Lateral Loading Conditions. *Appl. Sci.* **2022**, *12*, 6827. [CrossRef]

30. Malhotra, H.; Sanjay; Singh, K. Effect of Load Inclination on the Uplift Capacity of the Granular Anchor Pile Foundation in Cohesive Soil. *Arab. J. Geosci.* **2022**, *15*, 1–11. [CrossRef]

31. Yu, H.; Zhou, H.; Sheil, B.; Liu, H. Finite Element Modelling of Helical Pile Installation and Its Influence on Uplift Capacity in Strain Softening Clay. *Can. Geotech. J.* **2022**. [CrossRef]

32. Joseph, J.; Kumar, S.; Sawant, V.A.; Patel, J.B. An Experimental and Numerical Comparative Study on the Uplift Capacity of Single Granular Pile Anchor and Rough Pile in Sand. *Int. J. Geotech. Eng.* **2021**, *16*, 499–513. [CrossRef]

33. Alwalan, M.; Alnuaim, A. Axial Loading Effect on the Behavior of Large Helical Pile Groups in Sandy Soil. *Arab. J. Sci. Eng.* **2022**, *47*, 5017–5031. [CrossRef]

34. Mahmoudi-Mehrizi, M.E.; Ghanbari, A.; Sabermahani, M. The Study of Configuration Effect of Helical Anchor Group on Retaining Wall Displacement. *Geomech. Geoengin.* **2020**, *17*, 598–612. [CrossRef]

35. Lin, Y.; Xiao, J.; Le, C.; Zhang, P.; Chen, Q.; Ding, H. Bearing Characteristics of Helical Pile Foundations for Offshore Wind Turbines in Sandy Soil. *J. Mar. Sci. Eng.* **2022**, *10*, 889. [CrossRef]

36. Chaghameh, A.; Arjomand, M.; Adresi, M. Screw Pile and Its Application in Road's Subgrade Improvement. *Road* **2022**, *30*, 197–206. [CrossRef]

37. Joseph, J.; Kumar, S.; Patel, J.B.; Sawant, V.; Tandel, Y. Model Tests on Granular Pile Anchor and Helical Anchor: A Comparative Study. *Int. J. Geosynth. Ground Eng.* **2022**, *8*, 1–12. [CrossRef]

38. Akhtar, N.; Ishak, M.I.S.; Ahmad, M.I.; Umar, K.; Md Yusuff, M.S.; Anees, M.T.; Qadir, A.; Almanasir, Y.K.A. Modification of the Water Quality Index (Wqi) Process for Simple Calculation Using the Multi-Criteria Decision-Making (Mcdm) Method: A Review. *Water* **2021**, *13*, 905. [CrossRef]

39. Akhtar, N.; Syakir Ishak, M.I.; Bhawani, S.A.; Umar, K. Various Natural and Anthropogenic Factors Responsible for Water Quality Degradation: A Review. *Water* **2021**, *13*, 2660. [CrossRef]

40. Sabatini, P.J.; Pass, D.G.; Bachus, R.C.; Consultants, G. *Ground Anchors and Anchored Systems*; United States Federal Highway Administration, Office of Bridge Technology: Washington, DC, USA, 1999.

41. Al-Busoda, B.S.; Awn, S.H.A.; Abbase, H.O. Numerical Modeling of Retaining Wall Resting on Expansive Soil. *Geotech. Eng. J. Seags Agssea* **2017**, *48*, 116–121. Available online: https://www.researchgate.net/profile/Safa-Abid-Awn/publication/3315 22032_Numerical_Modeling_of_Retaining_Wall_Resting_on_Expansive_Soil/links/5c7e1bc392851c695054e094/Numerical-Modeling-of-Retaining-Wall-Resting-on-Expansive-Soil.pdf (accessed on 15 March 2022).

42. Al-Busoda, B.S.; Abbas, H.O. Numerical Simulation of Mitigation of Soil Swelling Problem under Communication Tower Using Helical Piles. *J. Geotech. Eng.* **2017**, *4*, 35–46. Available online: https://www.researchgate.net/profile/Hassan-Abbas-17/publication/331135065_Numerical_Simulation_of_Mitigation_of_Soil_Swelling_Problem_under_Communication_Tower_Using_Helical_Piles/links/5c670f874585156b57ffeab3/Numerical-Simulation-of-Mitigation-of-Soil-Swelling-Problem-under-Communication-Tower-Using-Helical-Piles.pdf (accessed on 15 March 2022).

43. Pérez, Z.A.; Schiavon, J.A.; de Tsuha, C.H.C.; Dias, D.; Thorel, L. Numerical and Experimental Study on Influence of Installation Effects on Behaviour of Helical Anchors in Very Dense Sand. *Can. Geotech. J.* **2017**, *55*, 1067–1080. [CrossRef]

44. Elsherbiny, Z.H.; El Naggar, M.H. Axial Compressive Capacity of Helical Piles from Field Tests and Numerical Study. *Can. Geotech. J.* **2013**, *50*, 1191–1203. [CrossRef]

45. George, B.E.; Banerjee, S.; Gandhi, S.R. Numerical Analysis of Helical Piles in Cohesionless Soil. *Int. J. Geotech. Eng.* **2017**, *14*, 361–375. [CrossRef]

46. Nowkandeh, M.J.; Choobbasti, A.J. Numerical Study of Single Helical Piles and Helical Pile Groups under Compressive Loading in Cohesive and Cohesionless Soils. *Bull. Eng. Geol. Environ.* **2021**, *80*, 4001–4023. [CrossRef]

47. Li, W.; Deng, L. Axial Load Tests and Numerical Modeling of Single-Helix Piles in Cohesive and Cohesionless Soils. *Acta Geotech.* **2019**, *14*, 461–475. [CrossRef]

48. Vignesh, V.; Muthukumar, M. Experimental and Numerical Study of Group Effect on the Behavior of Helical Piles in Soft Clays under Uplift and Lateral Loading. *Ocean Eng.* **2023**, *268*, 1–18. [CrossRef]

49. Garakani, A.A.; Serjoie, K.A. Ultimate Bearing Capacity of Helical Piles as Electric Transmission Tower Foundations in Unsaturated Soils: Analytical, Numerical, and Experimental Investigations. *Int. J. Geomech.* **2022**, *22*, 04022194. [CrossRef]

50. Cheng, P.; Guo, J.; Yao, K.; Chen, X. Numerical Investigation on Pullout Capacity of Helical Piles under Combined Loading in Spatially Random Clay. *Mar. Georesources Geotechnol.* **2022**, 1–14. [CrossRef]

51. Kranthikumar, A.; Sawant, V.A.; Shukla, S.K. Numerical Modeling of Granular Anchor Pile System in Loose Sandy Soil Subjected to Uplift Loading. *Int. J. Geosynth. Ground Eng.* **2016**, *2*, 1–7. [CrossRef]

52. Kranthikumar, A.; Sawant, V.A.; Kumar, P.; Shukla, S.K. Numerical and Experimental Investigations of Granular Anchor Piles in Loose Sandy Soil Subjected to Uplift Loading. *Int. J. Geomech.* **2017**, *17*, 04016059. [CrossRef]

53. Sharma, R.K. A Numerical Study of Granular Pile Anchors Subjected to Uplift Forces in Expansive Soils Using PLAXIS 3D. *Indian Geotech. J.* **2019**, *49*, 304–313. [CrossRef]

54. Malhotra, H.; Singh, S.K. Experimental and Numerical Studies on Pull-out Behavior of Granular Anchor Pile Foundation Embedded in Sandy Soil. *Arab. J. Sci. Eng.* **2021**, *46*, 4477–4487. [CrossRef]

55. Mahmoudi-Mehrizi, M.-E.; Ghanbari, A. A Review of the Advancement of Helical Foundations from 1990–2020 and the Barriers to Their Expansion in Developing Countries. *J. Eng. Geol.* **2021**, *14*, 37–84. [CrossRef]

56. Hosamo, H.H.; Nielsen, H.K.; Alnmr, A.N.; Svennevig, P.R.; Svidt, K. A Review of the Digital Twin Technology for Fault Detection in Buildings. *Front. Built Environ.* **2022**, *8*, 1–23. [CrossRef]

57. Perianes-Rodriguez, A.; Waltman, L.; Van Eck, N.J. Constructing Bibliometric Networks: A Comparison between Full and Fractional Counting. *J. Informetr.* **2016**, *10*, 1178–1195. [CrossRef]

58. Adem, H.H.; Vanapalli, S.K. Review of Methods for Predicting in Situ Volume Change Movement of Expansive Soil over Time. *J. Rock Mech. Geotech. Eng.* **2015**, *7*, 73–86. [CrossRef]

59. Nelson, J.; Miller, D.J. *Expansive Soils: Problems and Practice in Foundation and Pavement Engineering*; John Wiley & Sons: New York, NY, USA, 1997; ISBN 0471181145.

60. Holtz, R.D.; Kovacs, W.D.; Sheahan, T.C. *An Introduction to Geotechnical Engineering*; Prentice-Hall: Hoboken, NJ, USA, 1981; ISBN 0-13-484394-0.

61. Holtz, W.G. Expansive Clays-Properties and Problems. *Q. Colo. Sch. Mines* **1959**, *54*, 90–125.

62. Dakshanamurthy, V.; Raman, V. A Simple Method of Identifying an Expansive Soil. *Soils Found.* **1973**, *13*, 97–104. [CrossRef]

63. *AASHTO T258-81*; Standard Method of Test for Determining Expansive Soils. American Association of State and Highway Transportation Officials: Washington, DC, USA, 2018.

64. *ASTM D4546-14*; Standard Test Methods for One-Dimensional Swell or Collapse of Soils. ASTM International: West Conshohocken, PA, USA, 2014.
65. Chen, F.H. *Foundations on Expansive Soils*; Elsevier: New York, NY, USA, 1975; Volume 12, ISBN 044460166X.
66. Jones, L.D.; Jefferson, I. *Expansive Soils. Volume 1, Geotechnical Engineering Principles, Problematic Soils and Site Investigation*; ICE Publishing: London, UK, 2012; pp. 413–441.
67. Peck, R.B.; Hanson, W.E.; Thornburn, T.H. *Foundation Engineering*, 2nd ed.; Wiley: New York, NY, USA, 1974; ISBN 978-0-471-67585-3.
68. Murthy, V.N.S. *Geotechnical Engineering: Principles and Practices of Soil Mechanics and Foundation Engineering*; Marcel Dekker, Inc: New York, NY, USA; Basel, Switzerland, 2002; ISBN 0824708733.
69. Zumrawi, M.M.E.; Abdelmarouf, A.O.; Gameil, A.E.A. Damages of Buildings on Expansive Soils: Diagnosis and Avoidance. *Int. J. Multidiscip. Sci. Emerg. Res.* **2017**, *6*, 108–115.
70. Lytton, R.; Aubeny, C.; Bulut, R. *Design Procedure for Pavements on Expansive Soils: Volume 3*; Report 0–4; Texas Transportation Institute, Texas A & M University System: College Station, TX, USA, 2005.
71. Rao, M.R.; Rao, A.S.; Babu, R.D. Arresting Heave of Expansive Soil Beds with Lime-Stabilised Flyash Cushion. *J. Inst. Eng. Part CV Civ. Eng. Div.* **2007**, *87*, 13–17.
72. Murty, V.R.; Praveen, G.V. Use of Chemically Stabilized Soil as Cushion Material below Light Weight Structures Founded on Expansive Soils. *J. Mater. Civ. Eng.* **2008**, *20*, 392–400. [CrossRef]
73. Walsh, K.D.; Houston, S.; Houston, W.N.; Harraz, A.M. Finite Element Evaluation of Deep-Seated Swell. In Proceedings of the 4th Asia Pacific Conference on Unsaturated Soils, Newcastle, Australia, 23–25 November 2010; pp. 731–736.
74. Ahmed, A. Evaluation of Drying and Wetting Cycles with Soil Cushion to Mitigate the Potential of Expansive Soil in Upper Egypt. *Electron. J. Geotech. Eng.* **2009**, *15*, 1–11.
75. Srinivas, K.; Prasad, D.S.V.; Rao, E. A Study on Improvement of Expansive Soil by Using Cns (Cohesive Non Swelling) Layer. *Int. J. Innov. Res. Technol.* **2016**, *3*, 54–60. Available online: http://ijirt.org/Article?manuscript=143878 (accessed on 15 March 2022).
76. Gromko, G.J. Review of Expansive Soils. *J. Geotech. Eng. Div.* **1974**, *100*, 667–687. [CrossRef]
77. O'Neill, M.W.; Poormoayed, N. Methodology for Foundations on Expansive Clays. *J. Geotech. Eng. Div.* **1980**, *106*, 1345–1367. [CrossRef]
78. Prusinski, J.R.; Bhattacharja, S. Effectiveness of Portland Cement and Lime in Stabilizing Clay Soils. *Transp. Res. Rec.* **1999**, *1652*, 215–227. [CrossRef]
79. Moayed, R.Z.; Haratian, M.; Izadi, E. Improvement of Volume Change Characteristics of Saline Clayey Soils. *J. Appl. Sci.* **2011**, *11*, 76–85. [CrossRef]
80. Belabbaci, Z.; Mamoune, S.M.A.; Bekkouche, A. Stabilization of Expansive Soils with Milk of Lime: The Case of Clays of Tlemcen, Algeria. *Electron. J. Geotech. Eng.* **2012**, *17*, 1293–1304.
81. Al-Taie, A.; Disfani, M.M.; Evans, R.; Arulrajah, A.; Horpibulsuk, S. Swell-Shrink Cycles of Lime Stabilized Expansive Subgrade. *Procedia Eng.* **2016**, *143*, 615–622. [CrossRef]
82. Mahedi, M.; Cetin, B.; White, D.J. Cement, Lime, and Fly Ashes in Stabilizing Expansive Soils: Performance Evaluation and Comparison. *J. Mater. Civ. Eng.* **2020**, *32*, 1–16. [CrossRef]
83. Jeyapalan, J.K.; Rice, G.T.; Lytton, R.L. *State-of-the-Art Review of Expansive Soil Treatment Methods*; Texas A & M University: College Station, TX, USA, 1981.
84. Venkataramana, K. Building on Expansive Clays with Special Reference to Trinidad. *West Indian J. Eng.* **2003**, *25*, 43–53. Available online: https://sta.uwi.edu/eng/wije/vol2502_jan2003/documents/Buildingonexpansiveclays.pdf (accessed on 15 March 2022).
85. Dafalla, M.A.; Shamrani, M.A. Road Damage Due to Expansive Soils: Survey of the Phenomenon and Measures for Improvement. In Proceedings of the GeoHunan International Conference, Hunan, China, 16 May 2011; pp. 73–80.
86. Sorochan, E.A. Use of Piles in Expansive Soils. *Soil Mech. Found. Eng.* **1974**, *11*, 33–38. [CrossRef]
87. Aljorany, A.N.; Ibrahim, S.F.; Al-Adly, A.I. Heave Behavior of Granular Pile Anchor-Foundation System (GPA-Foundation System) in Expansive Soil. *J. Eng.* **2014**, *20*, 1–22. Available online: https://www.iasj.net/iasj/download/f80b6b01950d0ef7 (accessed on 15 March 2022).
88. Hugher, J.M.O.; Withers, N.J. Reinforcing of Soft Cohesive Soils with Stone Columns. *Ground Eng.* **1974**, *7*, 42–49.
89. Rao, A.S.; Phanikumar, B.R.; Babu, R.D.; Suresh, K. Pullout Behavior of Granular Pile-Anchors in Expansive Clay Beds in Situ. *J. Geotech. Geoenvironmental Eng.* **2007**, *133*, 531–538. [CrossRef]
90. Phanikumar, B.R.; Srirama Rao, A.; Suresh, K. Field Behaviour of Granular Pile-Anchors in Expansive Soils. *Proc. Inst. Civ. Eng. Ground Improv.* **2008**, *161*, 199–206. [CrossRef]
91. Phanikumar, B.R.; Sharma, R.S.; Rao, A.S.; Madhav, M.R. Granular Pile Anchor Foundation (GPAF) System for Improving the Engineering Behavior of Expansive Clay Beds. *Geotech. Test. J.* **2004**, *27*, 279–287. [CrossRef]
92. Cooper, M.R.; Rose, A.N. Stone Column Support for an Embankment on Deep Alluvial Soils. *Proc. Inst. Civ. Eng. Geotech. Eng.* **1999**, *137*, 15–25. [CrossRef]
93. Muir Wood, D.; Hu, W.; Nash, D.F.T. Group Effects in Stone Column Foundations: Model Tests. *Geotechnique* **2000**, *50*, 689–698. [CrossRef]
94. Vashishtha, H.R.; Sawant, V.A. An Experimental Investigation for Pullout Response of a Single Granular Pile Anchor in Clayey Soil. *Int. J. Geo-Eng.* **2021**, *12*, 1–19. [CrossRef]

95. Srirama Rao, A.; Phanikumar, B.R.; Suresh, K. Response of Granular Pile-Anchors under Compression. *Proc. Inst. Civ. Eng.-Ground Improv.* **2008**, *161*, 121–129. [CrossRef]
96. Ismail, M.; Shahin, M. Finite Element Analyses of Granular Pile Anchors as a Foundation Option for Reactive Soils. In Proceedings of the International Conference on Advances in Geotechnical Engineering, Perth, Australia, 7–9 November 2011; pp. 1047–1052.
97. Sivakumar, V.; O'Kelly, B.C.; Madhav, M.R.; Moorhead, C.; Rankin, B. Granular Anchors under Vertical Loading–Axial Pull. *Can. Geotech. J.* **2013**, *50*, 123–132. [CrossRef]
98. Krishna, P.H.; Murty, V.R. Pull-out Capacity of Granular Anchor Piles in Expansive Soils. *IOSR J. Mech. Civ. Eng.* **2013**, *5*, 24–31. [CrossRef]
99. Johnson, N.; Sandeep, M.N. Ground Improvement Using Granular Pile Anchor Foundation. *Procedia Technol.* **2016**, *24*, 263–270. [CrossRef]
100. Muthukumar, M.; Shukla, S.K. Comparative Study on the Behaviour of Granular Pile Anchors and Helical Pile Anchors in Expansive Soils Subjected to Swelling. *Int. J. Geotech. Eng.* **2017**, *14*, 49–54. [CrossRef]
101. Muthukumar, M.; Shukla, S.K. Swelling Behaviour of Expansive Clay Beds Reinforced with Encased Granular Pile Anchors. *Int. J. Geotech. Eng.* **2016**, *12*, 109–117. [CrossRef]
102. Abbas, H.O. Laboratory Study on Reinforced Expansive Soil with Granular Pile Anchors. *Int. J. Eng.* **2020**, *33*, 1167–1172. [CrossRef]
103. Sharma, A.; Sharma, R.K. An Experimental Study on Uplift Behaviour of Granular Anchor Pile in Stabilized Expansive Soil. *Int. J. Geotech. Eng.* **2021**, *15*, 950–963. [CrossRef]
104. Khan, H.A.; Gaddam, K. An Experimental Study on Heave and Uplift Behaviour of Granular Pile Anchor Foundation System. In Proceedings of the IOP Conference Series: Earth and Environmental Science, Surakarta, Indonesia, 24–25 August 2021; Volume 822, pp. 1–9.
105. Perko, H.A. *Helical Piles: A Practical Guide to Design and Installation*; John Wiley & Sons: New York, NY, USA, 2009; ISBN 0470404795.
106. Pack, J.S. Performance of Square Shaft Helical Pier Foundations in Swelling Soils. *Geotech. Pract. Publ.* **2007**, 76–85. [CrossRef]
107. Hoyt, R.M.; Clemence, S. Uplift Capacity of Helical Anchors in Soil. In Proceedings of the 12th International Conference on Soil Mechanics and Foundation Engineering, Rio de Janeiro, Brazil, 13–18 August 1989; pp. 1019–1022.
108. Ghaly, A.; Hanna, A. Experimental and Theoretical Studies on Installation Torque of Screw Anchors. *Can. Geotech. J.* **1991**, *28*, 353–364. [CrossRef]
109. Chao, K.C.; Nelson, J.D.; Overton, D.D. Factors Influencing Design of Deep Foundations on Expansive Soils. In Proceedings of the 5th Asia Pacific Conference on Unsaturated Soils, Pattaya, Thailand, 14–16 November 2011; Volume 2, pp. 829–834.
110. Al-Busoda, B.S.; Abbase, H.O. Mitigation of Expansive Soil Problems by Using Helical Piles with Additives. *J. Geotech. Eng.* **2015**, *2*, 30–40.
111. Al-Busoda, B.S.; Abbase, H.O. Helical Piles Embedded in Expansive Soil Overlaying Sandy Soil. *Al-Khwarizmi Eng. J.* **2016**, *12*, 19–25.
112. Albusoda, B.S.; Abbase, H.O. Performance Assessment of Single and Group of Helical Piles Embedded in Expansive Soil. *Int. J. Geo-Eng.* **2017**, *8*, 1–20. [CrossRef]
113. Mulyanda, D.; Iqbal, M.M.; Dewi, R. The Effect of Helical Size On Uplift Pile Capacity. *Int. J. Sci. Technol. Res.* **2020**, *9*, 4140–4145. Available online: https://repository.unsri.ac.id/49463/1/The-Effect-Of-Helical-Size-On-Uplift-Pile-Capacity.pdf (accessed on 15 March 2022).
114. Sangeetha, S.; Hari Krishna, P. Analysis of Heave Behaviour of Expansive Soil Provided with Granular Pile Anchors Using Plaxis. *Lect. Notes Civ. Eng.* **2020**, *55*, 391–404. [CrossRef]
115. Alsirawan, R.; Alnmr, A. Dynamic Behavior of Gravity Segmental Retaining Walls. *Pollack Period.* **2022**, *1*. [CrossRef]
116. Dysli, M. *Swiss Standard SN 670 010b, Characteristic Coefficients of Soils*; Strasse und Verkehr: Zurich, Switzerland, 2000.
117. Xiao, J.; Yang, H.; Zhang, J.; Tang, X. Properties of Drained Shear Strength of Expansive Soil Considering Low Stresses and Its Influencing Factors. *Int. J. Civ. Eng.* **2018**, *16*, 1389–1398. [CrossRef]
118. Adem, H.H.; Vanapalli, S.K. Elasticity Moduli of Expansive Soils from Dimensional Analysis. *Geotech. Res.* **2014**, *1*, 60–72. [CrossRef]
119. Pack, J.S. *Design and Inspection Guide for Helical Piles and Helical Tension Anchors*; Intermountainhelicalpiers Inc.: Denver, CO, USA, 2009; 194p.
120. Zhao, Q.; Wang, Y.; Tang, Y.; Ren, G.; Qiu, Z.; Luo, W.; Ye, Z. Numerical Analysis of the Installation Process of Screw Piles Based on the FEM-SPH Coupling Method. *Appl. Sci.* **2022**, *12*, 8508. [CrossRef]
121. Kaufmann, K.L.; Nielsen, B.N.; Augustesen, A.H. Finite Element Investigations on the Interaction between a Pile and Swelling Clay. *Uniw. Śląski* **2010**, *7*, 343–354.
122. Tripathy, S.; Subba Rao, K.S.; Fredlund, D.G. Water Content-Void Ratio Swell-Shrink Paths of Compacted Expansive Soils. *Can. Geotech. J.* **2011**, *39*, 938–959. [CrossRef]
123. Estabragh, A.R.; Parsaei, B.; Javadi, A.A. Laboratory Investigation of the Effect of Cyclic Wetting and Drying on the Behaviour of an Expansive Soil. *Soils Found.* **2015**, *55*, 304–314. [CrossRef]

124. Al-Shamrani, M.A.; Dhowian, A.W. Experimental Study of Lateral Restraint Effects on the Potential Heave of Expansive Soils. *Eng. Geol.* **2003**, *69*, 63–81. [CrossRef]
125. Thakur, V.; Narain Singh, D.; S Thakur, V.K.; Singh, D.N. Rapid Determination of Swelling Pressure of Clay Minerals. *J. Test. Eval.* **2005**, *33*, 239–245. [CrossRef]

 *sustainability*

Article

One-Dimensional Computational Model of Gyttja Clay for Settlement Prediction

Grzegorz Kacprzak, Artur Zbiciak, Kazimierz Józefiak, Paweł Nowak and Mateusz Frydrych *

Faculty of Civil Engineering, Warsaw University of Technology, 16 Armii Ludowej Ave., 00-637 Warsaw, Poland
* Correspondence: mateusz.frydrych.dokt@pw.edu.pl; Tel.: +48-22-2346543

Abstract: One of the most important subjects of geomechanics research is finding mathematical relationships which could correctly describe behavior of the soil under loading. Safety of every engineering structure depends strongly on accuracy and correctness of this description. As laboratory tests show, macroscopic properties of soil are complicated. Therefore, working out appropriate load-settlement relationships is considered to be a very difficult geomechanics tasks to solve. A majority of constitutive models proposed to date concern mineral soils and there is very little research related to modelling organic soil behavior under loading. In case of organic soils, due to their very complicated and composite structure, constitutive models are often formulated empirically based on laboratory tests of particular soils. The authors of this paper propose a 1-D rheological structure which accounts for complex behavior of soil related to the settlement process. The model simulates immediate reversible elastic settlement and plastic soil deformation as well as primary and secondary (creep effect) consolidation. Material parameters of the model were determined by a curve fitting procedure applied for a natural scale settlement test of plate foundation. The test was carried out in soil conditions connected with Eemian geological structure of Warsaw, i.e., Eemian glacial tunnel valley in Warsaw called Żoliborz Glacial Tunnel Valley filled with organic soils being up to 20 metres thick. This area has lately become an object of interest of investors as a site for building construction.

Keywords: soil rheology; settlement; creep; plate foundation; geotechnical engineering

Citation: Kacprzak, G.; Zbiciak, A.; Józefiak, K.; Nowak, P.; Frydrych, M. One-Dimensional Computational Model of Gyttja Clay for Settlement Prediction. *Sustainability* **2023**, *15*, 1759. https://doi.org/10.3390/su15031759

Academic Editors: Shuren Wang, Chen Cao and Hong-Wei Yang

Received: 12 December 2022
Revised: 11 January 2023
Accepted: 12 January 2023
Published: 17 January 2023

1. Introduction

One of the most important topics of geomechanics research is to find equations which will correctly describe the behavior of the loaded soil. Accuracy and correctness of description of loaded soil work, determines the safety of each structure. Due to a very complex and non-homogeneous structure of soil, the elaboration of accurate relationships load—settlement is considered to be one of the most difficult geomechanics task, especially in case of organic soils.

The necessity of erecting and maintaining building structures on an organic subsoil requires proper prediction of ground surface deformations with the use of geotechnical parameters. In the premises of Warsaw, the geological and engineering problems of organic soils are mostly related to organic carbonate deposits filling the glacial tunnel valley paleodoline the so-called Żoliborz Glacial Tunnel Valley, which extends meridionally Warsaw from the Żoliborz to the Okęcie District. Settlement of structures located in Warsaw within the glacial tunnel valley result primarily from settlement of gyttja, including its consolidation, which occur during the construction and operation phases of buildings. Gyttja, as a calcareous and organic mud, is formally classified as low-bearing capacity soil. This definition should rather refer to young Holocene organic lake sediments characterized by low geotechnical parameters with the texture of 'slime' or 'ooze' [1] as described by Hampus von Post in 1862. Geological history of 100 ka (kilo annum) changed geotechnical properties by consolidation and other different post-depositional processes sensu lato compacting scrutinized soil (i.e., moisture variation not caused by consolidation, cementation, time factor, etc.). A soil consolidation process is complicated and generally involves three stages: (i) initial compression, (ii) primary consolidation (pore pressure dissipation), (iii) secondary consolidation (creep rheological effect) [2]. All the three phases do not need to proceed one after another. Organic soils contain all three elements of total settlement in the initial phase of loading. A secondary consolidation phenomenon is still not fully understood.

The process of consolidation of highly compressible soils involves such phenomena as immediate deformation of the bubbles of gaseous water in the pores and deformation of the skeleton under the influence of effective stress [3–7]. This type of deformation is assumed to be elastic. The second one is related to the decrease volume of soil and gradual soil consolidation. The duration of consolidation under constant loading depends on soil permeability. In addition, there are secondary deformations, which are the result of long-term structural deformations of the soil (creep). The rate of these deformations depends on the rheological properties of the soil (viscosity); the greater the structural viscosity of the soil, the longer the skeleton creep process persists.

The results of research carried out by many researchers [3,6–9] indicate that the characteristics describing the process of organic soil consolidation is non-linear, which makes it difficult to use them in computational methods. The current state of research surrounds the topic presented in this article for several reasons. These are matters related to the very problem of consolidation based on non-classical rheological systems [10]. This non-linearity results not only from the change in the state of the soil, but also from the large anisotropy of stress in the subsoil and its variability in the deformation process [9]. In addition, the high compressibility of organic soils makes it necessary to take into account variable ground geometry during consolidation calculations, which leads to non-linear geometrical relationships in numerical solutions. The deformations which are more complicated than mineral soils, require the use of calculation methods based on complex soil models taking into account the different behavior of organic soils under load. Of great interest is the use of numerical analysis based on theoretical rheological models, the use of which provides adaptability to other geotechnical systems. A good example is the very fresh work on Non-Darcian flow and rheological consolidation of saturated clay [11]. Therefore, it is necessary to correctly estimate parameters or deformation characteristics of soils describing particular stages of the deformation process. The characteristics defining the deformation process influence the value of the parameter which depends on the stresses and time for a given type of soil. The values of deformation properties of the subsoil adopted for further analyses determine the stiffness of the subsoil under the direct foundation, as well as the stiffness of piles, displacement of columns or barettes included in the deep foundations.

Organic soils vary from the mineral soils primarily by a significant content of the organic substance (exceeding 2%) and, in most of cases, coloideality of the liquid phase. In this case study geological model covered gyttja with organic matter ranging mostly from 10% to 30%, with other components like calcium carbonate $CaCO_3$ (between 10–50% of the content) and other mineral (non-carbonate) and non-organic parts (in the range 28–68% of the content). Due to a very complex and non-homogeneous structure, in case of organic soils, the empirical models developed during the laboratory research of the soil samples formulation is the most common. The authors of this thesis decided to create a rheological model of the organic soil based on the results of a large scale test (scale 1:1), which is the sample loading of the foundation plate settled directly on the ground. The test was performed in soil conditions connected with Eemian geological structure of Warsaw, i.e., Eemian glacial tunnel valley in Warsaw called Żoliborz Glacial Tunnel Valley filled with organic soils with the thickness up to over 20 m. The theory of a time rate of one-dimensional consolidation was first proposed by Terzaghi [11]. It was based on the assumption that a relationship between effective stress and strain is linear elastic and it describes only the primary consolidation process. Thereafter, several investigators [12–17] used visco-elastic models to study one-dimensional consolidation, i. e. infinite soil layer of known thickness loaded at the top. For example Gibson & Lo [15] proposed a rheological structure for a soil skeleton which consisted of Hookean spring connected in series with a Kelvin-Voigt element where effective stress was calculated using Terzaghi's continuity equation. [17] used a Kelvin-Voigt element in effective stress space but with a non-linear dashpot. It is also possible to model soil settlement using more complicated rheological structures in total stress space where a certain dashpot accounts for Terzaghi's consolidation theory [2].

In this paper the authors propose a one-dimensional rheological model of plate foundations on gyttja clay during consolidation, defined in total stress space, taking into account all three phases of the process. The current state of knowledge describes the curves resulting from theoretical considerations, including using statistical analysis, for many reasons [18]. The choice of the serial connection of particular element was specified, as recalled, by the infiltration of consolidation phases. The model was not previously used by engineers for settlement prediction. The main feature of the model consists of an original set of non-linear explicit ODEs defining the evolution of the rheological structure. The set of constitutive relationships is strongly non-linear and should be integrated with the use of classical algorithms for solution of ordinary differential equation systems with respect to time.

2. Load Test in Natural Scale

In order to define model parameters of the investigated soil, a sample loading test was carried out. The test was performed on a concrete raft with the dimensions 5.0 m × 5.0 m × 0.30 m at depth of 3.0 m below the ground level, after the removal of ground layers on the non-reinforced subsoil. The geotechnical cross-section pertaining to the issue under consideration, along with the soil and water conditions, is shown in Figure 1. CPT soundings were particularly useful in correlating the results. Figure 1 shows the results of only the cone resistance for the CPT sounding and the pressure results for the DMT sounding, i.e., the most relevant results from the point of view of the conducted research. The authors decided to present only the most relevant data for the sake of legibility of the presented results.

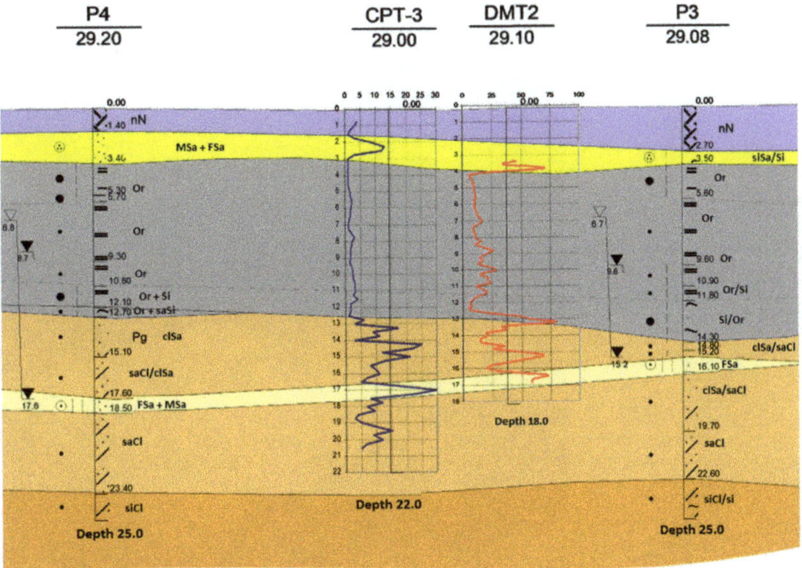

Figure 1. Geotechnical cross-section pertaining to the issue at hand.

The results of the CPT investigation are shown in Table 1. The documentation compiled was quite extensive, but only the most relevant information from the point of view of the article was selected, so the first culm of Table 1 shows the results for many soundings. S_u stands for shear strength in a undrained shear test.

Table 1. CPT probe results for the issue at hand.

CPT Number	*Depth*$[m]$	*LL*	*Su*$[kPa]$
CPT–S1	2.4–8.8	0.2	45
CPT–S2	3.0–13.0	0.2	50
CPT–S3	2.8–12.8	0.2	60
CPT–S4	3.0–13.0	0.25	55
CPT–S5	3.4–11.2	0.25	50
CPT–S6	3.0–9.6	0.22	50
CPT–S5b	11.2–13.8	−0.05	200
CPT–S6	3.0–9.6	0.22	50
CPT–S7	4.6–9.2	0.22	60
CPT–S8	3.2–14.2	0.15	70
CPT–S9	3.4–15.0	0.15	70
CPT–S10	3.3–9.0	0.18	60

The observations were conducted during a 1–1.5 month period. The sample loading was placed stage by stage, with the up-to-date measurements. A 300 t ballast was placed and, as a result, stress of 120 kPa was obtained. To control the settlement of concrete slab, 5 supervising benchmarks were stabilized in each corner (benchmark 1 to 4) and in the middle of the raft (benchmark 5). Positioning of the benchmarks is presented in Figures 2 and 3. The analysis carried out was very complicated, and measurements from the middle repertory (benchmark no. 5) were selected for simulation in order to keep the height of the results transparent.

Figure 2. Positioning of benchmarks no. 1–4.

Figure 3. Positioning of benchmark no. 5 and concrete plates view.

The measuring network consisted of 5 supervising benchmarks and 3 reference benchmarks. The measurement was executed using the precise levelling method, with the automatic leveler LEICA NA 3003 and the set of invar levelling rods. The initial measurement of the settlement of the concrete raft was taken before the plate was loaded. The following measurements were done successfully after loading the raft with a layer consisted of several concrete plates (Figure 3, right corner), and then during the removal of the plates. At the time of the measurement, the temperature was between 0 °C and 8 °C.

Calculations of the vertical settlement values were performed using a strict method submitting both the primary (initial) and secondary (up-to-date from the following days) measurements into equalization, conducting the analysis of the constancy of the reference benchmarks. The average error of evaluation of the benchmarks' settlement was given to indicate with what precision (accuracy) the slab settlement measurements were taken $m_{dH} \leq 0.5\ mm$.

3. Rheological Model of Settlement

3.1. Constitutive Differential Equations in 1D

The rheological structure, shown in Figure 4, was used to obtain constitutive differential equations. As in a settlement analysis vertical displacement of a foundation should be found, the authors assumed stress-displacement relation in the rheological structure instead of stress-strain relation. This resulted in obtaining the constitutive equations between vertical displacement and vertical total stress. In the model, there are four main elements connected in series. The displacements of every section are indicated in Figure 4.

The material parameters $C\alpha$ ($\alpha = 1, 2, 3$) are elastic proportionality constants. Their units are N/m^3 since a stress-displacement relation was chosen. Constants $C\alpha$ are introduced assuming that there is no constraints concerning lateral strains, i.e., boundary conditions are similar to Young's modulus definition. The material constants D_β ($\beta = 1, 2$) are soil viscosities and their units are $N \cdot s/m^3 = Pa \cdot s/m$. All the material parameters are defined in the total stress space.

The topmost spring element (displacement s^e) simulates the elastic behaviour according to the equation

$$s^e = \frac{\sigma}{C_1} \tag{1}$$

This element accounts for immediate reversible change in displacement after applying the load.

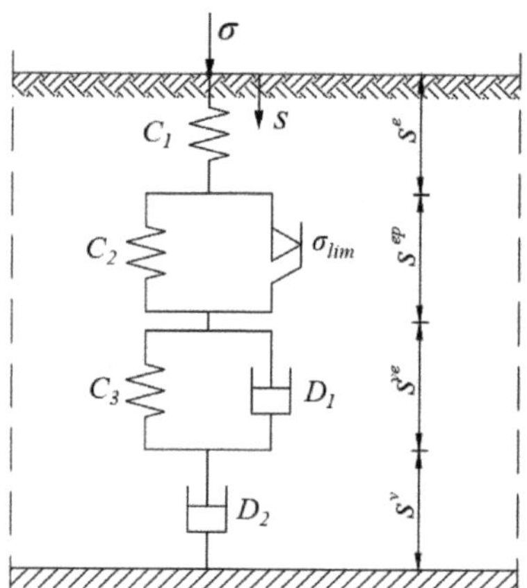

Figure 4. Soil settlement rheological structure.

The next section is an elastic-plastic element in which the displacement s^{ep} occurs only if the stress exceeds stress limit σ_{lim} [N/m^2]. This element accounts for plastic soil deformation. The following equation governs the elastic-plastic displacement rate $\dot{s}^{ep}$:

$$\dot{s}^{ep} = \begin{cases} \frac{C_1}{(C_1+C_2)\sigma^p}\left[\sigma^p\left(\dot{s} + \frac{C_3}{D_1}s^{ve} - \frac{\sigma}{D_{eq}}\right)\right]^+ & if \ |\sigma^p| \geq \sigma_{lim} \\ 0 & if \ |\sigma^p| < \sigma_{lim} \end{cases} \tag{2}$$

where

$$\sigma^p = \sigma - C_2 s^{ep} \tag{3}$$

and $[\cdot]^+$ denotes a projection onto the set of non-negative numbers

$$[z]^+ = \begin{cases} z & if \ z > 0 \\ 0 & if \ z \leq 0 \end{cases} \quad \forall z \in R \tag{4}$$

It should be emphasized that the model is strongly non-linear taking into account both viscoelastic and plastic properties of the soil material. The paper presents an original set of constitutive equations formulated in the explicit form via the non-linear ODE. Such equations cannot be found in the literature. Formulation of Equation (2) needs the notion of associated flow rule along with Kuhn-Tucker conditions [19]. In case of such approach a crucial point is to evaluate the so called Lagrange multiplier which defines the rate of the elastic-plastic displacement (denoted $\dot{s}^{ep}$ in the paper). Having the value of the Lagrange multiplier, the set of constitutive equations has a simple analytical form as presented in the paper.

Moving back to the analysis of the rheological scheme shown in Figure 4 let us note that the two bottom sections of the model contain dashpot elements which are related to the consolidation process. The bottom dashpot is responsible for irreversible creep displacement which is connected with secondary consolidation. The visco-elastic displacement rate $\dot{s}^{ve}$ and viscous displacement rate $\dot{s}^v$ can be calculated as in the case of Newtonian fluids

$$\dot{s}^{ve} = \frac{1}{D_1}(\sigma - C_3 s^{ve}) \tag{5}$$

$$\dot{s}^v = \frac{\sigma}{D_2} \tag{6}$$

The four displacement components (Figure 4) are connected in series and individual displacements should be added up to obtain the total displacement

$$s = s^e + s^{ep} + s^{ve} + s^v \tag{7}$$

Formula (2) for the case where $|\sigma^p| = \sigma_{lim}$ was obtained taking into account that during plastic yielding

$$C_1 s^e = \sigma_{lim} + s^{ep} C_2 \tag{8}$$

which implies

$$C_1 \dot{s}^e = \dot{s}^{ep} C_2 \tag{9}$$

and by substituting Equation (7), differentiated with respect to time. Variable σ^p was put into the square bracket to take into account loading and unloading cases using only one formula. Equations (1), (2), (5)–(7) form a differential-algebraic equation system which can be numerically solved with given material constants (C_α, D_β, σ_{lim}), loading function $\sigma(t)$ and initial conditions to obtain settlement values in time. The material parameters were determined in this paper based on the test loading experiment results. The initial conditions are

$$s(0) = 0;\ s^e(0) = 0;\ s^{ep}(0) = 0;\ s^{ve}(0) = 0 \tag{10}$$

An analytical solution of the system exists only for a simple instantaneous loading shown in Figure 5

$$\sigma(t) = \begin{cases} \frac{F_{max}}{A} & if\ t \geq 0\ and\ t < t_0 \\ 0 & otherwise \end{cases} \tag{11}$$

in the form [17]

$$s = \sigma_0 \varphi \text{ where } \varphi = \begin{cases} \varphi_1 & if\ t < t_0 \\ \varphi_2 & if\ t > t_0 \end{cases} \tag{12}$$

where settlement functions $\varphi\ \varphi_1$ and $\varphi\ \varphi_2$ depend on material parameters and are defined as follows:

$$\varphi_1 = \frac{1}{C_1} + \frac{A_1}{C_2} + \frac{t}{D_2} + \frac{1}{C_3}\left[1 - exp\left(-\frac{t}{\lambda}\right)\right] \tag{13}$$

$$\varphi_1 = \frac{A_2}{C_2} + \frac{t_0}{D_2} + \frac{1}{C_3}exp\left(-\frac{t}{\lambda}\right)\left[exp\left(\frac{t_0}{\lambda}\right) - 1\right] \tag{14}$$

where $\lambda = \frac{D_1}{C_3}$ and

$$A_1 = \left[1 - \frac{F_{gr}}{F}\right]^+ \tag{15}$$

$$A_2 = min\left(\frac{F_{gr}}{F},\ \left[\left[1 - \frac{F_{gr}}{F}\right]^+\right]\right) \tag{16}$$

and symbol $[\cdot]^+$ is defined in Equation (4).

For the loading shown in Figure 5, assuming A = 25 m² and F_{max} = 2512 kN and material parameters determined by optimization for the test loading experiment (Table 2), the model was solved numerically and using analytical formulae. There was no difference between a numerical and analytical solution. The results of this simple creep test are shown in Figure 6.

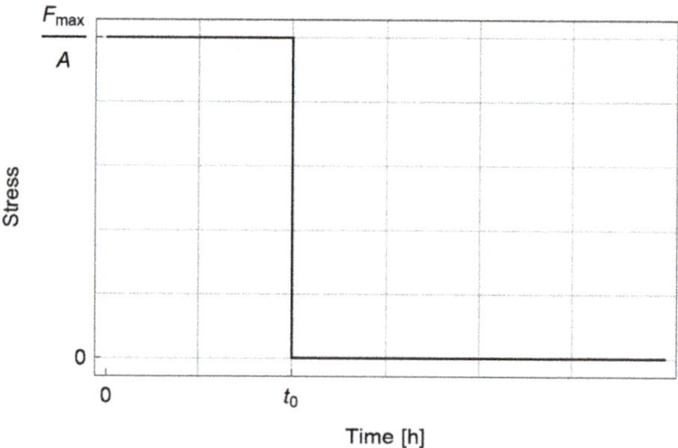

Figure 5. Instantaneous loading for creep test simulation.

Table 2. Material parameters of the settlement model determined based on the test loading experiment by curve fitting.

$C_1\left[\frac{kPa}{m}\right]$	$C_2\left[\frac{kPa}{m}\right]$	$C_3\left[\frac{kPa}{m}\right]$	$D_1\left[\frac{kPa}{m}\right]$	$D_2\left[\frac{kPa}{m}\right]$	$\sigma_{lim}[kPa]$
5495	25,145	5765	$4.12 \cdot 10^6$	$4 \cdot 10^7$	40

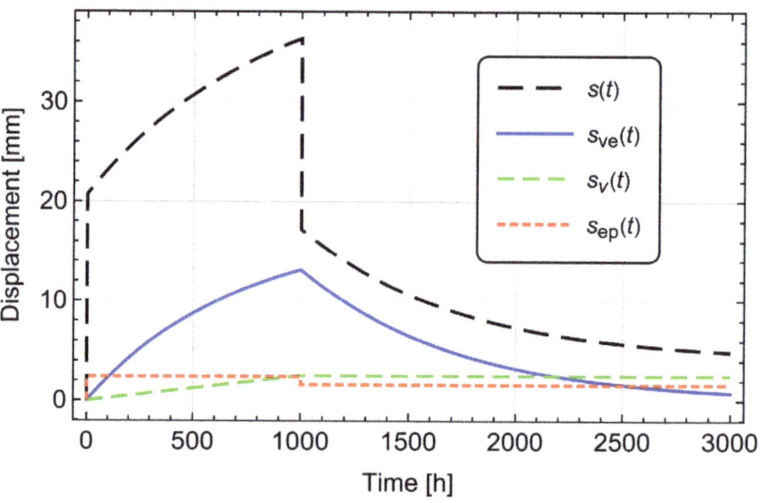

Figure 6. Creep test simulation results.

Material parameters of the model can be found by curve fitting to a plate foundation test loading experiment. Such a field test is carried out for a foundation of a certain shape and dimensions. In this paper, it has been done for a square plate with 5.0 m long sides. Material parameters of the model were determined by curve fitting procedure applied for settlement test results. To estimate model parameters for different foundation dimensions, the authors propose to use an elastic solution for deflection due to a uniformly loaded flexible area. It is a consolidation model in which there is only one spring element in effective stress space and the relationship for settlement can be written as [20,21]

$$s = \frac{q\omega B\left(1 - v^2\right)}{E} \tag{17}$$

where B denotes foundation width or diameter, ω is a coefficient that depends on a foundation shape and stiffness, E is effective soil deformation modulus related to normal consolidation or swelling line, and ν is Poisson's ratio. For deflection under the foundation center, ω ranges from 1.0 for a circular foundation to 4.0 for a rectangle plate for which $L/B = 100$. For a square foundation $\omega_{sq}\,1.12$. Having a set of parameters determined by curve fitting $\{C_\alpha^{cf}, D_\beta^{cf}, \sigma_{lim}^{cf}\}$ from a test loading experiment for a foundation characterized by ω^{cf} and B^{cf}, the model parameters of any foundation shape can be calculated as follows (see Equation (17)):

$$C_\alpha = B^{cf}\omega^{cf}\frac{C_\alpha^{cf}}{\omega B}, \quad \alpha = 1,2,3 \tag{18}$$

$$D_\beta = B^{cf}\omega^{cf}\frac{D_\beta^{cf}}{\omega B}, \quad \beta = 1,2 \tag{19}$$

$$\sigma_{lim} = B^{cf}\omega^{cf}\frac{\sigma_{lim}^{cf}}{\omega B} \tag{20}$$

For the experiment used for optimization in this paper $\omega^{cf} = \omega_{sq} = 1.12$ and $B^{cf} = 5.0\ m$.

3.2. Influence of the Width of Foundation on the Value of Settlements

In case of calculation of large plates, the definition of the depth of an active settlement zone becomes a problem.

According to the literature [21–25], the value of the settlement increases proportionally to the increase of the size of the foundation. While calculating the foundations of large dimensions (big footings, large foundation plates) the active zone theoretically becomes very large, which leads to difficult and time-consuming calculations. The solution of this problem may be a method of division of the plate into smaller sections and considering its settlement separately. In such case the influence on the settlement of the adjoining foundation should be taken into account. According to [26] there is a boundary dimension of the equivalent spread footing. When it is exceeded, the settlement of the foundation does not increase, and the equation for the calculation of the settlement of the foundation can be presented as:

$$s = \frac{\sigma_0^2}{E_0\gamma}\sqrt{\frac{3}{2\pi}}\left(\frac{\gamma B}{\sigma_0}\right)\left[2 - \sqrt[3]{\frac{3}{2\pi\beta}}\sqrt[3]{\frac{\gamma B}{\sigma_0}}\right] \tag{21}$$

Relationship (21) is the function of independent variable $\frac{\gamma B}{\sigma_0}$, the course of which is presented in Figure 7.

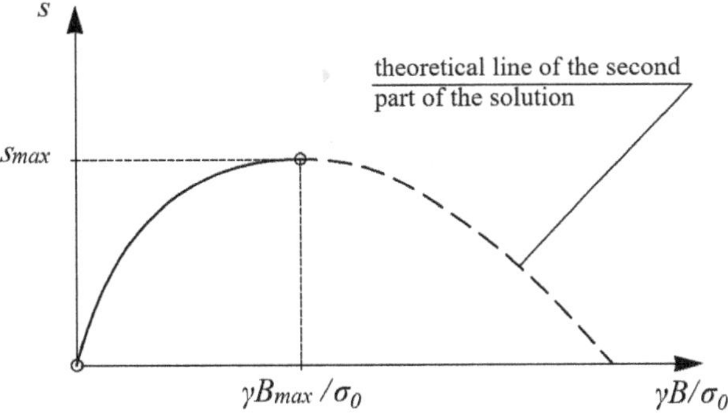

Figure 7. Graph of the settlement function $= s\left(\frac{\gamma B}{\sigma_0}\right)$.

From the graph, it can be read out that the settlement of the foundation increases with the growth of the dimension of foundation B. It can be indicated that function (21) reaches its maximum in point $\gamma B_{max}/\sigma_0$, which shows that in order to calculate the settlement of a large plate, it is enough to calculate the settlement for a given load, for a square slab with side dimension σ_0/γ. Therefore, the research was conducted in the natural scale in-situ on the square plate with the dimensions approximate to the equivalent square footing, resulting from the established range of loading σ_0 = 120 kPa and the soil weight γ = 21 kN/m^3. The dimension of the equivalent square footing obtained in order to calculate accurately the settlement of the large plate is $B_{max} = 120/21 = 5.7\ m$.

3.3. Influence of the Width of Foundation on the Value of Settlements

Material constants of the proposed settlement model were determined by optimization using data of the plate foundation test loading described in previous sections. Figure 8 shows the stress applied to the plate foundation in the course of the field test. Linear interpolation was assumed between test points. Having an interpolated function it was possible to numerically solve a differential-algebraic system of equations for the given values of material constants.

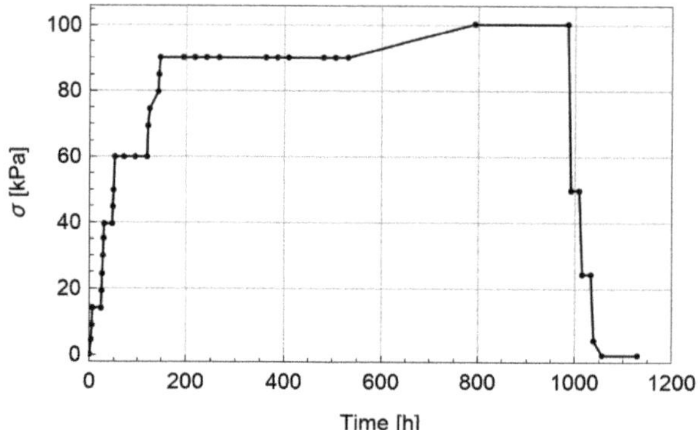

Figure 8. Stress under the plate foundation during the settlement field test, linearly interpolated between measurement points.

The optimal set of material parameters for the rheological model indicates such values of parameters $\widetilde{p}$ for which functional $F(p)$ reaches its minimum. Thus, the minimization problem can be stated as follows:

$$\widetilde{p} = \arg\min_{p\in\Omega} F(p) \tag{22}$$

where Ω denotes the set of admissible parameters' values.

The optimization problem defined by Equation (22) was solved using a least-squares method implemented in Mathematica software [27]. In this approach, the functional $F(p)$ was defined as follows:

$$F(p) := \Sigma_{j=1}^{N} \left| s_j^R(t) - s_j(t; p) \right|^2 \tag{23}$$

where $s_j(t; p)$ are the settlement values determined by the rheological model and s_j^R are test data and N is the number of data points. In our case N = 41 and $p = \{C_\alpha,\ D_\beta,\ \sigma_{lim}\}$.

Nelder-Mead nonlinear optimization technique was used [28]. However, it was not possible to find the global optimum of the functional with this method and only a local optimum could be found. Although in the case of this optimization there were many local minima, Nelder-Mead algorithm was used since it was computationally efficient. Many attempts with different initial constraints had to be made to obtain correct curve fitting results. In such a non-linear problem, where only a numerical solution of a differential-algebraic system was available, it was difficult to find a true global optimum. However, it was possible, in this case, to find a sufficient minimum (after rejecting incorrect local minima).

A determined set of parameters is shown in Table 2. Figures 9 and 10 show the optimization results and separate model displacement components, respectively. It can be observed that the model can simulate the consolidation process quite well, although some effort had to be made to find material parameters. The secondary effects are important and they take in a case of organic soil a very long time to occur. Therefore, to predict the long term settlement the proposed mathematical model needs to be validated for a wider period.

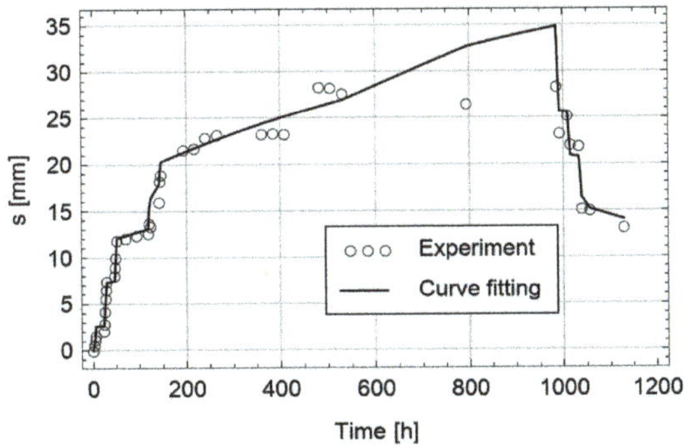

Figure 9. Curve fitting results. Total settlement curve.

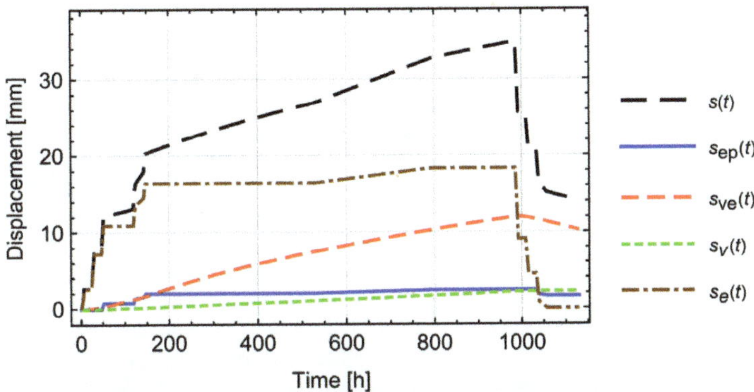

Figure 10. Separate displacement components of the rheological model.

4. Conclusions

The interpretation of the obtained values of C_1, C_2, C_3 defined as a stiffness modulus of the elastic subsoil by direct analogy to the Winkler—Zimmerman model (kN/m^3) must be carried out $k = 1/(1/C_1 + 1/C_2 + 1/C_3) = 2530$ kPa/m. This value is convergent to 4000 kPa/m as a result of field observation (120 kPa/0.03 m).

D_1 and D_2 constants define modified viscosity of Newton's liquid. In comparison with original viscosity unit expressed in kPa·s, D_1 and D_2 can be called the modified viscosity by analogy between a constant of Winkler-Zimmerman model (modulus of subgrade reaction k [kPa/m]) and the modulus of elasticity E expressed in kPa (by analogy to the modulus expressed in kPa/m in the discussed case the viscosity unit has been changed into kPa s/m). Observation of the results of fitting analysis of the D_1 and D_2 values is a prompt to think that there is a logic reference to a coefficient of soil consolidation C_v expressed in m^2/s. Assuming that the typical value of C_v equals 10^{-7} m^2/s [29] in the presented case the inverse value of C_v is compatible with D_1 and D_2 values: $D_1 = 4.12 \cdot 10^6$ kPa·h/m and $D_2 = 4.0 \cdot 10^7$ kPa h/m ($1/D = 1/D_1 + 1/D_2$ gives $3.6 \cdot 10^6$), respectively.

In the case of σ_{lim} treated as the shear strength in the undrained conditions (s_u or c_u according to the European Standard EN 1997), the values calculated in fitting analysis can be compared with the results of the triaxial tests $S_u = 1/2 \cdot (\sigma_1 - \sigma_3)$, which is the half values of the stress deviator q.

Figure 11 shows the results of the consolidated undrained triaxial test performed on the gyttja recovered from the depth of 6 m from the site of the real scale load test. Assuming that the submerged unit weight of gyttja is $\gamma' = 8$ kN/m^3 and the at-rest earth pressure coefficient K_0 of gyttja is 0.9, finally effective radial stress on the depth of 6 m is $\sigma_3' = 6$ m $\cdot$ 8 kN/m^3 $\cdot$ 0.9 = 43.2 kPa.

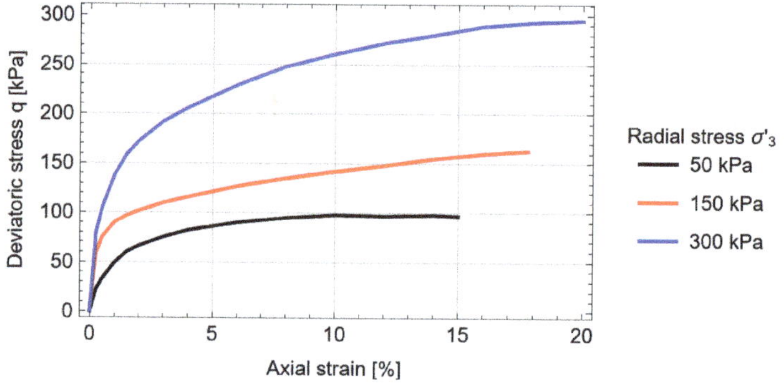

Figure 11. Triaxial compression test results for the gytias directly under the base plate.

The analysis of the results of triaxial test indicates satisfactory/good convergence of the shear strength s_u value from the laboratory tests with the value of limit stress σ_{lim} estimated in the curve fitting procedure. Taking into account half of the limit value of the deviator q for the radial stress $\sigma_3' = 50$ kPa (nearest to 43.2 kPa) the value of shear strength is $s_u = q/2 = 98/2 = 49$ kPa which is close to $\sigma_{lim} = 40$ kPa obtained in simulation.

It should be emphasized that the methodology of formulating constitutive relationships with the use of rheological schemes applied in this paper has led to their explicit form as a system of nonlinear differential equations. A similar approach was presented by the authors in [30–32], concerning not only soils, but also asphalt-aggregate mixtures. Such approach involves applying classical and non-classical rheological elements (springs, dashpots, sliders, etc.). The same elements can be used for the description of different phenomena occurring in soils and asphalt-aggregate mixtures. For example in soils primary and secondary consolidation can be modelled. On the other hand, in asphalt-aggregate mixtures rheological structures can be applied for modelling creep and relaxation phenomena as well as binders' properties called zero shear viscosity [33].

Proper parametrization of the rheological ground model is necessary also in order to analyze dynamical behavior of buildings considering soil-structure interaction. Presented approach to determining shear strength of soil can be useful for correct assessment of the control procedure of building—soil-foundation system with plastic behavior of soil [34–36].

The curve fitting of the model (see Table 2 and Figure 9) consists of iteratively solving the system of non-linear differential equations. For that purpose, Mathematica software was applied with the program's build-in ODEs solvers. These solvers are well suited for classical continuous problems. Therefore, special attention had to be put in order to solve a discontinuous differential equation as Equation (2). During calculations the authors performed a simplified sensitivity analysis which demonstrated that the small perturbations of rheological scheme parameters do not change the character of the solution significantly. In general, it was observed that the major changes (e.g., one order of magnitude) of plasticity (slider) and viscosity (dashpot) elements result in erroneous prediction of permanent settlement after unloading. An in-depth sensitivity analysis needs a lot more calculations. The work on this field is currently carried out by the authors.

Author Contributions: Conceptualization, G.K. and A.Z.; methodology, A.Z.; software, K.J.; validation, G.K. and K.J.; formal analysis, A.Z.; investigation, K.J.; resources, P.N.; data curation, K.J.; writing—original draft preparation, K.J. and M.F; writing—review and editing, G.K., A.Z., K.J., P.N. and M.F.; visualization, K.J., P.N. and M.F.; supervision, A.Z.; project administration, G.K., M.F. and P.N. All authors have read and agreed to the published version of the manuscript.

Funding: This paper was co-financed under the research grant of the Warsaw University of Technology supporting the scientific activity in the discipline of Civil Engineering and Transport.

Institutional Review Board Statement: Not applicable.

Informed Consent Statement: Not applicable.

Data Availability Statement: All data used in the article are given in the bibliography or own sources have been used in the case of photos.

Acknowledgments: The authors acknowledge the Warbud company for the opportunity to use the results of the field research.

Conflicts of Interest: The authors declare no conflict of interest.

References

1. Myślińska, E. *Organic Soils and Laboratory Methods of Research*; PWN: Warsaw, Poland, 2001. (In Polish)
2. Lee, K.I.; White, W.; Ingles, O.G. *Geotechnical Engineering*; Pitman Publishing Inc.: London, UK, 1983.
3. Berry, P.L.; Poskitt, T.J. The consolidation of peat. *Géotechnique* **1972**, *22*, 27–52. [CrossRef]
4. Koda, E. The Effect of Vertical Drainage on Accelerating the Consolidation of Organic Soils. Ph.D. Thesis, Warsaw University of Life Sciences, Warsaw, Poland, 1990. In Polish .
5. Haan, E.J.D. A compression model for non-brittle soft clays and peat. *Géotechnique* **1996**, *46*, 1–16. [CrossRef]
6. Sas, W. Modeling of Organic Soil Deformation Including Changes in the Properties of the Medium. Ph.D. Thesis, Warsaw University of Life Sciences, Warsaw, Poland, 2001. In Polish .
7. Szymański, A.; Sas, W. Deformation characteristics of organic soils. *Ann. Wars. Agric. Univ. Land Reclam.* **2001**, *32*, 117–126.
8. Edil, T.B.; Dhowian, A.W. Analysis of long term compression of peats. *Geotech. Eng.* **1979**, *10*, 153–178.
9. Wolski, W.; Larsson, R.; Szymański, A.; Lechowicz, Z.; Mirecki, J.; Garbulewski, K.; Hartlen, J.; Bergdahl, U. *Two Stage Constructed Embankments on Organic Soils*; Technical Report 32; Swedish Geotechnical Institute: Linköping, Sweden, 1988.
10. Wu, W.; Zong, M.; El Naggar, M.H.; Mei, G.; Liang, R. Analytical solution for one-dimensional consolidation of double-layered soil with exponentially time-growing drainage boundary. *Int. J. Distrib. Sens. Netw.* **2018**, *14*, 1550147718806716. [CrossRef]
11. Terzaghi, K. *Erdbaumechanik Auf Boden-Physikalischen Grundlagen*; Deuticke: Vienna, Austria, 1925.
12. Liu, Z.; Xia, Y.; Shi, M.; Zhang, J.; Zhu, X. Numerical Simulation and Experiment Study on the Characteristics of Non-Darcian Flow and Rheological Consolidation of Saturated Clay. *Water* **2019**, *11*, 1385. [CrossRef]
13. Taylor, D.W.; Merchant, W. A theory of clay consolidation accounting for secondary compression. *J. Math. Phys.* **1940**, *19*, 167–185. [CrossRef]
14. Taylor, D.W. *Research on Consolidation of Clays*; Technical Report 82; Massachusetts Institute of Technology: Cambridge, MA, USA, 1942.
15. Gibson, R.E.; Lo, K.Y. A Theory of Consolidation for Soils Exhibiting Secondary Compression. In *ACTA Politechnica Scandinavica*; Norwegian Geotechnical Institute: Oslo, Norway, 1961.
16. Schiffman, R.L.; Ladd, C.C.; Chen, A.T. The Secondary Consolidation of Clay. In *Symposium on Rheological Soil Mechanics*; Springer: Berlin/Heidelberg, Germany, 1964.
17. Barden, L. Primary and secondary consolidation of clay and peat. *Geotechnique* **1965**, *15*, 345–362. [CrossRef]
18. Olek, B.S. A Consolidation Curve Reproduction Based on Sigmoid Model: Evaluation and Statistical Assessment. *Materials* **2022**, *15*, 6188. [CrossRef] [PubMed]
19. Lubarda, V.A. *Elastoplasticity Theory*; CRC Press: Boca Raton, FL, USA, 2001. [CrossRef]
20. Nowacki, W. *The theory of Creep*; Arkady: Warsaw, Poland, 1963; In Polish .
21. Wiłun, Z. *An Outline Geotechnics*; WKŁ: Alphen, The Netherlands, 1987; In Polish .
22. Cernica, N.J. *Geotechnical Engineering: Foundation Design*; J. Wiley & Sons: Hoboken, NJ, USA, 1995.
23. Glazer, Z. *Soil Mechanics*; Wydawnictwa Geologiczne: Warszawa, Poland, 1985; In Polish .
24. Jeske, T.; Przedecki, T.; Rosiński, B. *Soil Mechanics*; PWN: Warsaw, Poland, 1966; In Polish .
25. Puła, O.; Rybak, C.; Sarniak, W. *Foundation Design*; Dolnoslaskie Wydawnictwo Edukacyjne: Wrocław, Poland, 2009; In Polish .

26. Meyer, Z. Określenie współczynnika podatności podłoża przy projektowaniu plyt fundamentowych w złożonych warunkach geotechnicznych. In Proceedings of the XXVIII Warsztaty Pracy Projektanta Konstrukcji, Wisła, Poland, 5 March 2013; Volume 1, pp. 343–392.
27. Wolfram Research, Inc. *Mathematica*, 10th ed.; Wolfram Research, Inc.: Champaign, IL, USA, 2014.
28. Nelder, J.A.; Mead, R. A Simplex Method for Function Minimization. *Comput. J.* **1965**, *7*, 308–313. [CrossRef]
29. Kacprzak, G. *Współpraca Fundamentu Płytowo-Palowego Z Podłozem Gruntowym*; Oficyna Wydawnicza Politechniki Warszawskiej: Warszawa, Poland, 2018; In Polish .
30. Józefiak, K.; Zbiciak, A. Secondary consolidation modelling by using rheological schemes. *MATEC Web Conf.* **2017**, *117*, 00069. [CrossRef]
31. Zbiciak, A.; Grzesikiewicz, W.; Wakulicz, A. One-dimensional rheological models of asphalt-aggregate mixtures. *Logistyka* **2010**, *6*, 3825–3832.
32. Zbiciak, A. Mathematical description of rheological properties of asphalt-aggregate mixes. *Bull. Pol. Acad. Sci. Tech. Sci.* **2013**, *61*, 65–72. [CrossRef]
33. Brzezinski, K. Zero shear viscosity estimation using a computer simulation of Van der Poel's nomograph. *J. Build. Chem.* **2016**, *1*, 10–17. [CrossRef]
34. Lamarque, C.H.; Kacprzak, G.; Awrejcewicz, J. Active control for a 2DOF of building including elastoplastic soil-structure coupling terms. In Proceedings of the DETC'01, 2001 ASME Design Engineering Technical Conferences, Pittsburgh, PA, USA, 9–12 September 2001.
35. Lamarque, C.H.; Awrejcewicz, J.; Kacprzak, G. Active control for a 2DOF mechanical system including elastoplastic terms. In Proceedings of the 10th International Conference on System-Modelling-Control, Zakopane, Poland, 21–25 May 2001; pp. 11–16.
36. Lamarque, C.H.; Awrejcewicz, J.; Kacprzak, G. Control of structures. Applied Mechanics. In Proceedings of the Americas: Proceedings of the Seventh PAN American Congress of Applied Mechanics, Temuco, Chile, 2–4 January 2002.

Article

Study on the Influence of Water Content on Mechanical Properties and Acoustic Emission Characteristics of Sandstone: Case Study from China Based on a Sandstone from the Nanyang Area

Xin Huang [1], Tong Wang [1], Yanbin Luo [2,*] and Jiaqi Guo [1,2,*]

[1]   School of Civil Engineering, Henan Polytechnic University, Jiaozuo 454000, China
[2]   School of Highway, Chang'an University, Xi'an 710064, China
*   Correspondence: lyb@chd.edu.cn (Y.L.); gjq519@163.com (J.G.)

Abstract: In order to study the influence of water content on the mechanical properties of sandstone and evolution of crack propagation, laboratory compression tests and Engineering Discrete Element Method (EDEM) numerical simulation of sandstone under different conditions were carried out by the RMT-150B rock mechanics test system. The sandstone samples were from Nanyang, Henan Province, containing a total of 12 rock samples. Under the confining pressure of 0, 5, 10, and 20 MPa, the rock samples with 0%, 1.81%, and 3.24% water content were tested. The findings demonstrated that as the sample's water content grew, the peak strain increased but the peak strength, elastic modulus, maximum energy rate of individual acoustic emission events, and cumulative acoustic emission energy rate all reduced. While the ratio of tensile cracks to shear cracks inside the rock samples rose with increasing water content, the failure mode of sandstone changes from shear failure to tensile failure with the increase of water content, but the sandstone specimens in the three conditions exhibited shear macroscopic fracture surfaces. Research results will provide reference for the safe construction of underground projects in water rich areas.

Keywords: sandstone; water content; mechanical properties; acoustic emission; failure mode; EDEM

Citation: Huang, X.; Wang, T.; Luo, Y.; Guo, J. Study on the Influence of Water Content on Mechanical Properties and Acoustic Emission Characteristics of Sandstone: Case Study from China Based on a Sandstone from the Nanyang Area. *Sustainability* **2023**, *15*, 552. https://doi.org/10.3390/su15010552

Academic Editors: Shuren Wang, Chen Cao and Hong-Wei Yang

Received: 22 November 2022
Revised: 17 December 2022
Accepted: 20 December 2022
Published: 28 December 2022

1. Introduction

In recent years, with the rapid development of underground engineering in China, tunnel collapse, water inrush, mud inrush, and other major engineering disasters caused by water rock interaction have been increasing, and the influence of water on rock stability has gradually been paid attention to [1]. Water content and surrounding rock pressure have a great impact on rock strength and failure deformation. In addition to softening the rock, water also dissolves some of its mineral constituents, enlarging the rock's interior fractures and altering its mechanical properties; this is especially true for sandstone, a rock with high porosity [2–4]. Therefore, the research on the influence of water content on the mechanical properties, failure characteristics, and internal damage evolution of rocks will provide references for the safe construction of underground projects in water-rich areas [5].

Domestic and foreign scholars have done a lot of experimental research on the impact of water on rock failure and have achieved fruitful results. K. Hashiba et al. [6] used uniaxial compression test to research the alternating loading rate of Mitsui Andesite under varying levels of water saturation, and the findings demonstrated that rock's uniaxial compressive strength rises as water saturation falls. Zhao et al. [7] conducted uniaxial compression test on red sandstone with different water content and compared the results by discrete element software. It was found that the strength and elastic modulus of rock decreased with the increase of water content. Feng et al. [8] carried out uniaxial compression tests of coal samples with different water content. Through the analysis of test results, they found that water softened the coal samples while weakening their brittleness, and the failure process was more moderate. Eunhye Kim [9] conducted static and rapid loading compression tests on sandstone with different water contents; rock strength and Young's modulus decrease with the increase of water content. Sun et al. [10] carried out triaxial loading tests of sandstone rock mass with different water content, analyzed the acoustic emission signals inside the rock during loading, and found that with the increase of moisture content, the strength of the rock gradually decreased, and the elasticity and accumulated damage of the rock decreased, resulting in the gradual decrease of the accumulated acoustic emission count and accumulated energy in the process of broken. Erguler Z [11] tested different types of rock and found that the average elastic modulus and tensile strength decreased significantly as the UCS water content decreased. Guo et al. [12] conducted uniaxial compression and conventional triaxial compression tests on limestone under different moisture content, and discovered that when water content increased, limestone's uniaxial compressive strength declined and its peak strain increased. Tan et al. [13] used PFC software to simulate the cyclic loading and unloading of coal samples under different confining pressures. By analyzing the change law of elastic modulus and the evolution law of plastic strain of coal samples during loading, they proposed that the increase of confining pressure is conducive to the suppression of damage and the increase of the sample's ability to withstand plastic strain.

Domestic and foreign scholars have made a lot of achievements in the study of the influence of water content on the mechanical properties and failure modes of rocks. However, the accuracy of the indoor test depends not only on the process of the test, but also on the monitoring means adopted in the test to a large extent. However, the monitoring of most studies is unilateral, and few studies have observed and analyzed the internal evolution process of rocks with different water contents during loading. In order to study the influence of water content on the mechanical properties of sandstone and evolution of crack propagation from many aspects. In this paper, a series of compression tests are carried out on sandstone samples with different water contents, and the acoustic emission system is used to monitor the process of broken rock and with the aid of engineering discrete element method (EDEM), simulation with uniaxial and triaxial compression tests of sandstone with different water content, and analyzes the damage evolution process of rock samples at the microcosmic level. Combined with the results of laboratory tests and numerical tests, water content has a significant negative effect on sandstone strength and will lead to the evolution of internal cracks from shear to tensile. The test results reveal the influence of water content on the mechanical properties, deformation characteristics, and failure characteristics of sandstone.

2. Materials and Methods

2.1. Sample Preparation

Sandstone samples are taken from Nanyang, Henan Province, China. The color is gray white and the appearance is rough. They are composed of various sand cementation. The rock mass is relatively complete and there are no obvious defects on the surface. Sampling and sample preparation are carried out in strict accordance with GB/T2356.1-2009 "rock physical and mechanical properties determination method", machining of field rock blocks into standard cylindrical specimens with a diameter of 50 mm and a height of 100 mm [14]. Before the test, the sound velocity of the obtained rock samples is calibrated, and the individuals with large wave speed deviation eliminated so as to minimize the influence of rock discreteness on the experimental results. The prepared standard samples were divided into three groups: dry group, natural group, and saturated group, respectively labeled D, N, and S. Compression tests were carried out with confining pressures of 0 MPa, 5 MPa, 10 MPa, and 20 MPa. Different confining pressures of dry group, natural group, and saturated group were D-0, D-5, D-10, D-20, N-0, N-5, N-10, N-20, S-0, S-5, S-10, and S-20. Sandstone sampling information and physical test equipment are shown in Figure 1.

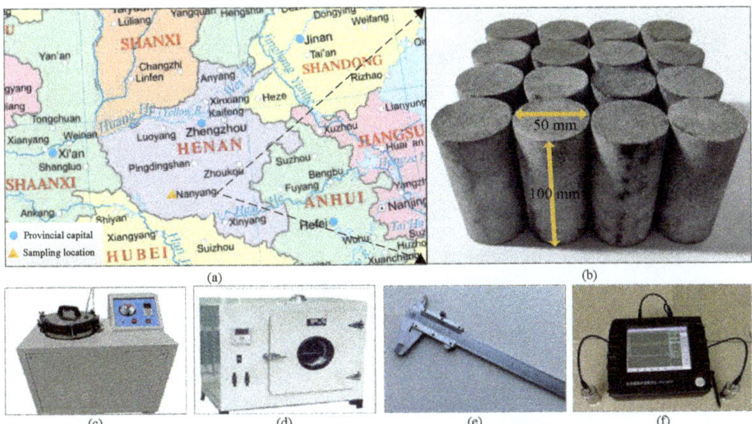

Figure 1. Sandstone sampling information and physical test equipment: (**a**) sampling location; (**b**) sandstone samples; (**c**) saturator; (**d**) dry oven; (**e**) saturation instrument; (**f**) acoustic test instrument.

2.2. Experimental Equipment and Methods

Dry, natural, and saturated specimens were prepared as follows, dry group: put the test sample into the dry oven for dry, set the dry temperature to 105 °C according to MT 224-1990 "Method for determination of permeability coefficient of coal and rock", take out, weigh, and record it after drying for 48 h, and then put it back into the dry oven. After 24 h, take out and weigh it again, compare the results of the two times [15]. If the quality continues to decline, repeat the above process. If the quality does not change twice, put the rock sample in the dry oven for cooling, and then wax seal it. Natural group: weigh and record after sample preparation, and wax seal until the test is carried out. Saturated group: place the sample in the saturation instrument, fill it with deionized water until the rock sample is completely submerged, seal and cover the vacuum water saturation instrument, pump it to the vacuum state, keep it for 48 h, take it out and weigh it, replace the rock sample after recording, take it out and weigh it again after 24 h, and compare the results of the two times. Repeat the technique outlined above if the quality keeps improving. If the quality does not change, remove any remaining water from the rock sample's surface and wax seal it. Calculate the water content of the sample according to Formula (1)

$$\omega = \frac{m_1}{m_0} - 1 \tag{1}$$

where: m_1 is the actual water mass in the rock sample, m_0 is the mass of dry rock sample, and the unit is g; ω is the water content of rock sample, unit is %.

In addition, the wave speed of sandstone samples in dry, natural, and saturated states were measured by acoustic wave tester. The water content, density, and wave speed of sandstone tests are shown in Table 1.

Table 1. Water content, density, and wave speed of sandstone.

Number	Moisture Condition	Water Content (%)	Density (kg/m^3)	Wave Speed (km/s)
D	Dry	0	23.553	3.378
N	Nature	1.81	23.845	3.521
S	Saturated	3.24	24.762	3.571

Uniaxial and triaxial compression tests were carried out on sandstone samples with different water content using a digitally controlled electro-hydraulic servo testing machine RMT-150B developed by Wuhan Institute of Rock and Soil Mechanics, Chinese Academy of

Sciences. The testing device's maximum axial load is 1000 kN, and its maximum confining pressure is 50 MPa. The displacement-controlled loading method was used during the uniaxial and triaxial compression tests, with the loading rate of 0.002 mm/s, and the rock sample was loaded once until it was destroyed. In the triaxial compression test, three sets of confining pressures—5 MPa, 10 MPa, and 20 MPa—were established. After, it was loaded at 0.5 MPa/s by means of stress loading, loaded to a specific confining pressure and then stabilized. The DS5-8 acoustic emission monitoring device was used to continuously track sandstone's acoustic emission events while it was being loaded. Vaseline was used to secure tight contact between the rock sample and the sensor and prevent signal transmission issues between the acoustic emission probe and the rock sample surface. The acquisition threshold for acoustic emission signals was set at 40 dB, and the sampling frequency was 5 MHz. In Figure 2, the test system is displayed [16,17].

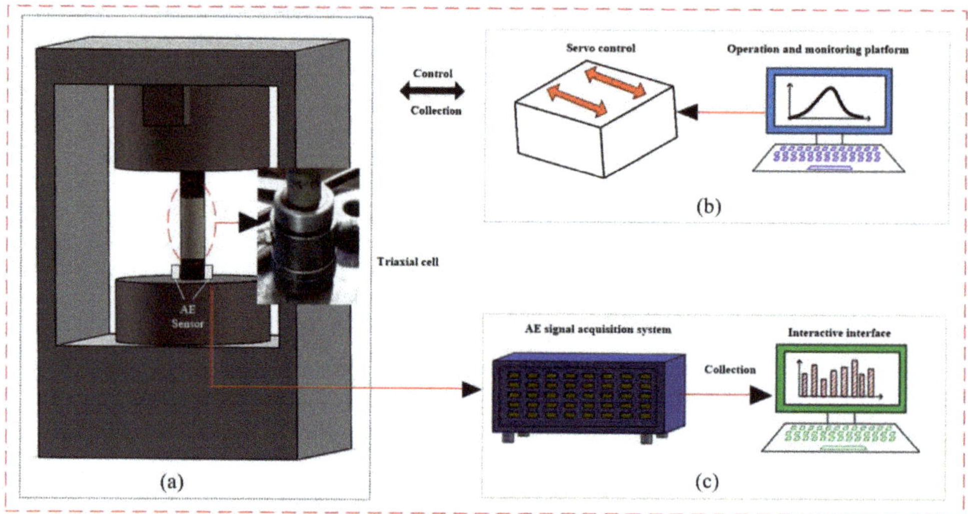

Figure 2. Test system: (**a**) RMT-150B rock mechanics test system; (**b**) rock mechanics control system; (**c**) acoustic emission synchronous detection system.

2.3. Establishment of Model and Determination of Parameters for Numerical Simulation

EDEM is a high-level discontinuous medium program software based on the discrete element method. It can be used to study the fracture problem of rock-like materials, which are essentially particle aggregates. It can reflect the crack evolution characteristics and failure mechanism of media under stress conditions. One advantage of EDEM is that it can simulate gravity accumulation when particles are accumulated, which makes internal pores generated in the sandstone model, so as to restore sandstone properties to the greatest extent, and it can reflect the compaction stage of sandstone well when simulating uniaxial compression [18,19].

First, a numerical analysis model was generated in the program, which is completely consistent with the size of sandstone samples. The model body was composed of particles, and the edges were constrained by setting "walls". Based on the Hertz Mindlin with Bonding and Hertz Mindlin with JKR particle contact model of EDEM software, the model particles were bonded to generate strength. After completion, the side wall was deleted, and the loading process was simulated by giving the top loading plate motion. The establishment process of the numerical model is shown in Figure 3. The loading speed was the same as the physical test, both of which were 0.001 mm/s.

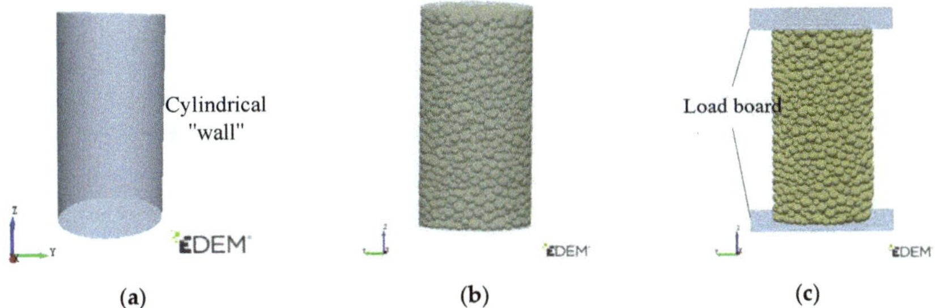

Figure 3. Establishment process of numerical model: (**a**) generate surrounding "walls"; (**b**) randomly generate particles; (**c**) calculation model.

Before the loading test, the microscopic parameters of the numerical model were calibrated first. Through the calculation method given by Shen [20], the mechanical parameters such as elastic modulus and Poisson's ratio measured in indoor tests were taken into account, and the mesoscopic parameters of the numerical simulation model were obtained to preliminarily establish the model body. Then, the simulation test was carried out at the same loading rate to obtain the uniaxial compressive strength, elastic modulus, and other results of the numerical model. The "trial and error method" was used to constantly adjust the microscopic parameters of the numerical model until the simulation calculation value was close to the real value obtained from the indoor test. The energy layer was applied to the particle surface to simulate the water content of the rock samples. This paper mainly studies the compressive strength of sandstone and does not discuss the tensile strength of rock. See Table 2 for the calibration results of relevant microscopic parameters.

Table 2. Microscopic parameters of sandstone model.

Number	Normal Contact Stiffness (N/m^3)	Tangential Contact Stiffness (N/m^3)	Critical Normal Stress (Pa)	Critical Shear Stress (Pa)	Surface Energy (J/m^2)
D	5.25×10^{12}	5.1×10^{12}	6.8×10^7	4.4×10^7	0
N	2×10^{12}	1.85×10^{12}	6.8×10^7	4.4×10^7	10
S	1.4×10^{12}	7.2×10^{11}	6.8×10^7	4.4×10^7	20

The numerical calculation method was used to simulate the rock, and the simulation results were mainly judged by the compressive strength and elastic modulus. Taking saturated rock sample as an example, Figure 4c shows the comparison between the stress-strain curves obtained from the test and the EDEM simulation results.

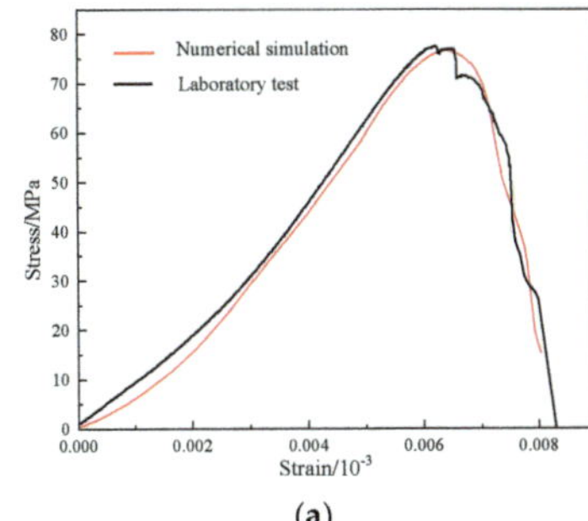

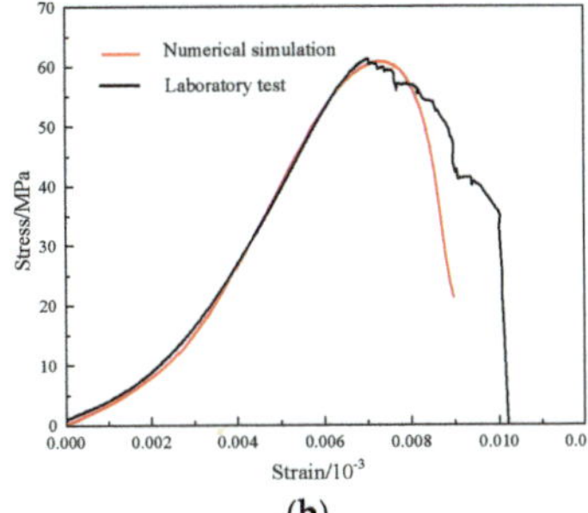

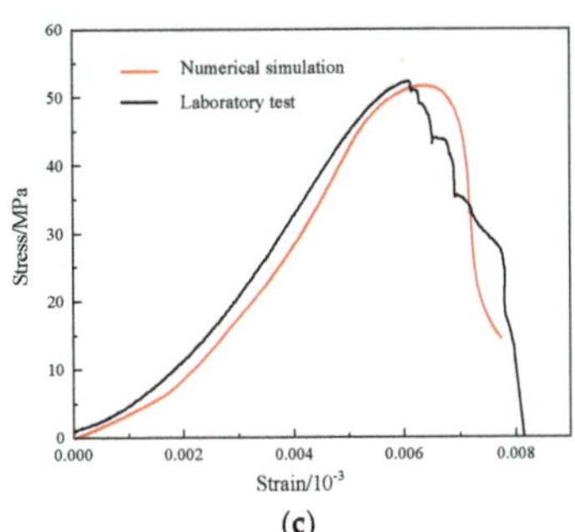

Figure 4. Stress-strain curves of sandstone laboratory test and numerical simulation: (**a**) dry sample; (**b**) nature sample; (**c**) saturated sample.

EDEM simulation results show that the peak strength of the material is 52.2 MPa and the elastic modulus is 11.078 GPa. The peak strength of rock material obtained by laboratory test was 51.335 MPa, and the elastic modulus was 10.418 GPa. The results show that there was a deviation of about 1.657% between the laboratory test and numerical simulation, and the difference of elastic modulus was 5.958%, which is within the allowable range.

3. Results and Discussion

3.1. Effect of Water Content on Stress-Strain Curve of Sandstone under Different Confining Pressures

Compression tests of sandstone under different confining pressure and water content were carried out, and the stress-strain curve is shown in Figure 5. With $\sigma_3 = 0$ MPa as an example, the effect of water content on the sandstone stress-strain curve at various points during loading was analyzed [21,22]. The stress-strain curve of the rock samples was significantly impacted by changes in water content, as illustrated in Figure 5a. The rock samples' peak strength in the dry state was the highest, measuring 77.40 MPa. The compaction stage was shorter, and the elastic stage was longer than that in the natural and saturated states. There was no obvious yielding stage. There was no warning before the failure. The stress after the peak decreased rapidly, and the bearing capacity was lost under a very short strain. The failure had obvious brittleness. The higher the water content, the more obvious the compaction stage of rock sample, the stress-strain curve at the initial stage of loading appeared obviously bending, the elastic stage became shorter, and there was an obvious yield stage. The curve obviously slowed down before reaching the peak, and there were obvious crack signs before failure. Peak rock strength of both natural and saturated rock samples considerably reduced as water content rose. Compared with dry rock sample, the peak strength of natural rock samples was 61.27 MPa, a decrease of 16.13 MPa, a decrease of 26.38%; the peak strength of the saturated rock samples was 52.24 MPa, a decrease of 25.16 MPa and 48.16%. Through comparison, it can be seen that water content significantly reduced the peak strength of sandstone.

The higher water content, the lower triaxial compressive strength of sandstone. When the confining pressure was 5 MPa, the compressive strength of sandstone in dry, natural, and saturated state was 107.07 MPa, 99.18 MPa, and 82.67 MPa, respectively. Compared with dry rock sample, the strength of natural rock samples decreased by nearly 7.96%, and the strength of saturated rock samples decreased by nearly 29.51%; when confining pressure was 10 MPa, the compressive strength of sandstone in dry, natural, and saturated state was 156.94 MPa, 126.62 MPa, and 115.57 MPa, respectively. Compared with dry rock sample, the strength of natural rock samples decreased by nearly 23.95%, and that of saturated rock samples decreased by nearly 35.80%; when confining pressure was 20 MPa, the compressive strength of sandstone in dry, natural, and saturated state was 211.86 MPa, 190.16 MPa, and 185.22 MPa, respectively. Compared with dry rock sample, the strength of natural rock samples decreased by nearly 11.41%, and that of saturated rock samples decreased by nearly 14%. Through comparison, it was found that the increase of water content significantly reduced the compressive strength of sandstone.

Figure 6 shows the fitting relation curve of sandstone peak strength with water content and confining pressure. Figure 6a illustrates that the peak strength of sandstone drops linearly as the water content rises. It can be concluded that confining pressure has little impact on the rate of decline of compressive strength of sandstone based on the influence coefficient of different water contents on the compressive strength of sandstone samples under four distinct types of confining pressure. Figure 6b illustrates that the increase in confining pressure improves the compressive strength of dry, natural, and saturated samples. Excellent linear fitting relationships exist between the compressive strength and confining pressure of samples with various water contents. Under all three circumstances, the fitting relationship coefficient (R^2) was higher than 0.9791. Comparing the slopes of the three curves, the slope of the dry state is significantly larger than that of the natural and saturated state, indicating that the increase of water content will reduce the sensitivity of confining pressure to strength gain.

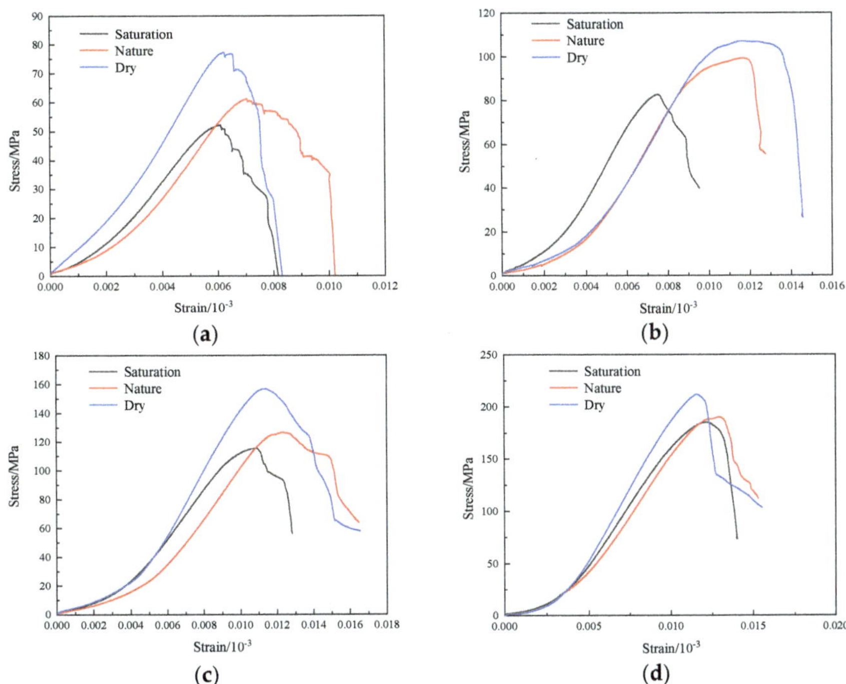

Figure 5. Stress-strain curves of sandstone samples with different water content under different confining pressure conditions: (**a**) $\sigma_3 = 0$ MPa (uniaxial compression); (**b**) $\sigma_3 = 5$ MPa; (**c**) $\sigma_3 = 10$ MPa; (**d**) $\sigma_3 = 20$ MPa.

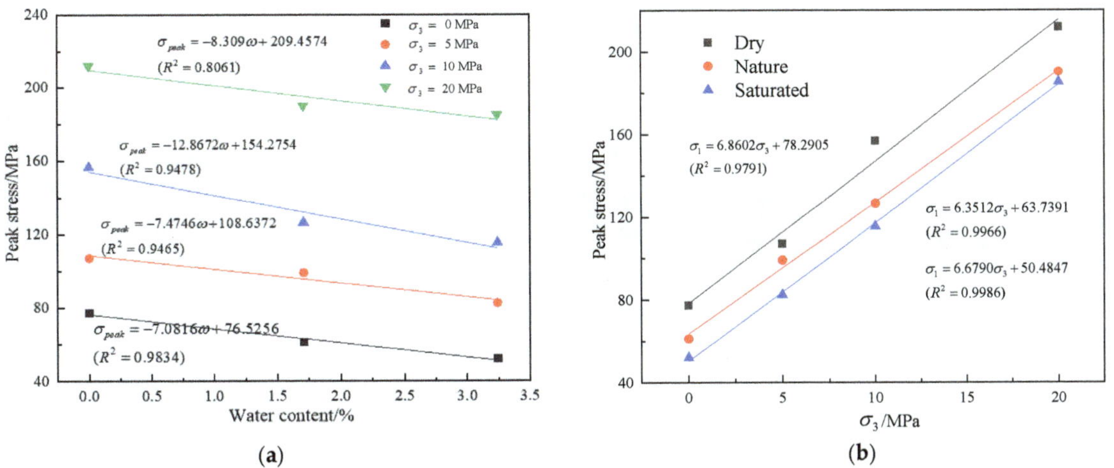

Figure 6. Fitting relation curve of sandstone peak strength with water content and confining pressure: (**a**) relationship between peak strength and water content; (**b**) relationship between peak strength and confining pressure.

3.2. Analysis of Acoustic Emission Characteristic

The acoustic emission signal of rock is that rock emits sound wave or ultrasonic wave in the process of deformation and fracture [23–25]. Analyzing the relationship between acoustic emission signals and crack propagation characteristics during rock fracture is helpful to study the internal crack evolution of sandstone with different water content under different confining pressures.

3.2.1. Ringing Count and Energy Rate Analysis

The stress in rock exceeds the maximum strength of rock, resulting in rock failure. Stress action mode and failure process on fracture surface during rock failure are called failure mechanism. Acoustic emission ringing count is a commonly used acoustic emission characterization parameter, which can reflect the number of energy release events during the formation and expansion of internal cracks in rock samples. The relationship between the stress-strain curve of dry and saturated rock samples and the acoustic emission ringing count accumulative count is shown in Figure 7.

Figure 7 illustrates that the cumulative ringing count curve exhibits an upward concave growth pattern and how the growth rate of the curve gradually rises as the external load continues to rise. At the low stress level, the curve growth rate is low, and at the high stress level, the curve growth rate is high. According to the characteristics of the stress–strain curve, the loading process can be divided into four stages: I compaction stage, II elasticity stage, III yield stage, and IV failure stage. The acoustic emission ring count characteristics of each stage are as follows [26]:

(1) I compaction stage: the primary microfractures inside the rock are closed under external loading, and the elastic strain energy is released due to the occlusion and friction of the particles near the primary and dissolved fractures, which generates a small amount of acoustic emission signal, and with the compaction of the rock samples skeleton in the loading process, the acoustic emission ring count is gradually reduced by the complete closure of the fractures and reaches the lowest at the end of stage I. The number of acoustic emission signals of water-saturated samples is significantly lower than that of dry sample, which is because the fracture is filled with water, the attenuation rate of acoustic emission signal in liquid medium is greater than that in gas medium, and water acts as a lubricant during particle occlusion and friction to reduce acoustic emission signal generation, resulting in the acoustic emission signal intensity in water-saturated condition being lower than that in dry condition. The larger the surrounding pressure, the more significant the effect.

(2) II elastic stage: the rock samples mainly underwent elastic deformation. With the increase of the load, the compacted rock samples gradually entered the linear elastic change stage, and a slight slip appeared between the cracks of the rock samples, but it had not yet reached the crack initiation stress of the rock. The acoustic emission ringing count as a whole remained at a low level, and the slope of the cumulative ringing count curve was basically unchanged; the ringing count of the dry sample was significantly higher than that of the saturated samples, and the effect of compaction pressure was minimal at this stage.

(3) III yield stage: the load stress exceeds the rock's crack initiation stress. At the early stage of this stage, in addition to the expansion of the original crack, a small number of new micro cracks are gradually generated in the samples, resulting in a slight increase in the acoustic emission ringing count. With the loading, new cracks are constantly generated and intersected in the rock samples, resulting in local fracture surfaces. The acoustic emission ringing count starts to increase significantly. With the decrease of water content and the increase of confining pressure, the ringing count value and the slope of the cumulative ringing count curve increase significantly.

(4) IV failure stage: the internal fracture of the rock intensifies, the local fracture on surface starts to penetrate, eventually forming a rupture surface accompanied and by a sharp increase in strong acoustic emission signals, and the peak value of acoustic

emission ringing count appears at this stage. The slope of the cumulative generation curve starts to surge.

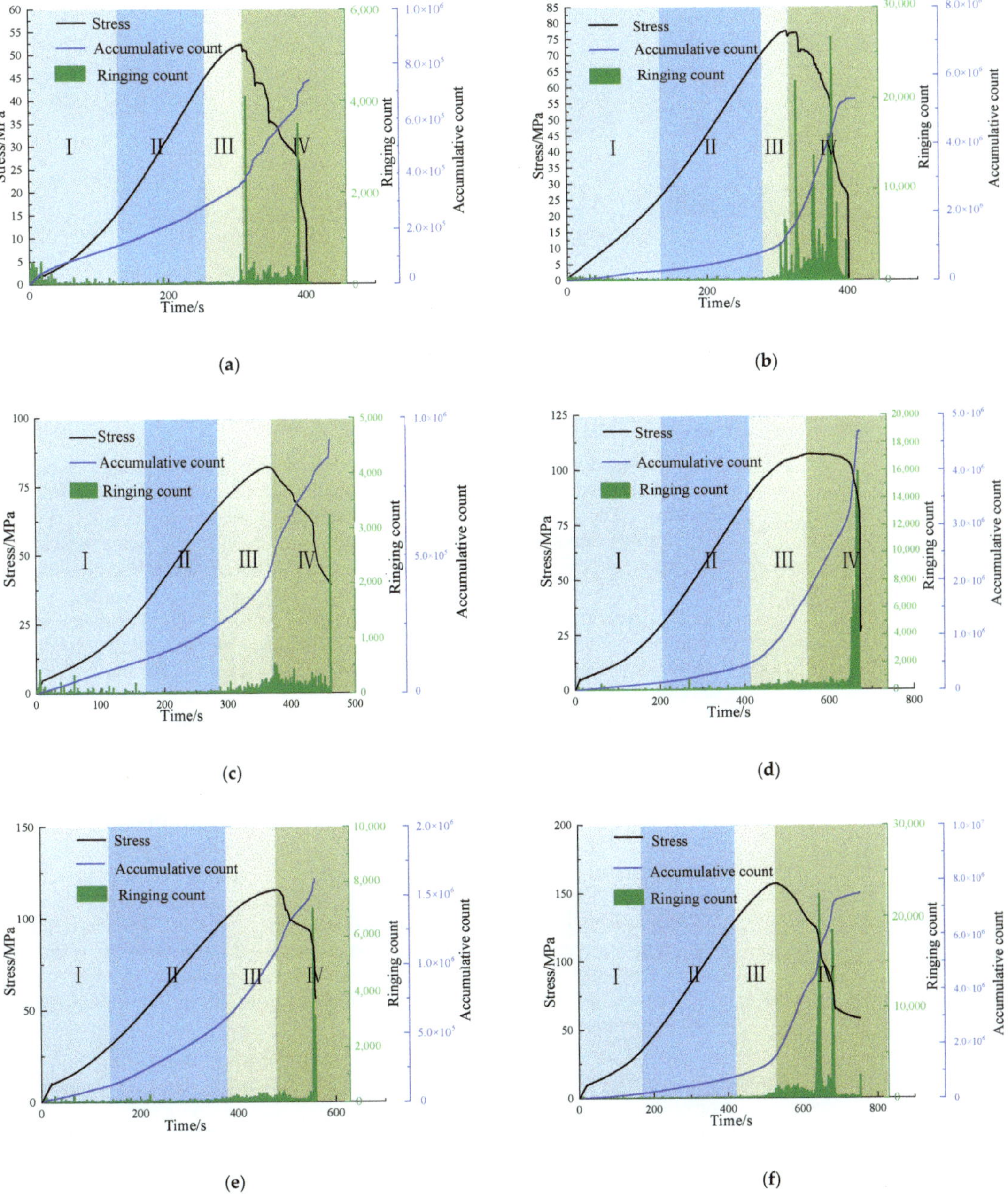

Figure 7. *Cont.*

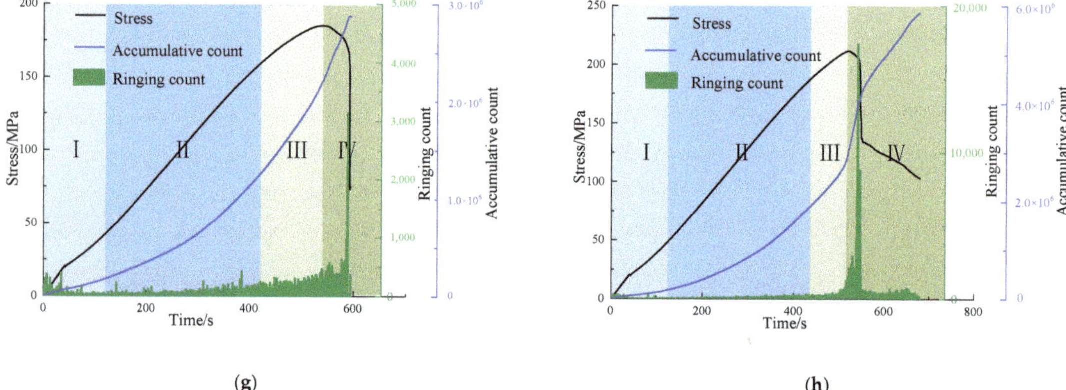

Figure 7. Overall stress-strain curves and ringing count of rock samples with different water content: (**a**) S-0; (**b**) D-0; (**c**) S-5; (**d**) D-5; (**e**) S-10; (**f**) D-10; (**g**) S-20; (**h**) D-20.

Comparing the test results of rock samples with different water contents, the following conclusions are obtained:

(1) The acoustic emission ring count of the dry sample is always bigger than that of the saturated sample in each corresponding stage when the confining pressure is the same. The maximum slope of the cumulative AE ringing count curve of the dry sample is always higher than that of the saturated sample, which indicates that the dry sample is more conducive to the generation of acoustic emission signals; another effect of water is reflected in the unstable fluctuation of acoustic emission signal. Under triaxial compression, the acoustic emission signal of water saturated samples fluctuates significantly at the yield stage, while the ringing count of dry sample increases steadily at this stage, which is the same as the research results of Yao [27].

(2) There are obvious differences in the characteristics of acoustic emission ringing count under the two loading modes. The acoustic emission ringing count of the samples under uniaxial compression is mostly concentrated in the failure stage, and less in the pre-peak stage. However, the ringing count of the samples under triaxial compression is also distributed in the yield stage, indicating that the intersection and penetration of cracks and the generation of local fracture surfaces of the samples before reaching the peak strength under uniaxial compression are lower than those under triaxial compression, Under triaxial compression, the energy of rock samples is released violently before the peak strength; at the same time, the slope of the cumulative AE ring count curve under uniaxial compression increased significantly at the peak point, while the point of sharp increase in the slope of the accumulative AE ring count curve under triaxial compression was located in the third stage, and the curve at the peak point was smoother.

AE energy rate is represented by wave carrying energy and duration, which can reflect the degree of fracture in the rock sample. The AE energy rate of the stress curve of sandstone samples with different water content is a comprehensive reflection of the energy and time of the monitored acoustic emission events, which can be used to characterize the degree of micro fracture in sandstone bodies. Overall stress-strain curves and energy rate and accumulative energy rate for rock samples with different water content are shown in Figure 8.

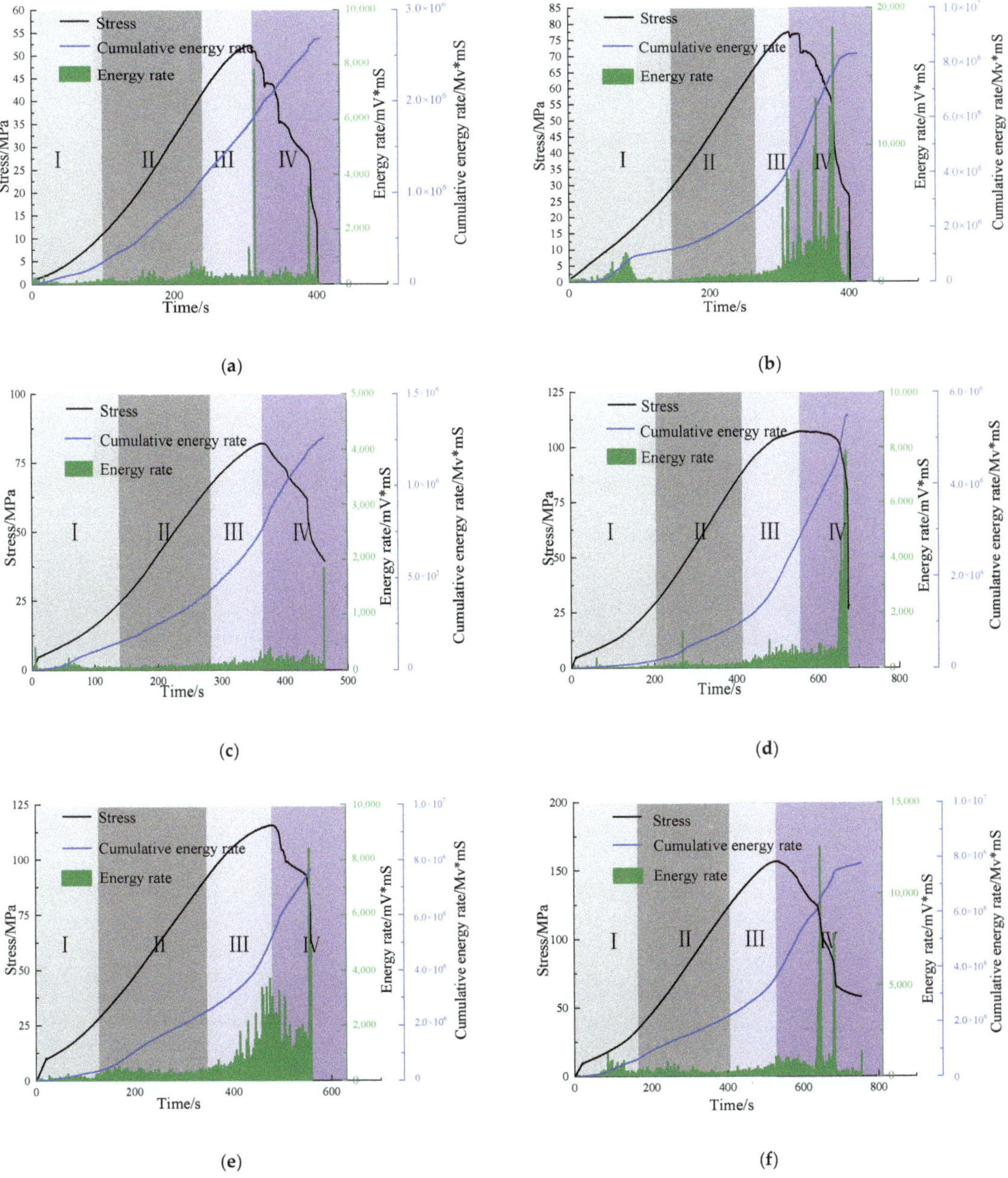

Figure 8. *Cont.*

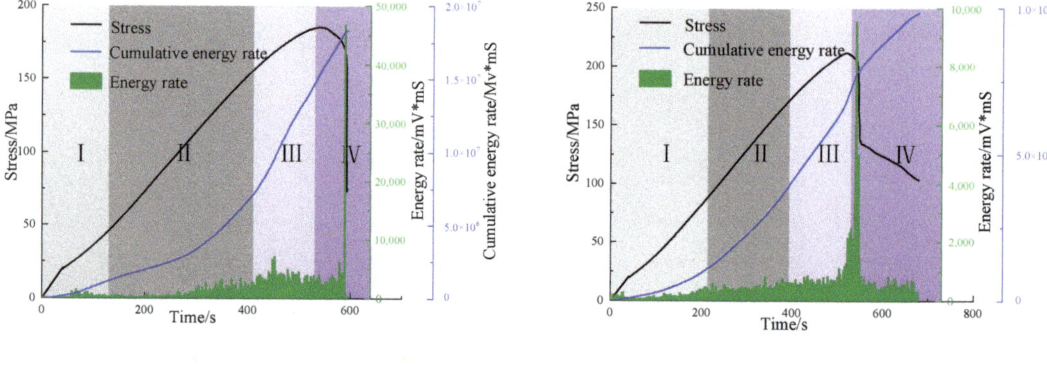

Figure 8. Overall stress-strain curves and energy rate, accumulative energy rate for rock samples with different water content: (**a**) S-0; (**b**) D-0; (**c**) S-5; (**d**) D-5; (**e**) S-10; (**f**) D-10; (**g**) S-20; (**h**) D-20.

Comparing Figures 7 and 8 shows that the acoustic emission energy characteristics are similar to the acoustic emission ringing count characteristics, and also have obvious stages. AE energy rate is lower in the compaction stage and elastic stage; at the early stage of the yield stage, the AE energy rate keeps increasing at a low speed in the later yield stage, the AE energy rate increases significantly; in the failure stage, the slope of the cumulative energy rate curve reaches the maximum at this stage. At the same time, there are many AE energy rate events at this stage, and the highest AE energy rate of the samples occurs at this stage.

Different from the distribution characteristics of ringing count, there is no obvious regularity in AE energy rate in the compaction stage, because AE energy rate reflects the energy released by an acoustic emission signal, and ringing count is the number of acoustic emission signals. Therefore, the ringing count in the compaction stage shows a gradual decrease with the loading.

3.2.2. Relationship between RA and AF Parameters and Failure Mode

Many researchers grouped the AE data into shear and tensile clusters and used the AE parameters to study the cracking mode of microcracks in experiments. The RA value can be obtained by dividing the rise time by the amplitude, and the unit is MS/v. The AF value is obtained by the ratio of the number of ringing times to the duration, in kHz. In general, when the AF value of acoustic emission signal is low and the RA value is high, shear fractures develop in rock samples; in contrast, tensile cracks are developed in the rock samples, as shown in Figure 9 [28]. According to the crack classification procedure of JCMS-III B5706 specification, it is recommended to use the RA-AF diagram shown in Figure 9 to partition the acoustic emission data to distinguish tensile cracks and shear cracks.

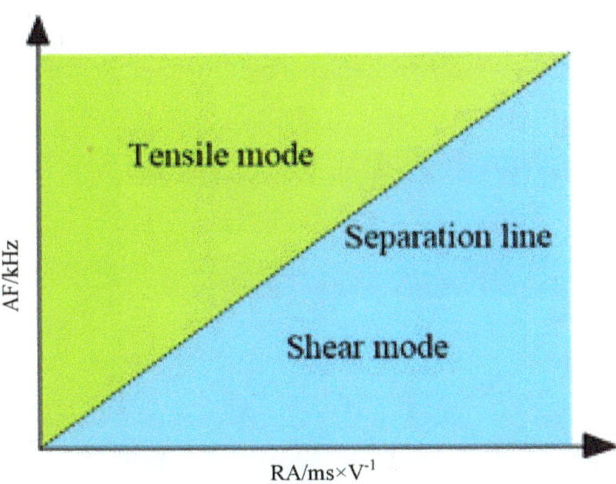

Figure 9. Failure mode classification of cracks according to RA-AF.

The Rise time/Amplitude (RA) value and Ring count/Duration (AF) value of sandstone samples is calculated and a scatter plot is drawn, as shown in Figure 10. AF and RA are distributed in [0, 200].

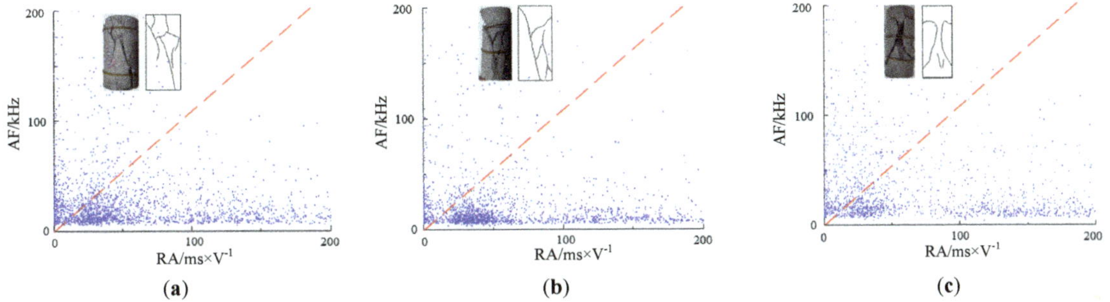

Figure 10. RA-AF distribution of different samples under uniaxial compression: (**a**) S-0; (**b**) N-0; (**c**) D-0.

Figure 10 illustrates the development of shear cracks and tensile cracks in the samples in the three states, with shear cracks serving as the predominant type. From the distribution characteristics of scatter points, there is little difference in the distribution range of tensile crack scattered points of samples in the three states, but the high-density distribution range of shear crack signal points of dry sample is larger, and with the decrease of water content, the overall proportion of scattered points below the separation line is gradually increasing. It shows that the lower the water content, the more favorable the formation of shear cracks, the more unfavorable the formation of tensile cracks.

Because there are many overlapping points in the scatter plot, and the skip point method is used in the drawing, in order to verify the above conclusions and further explore the failure mode and crack evolution of rock samples, the complete acoustic emission signals are statistically analyzed, as shown in Figure 11. From the quantity of AE signals, with the decrease of water content, the total number of the two types of cracks decreases, which corresponds to the gradual decline of the final damage and fragmentation of the water saturated rock samples to the dry sample in the test.

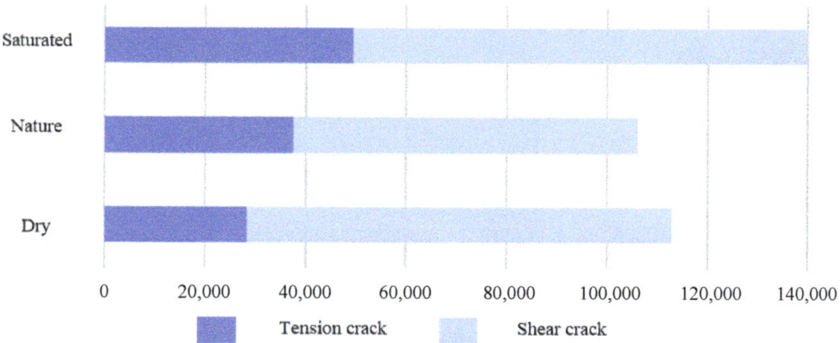

Figure 11. Statistics of total signal number of rock sample under different water content.

3.3. Influence of Water Content on Sandstone Failure Mode

The failure characteristics of rock samples are the final result of internal microcrack propagation, which contains rich information such as rock sample deformation, force chain evolution path, crack propagation results, etc. Analyzing the failure characteristics of rocks is of great significance for studying the influence of water content on sandstone properties [29].

Figure 12 shows the failure mode of rock samples with different water content. It can be seen from Figure 12 that the final failure mode of the dry sample under uniaxial compression presents a typical "X" conjugate shear failure. The two oblique main fracture surfaces are caused by shear slip and conjugate cross in the middle of the samples. After the failure, the rock samples have good integrity and a small number of fragments, as shown in Figure 12a; with the increase of the water content of the rock samples, the number of cracks increases, and the failure mode changes from the "X" conjugate shear failure in the dry state to the single slope shear failure. In the natural state, the rock samples finally present a "Y" shaped failure mode, and the degree of fragmentation of the rock samples obviously increases, as shown in Figure 12e; as the water content continues to increase, the fractures continue to increase, and the rock samples in the water saturated state finally presents a single slope shear failure mode, with the highest degree of fragmentation, as shown in Figure 12i. With the increase of water content, the ultimate fracture degree of sandstone increases, which is due to the role of water in weakening the cementation ability of rock particles. The uniaxial compression failure mode of rock samples in the three states is shear failure. The influence of water content on the failure mode of rock samples deduced from the final failure mode of rock samples is consistent with the change rule of water content on the failure mode of rock samples judged by RA-AF value in Section 3.2.2. Under the same water content, with the increase of confining pressure, the integrity of the rock samples is higher when it is destroyed, and the final failure form is shear failure, which indicates that confining pressure has an inhibitory effect on the failure of the rock samples.

3.4. Analysis of Numerical Simulation Results

Figure 13 is a comparison of the final failure characteristics of sandstone with different water content under EDEM simulation uniaxial compression and the final failure morphology of laboratory tests. When the dry sample is damaged (Figure 13a), the crack is a penetrating shear crack, and the failure form is shear failure; when the rock samples is damaged in the natural sample (Figure 13b), there are two vertical tensile cracks in the rock samples. At this time, the failure mode of the rock mass changes from shear failure to tensile failure; the saturated rock sample is damaged (Figure 13c). At this time, the tensile cracks of the samples are fully developed. The final failure form is typical tensile failure, and the degree of rock fragmentation is even greater than the former. The failure modes of the above rock samples under different water bearing conditions are consistent with the indoor test results.

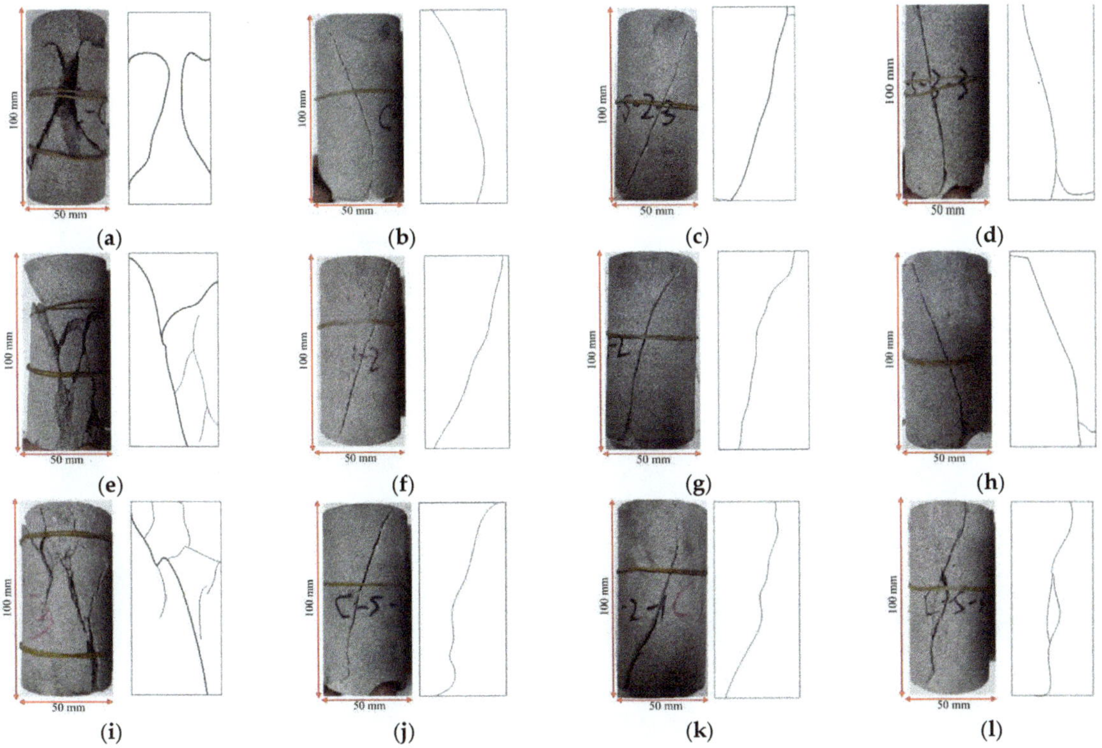

Figure 12. Failure modes of sandstone with different water content under different confining pressure: (**a**) D-0; (**b**) D-5; (**c**) D-10; (**d**) D-20; (**e**) N-0; (**f**) N-5; (**g**) N-10; (**h**) N-20; (**i**) S-0; (**j**) S-5; (**k**) S-10; (**l**) S-20.

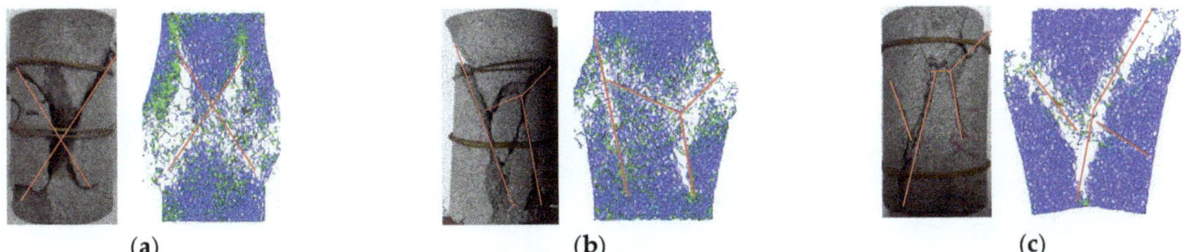

Figure 13. Uniaxial compression test results and simulation comparison of sandstone with different water content: (**a**) D-0; (**b**) N-0; (**c**) S-0.

Table 3 shows the crack propagation process of rock samples in different loading states. The stress distribution nephogram of bond at the boundary point from compaction stage to elastic stage, yield point, and peak point is shown in the table. By comparison, the bond stress of the three states rock samples at each stage point increases with the decrease of water content. Compared with other rock samples, the stress of the bond in the saturated rock sample is at a low level during the whole loading process. The cloud map is evenly blue, and a small part of the crack propagation zone has uneven stress. The cloud picture of the dry sample is uniform green during loading, and the bonding stress is always at a high level. The non-destructive areas at both ends of the rock samples are blue, and the stress is slightly lower. Table 3a,d,g rock samples are at boundary point from compaction stage to elastic stage, compared with the initial stress state, the bond of the rock samples is overall tighter without fracture; Table 3b,e,h is the yield point of the rock samples. The stress concentration occurs inside the rock samples, and the bond has been obviously connected, which corresponds to the internal cracks in the laboratory test. Table 3c,f,i is at the peak point of the rock samples. At this time, the broken bond has been connected, many cracks have appeared in the cloud map, and the edge has also fallen off, and the bearing capacity of the rock samples began to decline. Comparison Table 3a,d,g, compaction stage and elastic stage transition point, with the decrease of water content, the stress cloud of rock samples is more uniform, but the stress level increased significantly. At the peak point, with the increase of water content, the internal crack path of rock samples increases obviously.

Table 3. Bond stress nephogram of rock samples under different state.

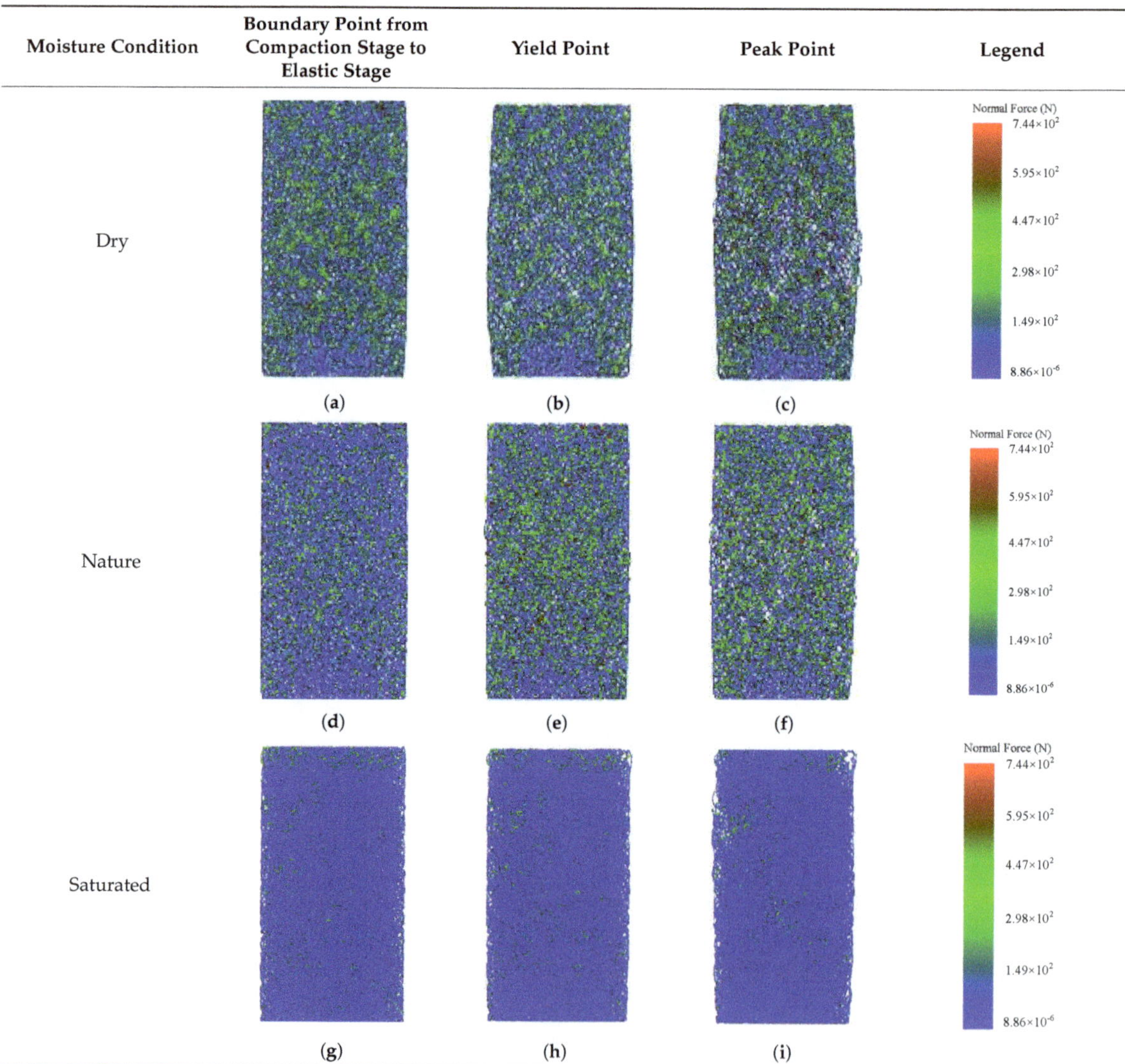

Moisture Condition	Boundary Point from Compaction Stage to Elastic Stage	Yield Point	Peak Point	Legend
Dry	(a)	(b)	(c)	
Nature	(d)	(e)	(f)	
Saturated	(g)	(h)	(i)	

Figure 14 shows the relationship between the amount of damage and the water content state during bond formation. It can be seen from the figure that with the increase of water content, the number of damaged bonds continues to rise. With the total number of particles unchanged, the total number of bonds originally generated was 97,572, and the number of bonds destroyed in the dry state was 34,076; the number of bonds broken in natural state is 38,396; the number of broken bonds in the saturated state is 44,636. With the increase of the water content, the number of broken bonds increases synchronously, which is the reason for the strength decline of the rock samples in the later loading process. This also corresponds well to the water eroding the rock mass inside the rock samples and reducing the adhesion between particles.

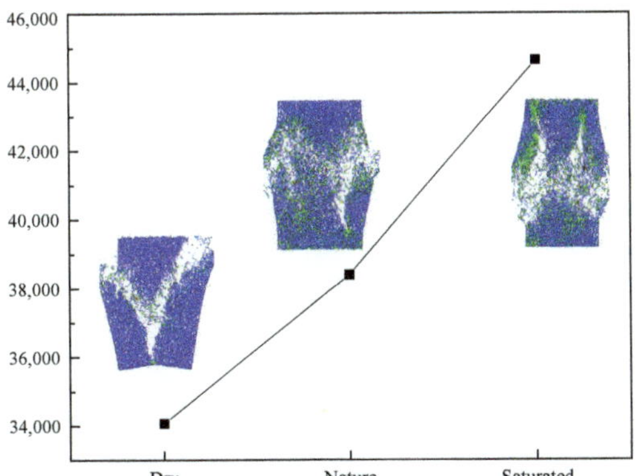

Figure 14. Relationship between the amount of damage bond and water content.

Figure 15 shows the relationship between the number of broken bonds and the stress-strain curve of sandstone under different water content. It can be seen from the diagram that the bonding of the three states did not break at the initial stage of loading (stage I); in the elastic stage (stage II), the fracture began to occur, and the number growth was stable and slow. With the continuous compression, the rock samples enter the pre-peak stage (stage III), the number of bond failure in the three states increases significantly, and the slope of the curve increases significantly. Eventually, the rock samples enters the failure stage (stage IV) and the number of broken bonds begins to increase dramatically, peaking at complete failure The time when the bond fracture first appears in the three different states is gradually advanced with the increase of water content, and the rising slope of the corresponding curves of the three states also decreases with the increase of water content after reaching the peak strength, which proves that water content is closely related to the weakening of sandstone strength and internal bond.

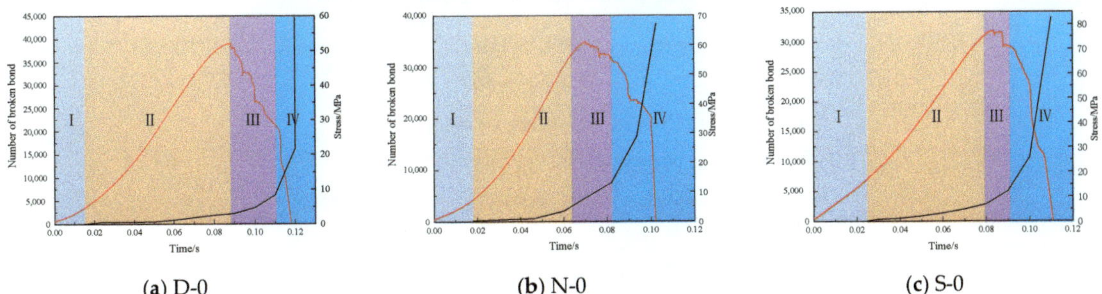

(**a**) D-0   (**b**) N-0   (**c**) S-0

Figure 15. Relationship between bond quantity and time of sandstone failure in different states: (**a**) D-0; (**b**) N-0; (**c**) S-0.

4. Conclusions

In this paper, the conventional uniaxial and triaxial tests of sandstone with different water content and confining pressures were carried out, and the laboratory tests are simulated by discrete element software to further study internal fracture evolution process of sandstone. The following conclusions are obtained:

(1) Compared with uniaxial compression results, the peak strength of sandstone decreases with the increase of water content. Water action will cause certain damage to the interior of sandstone samples, weaken the cementation between rock particles, and reduce the compressive strength and elastic modulus of sandstone samples. The higher the water content, the more complex the fracture development of sandstone samples, and the more broken the rock samples is. Therefore, water–rock interaction has a great influence on rock strength.

(2) The AE energy rate and ringing count of the rock samples during loading process show obvious "quiet period" to "frequent period" phased changes, and the stage boundary corresponds to the peak strength. With the decrease of the water content of sandstone samples, the proportion of frequent period decreases, and the maximum energy rate increases significantly. Acoustic emission is active in yield stage and failure stage, and a large amount of strain energy and acoustic emission signals are released from rock samples; with the increase of water content, the frequency of rock samples is obviously reduced, and the strain energy released by rock samples under load is reduced. These have reference significance for practical engineering risk prediction.

(3) Using discrete element method to analyze the destruction process of rock samples in detail, the increase of water content inside the rock mass is not conducive to the generation and maintenance of the number of connection bonds in the model body, and the number of destroyed bonds in dry sample at the same loading time is significantly less than that in water-saturated rock samples. It can be seen that the strength of sandstone increases with the decrease of water content. From the micro level analysis, water reduces the connection effect inside the rock mass, and this effect increases with the increase of water content. This result provides a new idea for the study of water–rock interaction in underground engineering by the discrete element method.

Author Contributions: Conceptualization, X.H. and Y.L.; methodology, T.W. and J.G; software, Y.L.; validation, T.W. and Y.L.; formal analysis, T.W. and J.G.; investigation, J.G.; resources, Y.L.; data curation, X.H. and T.W.; writing—original draft preparation, Y.L. and J.G.; writing—review and editing, T.W. and X.H.; supervision, X.H.; project administration, J.G.; funding acquisition, J.G. and X.H. All authors have read and agreed to the published version of the manuscript.

Funding: "This research was funded by the National Natural Science Foundation of China, grant number 52178388" and "The Opening Project of Key Laboratory of Highway Bridge and Tunnel of Shaanxi Province (Chang'an University), grant number 300102211517" and "The Key Scientific and Technological Project of Henan Province, grant number 212102310292" and "The Fundamental Research Funds for the Universities of Henan Province, grant number NSFRF210337".

Institutional Review Board Statement: Not applicable.

Informed Consent Statement: Not applicable.

Data Availability Statement: Not applicable.

Conflicts of Interest: The authors declare no conflict of interest.

References

1. Louis, N.Y.W.; Varun, M.; Liu, G. Water effects on rock strength and stiffness degradation. *Acta Geotech.* **2016**, *11*, 713–737.
2. Yilmaz, I. Influence of water content on the strength and deformability of gypsum. *Int. J. Rock Mech. Min. Sci.* **2009**, *47*, 342–347. [CrossRef]

3. Zhou, L.; Niu, C.Y.; Zhu, Z.M. Fracture properties and tensile strength of three typical sandstone materials under static and impact loads. *Geomech. Eng.* **2020**, *23*, 467–480.

4. Nara, Y.; Morimoto, K.; Yoneda, T. Effects of humidity and temperature on subcritical crack growth in sandstone. *Int. J. Solids Struct.* **2011**, *48*, 1130–1140. [CrossRef]

5. Cai, X.; Zhou, Z.; Liu, K.; Du, X.; Zang, H. Water-weakening effects on the mechanical behavior of different rock types: Phenomena and mechanisms. *Appl. Sci.* **2019**, *9*, 4450. [CrossRef]

6. Hashiba, K.; Fukui, K.; Kataoka, M. Effects of water saturation on the strength and loading-rate dependence of andesite. *Int. J. Rock Mech. Min. Sci.* **2019**, *117*, 142–149. [CrossRef]

7. Zhao, K.; Yang, D.; Zeng, P.; Huang, Z.; Wu, W.; Li, B.; Teng, T. Effect of water content on the failure pattern and acoustic emission characteristics of red sandstone. *Int. J. Rock Mech. Min. Sci.* **2021**, *142*, 1365–1609. [CrossRef]

8. Feng, G.R.; Wen, X.Z.; Guo, J. Study on influence of moisture content on coal sample AE properties and fragment distribution characteristics. *J. Cent. South Univ. (Sci. Technol.)* **2021**, *52*, 2910–2918.

9. Eunhye, K.; Hossein, C. Effect of water saturation and loading rate on the mechanical properties of Red and Buff Sandstones. *Int. J. Rock Mech. Min. Sci.* **2016**, *88*, 23–28.

10. Sun, X.; Xu, H.C.; Zheng, L.G.; He, M.; Gong, W. An experimental investigation on acoustic emission characteristics of sandstone rockburst with different moisture contents. *Sci. China Technol. Sci.* **2016**, *59*, 1549–1558. [CrossRef]

11. Erguler, Z.A.; Ulusay, R. Water-induced variations in mechanical properties of clay-bearing rocks. *Int. J. Rock Mech. Min. Sci.* **2009**, *46*, 355–370. [CrossRef]

12. Guo, J.Q.; Liu, X.L.; Qiao, C.S. Experimental study of mechanical properties and energy mechanism of karst limestone under natural and saturated states. *Chin. J. Rock Mech. Eng.* **2014**, *33*, 296–308.

13. Tan, P.; Rao, Q.H.; Li, Z. A new method for quantitative determination of PFC3D microscopic parameters considering fracture toughness. *J. Cent. South Univ. (Sci. Technol.)* **2021**, *52*, 2849–2866.

14. Zhu, J.; Deng, J.H.; Chen, F.; Ma, Y.; Yao, Y. Water-weakening effects on the strength of hard rocks at different loading rates: An experimental study. *Rock Mech. Rock Eng.* **2021**, *54*, 4347–4353. [CrossRef]

15. Vasarhelyi, B. Statistical analysis of the influence of water content on the strength of the miocene limestone. *Rock Mech. Rock Eng.* **2005**, *38*, 69–76. [CrossRef]

16. Yu, L.Q.; Yao, Q.L.; Xu, Q. Experimental and numerical simulation study on crack propagation of fractured fine sandstone under the influence of loading rate. *J. China Coal Soc.* **2021**, *46*, 3488–3501.

17. He, M.C.; Zhao, F.; Du, S.; Zheng, M.J. Rock burst characteristics based on experimental tests under different unloading rates. *Rock Soil Mech.* **2014**, *35*, 2737–2747.

18. Liu, J.; Li, J.T. Analysis on meso-damage characteristics of marble under triaxial cyclic loading and unloading based on particle flow simulation. *J. Cent. South Univ. (Sci. Technol.)* **2018**, *49*, 2797–2803.

19. Tian, W.L.; Yang, S.Q.; Fang, G. Particle flow simulation on mechanical behavior of coal specimen under triaxial cyclic loading and unloading. *J. China Coal Soc.* **2016**, *41*, 603–610.

20. Shen, H.H.; Zhang, H.; Fan, J.K. A rock modeling method of multi-parameters fitting in EDEM. *Rock Soil Mech.* **2021**, *42*, 2298–2310+2320.

21. Zhao, Y.C.; Yang, T.H.; Xu, T.; Zhang, P.; Shi, W. Mechanical and energy release characteristics of different water-bearing sandstones under uniaxial compression. *Int. J. Damage Mech.* **2018**, *27*, 640–656. [CrossRef]

22. Liu, H.L.; Zhu, W.C.; Yu, Y.J.; Xu, T.; Li, R.; Liu, X. Effect of water imbibition on uniaxial compression strength of sandstone. *Int. J. Rock Mech. Min. Sci.* **2020**, *127*, 104200. [CrossRef]

23. Wang, Q.S.; Chen, J.X.; Guo, J.Q.; Luo, Y.; Wang, H.; Liu, Q. Acoustic emission characteristics and energy mechanism in karst limestone failure under uniaxial and triaxial compression. *Bull. Eng. Geol. Environ.* **2019**, *78*, 1422–1427. [CrossRef]

24. Sherif, M.; Tanks, J.; Ozbulut, O. Acoustic emission analysis of cyclically loaded super elastic shape memory alloy fiber reinforced mortar beams. *Cem. Concr. Res.* **2017**, *95*, 178–187. [CrossRef]

25. Guo, J.; Feng, G.R.; Qi, T.Y. Dynamic Mechanical Behavior of Dry and Water Saturated Igneous Rock with Acoustic Emission Monitoring. *Shock. Vib.* **2018**, *2018*, 2348394. [CrossRef]

26. Chen, J.Q.; Li, T.B.; Wang, W.; Zhu, Z.; Chen, Z.; Tang, O. Weakening effects of the presence of water on the brittleness of hard sandstone. *Bull. Eng. Geol. Environ.* **2019**, *78*, 1471–1483. [CrossRef]

27. Yao, Q.L.; Chen, T.; Ju, M.H.; Liang, S.; Liu, Y.; Li, X. Effects of water intrusion on mechanical properties of and crack propagation in coal. *Rock Mech. Rock Eng.* **2016**, *49*, 4699–4709. [CrossRef]

28. Guo, J.Q.; Liu, P.F.; Fan, J.Q.; Zhang, H.Y. Study on the Mechanical Behavior and Acoustic Emission Properties of Granite under Triaxial Compression. *Geofluids* **2021**, *3954097*. [CrossRef]

29. Li, D.Y.; Sun, Z.; Zhu, Q.Q.; Peng, K. Triaxial loading and unloading tests on dry and saturated sandstone specimens. *Appl. Sci.* **2019**, *9*, 1689. [CrossRef]

MDPI AG
Grosspeteranlage 5
4052 Basel
Switzerland
Tel.: +41 61 683 77 34

Sustainability Editorial Office
E-mail: sustainability@mdpi.com
www.mdpi.com/journal/sustainability